Student Solutions Manual

to accompany

Beginning and Intermediate Algebra

Third Edition

Sherri Messersmith
College of DuPage

Student Solutions Manual to accompany
BEGINNING AND INTERMEDIATE ALGEBRA, THIRD EDITION
SHERRI MESSERSMITH

Published by McGraw-Hill Higher Education, an imprint of The McGraw-Hill Companies, Inc., 1221 Avenue of the Americas, New York, NY 10020. Copyright © 2012, 2009, and 2007 by The McGraw-Hill Companies, Inc. All rights reserved.

This book is printed on acid-free paper.

1 2 3 4 5 6 7 8 9 0 QDB/QDB 1 0 9 8 7 6 5 4 3 2 1

ISBN: 978–0–07–729702–2
MHID: 0–07–729702–4

Student Solutions Manual

Beginning and Intermediate Algebra
Third Edition

Table of Contents

Section 1.1: Exercises

1) a) $\dfrac{2}{5}$ b) $\dfrac{4}{6} = \dfrac{2}{3}$

   c) $\dfrac{4}{4} = 1$

3) $\dfrac{1}{2}$

5) a) $18 = 1 \cdot 18$
   $18 = 2 \cdot 9$
   $18 = 3 \cdot 6$
   The factors of 18 are 1, 2, 3, 6, 9, and 18.

   b) $40 = 1 \cdot 40$
   $40 = 2 \cdot 20$
   $40 = 4 \cdot 10$
   $40 = 5 \cdot 8$
   The factors of 40 are 1, 2, 4, 5, 8, 10, 20, and 40.

   c) $23 = 1 \cdot 23$
   The factors of 23 are 1 and 23.

7) a) Composite   b) Composite

   c) Prime

9) Composite. It is divisible by 2 and has other factors as well.

11) a) $18 = 2 \cdot 3 \cdot 3$ b) $54 = 2 \cdot 3 \cdot 3 \cdot 3$

    c) $42 = 2 \cdot 3 \cdot 7$ d) $150 = 2 \cdot 3 \cdot 5 \cdot 5$

13) a) $\dfrac{9}{12} = \dfrac{3 \cdot \cancel{3}}{2 \cdot 2 \cdot \cancel{3}} = \dfrac{3}{4}$

    b) $\dfrac{54}{72} = \dfrac{\cancel{2} \cdot \cancel{3} \cdot \cancel{3} \cdot 3}{\cancel{2} \cdot 2 \cdot 2 \cdot \cancel{3} \cdot \cancel{3}} = \dfrac{3}{4}$

c) $\dfrac{84}{35} = \dfrac{2 \cdot 2 \cdot 3 \cdot \cancel{7}}{5 \cdot \cancel{7}} = \dfrac{12}{5}$ or $2\dfrac{2}{5}$

d) $\dfrac{120}{280} = \dfrac{\cancel{2} \cdot \cancel{2} \cdot \cancel{2} \cdot 3 \cdot \cancel{5}}{\cancel{2} \cdot \cancel{2} \cdot \cancel{2} \cdot \cancel{5} \cdot 7} = \dfrac{3}{7}$

15) a) $\dfrac{2}{7} \cdot \dfrac{3}{5} = \dfrac{2 \cdot 3}{7 \cdot 5} = \dfrac{6}{35}$

    b) $\dfrac{\overset{5}{\cancel{15}}}{\underset{13}{\cancel{26}}} \cdot \dfrac{\overset{2}{\cancel{4}}}{\underset{3}{\cancel{9}}} = \dfrac{5 \cdot 2}{13 \cdot 3} = \dfrac{10}{39}$

    c) $\dfrac{1}{\underset{1}{\cancel{2}}} \cdot \dfrac{\overset{7}{\cancel{14}}}{15} = \dfrac{1 \cdot 7}{1 \cdot 15} = \dfrac{7}{15}$

    d) $\dfrac{\overset{6}{\cancel{42}}}{\underset{5}{\cancel{55}}} \cdot \dfrac{\overset{2}{\cancel{22}}}{\underset{5}{\cancel{35}}} = \dfrac{6 \cdot 2}{5 \cdot 5} = \dfrac{12}{25}$

    e) $4 \cdot \dfrac{1}{8} = \dfrac{\overset{1}{\cancel{4}}}{1} \cdot \dfrac{1}{\underset{2}{\cancel{8}}} = \dfrac{1 \cdot 1}{2} = \dfrac{1}{2}$

    f) $6\dfrac{1}{8} \cdot \dfrac{2}{7} = \dfrac{\overset{7}{\cancel{49}}}{\underset{4}{\cancel{8}}} \cdot \dfrac{\overset{1}{\cancel{2}}}{\underset{1}{\cancel{7}}} = \dfrac{7 \cdot 1}{4 \cdot 1} = \dfrac{7}{4}$ or $1\dfrac{3}{4}$

17) She multiplied the whole numbers and multiplied the fractions. She should have converted the mixed numbers to improper fractions before multiplying.
$5\dfrac{1}{2} \cdot 2\dfrac{1}{3} = \dfrac{11}{2} \cdot \dfrac{7}{3} = \dfrac{11 \cdot 7}{2 \cdot 3} = \dfrac{77}{6}$ or $12\dfrac{5}{6}$

19) a) $\dfrac{1}{42} \div \dfrac{2}{7} = \dfrac{1}{\underset{6}{\cancel{42}}} \cdot \dfrac{\overset{1}{\cancel{7}}}{2} = \dfrac{1 \cdot 1}{6 \cdot 2} = \dfrac{1}{12}$

    b) $\dfrac{3}{11} \div \dfrac{4}{5} = \dfrac{3}{11} \cdot \dfrac{5}{4} = \dfrac{3 \cdot 5}{11 \cdot 4} = \dfrac{15}{44}$

c) $\dfrac{18}{35} \div \dfrac{9}{10} = \dfrac{\overset{2}{\cancel{18}}}{\underset{7}{\cancel{35}}} \cdot \dfrac{\overset{2}{\cancel{10}}}{\underset{1}{\cancel{9}}} = \dfrac{2 \cdot 2}{7 \cdot 1} = \dfrac{4}{7}$

d) $\dfrac{14}{15} \div \dfrac{2}{15} = \dfrac{\overset{7}{\cancel{14}}}{\underset{1}{\cancel{15}}} \cdot \dfrac{\overset{1}{\cancel{15}}}{\underset{1}{\cancel{2}}} = \dfrac{7 \cdot 1}{1 \cdot 1} = 7$

e) $6\dfrac{2}{5} \div 1\dfrac{13}{15} = \dfrac{32}{5} \div \dfrac{28}{15}$

$= \dfrac{\overset{8}{\cancel{32}}}{\underset{1}{\cancel{5}}} \cdot \dfrac{\overset{3}{\cancel{15}}}{\underset{7}{\cancel{28}}}$

$= \dfrac{8 \cdot 3}{1 \cdot 7}$

$= \dfrac{24}{7} \text{ or } 3\dfrac{3}{7}$

f) $\dfrac{4}{7} \div 8 = \dfrac{\overset{1}{\cancel{4}}}{7} \cdot \dfrac{1}{\underset{2}{\cancel{8}}} = \dfrac{1 \cdot 1}{7 \cdot 2} = \dfrac{1}{14}$

21) $10 = 2 \cdot 5, \ 15 = 3 \cdot 5$

LCM of 10 and 15 is $2 \cdot 3 \cdot 5 = 30$

23) a) $10 = 2 \cdot 5, \ 30 = 2 \cdot 3 \cdot 5$

LCD of $\dfrac{9}{10}$ and $\dfrac{5}{12}$ is $2 \cdot 3 \cdot 5 = 30$

b) $8 = 2 \cdot 2 \cdot 2, \ 12 = 2 \cdot 2 \cdot 3$

LCD of $\dfrac{7}{8}$ and $\dfrac{5}{12}$ is $2 \cdot 2 \cdot 2 \cdot 3 = 24$

c) $9 = 3 \cdot 3, \ 6 = 2 \cdot 3, \ 4 = 2 \cdot 2$

LCD of $\dfrac{4}{9}, \dfrac{1}{6}$, and $\dfrac{3}{4}$ is $2 \cdot 2 \cdot 3 \cdot 3 = 36$

25) a) $\dfrac{6}{11} + \dfrac{2}{11} = \dfrac{6+2}{11} = \dfrac{8}{11}$

b) $\dfrac{19}{20} - \dfrac{7}{20} = \dfrac{19-7}{20} = \dfrac{12}{20} = \dfrac{3}{5}$

c) $\dfrac{4}{25} + \dfrac{2}{25} + \dfrac{9}{25} = \dfrac{4+2+9}{25}$

$= \dfrac{15}{25} = \dfrac{3}{5}$

d) $\dfrac{2}{9} + \dfrac{1}{6} = \dfrac{4}{18} + \dfrac{3}{18} = \dfrac{4+3}{18} = \dfrac{7}{18}$

e) $\dfrac{3}{5} + \dfrac{11}{30} = \dfrac{18}{30} + \dfrac{11}{30} = \dfrac{18+11}{30} = \dfrac{29}{30}$

f) $\dfrac{13}{18} - \dfrac{2}{3} = \dfrac{13}{18} - \dfrac{12}{18} = \dfrac{13-12}{18} = \dfrac{1}{18}$

g) $\dfrac{4}{7} + \dfrac{5}{9} = \dfrac{36}{63} + \dfrac{35}{63} = \dfrac{36+35}{63}$

$= \dfrac{71}{63} \text{ or } 1\dfrac{8}{63}$

h) $\dfrac{5}{6} - \dfrac{1}{4} = \dfrac{10}{12} - \dfrac{3}{12} = \dfrac{10-3}{12} = \dfrac{7}{12}$

i) $\dfrac{3}{10} + \dfrac{7}{20} + \dfrac{3}{4} = \dfrac{6}{20} + \dfrac{7}{20} + \dfrac{15}{20}$

$= \dfrac{6+7+15}{20}$

$= \dfrac{28}{20} = \dfrac{7}{5} \text{ or } 1\dfrac{2}{5}$

j) $\dfrac{1}{6} + \dfrac{2}{9} + \dfrac{10}{27} = \dfrac{9}{54} + \dfrac{12}{54} + \dfrac{20}{54}$

$= \dfrac{9+12+20}{54}$

$= \dfrac{41}{54}$

27) a) $8\dfrac{5}{11}+6\dfrac{2}{11}=14\dfrac{5+2}{11}=14\dfrac{7}{11}$

b) $2\dfrac{1}{10}+9\dfrac{3}{10}=11\dfrac{1+3}{10}$
$=11\dfrac{4}{10}=11\dfrac{2}{5}$

c) $7\dfrac{11}{12}-1\dfrac{5}{12}=6\dfrac{11-5}{12}=6\dfrac{6}{12}=6\dfrac{1}{2}$

d) $3\dfrac{1}{5}+2\dfrac{1}{4}=3\dfrac{4}{20}+2\dfrac{5}{20}$
$=5\dfrac{4+5}{20}=5\dfrac{9}{20}$

e) $5\dfrac{2}{3}-4\dfrac{4}{15}=5\dfrac{10}{15}-4\dfrac{4}{15}$
$=1\dfrac{10-4}{15}$
$=1\dfrac{6}{15}=1\dfrac{2}{5}$

f) $9\dfrac{5}{8}-5\dfrac{3}{10}=9\dfrac{25}{40}-5\dfrac{12}{40}$
$=4\dfrac{25-12}{40}=4\dfrac{13}{40}$

g) $4\dfrac{3}{7}+6\dfrac{3}{4}=4\dfrac{12}{28}+6\dfrac{21}{28}$
$=10\dfrac{12+21}{28}$
$=10\dfrac{33}{28}=11\dfrac{5}{28}$

h) $7\dfrac{13}{20}+\dfrac{4}{5}=7\dfrac{13}{20}+\dfrac{16}{20}$
$=7\dfrac{13+16}{20}$
$=7\dfrac{29}{20}=8\dfrac{9}{20}$

29) $7\div1\dfrac{2}{3}=7\div\dfrac{5}{3}=\dfrac{7}{1}\cdot\dfrac{3}{5}=\dfrac{21}{5}=4\dfrac{1}{5}$
Alex can make 4 whole bears.
Determine amount of fabric used to make 4 bears:
Amount of fabric remaining
$= 7$ yds $- \dfrac{20}{3}$ yds
$= \dfrac{21}{3}$ yds $- \dfrac{20}{3}$ yds
$= \dfrac{1}{3}$ yd left over

30)

31) $175\cdot\dfrac{2}{7}=\dfrac{\overset{25}{\cancel{175}}}{1}\cdot\dfrac{2}{\cancel{7}}=50$ hits

33) Add the width of the frame twice to each dimension since it is being added to both sides of the picture.
$18\dfrac{3}{8}+2\dfrac{1}{8}+2\dfrac{1}{8}=22\dfrac{3+1+1}{8}$
$=22\dfrac{5}{8}$
$12\dfrac{1}{4}+2\dfrac{1}{8}+2\dfrac{1}{8}=2\dfrac{2}{8}+2\dfrac{1}{8}+2\dfrac{1}{8}$
$=16\dfrac{2+1+1}{8}$
$=16\dfrac{4}{8}=16\dfrac{1}{2}$
$16\dfrac{1}{2}$ in. by $22\dfrac{5}{8}$ in.

35) $\dfrac{2}{3}+1\dfrac{1}{4}+1\dfrac{1}{2}=\dfrac{8}{12}+1\dfrac{3}{12}+1\dfrac{6}{12}$

$\qquad =2\dfrac{8+3+6}{12}$

$\qquad =2\dfrac{17}{12}$

$\qquad =3\dfrac{5}{12}$ cups

37) $16\dfrac{3}{4}-11\dfrac{3}{5}=16\dfrac{15}{20}-11\dfrac{12}{20}$

$\qquad =5\dfrac{15-12}{20}$

$\qquad =5\dfrac{3}{20}$ gallons

39) $42\cdot\dfrac{5}{6}=\dfrac{\overset{7}{\cancel{42}}}{1}\cdot\dfrac{5}{\underset{1}{\cancel{6}}}=35$ problems

41) Add the total length welded so far:

$14\dfrac{1}{6}+10\dfrac{3}{4}=14\dfrac{2}{12}+10\dfrac{9}{12}$

$\qquad =24\dfrac{2+9}{12}=24\dfrac{11}{12}$

Subtract the total from the desired length:

$32\dfrac{7}{8}-24\dfrac{11}{12}=32\dfrac{21}{24}-24\dfrac{22}{24}$

$\qquad =31\dfrac{45}{24}-24\dfrac{22}{24}$

$\qquad =7\dfrac{45-22}{24}$

$\qquad =7\dfrac{23}{24}$ in.

43) $400\cdot\dfrac{3}{5}=\dfrac{\overset{8}{\cancel{400}}}{1}\cdot\dfrac{3}{\underset{1}{\cancel{5}}}=240$ students

Section 1.2: Exercises

1) a) base: 6; exponent: 4

   b) base: 2; exponent: 3

   c) base: $\dfrac{9}{8}$; exponent: 5

3) a) $9\cdot9\cdot9\cdot9=9^4$

   b) $2\cdot2\cdot2\cdot2\cdot2\cdot2\cdot2\cdot2=2^8$

   c) $\dfrac{1}{4}\cdot\dfrac{1}{4}\cdot\dfrac{1}{4}=\left(\dfrac{1}{4}\right)^3$

5) a) $8^2=8\cdot8=64$

   b) $(11)^2=11\cdot11=121$

   c) $2^4=2\cdot2\cdot2\cdot2=16$

   d) $5^3=5\cdot5\cdot5=125$

   e) $3^4=3\cdot3\cdot3\cdot3=81$

   f) $(12)^2=12\cdot12=144$

   g) $1^2=1\cdot1=1$

   h) $\left(\dfrac{3}{10}\right)^2=\dfrac{3}{10}\cdot\dfrac{3}{10}=\dfrac{9}{100}$

   i) $\left(\dfrac{1}{2}\right)^6=\dfrac{1}{2}\cdot\dfrac{1}{2}\cdot\dfrac{1}{2}\cdot\dfrac{1}{2}\cdot\dfrac{1}{2}\cdot\dfrac{1}{2}=\dfrac{1}{64}$

   j) $(0.3)^2=(0.3)\cdot(0.3)=0.09$

7) $(0.5)^2=0.5\cdot0.5=0.25$ or

   $(0.5)^2=\left(\dfrac{1}{2}\right)^2=\dfrac{1}{4}$

9) Answers may vary.

11) $17 - 2 + 4 = 15 + 4 = 19$

13) $48 \div 2 + 14 = 24 + 14 = 38$

15) $20 - 3 \cdot 2 + 9 = 20 - 6 + 9$
$= 14 + 9 = 23$

17) $8 + 12 \cdot \dfrac{3}{4} = 8 + 9 = 17$

19) $\dfrac{2}{5} \cdot \dfrac{1}{8} + \dfrac{2}{3} \cdot \dfrac{9}{10} = \dfrac{\cancel{2}}{5} \cdot \dfrac{1}{\cancel{8}^{\,4}} + \dfrac{\cancel{2}^{\,1}}{\cancel{3}^{\,1}} \cdot \dfrac{\cancel{9}^{\,3}}{\cancel{10}^{\,5}}$

$= \dfrac{1}{20} + \dfrac{3}{5} = \dfrac{1}{20} + \dfrac{12}{20}$

$= \dfrac{13}{20}$

21) $2 \cdot \dfrac{3}{4} - \left(\dfrac{2}{3}\right)^2 = 2 \cdot \dfrac{3}{4} - \dfrac{4}{9}$

$= \dfrac{3}{2} - \dfrac{4}{9}$

$= \dfrac{27}{18} - \dfrac{8}{18}$

$= \dfrac{19}{18}$ or $1\dfrac{1}{18}$

23) $25 - 11 \cdot 2 + 1 = 25 - 22 + 1$
$= 3 + 1 = 4$

25) $39 - 3(9 - 7)^3 = 39 - 3(2)^3$
$= 39 - 3(8)$
$= 39 - 24 = 15$

27) $60 \div 15 + 5 \cdot 3 = 4 + 15 = 19$

29) $7\left[45 \div (19 - 10)\right] + 2$
$= 7\left[45 \div (9)\right] + 2$
$= 7 \cdot 5 + 2$
$= 35 + 2 = 37$

31) $1 + 2\left[(3 + 2)^3 \div (11 - 6)^2\right]$
$= 1 + 2\left[(5)^3 \div (5)^2\right]$
$= 1 + 2[125 \div 25]$
$= 1 + 2[5]$
$= 1 + 10 = 11$

33) $\dfrac{4(7 - 2)^2}{12^2 - 8 \cdot 3} = \dfrac{4 \cdot (5)^2}{144 - 24} = \dfrac{4 \cdot 25}{120}$

$= \dfrac{100}{120} = \dfrac{10}{12} = \dfrac{5}{6}$

35) $\dfrac{4(9 - 6)^3}{(2)^2 + 3 \cdot 8} = \dfrac{4(3)^3}{4 + 24}$

$= \dfrac{\cancel{4}^{\,1}(27)}{\cancel{28}_{\,7}}$

$= \dfrac{27}{7}$ or $3\dfrac{6}{7}$

Section 1.3: Exercises

1) Acute

3) Straight

5) Supplementary; complementary

7) $90° - 59° = 31°$

9) $90° - 12° = 78°$

11) $180° - 143° = 37°$

13) $180° - 38° = 142°$

15)   Angle Measure   Reason

$m\angle A = 180° - 31° = 149°$ supplementary

$m\angle B = 31°$ vertical

$m\angle C = m\angle A = 149°$ vertical

17) 180

19) the sum of the two known angles

$119° + 22° = 141°$
the measure of the unknown angle

$180° - 141° = 39°$
The triangle is obtuse since it contains one obtuse angle.

21) the sum of the two known angles

$90° + 51° = 141°$
the measure of the unknown angle

$180° - 141° = 39°$
The triangle is right since it contains one right angle.

23) Equilateral

25) Isosceles

27) True

29) $A = l \cdot w$ $P = 2l + 2w$

$\quad = 10 \cdot 8$ $= 2 \cdot 10 + 2 \cdot 8$

$\quad = 80 \text{ ft}^2$ $= 20 + 16 = 36 \text{ ft}$

31) $A = \dfrac{1}{2}bh$

$\quad = \dfrac{1}{2}(14 \text{ cm})(6 \text{ cm})$

$\quad = 7 \cdot 6 = 42 \text{ cm}^2$

$P = a + b + c$

$\quad = 8 \text{ cm} + 14\text{cm} + 7.25 \text{ cm} = 29.25 \text{ cm}$

33) $A = s^2$

$\quad = (6.5 \text{ mi})^2 = 42.25 \text{ mi}^2$

$P = 4s = 4 \cdot 6.5 \text{ mi} = 26 \text{ miles}$

35) $A = \dfrac{1}{2}h(b_1 + b_2)$

$\quad = \dfrac{1}{2} \cdot 12(11 + 16)$

$\quad = 6 \cdot 27 = 162 \text{ in}^2$

$P = a + b + c + d$

$\quad = 13 \text{ in} + 11 \text{ in} + 12 \text{ in} + 16 \text{ in}$

$\quad = 52 \text{ in}$

37) a) $A = \pi r^2$ b) $C = 2\pi r$

$\quad = \pi(5 \text{ in})^2$ $= 2\pi(5 \text{ in})$

$\quad = 25\pi \text{ in}^2$ $= 10\pi \text{ in}$

$\quad \approx 79.5 \text{ in}^2$ $\approx 31.4 \text{ in}$

39) a) $A = \pi r^2$

$\quad = \pi(2.5 \text{ m})^2$

$\quad = 6.25\pi \text{ m}^2 \approx 19.625 \text{ m}^2$

b) $C = 2\pi r$

$\quad = 2\pi(2.5 \text{ m})$

$\quad = 5\pi \text{ m} \approx 15.7 \text{ m}$

41) $A = \pi r^2$ $C = 2\pi r$

$\quad = \pi\left(\dfrac{1}{2} \text{ m}\right)^2$ $= 2\pi\left(\dfrac{1}{2} \text{ m}\right)$

$\quad = \dfrac{1}{4}\pi \text{ m}^2$ $= \pi \text{ m}$

43) $A = \pi r^2$ $C = 2\pi r$

$\quad = \pi(7 \text{ ft})^2$ $= 2\pi(7 \text{ ft})$

$\quad = 49\pi \text{ ft}^2$ $= 14\pi \text{ ft}$

45) $A = (23 \text{ m})(13 \text{ m}) + (11 \text{ m})(7 \text{ m})$

    $= 299 \text{ m}^2 + 77 \text{ m}^2$

    $= 376 \text{ m}^2$

    $P = 2(23 \text{ m}) + 2(20 \text{ m})$

    $= 46 \text{ m} + 40 \text{ m} = 86 \text{ m}$

47) $A = (20.5 \text{ in})(4.8 \text{ in}) + (5.7 \text{ in})(8.4 \text{ in})$

    $+ (11.2 \text{ in})(4.9 \text{ in})$

    $= 98.40 \text{ in}^2 + 47.88 \text{ in}^2 + 54.88 \text{ in}^2$

    $= 201.16 \text{ in}^2$

    $P = 2(20.5 \text{ in}) + 2(13.2 \text{ in})$

    $= 41 \text{ in} + 26.4 \text{ in} = 67.4 \text{ in}$

49) $A = (14 \text{ in})(12 \text{ in}) - (10 \text{ in})(8 \text{ in})$

    $= 168 \text{ in}^2 - 80 \text{ in}^2 = 88 \text{ in}^2$

51) $A = (4 \text{ ft})(7 \text{ ft}) - (1.5 \text{ ft})^2$

    $= 28 \text{ ft}^2 - 2.25 \text{ ft}^2 = 25.75 \text{ ft}^2$

53) $A = (16 \text{ cm})^2 - 3.14(5 \text{ cm})^2$

    $= 256 \text{ cm}^2 - 78.5 \text{ cm}^2 = 177.5 \text{ cm}^2$

55) $V = lwh = (7 \text{ m})(5 \text{ m})(2 \text{ m}) = 70 \text{ m}^3$

57) a) $V = \dfrac{4}{3}\pi r^3$

    $= \dfrac{4}{3}\pi(6 \text{ in})^3 = \dfrac{4}{3}\pi(216 \text{ in}^3)$

    $= 4\pi \cdot 72 \text{ in}^3 = 288\pi \text{ in}^3$

59) $V = \dfrac{4}{3}\pi r^3$

    $= \dfrac{4}{3}\pi(5 \text{ ft})^3 = \dfrac{4}{3}\pi(125 \text{ ft}^3) = \dfrac{500}{3}\pi \text{ ft}^3$

61) $V = \pi r^2 h$

    $= \pi(4 \text{ cm})^2(8.5 \text{ cm})$

    $= \pi(16 \text{ cm}^2)(8.5 \text{ cm})$

    $= 136\pi \text{ cm}^3$

63) a) $A = (9 \text{ ft})(6.5 \text{ ft}) = 58.5 \text{ ft}^2$

    $58.5 \ \cancel{\text{ft}^2} \cdot \left(\dfrac{\$20}{\cancel{\text{ft}^2}}\right) = \$1170$

    b) No, it would cost
       $\$1170$ to use this glass.

65) a) $V = \pi(3 \text{ ft})^2(8 \text{ ft})$

    $\approx (3.14)(9 \text{ ft}^2)(8 \text{ ft})$

    $\approx 226.08 \text{ ft}^3$

    b) $\text{Capacity} = (7.48)(226.08)$

       $\approx 1691 \text{ gal}$

67) a) $C = 2\pi(10 \text{ in}) = 20(3.14) \text{ in}$

    $\approx 62.8 \text{ in}$

    b) $A = \pi(10 \text{ in})^2 \approx 3.14(100 \text{ in}^2)$

    $\approx 314 \text{ in}^2$

69) $V = (30 \text{ ft})(19 \text{ ft})(1.5 \text{ ft})$

    $= 855 \text{ ft}^3$

  $\text{Capacity} = (855)(7.48) = 6395.4 \text{ gal}$

71) First, find the area of the countertop;
It is made of three sections

$$A = 2\frac{1}{4}\left(10\frac{1}{4} - 2\frac{3}{4}\right)$$

$$+2\frac{1}{4}\left(9\frac{1}{6} - 2\frac{1}{2} - 2 \cdot 2\frac{1}{4}\right) + 2\frac{1}{4}\left(6\frac{5}{6}\right)$$

$$= 2\frac{1}{4}\left(7\frac{1}{2} + 6\frac{5}{6} + 2\frac{1}{6}\right)$$

$$= 2\frac{1}{4}\left(16\frac{1}{2}\right) = 37\frac{1}{8}$$

The area of the countertop is $37\frac{1}{8}$ft².

To find the total cost, multiply the area
by the unit cost.

$$\text{Cost} = 37\frac{1}{8}\text{ft}^2 \cdot (\$80.00) = \$2970.00$$

No, she cannot afford her first
choice granite.
The granite countertop would
cost $2970.00

73) $C = 2\pi r$

$$= 2\pi(4.6\text{ in})$$

$$= 2(3.14)(4.6\text{ in})$$

$$= 28.9\text{ in}$$

75) a) $A = \frac{1}{2}h(b_1 + b_2)$

$$= \frac{1}{2}(4\text{ft})(14\text{ft}+8\text{ft})$$

$$= 2\text{ft}(22\text{ft}) = 44\text{ft}^2$$

b) First, find the perimeter
of the garden
$P = 14\text{ft}+5\text{ft}+8\text{ft}+5\text{ft} = 32\text{ft}$
To find the total cost, multiply
the perimeter by unit cost
$\text{Cost} = 32\text{ft} \cdot \$32.50/\text{ft} = \$752$

77) $A = \frac{1}{2}bh = \frac{1}{2}(15.6\text{in})(18\text{in})$

$$= 140.4\text{in}^2$$

Use 3.14 for π

$$V = \frac{1}{3}\pi r^2 h = \frac{1}{3}(3.14)(12\text{ft}^2)(8\text{ft})$$

$$= 1205.76\text{ft}^3$$

Section 1.4: Exercises

1) Answers may vary.

3) a) 17 b) 17,0 c) 17,0,−25

d) $17, 3.8, \frac{4}{5}, 0, -25, 6.\overline{7}, -2\frac{1}{8}$

e) $\sqrt{10}, 9.721983...$
f) all numbers in the set

5) true

7) false

9) true

11)

13)

15) The distance of the number from zero

17) −8

19) 15

21) $\frac{3}{4}$

23) $|-10| = 10$

25) $\left|\frac{9}{4}\right| = \frac{9}{4}$

27) $-|-14| = -14$

29) $|17 - 4| = |13| = 13$

31) $-\left|-4\dfrac{1}{7}\right| = -4\dfrac{1}{7}$

33) $-10, -2, 0, \dfrac{9}{10}, 3.8, 7$

35) $-6.51, -6.5, -5, 2, 7\dfrac{1}{3}, 7\dfrac{5}{6}$

37) true

39) false

41) false

43) false

45) -53

47) 14 million

49) $-419,000$

Section 1.5: Exercises

1) Answers may vary.

3) Answers may vary.

5) $6 - 11 = -5$

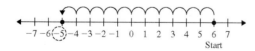

7) $-2 + (-7) = -9$

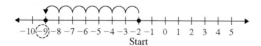

9) $8 + (-15) = -7$

11) $-3 - 11 = -14$

13) $-31 + 54 = 23$

15) $-26 - (-15) = -26 + 15 = -11$

17) $-352 - 498 = -850$

19) $-\dfrac{7}{12} + \dfrac{3}{4} = -\dfrac{7}{12} + \dfrac{9}{12} = \dfrac{2}{12} = \dfrac{1}{6}$

21) $-\dfrac{1}{6} - \dfrac{7}{8} = -\dfrac{4}{24} + \left(-\dfrac{21}{24}\right) = -\dfrac{25}{24}\ or -1\dfrac{1}{24}$

23) $-\dfrac{4}{9} - \left(-\dfrac{4}{15}\right) = -\dfrac{20}{45} + \dfrac{12}{45} = -\dfrac{8}{45}$

25) $19.4 + (-16.7) = 2.7$

27) $-25.8 - (-16.57) = -25.8 + 16.57 = -9.23$

29) $9 - (5 - 11) = 9 - (-6) = 9 + 6 = 15$

31) $-1 + (-6 - 4) = -1 + (-10) = -11$

33) $(-3 - 1) - (-8 + 6) = (-4) - (-2)$
$= -4 + 2 = -2$

35) $-16 + 4 + 3 - 10 = -12 + 3 - 10$
$= -9 - 10 = -19$

37) $5 - (-30) - 14 + 2 = 5 + 30 - 14 + 2$
$= 35 - 14 + 2$
$= 21 + 2 = 23$

Chapter 1: The Real Number System and Geometry

39) $\dfrac{4}{9} - \left(\dfrac{2}{3} + \dfrac{5}{6}\right) = \dfrac{8}{18} - \left(\dfrac{12}{18} + \dfrac{15}{18}\right)$

$\qquad\qquad = \dfrac{8}{18} - \left(\dfrac{27}{18}\right)$

$\qquad\qquad = -\dfrac{19}{18} \text{ or } -1\dfrac{1}{18}$

41) $\left(\dfrac{1}{8} - \dfrac{1}{2}\right) + \left(\dfrac{3}{4} - \dfrac{1}{6}\right)$

$\qquad\qquad = \left(\dfrac{3}{24} - \dfrac{12}{24}\right) + \left(\dfrac{18}{24} - \dfrac{4}{24}\right)$

$\qquad\qquad = -\dfrac{9}{24} + \dfrac{14}{24} = \dfrac{5}{24}$

43) $(2.7 + 3.8) - (1.4 - 6.9) = 6.5 - (-5.5)$

$\qquad\qquad\qquad\qquad = 6.5 + 5.5 = 12$

45) $|7 - 11| + |6 + (-13)| = |-4| + |-7|$

$\qquad\qquad\qquad\quad = 4 + 7 = 11$

47) $-|-2 - (-3)| - 2|-5 + 8|$

$\quad = -|-2 + 3| - 2|3|$

$\quad = -|1| - 2(3)$

$\quad = -1 - 6 = -7$

49) false   51) false   53) true

55) $-18 + 6 = -12$
His score in the 2005 Masters is -12.

57) $6,110,000 - 5,790,000 = 320,000$
The carbon emissions in China were 320,000 thousand metric tons more than those of the United States.

59) $881,566 + 45,407 = 926,973$
There were 926,973 flights at O'Hare in 2007.

61) a) $1745 - 1748 = -3$

b) $1611 - 1745 = -134$

c) $1480 - 1611 = -131$

d) $1480 - 1763 = -283$

63) a) $2.1 - 2.8 = -0.7$

b) $2.5 - 2.1 = 0.4$

c) $2.6 - 2.5 = 0.1$

d) $2.6 - 2.8 = -0.2$

65) $5 + 7; 12$

67) $10 - 16; -6$

69) $-21 + 13; -8$

71) $-20 + 30; 10$

73) $-37 + 22; -15$

75) $23 - 19; 4$

77) $(-5 + 11) - 18 = 6 - 18 = -12$

Section 1.6: Exercises

1) negative

3) $-8 \cdot 7 = -56$

5) $-15 \cdot (-3) = 45$

7) $-4 \cdot 3 \cdot (-7) = -12 \cdot (-7) = 84$

9) $\dfrac{4}{33} \cdot \left(-\dfrac{11}{10}\right) = \dfrac{\overset{}{\cancel{4}}}{\underset{3}{\cancel{33}}} \cdot \left(-\dfrac{\overset{1}{\cancel{11}}}{\underset{5}{\cancel{10}}}\right) = -\dfrac{2}{15}$

11) $(-0.5)(-2.8) = 1.4$

10

13) $-9 \cdot (-5) \cdot (-1) \cdot (-3) = 45 \cdot 3 = 135$

15) $\dfrac{3}{10} \cdot (-7) \cdot (8) \cdot (-1) \cdot (-5) = \dfrac{3}{\cancel{10}_{1}} \cdot \dfrac{\cancel{280}^{-28}}{1}$

$$= -84$$

17) when k is negative

19) when $k \neq 0$

21) $(-6)^2 = 36$

23) $-5^3 = -125$

25) $(-1)^5 = -1$

27) $-7^2 = -49$

29) $-2^5 = -32$

31) positive

33) $-50 \div (-5) = 10$

35) $\dfrac{64}{-16} = -4$

37) $\dfrac{-2.4}{0.3} = -8$

39) $-\dfrac{12}{13} \div (-\dfrac{6}{5}) = \dfrac{10}{13}$

$-\dfrac{0}{7} = 0$

41) $\dfrac{270}{-180} = -\dfrac{3}{2} \ or \ -1\dfrac{1}{2}$

43) $7 + 8(-5) = 7 + (-40) = -33$

45) $(9-14)^2 - (-3)(6) = (-5)^2 - (-18)$

$$= 25 + 18$$

$$= 43$$

47) $10 - 2(1-4)^3 \div 9$

$$= 10 - 2(-3)^3 \div 9$$

$$= 10 - 2(-27) \div 9$$

$$= 10 - (-54) \div 9$$

$$= 10 + 6 = 16$$

49) $\left(-\dfrac{3}{4}\right)(8) - 2\left[7 - (-3)(-6)\right]$

$$= -6 - 2[7-18]$$

$$= -6 - 2[-11]$$

$$= -6 + 22 = 16$$

51) $\dfrac{-46 - 3(-12)}{(-5)(-2)(-4)} = \dfrac{-46 + 36}{-40}$

$$= \dfrac{-10}{-40} = \dfrac{1}{4}$$

53) $-12 \cdot 6 = -72$

57) $9 + (-7)(-5) = 9 + 35 = 44$

55) $\dfrac{63}{-9} + 7 = -7 + 7 = 0$

61) $(-4)(-8) - 19 = 32 - 19 = 13$

63) $\dfrac{-100}{4} - (-7+2) = -25 - (-7+2)$

$$= -25 - (-5) = -25 + 5 = -20$$

65) $2\lceil 18 + (-31) \rceil = 2[-13] = -26$

67) $\dfrac{2}{3}(-27) = \dfrac{2}{\cancel{3}}\left(-\cancel{27}^{9}\right) = -18$

Chapter 1: The Real Number System and Geometry

69) $12(-5)+\dfrac{1}{2}(36)=-60+18=-42$

Section 1.7: Exercises

1) The constant is 4.

Term	Coeff.
$7p^2$	7
$-6p$	-6
4	4

3) The constant is 11.

Term	Coeff.
x^2y^2	1
$2xy$	2
$-y$	-1
11	11

5) The constant is -1.

Term	Coeff.
$-2g^5$	-2
$\dfrac{g^4}{5}$	$\dfrac{1}{5}$
$3.8g^2$	3.8
g	1
-1	-1

7) a) $4c+3$ when $c=2$

$$4(2)+3=8+3=11$$

b) $4c+3$ when $c=-5$

$$4(-5)+3=-20+3=-17$$

9) $x+4y=3+4(-5)=3-20=-17$

11) $z^2-xy-19$

$$=(-2)^2-(3)(-5)-19$$
$$=4-(-15)-19=4+15-19$$
$$=19-19=0$$

13) $\dfrac{x^3}{2y+1}=\dfrac{(3)^3}{(2)(-5)+1}$

$$=\dfrac{27}{-10+1}=\dfrac{27}{-9}=-3$$

15) $\dfrac{z^2-y^2}{2y-4(x+z)}$

$$=\dfrac{(-2)^2-(-5)^2}{2(-5)-4(3-2)}$$
$$=\dfrac{4-25}{-10-4(1)}$$
$$=\dfrac{-21}{-14}=\dfrac{\overset{3}{\cancel{21}}}{\underset{2}{\cancel{14}}}=\dfrac{3}{2}$$

17) No, the exponents are different.

19) Yes. Both are a^3b terms

21) 1

23) -5

25) distributive

27) identity

29) commutative

31) associative

33) $19+p$

35) $(8+1)+9$

37) $3k-21$

39) y

41) No. Subtraction is not commutative.

43) $2(1+9) = 2\cdot 1 + 2\cdot 9 = 2 + 18 = 20$

45) $-2(5+7) = -2\cdot 5 + (-2)\cdot 7$
$$= -10 + (-14) = -24$$

47) $4(8-3) = 4\cdot 8 + 4\cdot(-3) = 32 + (-12) = 20$

49) $-(10-4) = -10 + 4 = -6$

51) $8(y+3) = 8y + 8\cdot 3 = 8y + 24$

53) $-10(z+6) = -10z + (-10)\cdot 6 = 10z - 60$

55) $-3(x-4y-6)$
$$= -3x + (-3)\cdot(-4y) + (-3)\cdot(-6)$$
$$= -3x + 12y + 18$$

57) $-(-8c + 9d - 14) = 8c - 9d + 14$

59) $10p + 9 + 14p - 2 = 10p + 14p + 9 - 2$
$$= 24p + 7$$

61) $-18y^2 - 2y^2 + 19 + y^2 - 2 + 13$
$$= -18y^2 - 2y^2 + y^2 + 19 - 2 + 13$$
$$= -19y^2 + 30$$

63) $\dfrac{4}{9} + 3r - \dfrac{2}{3} + \dfrac{1}{5}r = 3r + \dfrac{1}{5}r + \dfrac{4}{9} - \dfrac{2}{3}$
$$= \dfrac{15}{5}r + \dfrac{1}{5}r + \dfrac{4}{9} - \dfrac{6}{9}$$
$$= \dfrac{16}{5}r - \dfrac{2}{9}$$

65) $2(3w+5) + w = 6w + 10 + w$
$$= 6w + w + 10$$
$$= 7w + 10$$

67) $9 - 4(3-x) - 4x + 3$
$$= 9 - 12 + 4x - 4x + 3$$
$$= 4x - 4x + 9 - 12 + 3$$
$$= 0x + 0 = 0$$

69) $3g - (8g+3) + 5 = 3g - 8g - 3 + 5$
$$= -5g + 2$$

71) $-5(t-2) - (10-2t)$
$$= -5t + 10 - 10 + 2t$$
$$= -5t + 2t + 10 - 10$$
$$= -3t$$

73) $3\big[2(5x+7) - 11\big] + 4(7+x)$
$$= 3\big[10x + 14 - 11\big] + 28 + 4x$$
$$= 3\big[10x + 3\big] + 28 + 4x$$
$$= 30x + 9 + 28 + 4x$$
$$= 30x + 4x + 9 + 28$$
$$= 34x + 37$$

75) $\dfrac{4}{5}(2z+10) - \dfrac{1}{2}(z+3)$
$$= \dfrac{8}{5}z + 8 - \dfrac{1}{2}z - \dfrac{3}{2}$$
$$= \dfrac{8}{5}z - \dfrac{1}{2}z + 8 - \dfrac{3}{2}$$
$$= \dfrac{16}{10}z - \dfrac{5}{10}z + \dfrac{16}{2} - \dfrac{3}{2}$$
$$= \dfrac{11}{10}z + \dfrac{13}{2}$$

Chapter 1: The Real Number System and Geometry

77) $1+\dfrac{3}{4}(10t-3)+\dfrac{5}{8}\left(t+\dfrac{1}{10}\right)$

$\qquad = 1+\dfrac{15}{2}t-\dfrac{9}{4}+\dfrac{5}{8}t+\dfrac{1}{16}$

$\qquad = \dfrac{15}{2}t+\dfrac{5}{8}t+1-\dfrac{9}{4}+\dfrac{1}{16}$

$\qquad = \dfrac{60}{8}t+\dfrac{5}{8}t+\dfrac{16}{16}-\dfrac{36}{16}+\dfrac{1}{16}$

$\qquad = \dfrac{65}{8}t-\dfrac{19}{16}$

79) $2.5(x-4)-1.2(3x+8)$

$\qquad = 2.5x-10-3.6x-9.6$

$\qquad = 2.5x-3.6x-10-9.6$

$\qquad = -1.1x-19.6$

81) $x+18$

83) $x-6$

85) $x-3$

87) $12+2x$

89) $(3+2x)-7=2x+3-7=2x-4$

91) $(x+15)-5=x+15-5=x+10$

Chapter 1 Review

1) a) $16=1\cdot16$

$\qquad 16=2\cdot8$

$\qquad 16=4\cdot4$

$\qquad$ The factors of 16 are 1, 2, 4, 8, 16.

b) $37=1\cdot37$

$\qquad$ The factors of 37 are 1 and 37.

3) a) $\dfrac{12}{30}=\dfrac{12\div6}{30\div6}=\dfrac{2}{5}$

b) $\dfrac{414}{702}=\dfrac{414\div18}{702\div18}=\dfrac{23}{39}$

5) $\dfrac{45}{64}\cdot\dfrac{32}{75}=\dfrac{\overset{3}{\cancel{45}}}{\underset{2}{\cancel{64}}}\cdot\dfrac{\overset{1}{\cancel{32}}}{\underset{5}{\cancel{75}}}=\dfrac{3}{10}$

7) $35\div\dfrac{7}{8}=35\cdot\dfrac{8}{7}=\overset{5}{\cancel{35}}\cdot\dfrac{8}{\underset{1}{\cancel{7}}}=40$

9) $\dfrac{30}{49}\div2\dfrac{6}{7}=\dfrac{30}{49}\div\dfrac{20}{7}$

$\qquad = \dfrac{30}{49}\cdot\dfrac{7}{20}$

$\qquad = \dfrac{\overset{3}{\cancel{30}}}{\underset{7}{\cancel{49}}}\cdot\dfrac{\overset{1}{\cancel{7}}}{\underset{2}{\cancel{20}}}=\dfrac{3}{14}$

11) $\dfrac{2}{3}+\dfrac{1}{4}=\dfrac{8}{12}+\dfrac{3}{12}=\dfrac{11}{12}$

13) $\dfrac{1}{5}+\dfrac{1}{3}+\dfrac{1}{6}=\dfrac{6}{30}+\dfrac{10}{30}+\dfrac{5}{30}=\dfrac{21}{30}=\dfrac{7}{10}$

15) $\dfrac{5}{8}-\dfrac{2}{7}=\dfrac{35}{56}-\dfrac{16}{56}=\dfrac{19}{56}$

17) $9\dfrac{3}{8}-2\dfrac{5}{6}=9\dfrac{9}{24}-2\dfrac{20}{24}$

$\qquad = 8\dfrac{33}{24}-2\dfrac{20}{24}$

$\qquad = 6\dfrac{33-20}{24}=6\dfrac{13}{24}$

19) $3^4=3\cdot3\cdot3\cdot3=81$

21) $\left(\dfrac{3}{4}\right)^3=\left(\dfrac{3}{4}\right)\cdot\left(\dfrac{3}{4}\right)\cdot\left(\dfrac{3}{4}\right)=\dfrac{27}{64}$

23) $13 - 7 + 4 = 6 + 4 = 10$

25) $\dfrac{12 - 56 \div 8}{(1+5)^2 - 2^4} = \dfrac{12 - 7}{(6)^2 - 2^4}$

$= \dfrac{5}{36 - 16}$

$= \dfrac{5}{20} = \dfrac{1}{4}$

27) $180° - 78° = 102°$

29) $A = lw$

$= \left(3\dfrac{1}{2} \text{ mi}\right)\left(1\dfrac{7}{8} \text{ mi}\right)$

$= \left(\dfrac{7}{2} \text{ mi}\right)\left(\dfrac{15}{8} \text{ mi}\right)$

$= \dfrac{105}{16} \text{ mi}^2 = 6\dfrac{9}{16} \text{ mi}^2$

$P = 2l + 2w$

$= 2\left(3\dfrac{1}{2} \text{ mi}\right) + 2\left(1\dfrac{7}{8} \text{ mi}\right)$

$= 7 \text{ mi} + \dfrac{15}{4} \text{ mi} = 7\text{mi} + 3\dfrac{3}{4}\text{mi}$

$= 10\dfrac{3}{4} \text{ mi}$

31) $A = \dfrac{1}{2}h(b_1 + b_2)$

$= \dfrac{1}{2}(5 \text{ in})(6 \text{ in} + 8 \text{ in})$

$= 35 \text{ in}^2$

$P = a + c + b_1 + b_2$

$= 13 \text{ in} + 5 \text{ in} + 6 \text{ in} + 8 \text{ in}$

$= 32 \text{ in}$

33) a) $A = \pi r^2$

$= \pi (3 \text{ in})^2$

$= 9\pi \text{ in}^2 \approx 28.26 \text{ in}^2$

b) $C = 2\pi r$

$= 2\pi (3 \text{ in})$

$= 6\pi \text{ in} \approx 18.84 \text{ in}$

35) $A = \pi (13 \text{ cm})^2 - \dfrac{1}{2}(20 \text{ cm})(17 \text{ cm})$

$= 169(3.14) \text{ cm}^2 - 170 \text{ cm}^2$

$= 530.66 \text{ cm}^2 - 170 \text{ cm}^2$

$= 360.66 \text{ cm}^2$

37) $V = \pi r^2 h$

$= \pi (1 \text{ ft})^2 (1.3 \text{ ft}) = 1.3\pi \text{ ft}^3$

39) $V = s^3 = \left(2\dfrac{1}{2} \text{ in}\right)^3$

$= \left(\dfrac{5}{2} \text{ in}\right)^3 = \dfrac{125}{8} \text{ in}^3 \text{ or } 15\dfrac{5}{8} \text{ in}^3$

41) a) $\{-16, 0, 4\}$

b) $\left\{\dfrac{7}{16}, -16, 0, 3.\overline{2}, 8.5, 4\right\}$

c) $\{4\}$

d) $\{0, 4\}$

e) $\left\{\sqrt{31}, 6.01832...\right\}$

43) a) $|-18| = 18$ b) $-|-7| = -7$

45) $-21 - (-40) = -21 + 40 = 19$

47) $\dfrac{5}{12} - \dfrac{5}{8} = \dfrac{5}{12} \cdot \dfrac{2}{2} - \dfrac{5}{8} \cdot \dfrac{3}{3} = \dfrac{10-15}{24} = -\dfrac{5}{24}$

49) $\left(-\dfrac{3}{2}\right)(8) = \left(-\dfrac{3}{\cancel{2}_{1}}\right)\left(\dfrac{\cancel{8}^{4}}{}\right) = -12$

51) $(-4)(3)(-2)(-1)(-3) = 72$

53) $-108 \div 9 = -12$

55) $-3\dfrac{1}{8} \div \left(-\dfrac{5}{6}\right) = \dfrac{\cancel{25}^{5}}{\cancel{8}_{4}} \cdot \dfrac{\cancel{6}^{3}}{\cancel{5}_{1}} = \dfrac{15}{4} = 3\dfrac{3}{4}$

57) $-6^2 = -36$

59) $(-2)^6 = 64$

61) $3^3 = 27$

63) $56 \div (-7) - 1 = -8 - 1 = -9$

65) $-11 + 4 \cdot 3 + (-8+6)^5 = -11 + 4 \cdot 3 + (-2)^5$
$$= -11 + 12 + (-32)$$
$$= 1 + (-32)$$
$$= -31$$

67) $\dfrac{-120}{-3}; \ 40$

69) $(-4) \cdot 7 - 15 = -28 - 15 = -43$

71) The constant is 14

Term	Coeff.
$5z^4$	5
$-8z^3$	-8
$\dfrac{3}{5}z^2$	$\dfrac{3}{5}$
$-z$	-1
14	14

73) $\dfrac{2a+b}{a^3 - b^2}$ when $a = -3$ and $b = 5$

$$\dfrac{2(-3)+5}{(-3)^3 - (5)^2} = \dfrac{-6+5}{-27-25} = \dfrac{-1}{-52} = \dfrac{1}{52}$$

75) inverse

77) distributive

79) $7(3-9) = 7 \cdot 3 - 7 \cdot 9 = 21 - 63 = -42$

81) $-(15-3) = -15 + 3 = -12$

83) $9m - 14 + 3m + 4 = 12m - 10$

85) $15y^2 + 8y - 4 + 2y^2 - 11y + 1$
$$= 15y^2 + 2y^2 + 8y - 11y - 4 + 1$$
$$= 17y^2 - 3y - 3$$

87) $\frac{3}{2}(5n-4)+\frac{1}{4}(n+6)$

$$=\frac{15}{2}n-6+\frac{1}{4}n+\frac{3}{2}$$

$$=\frac{15}{2}n+\frac{1}{4}n-6+\frac{3}{2}$$

$$=\frac{30}{4}n+\frac{1}{4}n-\frac{12}{2}+\frac{3}{2}$$

$$=\frac{31}{4}n-\frac{9}{2}$$

Chapter 1 Test

1) $210=2\cdot3\cdot5\cdot7$

3) $\frac{7}{16}\cdot\frac{10}{21}=\frac{\cancel{7}^{1}}{\cancel{16}_{8}}\cdot\frac{\cancel{10}^{5}}{\cancel{21}_{3}}=\frac{5}{24}$

5) $10\frac{2}{3}-3\frac{1}{4}=10\frac{8}{12}-3\frac{3}{12}$

$$=7\frac{8-3}{12}=7\frac{5}{12}$$

7) $\frac{3}{5}-\frac{17}{20}=\frac{12}{20}-\frac{17}{20}=-\frac{\cancel{5}^{1}}{\cancel{20}_{4}}=-\frac{1}{4}$

9) $16+8\div2=16+4=20$

11) $-15\cdot(-4)=60$

13) $23-6\left[-4+(9-11)^{4}\right]$

$$=23-6\left[-4+(-2)^{4}\right]$$

$$=23-6[-4+16]=23-6[12]$$

$$=23-72=-49$$

15) $14,693\text{ ft}-(-518\text{ ft})$

$=14,693\text{ ft}+518\text{ ft}=15,211\text{ ft}$

17) $180°-31°=149°$

19) a) $A=\frac{1}{2}bh=\frac{1}{2}(6\text{ mm})(3\text{ mm})$

$$=9\text{ mm}^2$$

$P=a+b+c$

$=5\text{ mm}+6\text{ mm}+3.6\text{ mm}$

$=14.6\text{ mm}$

b) $A=lw=(15\text{ cm})(7\text{ cm})$

$$=105\text{ cm}^2$$

$P=2l+2w$

$=2(15\text{ cm})+2(7\text{ cm})$

$=30\text{ cm}+14\text{ cm}=44\text{ cm}$

c) $A=l\cdot w+2\cdot l\cdot w$

$(16\text{in})((14-4)\text{in})+2(5\text{in})(4\text{in})$

$=160\text{ in}^2+40\text{ in}^2=200\text{ in}^2$

$P=(5+4+6+4+5+14+16+14)\text{ in}$

$=68\text{ in}$

21) a) $A=\pi r^2=\pi(9\text{ ft})^2=81\pi\text{ ft}^2$

b) $A=\pi r^2\approx3.14(81)\text{ ft}^3$

$=254.34\text{ ft}^3$

23)

Chapter 1: The Real Number System and Geometry

25) The constant is -10.

Term	Coeff.
$4p^3$	4
$-p^2$	-1
$\frac{1}{3}p$	$\frac{1}{3}$
-10	-10

27) a) commutative

b) associative

c) inverse

d) distributive

29) a) $\left(-8k^2+3k-5\right)+\left(2k^2+k-9\right)$

$$=-8k^2+2k^2+3k+k-5-9$$
$$=-6k^2+4k-14$$

b) $\frac{4}{3}(6c-5)-\frac{1}{2}(4c+3)$

$$=8c-\frac{20}{3}-2c-\frac{3}{2}$$
$$=8c-2c-\frac{20}{3}-\frac{3}{2}$$
$$=8c-2c-\frac{40}{6}-\frac{9}{6}=6c-\frac{49}{6}$$

Section 2.1A: Exercises

1) $9 \cdot 9 \cdot 9 \cdot 9 \cdot 9 \cdot 9 = 9^6$

3) $\left(\dfrac{1}{7}\right)\left(\dfrac{1}{7}\right)\left(\dfrac{1}{7}\right)\left(\dfrac{1}{7}\right) = \left(\dfrac{1}{7}\right)^4$

5) $(-5)(-5)(-5)(-5)(-5)(-5)(-5)$
$= (-5)^7$

7) $(-3y)(-3y)(-3y)(-3y)$
$\cdot(-3y)(-3y)(-3y)(-3y)$
$= (-3y)^8$

9) base: 6; exponent: 8

11) base: 0.05; exponent: 7

13) base: -8; exponent: 5

15) base: $9x$; exponent: 8

17) base: $-11a$; exponent: 2

19) base: p; exponent: 4

21) base: y; exponent: 2

23) $(3+4)^2 = 7^2 = 49$, $3^2 + 4^2 = 9 + 16 = 25$.
"adding then squaring" and "squaring
then adding" are not the same

25) Answers may vary.

27) No. $3t^4 = 3 \cdot t^4$
$(3t)^4 = 3^4 \cdot t^4 = 81t^4$

29) $2^5 = 32$

31) $(11)^2 = 121$ 32)

33) $(-2)^4 = 16$ 34)

35) $-3^4 = -81$ 36)

37) $-2^3 = -8$ 38)

39) $\left(\dfrac{1}{5}\right)^3 = \dfrac{1}{125}$

41) $2^2 \cdot 2^3 = 2^{2+3} = 2^5 = 32$

43) $3^2 \cdot 3^2 = 3^{2+2} = 3^4 = 81$

45) $5^2 \cdot 2^3 = 25 \cdot 8 = 200$

47) $\left(\dfrac{1}{2}\right)^4 \cdot \left(\dfrac{1}{2}\right)^2 = \left(\dfrac{1}{2}\right)^{4+2} = \left(\dfrac{1}{2}\right)^6 = \dfrac{1}{64}$

49) $8^3 \cdot 8^9 = 8^{3+9} = 8^{12}$

51) $5^2 \cdot 5^4 \cdot 5^5 = 5^{2+4+5} = 5^{11}$

53) $(-7)^2 \cdot (-7)^3 \cdot (-7)^3 = (-7)^{2+3+3}$
$= (-7)^8$

55) $b^2 \cdot b^4 = b^{2+4} = b^6$

57) $k \cdot k^2 \cdot k^3 = k^{1+2+3} = k^6$

59) $8y^3 \cdot y^2 = 8\left(y^3 \cdot y^2\right) = 8y^5$

61) $\left(9m^4\right)\left(6m^{11}\right) = (9 \cdot 6)\left(m^4 \cdot m^{11}\right)$
$= 54m^{15}$

63) $(-6r)\left(7r^4\right) = (-6 \cdot 7)\left(r \cdot r^4\right) = -42r^5$

Chapter 2: The Rules of Exponents

65) $\left(-7t^6\right)\left(t^3\right)\left(-4t^7\right)$

$\quad =\left[-7\cdot1\cdot(-4)\right]\left(t^6\cdot t^3\cdot t^7\right)$

$\quad =28t^{16}$

67) $\left(\dfrac{5}{3}x^2\right)(12x)\left(-2x^3\right)$

$\quad =\left[\dfrac{5}{3}\cdot12\cdot(-2)\right]\left(x^2\cdot x\cdot x^3\right)$

$\quad =-40x^6$

69) $\left(\dfrac{8}{21}b\right)\left(-6b^8\right)\left(-\dfrac{7}{2}b^6\right)$

$\quad =\left[\dfrac{8}{21}\cdot(-6)\cdot\left(-\dfrac{7}{2}\right)\right]\left(b\cdot b^8\cdot b^6\right)$

$\quad =8b^{15}$

71) $\left(y^3\right)^4=y^{3\cdot4}=y^{12}$

73) $\left(w^{11}\right)^7=w^{11\cdot7}=w^{77}$

75) $\left(3^3\right)^2=3^{3\cdot2}=3^6=729$

77) $\left(-5^3\right)^2=(-5)^{3\cdot2}=(-5)^6$

79) $\left(\dfrac{1}{3}\right)^4=\dfrac{1}{3^4}=\dfrac{1}{81}$

81) $\left(\dfrac{6}{a}\right)^2=\dfrac{6^2}{a^2}=\dfrac{36}{a^2}$

83) $\left(\dfrac{m}{n}\right)^5=\dfrac{m^5}{n^5}$

85) $(10y)^4=10^4y^4=10,000y^4$

87) $(-3p)^4=(-3)^4\cdot p^4=81p^4$

89) $(-4ab)^3=(-4)^3\cdot a^3\cdot b^3=-64a^3b^3$

91) $6(xy)^3=6\cdot x^3\cdot y^3=6x^3y^3$

93) $-9(tu)^4=-9\cdot t^4\cdot u^4=-9t^4u^4$

95) a) $\quad A=lw$

$\qquad =(3w)(w)$

$\qquad =3(w\cdot w)$

$\qquad =3w^2$ sq units

$\quad P=2l+2w$

$\qquad =2(3w)+2(w)$

$\qquad =6w+2w$

$\qquad =8w$ units

b) $\quad A=lw$

$\qquad =\left(5k^3\right)\left(k^2\right)$

$\qquad =5\left(k^3\cdot k^2\right)$

$\qquad =5k^5$ sq units

$\quad P=2l+2w$

$\qquad =2\left(5k^3\right)+2\left(k^2\right)$

$\qquad =10k^3+2k^2$ units

97) $A=\dfrac{1}{2}bh$

$\quad =\dfrac{1}{2}(x)\left(\dfrac{3}{4}x\right)$

$\quad =\left(\dfrac{1}{2}\cdot\dfrac{3}{4}\right)(x\cdot x)$

$\quad =\dfrac{3}{8}x^2$ sq units

Section 2.1B: Exercises

1) operations

3) $\left(k^9\right)^2 \left(k^3\right)^2 = \left(k^{18}\right)\left(k^6\right) = k^{24}$

5) $\left(5z^4\right)^2 \left(2z^6\right)^3 = \left(5^2\right)\left(z^4\right)^2\left(2^3\right)\left(z^6\right)^3$
$$= \left(25z^8\right)\left(8z^{18}\right)$$
$$= 200z^{26}$$

7) $6ab\left(-a^{10}b^2\right)^3$
$$= 6ab \cdot (-1)^3\left(a^{10}\right)^3\left(b^2\right)^3$$
$$= 6ab \cdot \left(-1a^{30}b^6\right)$$
$$= -6a^{31}b^7$$

9) $(9+2)^2 = 11^2 = 121$

11) $\left(-4t^6u^2\right)^3 \left(u^4\right)^5$
$$= (-4)^3\left(t^6\right)^3\left(u^2\right)^3 \cdot u^{20}$$
$$= \left(-64t^{18}u^6\right) \cdot u^{20}$$
$$= -64t^{18}u^{26}$$

13) $8\left(6k^7l^2\right)^2 = 8 \cdot \left(6^2\right)\left(k^7\right)^2\left(l^2\right)^2$
$$= 8 \cdot (36)k^{14}l^4$$
$$= 288k^{14}l^4$$

15) $\left(\dfrac{3}{g^5}\right)^3 \left(\dfrac{1}{6}\right)^2 = \dfrac{3^3}{\left(g^5\right)^3} \cdot \dfrac{1}{6^2} = \dfrac{27}{g^{15}} \cdot \dfrac{1}{36}$
$$= \dfrac{27}{36g^{15}} = \dfrac{\overset{3}{\cancel{27}}}{\underset{4}{\cancel{36}}\, g^{15}}$$
$$= \dfrac{3}{4g^{15}}$$

17) $\left(\dfrac{7}{8}n^2\right)^2 \left(-4n^9\right)^2$
$$= \left(\dfrac{7^2}{8^2}\right)\left(n^2\right)^2 (-4)^2\left(n^9\right)^2$$
$$= \dfrac{49}{64}\left(n^4\right)(16)\left(n^{18}\right)$$
$$= \dfrac{49}{\underset{4}{\cancel{64}}}\left(n^4\right)\left(\overset{1}{\cancel{16}}\right)\left(n^{18}\right)$$
$$= \dfrac{49}{4}n^{22}$$

19) $h^4\left(10h^3\right)^2\left(-3h^9\right)^2$
$$= h^4\left(10^2\right)\left(h^3\right)^2(-3)^2\left(h^9\right)^2$$
$$= h^4(100)\left(h^6\right)(9)\left(h^{18}\right)$$
$$= 900h^{28}$$

21) $3w^{11}\left(7w^2\right)^2\left(-w^6\right)^5$
$$= 3w^{11}\left(7^2\right)\left(w^2\right)^2(-1)^5\left(w^6\right)^5$$
$$= 3w^{11}(49)\left(w^4\right)(-1)\left(w^{30}\right)$$
$$= -147w^{45}$$

23) $\dfrac{\left(12x^3\right)^2}{\left(10y^5\right)^2} = \dfrac{\left(12^2\right)\left(x^3\right)^2}{\left(10^2\right)\left(y^5\right)^2}$
$$= \dfrac{144x^6}{100y^{10}}$$
$$= \dfrac{36x^6}{25y^{10}}$$

25) $\dfrac{\left(4d^9\right)^2}{\left(-2c^5\right)^6} = \dfrac{\left(4^2\right)\left(d^9\right)^2}{\left(-2\right)^6\left(c^5\right)^6}$

$= \dfrac{16d^{18}}{64c^{30}}$

$= \dfrac{d^{18}}{4c^{30}}$

27) $\dfrac{8\left(a^4b^7\right)^9}{\left(6c\right)^2} = \dfrac{8\left(a^4\right)^9\left(b^7\right)^9}{\left(6^2\right)\left(c^2\right)}$

$= \dfrac{8a^{36}b^{63}}{36c^2}$

$= \dfrac{2a^{36}b^{63}}{9c^2}$

29) $\dfrac{r^4\left(r^5\right)^7}{2t\left(11t^2\right)^2} = \dfrac{r^4\left(r^{35}\right)}{2t\left(11^2\right)\left(t^2\right)^2}$

$= \dfrac{r^{39}}{2t\left(121t^4\right)}$

$= \dfrac{r^{39}}{242t^5}$

31) $\left(\dfrac{4}{9}x^3y\right)^2\left(\dfrac{3}{2}x^6y^4\right)^3$

$= \left(\dfrac{4^2}{9^2}\right)\left(x^3\right)^2\left(y^2\right)\left(\dfrac{3^3}{2^3}\right)\left(x^6\right)^3\left(y^4\right)^3$

$= \left(\dfrac{16}{81}x^6y^2\right)\left(\dfrac{27}{8}x^{18}y^{12}\right)$

$= \left(\dfrac{\overset{2}{\cancel{16}}}{\underset{3}{\cancel{81}}}x^6y^2\right)\left(\dfrac{\overset{1}{\cancel{27}}}{\underset{1}{\cancel{8}}}x^{18}y^{12}\right)$

$= \dfrac{2}{3}x^{24}y^{14}$

33) $\left(-\dfrac{2}{5}c^9d^2\right)^3\left(\dfrac{5}{4}cd^6\right)^2$

$= (-1)^3\left(\dfrac{2^3}{5^3}\right)\left(c^9\right)^3\left(d^2\right)^3\left(\dfrac{5^2}{4^2}\right)\left(c^2\right)\left(d^6\right)^2$

$= \left(-\dfrac{8}{125}c^{27}d^6\right)\left(\dfrac{25}{16}c^2d^{12}\right)$

$= \left(-\dfrac{\overset{1}{\cancel{8}}}{\underset{5}{\cancel{125}}}c^{27}d^6\right)\left(\dfrac{\overset{1}{\cancel{25}}}{\underset{2}{\cancel{16}}}c^2d^{12}\right)$

$= -\dfrac{1}{10}c^{29}d^{18}$

35) $\left(\dfrac{5x^5y^2}{z^4}\right)^3 = \dfrac{\left(5^3\right)\left(x^5\right)^3\left(y^2\right)^3}{\left(z^4\right)^3} = \dfrac{125x^{15}y^6}{z^{12}}$

37) $\left(-\dfrac{3t^4u^9}{2v^7}\right)^4 = (-1)^4 \cdot \dfrac{\left(3^4\right)\left(t^4\right)^4\left(u^9\right)^4}{\left(2^4\right)\left(v^7\right)^4}$

$= \dfrac{81t^{16}u^{36}}{16v^{28}}$

39) $\left(\dfrac{12w^5}{4x^3y^6}\right)^2 = \left(\dfrac{\overset{3}{\cancel{12}}w^5}{\underset{1}{\cancel{4}}x^3y^6}\right)^2$

$= \left(\dfrac{3w^5}{x^3y^6}\right)^2 = \dfrac{\left(3\right)^2\left(w^5\right)^2}{\left(x^3\right)^2\left(y^6\right)^2}$

$= \dfrac{9w^{10}}{x^6y^{12}}$

41) a) $P = 4\left(5l^2 \text{ units}\right) = 20l^2 \text{ units}$

b) $A = \left(5l^2 \text{ units}\right)^2$

$= \left(5^2\right)\left(l^2\right)^2 \text{ sq units}$

$= 25l^4 \text{ sq units}$

43) a) $A = (x \text{ units})\left(\dfrac{3}{8} x \text{ units}\right)$

$\quad\quad = \dfrac{3}{8} x^2 \text{ sq. units}$

b) $P = 2(x \text{ units}) + 2\left(\dfrac{3}{8} x \text{ units}\right)$

$\quad\quad = 2x \text{ units} + \dfrac{3}{4} x \text{ units}$

$\quad\quad = \dfrac{8}{4} x \text{ units} + \dfrac{3}{4} x \text{ units}$

$\quad\quad = \dfrac{11}{4} x \text{ units}$

Section 2.2A: Exercises

1)   false

3)   true

5) $2^0 = 1$

7) $-5^0 = -1 \cdot 5^0 = -1 \cdot 1 = -1$

9) $0^8 = 0$

11) $(5)^0 + (-5)^0 = 1 + 1 = 2$

13) $6^{-2} = \left(\dfrac{1}{6}\right)^2 = \dfrac{1^2}{6^2} = \dfrac{1}{36}$

15) $2^{-4} = \left(\dfrac{1}{2}\right)^4 = \dfrac{1^4}{2^4} = \dfrac{1}{16}$

17) $5^{-3} = \left(\dfrac{1}{5}\right)^3 = \dfrac{1^3}{5^3} = \dfrac{1}{125}$

19) $\left(\dfrac{1}{8}\right)^{-2} = 8^2 = 64$

21) $\left(\dfrac{1}{2}\right)^{-5} = 2^5 = 32$

23) $\left(\dfrac{4}{3}\right)^{-3} = \left(\dfrac{3}{4}\right)^3 = \dfrac{3^3}{4^3} = \dfrac{27}{64}$

25) $\left(\dfrac{9}{7}\right)^{-2} = \left(\dfrac{7}{9}\right)^2 = \dfrac{7^2}{9^2} = \dfrac{49}{81}$

27) $\left(-\dfrac{1}{4}\right)^{-3} = \left(\dfrac{4}{-1}\right)^3 = \dfrac{4^3}{(-1)^3} = \dfrac{64}{-1} = -64$

29) $\left(-\dfrac{3}{8}\right)^{-2} = \left(\dfrac{8}{-3}\right)^2 = \dfrac{8^2}{(-3)^2} = \dfrac{64}{9}$

31) $-2^{-6} = -\left(\dfrac{1}{2}\right)^6 = -\dfrac{1^6}{2^6} = -\dfrac{1}{64}$

33) $-1^{-5} = -(1)^5 = -1$

$\quad\quad -9^{-2} = -\left(\dfrac{1}{9}\right)^2 = -\dfrac{1^2}{9^2} = -\dfrac{1}{81}$

35) $2^{-3} - 4^{-2} = \left(\dfrac{1}{2}\right)^3 - \left(\dfrac{1}{4}\right)^2$

$\quad\quad = \dfrac{1^3}{2^3} - \dfrac{1^2}{4^2} = \dfrac{1}{8} - \dfrac{1}{16}$

$\quad\quad = \dfrac{2}{16} - \dfrac{1}{16} = \dfrac{1}{16}$

37) $2^{-2} + 3^{-2} = \left(\dfrac{1}{2}\right)^2 + \left(\dfrac{1}{3}\right)^2$

$\quad\quad = \dfrac{1^2}{2^2} + \dfrac{1^2}{3^2} = \dfrac{1}{4} + \dfrac{1}{9}$

$\quad\quad = \dfrac{9}{36} + \dfrac{4}{36} = \dfrac{13}{36}$

39) $-9^{-2} + 3^{-3} + (-7)^0$

$$= -\left(\frac{1}{9}\right)^2 + \left(\frac{1}{3}\right)^3 + 1$$

$$= -\frac{1^2}{9^2} + \frac{1^3}{3^3} + 1 = -\frac{1}{81} + \frac{1}{27} + 1$$

$$= -\frac{1}{81} + \frac{3}{81} + \frac{81}{81} = \frac{83}{81}$$

Section 2.2B: Exercises

1) a) w b) n c) $2p$ d) c

3) $r^0 = 1$

5) $-2k^0 = -2 \cdot 1 = -2$

7) $x^0 + (2x)^0 = 1 + 1 = 2$

9) $d^{-3} = \left(\frac{1}{d}\right)^3 = \frac{1^3}{d^3} = \frac{1}{d^3}$

11) $p^{-1} = \left(\frac{1}{p}\right)^1 = \frac{1}{p}$

13) $\left(\frac{a^{-10}}{b^{-3}}\right) = \frac{b^3}{a^{10}}$

15) $\frac{y^{-8}}{x^{-5}} = \frac{x^5}{y^8}$

17) $\frac{t^5}{8u^{-3}} = \frac{t^5 u^3}{8}$

19) $5m^6 n^{-2} = \frac{5m^6}{n^2}$

21) $\frac{2}{t^{-11} u^{-5}} = 2t^{11} u^5$

23) $\frac{8a^6 b^{-1}}{5c^{-10} d} = \frac{8a^6 c^{10}}{5bd}$

25) $\frac{2z^4}{x^{-7} y^{-6}} = 2x^7 y^6 z^4$

27) $\left(\frac{a}{6}\right)^{-2} = \left(\frac{6}{a}\right)^2 = \frac{6^2}{a^2} = \frac{36}{a^2}$

29) $\left(\frac{2n}{q}\right)^{-5} = \left(\frac{q}{2n}\right)^5 = \frac{q^5}{2^5 n^5} = \frac{q^5}{32n^5}$

31) $\left(\frac{12b}{cd}\right)^{-2} = \left(\frac{cd}{12b}\right)^2 = \frac{c^2 d^2}{12^2 b^2} = \frac{c^2 d^2}{144 b^2}$

33) $-9k^{-2} = -9 \cdot \frac{1}{k^2} = -\frac{9}{k^2}$

35) $3t^{-3} = 3 \cdot \frac{1}{t^3} = \frac{3}{t^3}$

37) $-m^{-9} = -1 \cdot \frac{1}{m^9} = -\frac{1}{m^9}$

39) $\left(\frac{1}{z}\right)^{-10} = z^{10}$

41) $\left(\frac{1}{j}\right)^{-1} = j$

43) $5\left(\frac{1}{n}\right)^{-2} = 5n^2$

45) $c\left(\frac{1}{d}\right)^{-3} = cd^3$

Section 2.3: Exercises

1) You must subtract the denominator's exponent from the numerator's exponent; a^2

3) $\dfrac{d^{10}}{d^5} = d^{10-5} = d^5$

5) $\dfrac{m^9}{m^5} = m^{9-5} = m^4$

7) $\dfrac{8t^{15}}{t^8} = 8t^{15-8} = 8t^7$

9) $\dfrac{6^{12}}{6^{10}} = 6^{12-10} = 6^2 = 36$

11) $\dfrac{3^{12}}{3^8} = 3^{12-8} = 3^4 = 81$

13) $\dfrac{2^5}{2^9} = 2^{5-9} = 2^{-4} = \left(\dfrac{1}{2}\right)^4 = \dfrac{1}{16}$

15) $\dfrac{5^6}{5^9} = 5^{6-9} = 5^{-3} = \left(\dfrac{1}{5}\right)^3 = \dfrac{1}{125}$

17) $\dfrac{10d^4}{d^2} = 10d^{4-2} = 10d^2$

19) $\dfrac{20c^{11}}{30c^6} = \dfrac{\overset{2}{\cancel{20}}}{\underset{3}{\cancel{30}}}c^{11-6} = \dfrac{2}{3}c^5$

21) $\dfrac{y^3}{y^8} = y^{3-8} = y^{-5} = \dfrac{1}{y^5}$

23) $\dfrac{x^{-3}}{x^6} = x^{-3-6} = x^{-9} = \dfrac{1}{x^9}$

25) $\dfrac{t^{-6}}{t^{-3}} = t^{-6+3} = t^{-3} = \dfrac{1}{t^3}$

27) $\dfrac{a^{-1}}{a^9} = a^{-1-9} = a^{-10} = \dfrac{1}{a^{10}}$

29) $\dfrac{t^4}{t} = t^{4-1} = t^3$

31) $\dfrac{15w^2}{w^{10}} = 15w^{2-10} = 15w^{-8} = \dfrac{15}{w^8}$

33) $\dfrac{-6k}{k^4} = -6k^{1-4} = -6k^{-3} = -\dfrac{6}{k^3}$

35) $\dfrac{a^4b^9}{ab^2} = a^{4-1}b^{9-2} = a^3b^7$

37) $\dfrac{10k^{-2}l^{-6}}{15k^{-5}l^2} = \dfrac{\overset{2}{\cancel{10}}k^{-2+5}}{\underset{3}{\cancel{15}}l^{2+6}} = \dfrac{2k^3}{3l^8}$

39) $\dfrac{300x^7y^3}{30x^{12}y^8} = \dfrac{\overset{10}{\cancel{300}}x^7y^3}{\cancel{30}x^{12}y^8}$

$\qquad = \dfrac{10}{x^{12-7}y^{8-3}}$

$\qquad = \dfrac{10}{x^5y^5}$

41) $\dfrac{6v^{-1}w}{54v^2w^{-5}} = \dfrac{\overset{1}{\cancel{6}}v^{-1}w}{\underset{9}{\cancel{54}}v^2w^{-5}}$

$\qquad = \dfrac{1}{9}v^{-1-2}w^{1+5}$

$\qquad = \dfrac{1}{9}v^{-3}w^6 = \dfrac{w^6}{9v^3}$

43) $\dfrac{3c^5 d^{-2}}{8cd^{-3}} = \dfrac{3c^{5-1} d^{-2+3}}{8}$

$= \dfrac{3}{8} c^4 d$

45) $\dfrac{(x+y)^9}{(x+y)^2} = (x+y)^{9-2} = (x+y)^7$

47) $\dfrac{(c+d)^{-5}}{(c+d)^{-11}} = (c+d)^{-5+11} = (c+d)^6$

Putting It All Together

1) $\left(\dfrac{2}{3}\right)^4 = \dfrac{2^4}{3^4} = \dfrac{16}{81}$

3) $\dfrac{3^9}{3^5 \cdot 3^4} = \dfrac{3^9}{3^{5+4}} = \dfrac{3^9}{3^9} = 3^{9-9} = 3^0 = 1$

5) $\left(\dfrac{10}{3}\right)^{-2} = \left(\dfrac{3}{10}\right)^2 = \dfrac{3^2}{10^2} = \dfrac{9}{100}$

7) $(9-6)^2 = 3^2 = 9$

9) $10^{-2} = \dfrac{1}{10^2} = \dfrac{1}{100}$

11) $\dfrac{2^7}{2^{12}} = 2^{7-12} = 2^{-5} = \dfrac{1}{2^5} = \dfrac{1}{32}$

13) $\left(-\dfrac{5}{3}\right)^{-7} \cdot \left(-\dfrac{5}{3}\right)^4 = \left(-\dfrac{5}{3}\right)^{-7+4}$

$= \left(-\dfrac{5}{3}\right)^{-3} = \left(-\dfrac{3}{5}\right)^3$

$= -\dfrac{27}{125}$

15) $3^{-2} - 12^{-1} = \left(\dfrac{1}{3}\right)^2 - \dfrac{1}{12}$

$= \dfrac{1}{9} - \dfrac{1}{12}$

$= \dfrac{4}{36} - \dfrac{3}{36} = \dfrac{1}{36}$

17) $-10\left(-3g^4\right)^3 = -10 \cdot (-27) g^{12}$

$= 270 g^{12}$

19) $\dfrac{33s}{s^{12}} = 33s^{1-12} = 33s^{-11} = \dfrac{33}{s^{11}}$

21) $\left(\dfrac{2xy^4}{3x^{-9} y^{-2}}\right)^4 = \left(\dfrac{2}{3} x^{1-(-9)} y^{4-(-2)}\right)^4$

$= \left(\dfrac{2}{3} x^{1+9} y^{4+2}\right)^4$

$= \left(\dfrac{2}{3} x^{10} y^6\right)^4 = \dfrac{16}{81} x^{40} y^{24}$

23) $\left(\dfrac{9m^8}{n^3}\right)^{-2} = \left(\dfrac{n^3}{9m^8}\right)^2 = \dfrac{n^6}{81m^{16}}$

25) $\left(-b^5\right)^3 = -b^{15}$

27) $\left(-3m^5 n^2\right)^3 = -27m^{15} n^6$

29) $\left(-\dfrac{9}{4} z^5\right)\left(\dfrac{8}{3} z^{-2}\right) = \left(-\dfrac{\cancel{9}^{\,3}}{\cancel{4}_{\,1}} z^5\right)\left(\dfrac{\cancel{8}^{\,2}}{\cancel{3}_{\,1}} z^{-2}\right)$

$= -6z^{5-2} = -6z^3$

31) $\left(\dfrac{s^7}{t^3}\right)^{-6} = \left(\dfrac{t^3}{s^7}\right)^6 = \dfrac{t^{18}}{s^{42}}$

33) $(-ab^3c^5)^2 \left(\dfrac{a^4}{bc}\right)^3 = \left(a^2b^6c^{10}\right)\left(\dfrac{a^{12}}{b^3c^3}\right)$

$\qquad = a^{2+12}b^{6-3}c^{10-3}$

$\qquad = a^{14}b^3c^7$

35) $\left(\dfrac{48u^{-7}v^2}{36u^3v^{-5}}\right)^{-3} = \left(\dfrac{36u^3v^{-5}}{48u^{-7}v^2}\right)^3$

$\qquad = \left(\dfrac{\overset{3}{\cancel{36}}\ u^3v^{-5}}{\underset{4}{\cancel{48}}\ u^{-7}v^2}\right)^3$

$\qquad = \left(\dfrac{3}{4}u^{3+7}v^{-7}\right)^3$

$\qquad = \left(\dfrac{3}{4}u^{10}v^{-7}\right)^3$

$\qquad = \dfrac{27}{64}u^{30}v^{-21} = \dfrac{27u^{30}}{64v^{21}}$

37) $\left(\dfrac{-3t^4u}{t^2u^{-4}}\right)^3 = \left(-3t^{4-2}u^{1+4}\right)^3$

$\qquad = \left(-3t^2u^5\right)^3 = -27t^6u^{15}$

39) $\left(h^{-3}\right)^6 = h^{-18} = \dfrac{1}{h^{18}}$

41) $\left(\dfrac{h}{2}\right)^4 = \dfrac{h^4}{16}$

43) $-7c^4\left(-2c^2\right)^3 = -7c^4 \cdot (-8)c^6 = 56c^{10}$

45) $\left(12a^7\right)^{-1}\left(6a\right)^2 = \dfrac{(6a)^2}{\left(12a^7\right)} = \dfrac{36a^2}{12a^7}$

$\qquad = \dfrac{\overset{3}{\cancel{36}}\ a^2}{\underset{1}{\cancel{12}}a^7} = 3a^{2-7}$

$\qquad = 3a^{-5} = \dfrac{3}{a^5}$

47) $\left(\dfrac{9}{20}r^4\right)\left(4r^{-3}\right)\left(\dfrac{2}{33}r^9\right)$

$\qquad = \dfrac{9}{20}\cdot 4\cdot \dfrac{2}{33}r^4r^{-3}r^9$

$\qquad = \dfrac{\overset{3}{\cancel{9}}\cdot \overset{1}{\cancel{4}}\cdot 2}{\underset{5}{\cancel{20}}\cdot 1\cdot \underset{11}{\cancel{33}}}r^{4-3+9}$

$\qquad = \dfrac{6}{55}r^{10}$

49) $\dfrac{(a^2b^{-5}c)^{-3}}{(a^4b^{-3}c)^{-2}} = \dfrac{a^{-6}b^{15}c^{-3}}{a^{-8}b^6c^{-2}} = \dfrac{a^{-6+8}b^{15-6}}{c^{-2+3}} = \dfrac{a^2b^9}{c}$

51) $\dfrac{(2mn^{-2})^3(5m^2n^{-3})^{-1}}{(3m^{-3}n^3)^{-2}}$

$\qquad = \dfrac{(2mn^{-2})^3(3m^{-3}n^3)^2}{(5m^2n^{-3})^1}$

$\qquad = \dfrac{8m^3n^{-6}9m^{-6}n^6}{5m^2n^{-3}}$

$\qquad = \dfrac{72}{5}m^{3-6-2}n^{-6+6+3} = \dfrac{72}{5}m^{-5}n^3$

$\qquad = \dfrac{72n^3}{5m^5}$

53) $\left(\dfrac{4n^{-3}m}{n^8m^2}\right)^0 = 1$

55) $\left(\dfrac{49c^4d^8}{21c^4d^5}\right)^{-2} = \left(\dfrac{21c^4d^5}{49c^4d^8}\right)^2$

$\qquad = \left(\dfrac{\overset{3}{\cancel{21}}}{\underset{7}{\cancel{49}}}c^{4-4}d^{5-8}\right)^2$

$\qquad = \left(\dfrac{3}{7}c^0d^{-3}\right)^2 = \left(\dfrac{3}{7d^3}\right)^2 = \dfrac{9}{49d^6}$

57) $(p^{2c})^6 = p^{12c}$

59) $y^m \cdot y^{3m} = y^{4m}$

61) $t^{5b} \cdot t^{-8b} = t^{-3b} = \dfrac{1}{t^{3b}}$

63) $\dfrac{25c^{2x}}{40c^{9x}} = \dfrac{\overset{5}{\cancel{25}}}{\underset{8}{\cancel{40}}\, c^{9x-2x}} = \dfrac{5}{8c^{7x}}$

Section 2.4: Exercises

1) yes 3) no

5) no

7) yes

9) Answers may vary.

11) Answers may vary.

13) $71.765 \times 10^2 : 71\underset{\smile\smile}{76}.5 = 7176.5$

15) $40.6 \times 10^{-3} : 0.0\underset{\smile\smile}{40}6 = 0.0406$

17) $1,200,006 \times 10^{-7} : 0.1\underset{\smile\smile\smile\smile\smile}{200006} = 0.1200006$

19) $-6.8 \times 10^{-5} : -0\underset{\smile\smile\smile\smile}{0006}.8 = -0.000068$

21) $-5.26 \times 10^4 : -5.2\underset{\smile\smile}{600} = -52,600$

23) $8 \times 10^{-6} : \underset{\smile\smile\smile\smile\smile}{000008}. = 0.000008$

25) $6.021967 \times 10^5 : 6.02\underset{\smile}{1}967$
$= 602,196.7$

27) $3 \times 10^6 : 3.\underset{\smile\smile\smile\smile\smile\smile}{000000} = 3,000,000$

29) $-7.44 \times 10^{-4} : -0\underset{\smile\smile\smile}{007}.44 = -0.000744$

31) $2.4428 \times 10^7 : 2.\underset{\smile\smile\smile\smile\smile\smile}{4428000} = 24,428,000$

33) $2.5 \times 10^{-11} : \underset{\smile\smile\smile\smile\smile\smile\smile\smile\smile}{00000000025}.0$
$= 0.000000000025\,meters$

35) $2110.5 = 21\underset{\smile\smile}{10}.5 = 2.1105 \times 10^3$

37) $0.000096 = 0.\underset{\smile\smile\smile\smile\smile}{00009}6 = 9.6 \times 10^{-5}$

39) $-7,000,000 = -7,\underset{\smile\smile\smile}{000},000. = -7 \times 10^6$

41) $3400 = 3\underset{\smile\smile\smile}{400}. = 3.4 \times 10^3$

43) $0.0008 = 0.\underset{\smile\smile\smile}{000}8 = 8 \times 10^{-4}$

45) $-0.076 = -0.0\underset{\smile\smile}{76} = -7.6 \times 10^{-2}$

47) $6000 = 6\underset{\smile\smile\smile}{000}. = 6 \times 10^3$

49) $380,800,000 \text{ kg} = 3.808 \times 10^8 \text{ kg}$

51) $0.00000001 \text{ cm} = 1 \times 10^{-8} \text{ cm}$

53) $\dfrac{6 \times 10^9}{2 \times 10^5} = \dfrac{6}{2} \times \dfrac{10^9}{10^5} = 3 \times 10^4 = 30,000$

55) $(2.3 \times 10^3)(3 \times 10^2)$
$= (2.3 \times 3)(10^3 \times 10^2)$
$= 6.9 \times 10^5$
$= 690,000$

57) $\dfrac{8.4 \times 10^{12}}{-7 \times 10^9} = -\dfrac{8.4}{7} \times \dfrac{10^{12}}{10^9}$
$= -1.2 \times 10^3 = -1200$

59) $\left(-1.5\times10^{-8}\right)\left(4\times10^{6}\right)$

$=\left(-1.5\times4\right)\left(10^{-8}\times10^{6}\right)$

$=-6.0\times10^{-2}$

$=-0.06$

61) $\dfrac{-3\times10^{5}}{6\times10^{8}}=-\dfrac{3}{6}\times\dfrac{10^{5}}{10^{8}}$

$=-0.5\times10^{-3}=-0.0005$

63) $\left(9.75\times10^{4}\right)+\left(6.25\times10^{4}\right)$

$=\left(9.75+6.25\right)10^{4}$

$=16\times10^{4}$

$=160,000$

65) $\left(3.19\times10^{-5}\right)+\left(9.2\times10^{-5}\right)$

$=\left(3.19+9.2\right)10^{-5}$

$=12.39\times10^{-5}$

$=0.0001239$

67) $365\left(1.44\times10^{7}\right)$

$=\left(365\cdot1.44\right)10^{7}$

$=525.6\times10^{7}$

$=5,256,000,000$ particles

69) $\dfrac{2.21\times10^{10}\text{ lb}}{1,300,000\text{ cow}}=\dfrac{2.21\times10^{10}\text{ lb}}{1.3\times10^{6}\text{ cow}}$

$=\dfrac{2.21}{1.3}\times\dfrac{10^{10}\text{ lb}}{10^{6}\text{ cow}}$

$=1.7\times10^{4}\dfrac{\text{lb}}{\text{cow}}$

$=17,000\dfrac{\text{lb}}{\text{cow}}$

71) First determine the area of the photo.

$A=4\text{ in}\cdot6\text{ in}=24\text{ in}^{2}$

Then multiply by the rate.

$24\text{ in}^{2}\cdot\dfrac{1.1\times10^{6}\text{ droplets}}{\text{in}^{2}}$

$=2.4\times10\text{ in}^{2}\cdot\dfrac{1.1\times10^{6}\text{ droplets}}{\text{in}^{2}}$

$=\left(2.4\times1.1\right)\left(10\times10^{6}\right)$ droplets

$=2.64\times10^{7}$ droplets

$=26,400,000$ droplets

73) $\left(\begin{array}{l}\text{money spent by}\\\text{average household}\\\text{on food}\end{array}\right)=\dfrac{\left(\begin{array}{l}\text{total \$ spent}\\\text{on food}\end{array}\right)}{\left(\begin{array}{l}\text{\# of house}\\\text{holds}\end{array}\right)}$

$=\dfrac{7.3\times10^{11}}{120,000,000}=\$6083$

75) $\left(\text{distance traveled}\right)=\left(\text{speed}\right)\left(\text{time}\right)$

$=\left(7800\text{ m/s}\right)\left(2\text{ days}\right)$

$=\left(7800\right)\left(2\cdot24\cdot60\cdot60\text{ seconds}\right)$

$=\left(7.8\times10^{3}\right)\left(1.728\times10^{5}\right)$

$=13.4784\times10^{8}$

$=1.34784\times10^{9}\text{ m}$

77) $\left(\begin{array}{l}\text{\# of metric tons of}\\\text{carbon emissions}\\\text{per person}\end{array}\right)=$

$\dfrac{\left(\begin{array}{l}\text{total \# of metric tons}\\\text{of carbon emissions}\end{array}\right)}{\left(\begin{array}{l}\text{total US}\\\text{population}\end{array}\right)}$

$=\dfrac{6\times10^{9}}{300,000,000}=20\text{ metric tons.}$

Chapter 2: The Rules of Exponents

Chapter 2 Review
(2.1A)

1) a) $8 \cdot 8 \cdot 8 \cdot 8 \cdot 8 \cdot 8 = 8^6$

b) $(-7)(-7)(-7)(-7) = (-7)^4$

3) a) $2^3 \cdot 2^2 = 2^{3+2} = 2^5 = 32$

b) $\left(\dfrac{1}{3}\right)^2 \cdot \left(\dfrac{1}{3}\right) = \left(\dfrac{1}{3}\right)^{2+1} = \left(\dfrac{1}{3}\right)^3 = \dfrac{1}{27}$

c) $\left(7^3\right)^4 = 7^{3 \cdot 4} = 7^{12}$

d) $\left(k^5\right)^6 = k^{5 \cdot 6} = k^{30}$

5) a) $(5y)^3 = 5^3 y^3 = 125y^3$

b) $\left(-7m^4\right)\left(2m^{12}\right) = -14m^{4+12}$
$$= -14m^{16}$$

c) $\left(\dfrac{a}{b}\right)^6 = \dfrac{a^6}{b^6}$

d) $6(xy)^2 = 6x^2 y^2$

e) $\left(\dfrac{10}{9}c^4\right)(2c)\left(\dfrac{15}{4}c^3\right)$
$$= \dfrac{10}{9} \cdot 2 \cdot \dfrac{15}{4} c^4 c c^3$$
$$= \dfrac{\cancel{10}^5}{\cancel{9}_3} \cdot \cancel{2}^1 \cdot \dfrac{\cancel{15}^5}{\cancel{4}_1} c^{4+1+3}$$
$$= \dfrac{25}{3} c^8$$

7) a) $\left(z^5\right)^2 \left(z^3\right)^4 = z^{10} z^{12} = z^{22}$

b) $-2\left(3c^5 d^8\right)^2 = -2\left(9c^{10} d^{16}\right)$
$$= -18c^{10} d^{16}$$

c) $(9-4)^3 = 5^3 = 125$

d) $\dfrac{\left(10t^3\right)^2}{\left(2u^7\right)^3} = \dfrac{100t^6}{8u^{21}} = \dfrac{25t^6}{2u^{21}}$

9) a) $8^0 = 1$ b) $-3^0 = -1$

c) $9^{-1} = \dfrac{1}{9}$

d) $3^{-2} - 2^{-2} = \left(\dfrac{1}{3}\right)^2 - \left(\dfrac{1}{2}\right)^2$
$$= \dfrac{1}{9} - \dfrac{1}{4}$$
$$= \dfrac{4}{36} - \dfrac{9}{36} = -\dfrac{5}{36}$$

e) $\left(\dfrac{4}{5}\right)^{-3} = \left(\dfrac{5}{4}\right)^3 = \dfrac{125}{64}$

11) a) $v^{-9} = \dfrac{1}{v^9}$ b) $\left(\dfrac{9}{c}\right)^{-2} = \left(\dfrac{c}{9}\right)^2 = \dfrac{c^2}{81}$

c) $\left(\dfrac{1}{y}\right)^{-8} = y^8$ d) $-7k^{-9} = -\dfrac{7}{k^9}$

e) $\dfrac{19z^{-4}}{a^{-1}} = \dfrac{19a}{z^4}$ f) $20m^{-6}n^5 = \dfrac{20n^5}{m^6}$

g) $\left(\dfrac{2j}{k}\right)^{-5} = \left(\dfrac{k}{2j}\right)^5 = \dfrac{k^5}{32j^5}$

13) a) $\dfrac{3^8}{3^6} = 3^{8-6} = 3^2 = 9$

f) $\left(\dfrac{14m^5 n^5}{7m^4 n}\right)^3 = \left(2m^{5-4} n^{5-1}\right)^3$

$= \left(2mn^4\right)^3$

$= 8m^3 n^{12}$

b) $\dfrac{r^{11}}{r^3} = r^{11-3} = r^8$

c) $\dfrac{48t^{-2}}{32t^3} = \dfrac{3}{2} t^{-2-3} = \dfrac{3}{2} t^{-5} = \dfrac{3}{2t^5}$

g) $\left(\dfrac{3k^{-1}t}{5k^{-7}t^4}\right)^{-3} = \left(\dfrac{3}{5} k^{-1-(-7)} t^{1-4}\right)^{-3}$

$= \left(\dfrac{3}{5} k^6 t^{-3}\right)^{-3}$

d) $\dfrac{21xy^2}{35x^{-6} y^3} = \dfrac{3}{5} x^{1-(-6)} y^{2-3}$

$= \dfrac{3}{5} x^7 y^{-1} = \dfrac{3x^7}{5y}$

$= \left(\dfrac{3k^6}{5t^3}\right)^{-3}$

$= \left(\dfrac{5t^3}{3k^6}\right)^{3} = \dfrac{125t^9}{27k^{18}}$

15) a) $\left(-3s^4 t^5\right)^4 = 81s^{16} t^{20}$

b) $\dfrac{\left(2a^6\right)^5}{\left(4a^7\right)^2} = \dfrac{32a^{30}}{16a^{14}} = 2a^{30-14} = 2a^{16}$

h) $\left(\dfrac{40}{21} x^{10}\right)\left(3x^{-12}\right)\left(\dfrac{49}{20} x^2\right)$

$= \dfrac{40}{21} \cdot 3 \cdot \dfrac{49}{20} x^{10} x^{-12} x^2$

c) $\left(\dfrac{z^4}{y^3}\right)^{-6} = \left(\dfrac{y^3}{z^4}\right)^6 = \dfrac{y^{18}}{z^{24}}$

$= \dfrac{\overset{2}{\cancel{40}} \cdot \overset{1}{\cancel{3}} \cdot \overset{7}{\cancel{49}}}{\underset{1}{\cancel{21}} \cdot 1 \cdot \underset{1}{\cancel{20}}} x^{10-12+2} = 14x^0 = 14$

d) $\left(-x^3 y\right)^5 \left(6x^{-2} y^3\right)^2$

$= \left(-x^{15} y^5\right)\left(36x^{-4} y^6\right)$

$= -36x^{11} y^{11}$

17) a) $y^{3k} \cdot y^{7k} = y^{3k+7k} = y^{10k}$

b) $\left(x^{5p}\right)^2 = x^{2 \cdot 5p} = x^{10p}$

e) $\left(\dfrac{cd^{-4}}{c^8 d^{-9}}\right)^5 = \left(c^{1-8} d^{-4-(-9)}\right)^5$

$= \left(c^{-7} d^5\right)^5$

c) $\dfrac{z^{12c}}{z^{5c}} = z^{12c-5c} = z^{7c}$

$= c^{-35} d^{25}$

$= \dfrac{d^{25}}{c^{35}}$

d) $\dfrac{t^{6d}}{t^{11d}} = t^{6d-11d} = t^{-5d} = \dfrac{1}{t^{5d}}$

19) $-4.185 \times 10^2 = -418.5$

21) $6.7 \times 10^{-4} = 0.00067$

23) $2 \times 10^4 = 20,000$

25) $0.0000575 = 5.75 \times 10^{-5}$

27) $32,000,000 = 3.2 \times 10^7$

29) $178,000 = 1.78 \times 10^5$

31) $0.0009315 = 9.315 \times 10^{-4}$

33) $\dfrac{8 \times 10^6}{2 \times 10^{13}} = \dfrac{8}{2} \times \dfrac{10^6}{10^{13}}$

$\qquad = 4 \times 10^{-7}$

$\qquad = 0.0000004$

35) $\left(9 \times 10^{-8}\right)\left(4 \times 10^7\right)$

$\qquad = (9 \times 4)\left(10^{-8} \times 10^7\right)$

$\qquad = 36 \times 10^{-1}$

$\qquad = 3.6$

37) $\dfrac{-3 \times 10^{10}}{-4 \times 10^6} = \dfrac{3}{4} \times \dfrac{10^{10}}{10^6}$

$\qquad = .75 \times 10^4$

$\qquad = 7500$

39) $\dfrac{2.4 \times 10^5}{8} = \dfrac{2.4}{8} \times 10^5$

$\qquad = 0.3 \times 10^5$

$\qquad = 30,000$ quills

41) $\dfrac{2.99 \times 10^{-23} \text{ g}}{\text{molecule}} \cdot 100,000,000 \text{ molecules}$

$\qquad = 2.99 \times 10^{-23} \cdot 1 \times 10^8$ g

$\qquad = 2.99 \left(10^{-23} \times 10^8\right)$ g

$\qquad = 2.99 \times 10^{-15}$ g

$\qquad = 0.00000000000000299$ g

43) $\left(\dfrac{143,000 \text{ visits}}{\text{second}}\right) \text{x} \left(\dfrac{60 \text{ seconds}}{\text{minute}}\right) \text{x} \left(3 \text{ minutes}\right)$

$= \left(1.43\text{x}10^5\right)(6\text{x}10)(3)$ visits

$= (1.43 \cdot 6 \cdot 3)\left(10^5 \cdot 10\right)$ visits

$= 25.74 \times 10^6$ visits

$= 25,740,000$ visits

Chapter 2 Test

1) $(-3)(-3)(-3) = (-3)^3$

3) $5^2 \cdot 5 = 5^3 = 125$

5) $\left(8^3\right)^{12} = 8^{36}$

7) $3^4 = 81$

9) $2^{-5} = \left(\dfrac{1}{2}\right)^5 = \dfrac{1}{32}$

11) $\left(-\dfrac{3}{4}\right)^3 = -\dfrac{27}{64}$

13) $\left(5n^6\right)^3 = 125n^{18}$

15) $\dfrac{m^{10}}{m^4} = m^{10-4} = m^6$

17) $\left(\dfrac{-12t^{-6}u^8}{4t^5u^{-1}}\right)^{-3} = \left(-3t^{-6-5}u^{8-(-1)}\right)^{-3}$

$\qquad = \left(-3t^{-11}u^9\right)^{-3}$

$\qquad = \left(\dfrac{-3u^9}{t^{11}}\right)^{-3}$

$\qquad = \left(-\dfrac{t^{11}}{3u^9}\right)^3 = -\dfrac{t^{33}}{27u^{27}}$

19) $\left(\dfrac{(9x^2y^{-2})^3}{4xy}\right)^0 = 1$

b) First, find the area and then multiply by the cost per yd^2.

$$A = \left(53\dfrac{1}{3}\ yd\right)(120\ yd)$$

$$= \left(\dfrac{\cancel{160}}{\cancel{3}}_{1}\ yd\right)\left(\cancel{120}^{40}\ yd\right)$$

$$= 6400\ yd^2$$

21) $\dfrac{12a^4b^{-3}}{20c^{-2}d^3} = \dfrac{\cancel{12}^3 a^4b^{-3}}{\cancel{20}_5 c^{-2}d^3} = \dfrac{3a^4c^2}{5b^3d^3}$

$$6400\ yd^2 \cdot \dfrac{\$1.80}{yd^2} = \$11{,}520$$

23) $t^{10k} \cdot t^{3k} = t^{10k+3k} = t^{13k}$

11) $V = \dfrac{4}{3}\pi r^3$ where r is the radius of the sphere

25) $0.000165 = 1.65 \times 10^{-4}$

27) $2.18 \times 10^7 = 21{,}800{,}000$

13) $4x^3 + 2x - 3$, when $x = 4$
$= 4(4)^3 + 2(4) - 3$
$= 4(64) + 8 - 3$
$= 256 + 8 - 3$
$= 261$

Cumulative Review: Chapters 1-2

1) $\dfrac{90}{150} = \dfrac{90 \div 30}{150 \div 30} = \dfrac{3}{5}$

3) $\dfrac{4}{15} \div \dfrac{20}{21} = \dfrac{4}{15} \cdot \dfrac{21}{20} = \dfrac{\cancel{4}}{\cancel{15}_5} \cdot \dfrac{\cancel{21}^7}{\cancel{20}_5} = \dfrac{7}{25}$

15) $5(t^2 + 7t - 3) - 2(4t^2 - t + 5)$
$= 5t^2 + 35t - 15 - 8t^2 + 2t - 10$
$= 5t^2 - 8t^2 + 35t + 2t - 15 - 10$
$= -3t^2 + 37t - 25$

5) $-26 + 5 - 7 = -21 - 7 = -28$

7) $(-1)^2 = -1$

9) a) $P = 2\left(53\dfrac{1}{3}\ yd\right) + 2(120\ yd)$

$$= 2\left(\dfrac{160}{3}\ yd\right) + 240\ yd$$

$$= \dfrac{320}{3}\ yd + 240\ yd$$

$$= \dfrac{320}{3}\ yd + \dfrac{720}{3}\ yd$$

$$= \dfrac{1040}{3}\ yd = 346\dfrac{2}{3}\ yd$$

17) $4^3 \cdot 4^7 = 4^{10}$

19) $\left(\dfrac{32x^3}{8x^{-2}}\right)^{-1} = \left(\dfrac{32}{8}x^{3-(-2)}\right)^{-1} = \left(4x^5\right)^{-1}$

$$= \dfrac{1}{4x^5}$$

21) $\left(4z^3\right)\left(-7z^5\right) = -28z^8$

23) $\left(-2a^{-6}b\right)^5 = (-2)^5(a^{-6})^5(b^5)$

$$= -32a^{-30}b^5 = \frac{-32b^5}{a^{30}}$$

25) $\left(6.2\times10^5\right)\left(9.4\times10^{-2}\right)$

$= \left(6.2\times9.4\right)\left(10^5\times10^{-2}\right)$

$= 58.28\times10^3$

$= 58,280$

Chapter 3: Linear Equations and Inequalities

Section 3.1 Exercises

1) Equation

3) Expression

5) No, it is an expression.

7) b, d

9) No

$$a - 4 = -9$$
$$5 - 4 = -9$$
$$1 \neq -9$$

11) Yes

$$-8p = 12$$
$$-8\left(-\frac{3}{2}\right) = 12$$
$$\overset{4}{\cancel{-8}}\left(\frac{3}{\underset{1}{\cancel{-2}}}\right) = 12$$
$$12 = 12$$

13) Yes

$$10 - 2(3y - 1) + y = -8$$
$$10 - 2(3(4) - 1) + 4 = -8$$
$$10 - 2(12 - 1) + 4 = -8$$
$$10 - 2(11) + 4 = -8$$
$$10 - 22 + 4 = -8$$
$$-8 = -8$$

15)
$$r - 6 = 11$$
$$r - 6 + 6 = 11 + 6$$
$$r = 17$$
The solution set is $\{17\}$.

17)
$$b + 10 = 4$$
$$b + 10 - 10 = 4 - 10$$
$$b = -6$$
The solution set is $\{-6\}$.

19)
$$-16 = k - 12$$
$$-16 + 12 = k - 12 + 12$$
$$-4 = k$$
The solution set is $\{-4\}$.

21)
$$a + \frac{5}{8} = \frac{1}{2}$$
$$a + \frac{5}{8} - \frac{5}{8} = \frac{1}{2} - \frac{5}{8}$$
$$a = \frac{4}{8} - \frac{5}{8}$$
$$a = -\frac{1}{8}$$
The solution set is $\left\{-\frac{1}{8}\right\}$.

23)
$$13.1 = v + 7.2$$
$$13.1 - 7.2 = v + 7.2 - 7.2$$
$$5.9 = v$$
$$v = 5.9$$
The solution set is $\{5.9\}$.

25) Answers may vary.

27)
$$3y = 30$$
$$\frac{\overset{1}{\cancel{3}} y}{\underset{1}{\cancel{3}}} = \frac{30}{3}$$
$$y = 10$$
The solution set is $\{10\}$.

29) $-5z = 35$

$$\frac{\cancel{-5}z}{\cancel{-5}} = \frac{35}{-5}$$

$$z = -7$$

The solution set is $\{-7\}$.

31) $-56 = -7v$

$$\frac{-56}{-7} = \frac{\cancel{-7}v}{\cancel{-7}}$$

$$8 = v$$

The solution set is $\{8\}$.

33) $\dfrac{a}{4} = 12$

$$4 \cdot \frac{a}{4} = 4 \cdot 12$$

$$1a = 48$$

$$a = 48$$

The solution set is $\{48\}$.

35) $-6 = \dfrac{k}{8}$

$$-6 \cdot 8 = \frac{k}{8} \cdot 8$$

$$-48 = 1k$$

$$-48 = k$$

The solution set is $\{-48\}$.

37) $\dfrac{2}{3}g = -10$

$$\frac{3}{2} \cdot \frac{2}{3}g = \frac{3}{2} \cdot (-10)$$

$$1g = \frac{3}{\cancel{2}} \cdot \left(-\cancel{10}^{5}\right)$$

$$g = -15$$

The solution set is $\{-15\}$.

39) $-\dfrac{5}{3}d = -30$

$$-\frac{3}{5} \cdot \left(-\frac{5}{3}\right)d = -\frac{3}{5} \cdot (-30)$$

$$1d = \frac{3}{\cancel{5}} \cdot \left(\cancel{30}^{6}\right)$$

$$d = 18$$

The solution set is $\{18\}$.

41) $\dfrac{11}{15} = \dfrac{1}{3}y$

$$3 \cdot \frac{11}{15} = 3 \cdot \frac{1}{3}y$$

$$\cancel{3}^{1} \cdot \frac{11}{\cancel{15}_{5}} = 1y$$

$$\frac{11}{5} = y$$

The solution set is $\left\{\dfrac{11}{5}\right\}$.

43) $0.5q = 6$

$$\frac{0.5q}{0.5} = \frac{6}{0.5}$$

$$q = 12$$

The solution set is $\{12\}$.

45) $-w = -7$

$$\frac{-w}{-1} = \frac{-7}{-1}$$

$$w = 7$$

The solution set is $\{7\}$.

$-12d = 0$

$$\frac{-12d}{-12} = \frac{0}{-12}$$

$$d = 0$$

The solution set is $\{0\}$.

49) $\quad 3x - 7 = 17$

$3x - 7 + 7 = 17 + 7$

$3x = 24$

$\dfrac{\cancel{3}x}{\cancel{3}} = \dfrac{24}{3}$

$x = 8$

The solution set is $\{8\}$.

51) $\quad 7c + 4 = 18$

$7c + 4 - 4 = 18 - 4$

$7c = 14$

$\dfrac{\cancel{7}c}{\cancel{7}} = \dfrac{14}{7}$

$c = 2$

The solution set is $\{2\}$.

53) $\quad 8d - 15 = -15$

$8d - 15 + 15 = -15 + 15$

$8d = 0$

$\dfrac{\cancel{8}d}{\cancel{8}} = \dfrac{0}{8}$

$d = 0$

The solution set is $\{0\}$.

55) $\quad -11 = 5t - 9$

$-11 + 9 = 5t - 9 + 9$

$-2 = 5t$

$\dfrac{-2}{5} = \dfrac{\cancel{5}t}{\cancel{5}}$

$-\dfrac{2}{5} = t$

The solution set is $\left\{-\dfrac{2}{5}\right\}$.

57) $\quad 10 = 3 - 7y$

$10 - 3 = 3 - 3 - 7y$

$7 = -7y$

$\dfrac{7}{-7} = \dfrac{\cancel{-7}y}{\cancel{-7}}$

$-1 = y$

The solution set is $\{-1\}$.

59) $\quad \dfrac{4}{9}w - 11 = 1$

$\dfrac{4}{9}w - 11 + 11 = 1 + 11$

$\dfrac{4}{9}w = 12$

$\dfrac{9}{4} \cdot \dfrac{4}{9}w = \dfrac{9}{4} \cdot 12$

$1w = \dfrac{9}{\cancel{4}} \cdot \cancel{12}^{\,3}$

$w = 27$

The solution set is $\{27\}$.

61) $\quad \dfrac{10}{7}m + 3 = 1$

$\dfrac{10}{7}m + 3 - 3 = 1 - 3$

$\dfrac{10}{7}m = -2$

$\dfrac{7}{10} \cdot \dfrac{10}{7}m = \dfrac{7}{10} \cdot -2$

$1m = \dfrac{7}{\cancel{10}} \cdot -\cancel{2}^{\,1}$

$m = -\dfrac{7}{5}$

The solution set is $\left\{-\dfrac{7}{5}\right\}$.

63) $\dfrac{1}{2}d + 7 = 12$

$\dfrac{1}{2}d + 7 - 7 = 12 - 7$

$\dfrac{1}{2}d = 5$

$\dfrac{2}{1} \cdot \dfrac{1}{2}d = 5 \cdot \dfrac{2}{1}$

$d = 10$

The solution set is $\{10\}$.

65) $2 - \dfrac{5}{6}t = -2$

$2 - \dfrac{5}{6}t - 2 = -2 - 2$

$-\dfrac{5}{6}t = -4$

$\left(-\dfrac{6}{5}\right)\left(-\dfrac{5}{6}\right)t = \left(-\dfrac{6}{5}\right)(-4)$

$t = \dfrac{24}{5}$

The solution set is $\left\{\dfrac{24}{5}\right\}$.

67) $-\dfrac{1}{6}z + \dfrac{1}{2} = \dfrac{3}{4}$

$-\dfrac{1}{6}z + \dfrac{1}{2} - \dfrac{1}{2} = \dfrac{3}{4} - \dfrac{1}{2}$

$-\dfrac{1}{6}z = \dfrac{3}{4} - \dfrac{2}{4}$

$-\dfrac{1}{6}z = \dfrac{1}{4}$

$-6 \cdot \left(-\dfrac{1}{6}z\right) = -6 \cdot \dfrac{1}{4}$

$1z = -\overset{3}{\cancel{6}} \cdot \dfrac{1}{\underset{2}{\cancel{4}}}$

$z = -\dfrac{3}{2}$

The solution set is $\left\{-\dfrac{3}{2}\right\}$.

69) $5 - 0.4p = 2.6$

$5 - 5 - 0.4p = 2.6 - 5$

$-0.4p = -2.4$

$\dfrac{\overset{1}{\cancel{-0.4}}\,p}{\underset{1}{\cancel{-0.4}}} = \dfrac{-2.4}{-0.4}$

$1p = 6$

$p = 6$

The solution set is $\{6\}$.

71) $4.3a + 1.98 = -14.36$

$4.3a + 1.98 - 1.98 = -14.36 - 1.98$

$4.3a = -16.34$

$\dfrac{\overset{1}{\cancel{4.3}}a}{\underset{1}{\cancel{4.3}}} = \dfrac{-16.34}{4.3}$

$1a = -3.8$

$a = -3.8$

The solution set is $\{-3.8\}$.

73) $3x + 7 + 5x + 4 = 27$

$8x + 11 = 27$ Combine like terms.

$8x + 11 - 11 = 27 - 11$ Subtraction Prop. of Equality.

$8x = 16$ Combine like terms.

$\dfrac{\overset{1}{\cancel{8}}x}{\underset{1}{\cancel{8}}} = \dfrac{16}{8}$ Division Prop. Equality.

$x = 2$ Simplify.

The solution set is $\{2\}$.

75) $10v+9-2v+16=1$

$8v+25=1$

$8v+25-25=1-25$

$8v=-24$

$$\frac{\overset{1}{\cancel{8}}v}{\underset{1}{\cancel{8}}}=\frac{-24}{8}$$

$v=-3$

The solution set is $\{-3\}$.

77) $5-3m+9m+10-7m=-4$

$-m+15=-4$

$-m+15-15=-4-15$

$-m=-19$

$$\frac{-m}{-1}=\frac{-19}{-1}$$

$m=19$

The solution set is $\{19\}$.

79) $5=-12p+7+4p-12$

$5=-8p-5$

$5+5=-8p-5+5$

$10=-8p$

$$\frac{10}{-8}=\frac{\cancel{-8}p}{\underset{1}{\cancel{-8}}}$$

$-\dfrac{5}{4}=p$

The solution set is $\left\{-\dfrac{5}{4}\right\}$.

81) $2(5x+3)-3x+4=-11$

$10x+6-3x+4=-11$

$7x+10=-11$

$7x+10-10=-11-10$

$7x=-21$

$$\frac{\cancel{7}x}{\underset{1}{\cancel{7}}}=\frac{-21}{7}$$

$x=-3$

The solution set is $\{-3\}$.

83) $\quad -12=7(2a-3)-(8a-9)$

$-12=14a-8a-21+9$

$-12=6a-12$

$-12+12=6a-12+12$

$0=6a$

$$\frac{0}{6}=\frac{\overset{1}{\cancel{6}}a}{\underset{1}{\cancel{6}}}$$

$0=a$

The solution set is $\{0\}$.

85) $\dfrac{1}{3}(3w+4)-\dfrac{2}{3}=-\dfrac{1}{3}$

$w+\dfrac{4}{3}-\dfrac{2}{3}=-\dfrac{1}{3}$

$w+\dfrac{2}{3}=-\dfrac{1}{3}$

$w=-\dfrac{1}{3}-\dfrac{2}{3}$

$w=-1$

The solution set is $\{-1\}$.

87) $\dfrac{1}{2}(c-2)+\dfrac{1}{4}(2c+1)=\dfrac{5}{4}$

$\qquad \dfrac{1}{2}c-1+\dfrac{1}{2}c+\dfrac{1}{4}=\dfrac{5}{4}$

$\qquad\qquad\qquad c-\dfrac{3}{4}=\dfrac{5}{4}$

$\qquad\qquad\qquad c=\dfrac{5}{4}+\dfrac{3}{4}$

$\qquad\qquad\qquad c=\dfrac{8}{4}$

$\qquad\qquad\qquad c=2$

The solution set is $\{2\}$.

89) $\qquad \dfrac{4}{3}(t+1)-\dfrac{1}{6}(4t-3)=2$

$\qquad \dfrac{4}{3}t+\dfrac{4}{3}-\dfrac{2}{3}t+\dfrac{1}{2}=2$

$\dfrac{4}{3}t+\dfrac{2}{2}\cdot\dfrac{4}{3}-\dfrac{2}{3}t+\dfrac{3}{3}\cdot\dfrac{1}{2}=2$

$\qquad \dfrac{4}{3}t-\dfrac{2}{3}t+\dfrac{8}{6}+\dfrac{3}{6}=2$

$\qquad\qquad\qquad \dfrac{2}{3}t+\dfrac{11}{6}=2$

$\qquad\qquad\qquad \dfrac{2}{3}t=\dfrac{12}{6}-\dfrac{11}{6}$

$\qquad\qquad\qquad \dfrac{2}{3}t=\dfrac{1}{6}$

$\qquad\qquad \dfrac{3}{2}\cdot\dfrac{2}{3}t=\dfrac{1}{6}\cdot\dfrac{3}{2}$

$\qquad\qquad\qquad t=\dfrac{1}{4}$

The solution set is $\left\{\dfrac{1}{4}\right\}$.

Section 3.2 Exercises

1) Step1:Clear parentheses and combine like terms on each side of the equation.
Step2: Isolate the variable.
Step3:Solve for the variable.
Step4:Check the solution.

3) $\qquad 2y+7=5y-2$

$\qquad 2y-2y+7=5y-2y-2$

$\qquad\qquad\quad 7=3y-2$

$\qquad\qquad 7+2=3y-2+2$

$\qquad\qquad\quad 9=3y$

$\qquad\qquad\quad \dfrac{9}{3}=\dfrac{3y}{3}$

$\qquad\qquad\quad 3=y$

The solution set is $\{3\}$.

5) $\qquad 6-7p=2p+33$

$\qquad 6-7p+7p=2p+7p+33$

$\qquad\qquad\qquad 6=9p+33$

$\qquad\qquad 6-33=9p+33-33$

$\qquad\qquad\qquad -27=9p$

$\qquad\qquad\qquad \dfrac{-27}{9}=\dfrac{9p}{9}$

$\qquad\qquad\qquad -3=p$

The solution set is $\{-3\}$.

7) $-8x+6-2x+11=3+3x-7x$

$-10x+17=3-4x$

$-10x+10x+17=3-4x+10x$

$17=3+6x$

$17-3=3-3+6x$

$14=6x$

$\dfrac{14}{6}=\dfrac{6x}{6}$

$\dfrac{7}{3}=x$

The solution set is $\left\{\dfrac{7}{3}\right\}$.

9) $18-h+5h-11=9h+19-3h$

$7+4h=6h+19$

$7+4h-4h=6h+19-4h$

$7=2h+19$

$7-19=2h+19-19$

$-12=2h$

$\dfrac{-12}{2}=\dfrac{2h}{2}$

$-6=h$

The solution set is $\{-6\}$.

11) $4(2t+5)-7=5(t+5)$

$8t+20-7=5t+25$

$8t+13=5t+25$

$8t-5t+13=5t-5t+25$

$3t+13=25$

$3t+13-13=25-13$

$3t=12$

$\dfrac{3t}{3}=\dfrac{12}{3}$

$t=4$

The solution set is $\{4\}$.

13) $2(1-8c)=5-3(6c+1)+4c$

$2-16c=5-18c-3+4c$

$2-16c=2-14c$

$2-16c-2=2-14c-2$

$-16c=-14c$

$-16c+16c=-14c+16c$

$0=2c$

$0=c$

The solution set is $\{0\}$

15) $2(6d+5)=16-(7d-4)+11d$

$12d+10=16-7d+4+11d$

$12d+10=20+4d$

$12d+10-4d=20+4d-4d$

$8d+10=20$

$8d+10-10=20-10$

$8d=10$

$\dfrac{8d}{8}=\dfrac{\overset{5}{\cancel{10}}}{\underset{4}{\cancel{8}}}$

$d=\dfrac{5}{4}$

The solution set is $\left\{\dfrac{5}{4}\right\}$.

17) Eliminate fractions by multiplying both sides of the equation by the LCD of all the fractions in the equation.

19) Multiply both sides of the equation by 8.

21) $\dfrac{3}{8}x - \dfrac{1}{2} = \dfrac{1}{8}x + \dfrac{3}{4}$

$$8\left(\dfrac{3}{8}x - \dfrac{1}{2}\right) = 8\left(\dfrac{1}{8}x + \dfrac{3}{4}\right)$$

$$8\left(\dfrac{3}{8}x\right) - 8\left(\dfrac{1}{2}\right) = 8\left(\dfrac{1}{8}x\right) + 8\left(\dfrac{3}{4}\right)$$

$$3x - 4 = x + 6$$

$$3x - x - 4 = x - x + 6$$

$$2x - 4 = 6$$

$$2x - 4 + 4 = 6 + 4$$

$$2x = 10$$

$$\dfrac{2x}{2} = \dfrac{10}{2}$$

$$x = 5$$

The solution set is $\{5\}$.

23) $\dfrac{2}{3}d - 1 = \dfrac{1}{5}d + \dfrac{2}{5}$

$$15\left(\dfrac{2}{3}d - 1\right) = 15\left(\dfrac{1}{5}d + \dfrac{2}{5}\right)$$

$$15\left(\dfrac{2}{3}d\right) - 15(1) = 15\left(\dfrac{1}{5}d\right) + 15\left(\dfrac{2}{5}\right)$$

$$10d - 15 = 3d + 6$$

$$10d - 3d - 15 = 3d - 3d + 6$$

$$7d - 15 = 6$$

$$7d - 15 + 15 = 6 + 15$$

$$7d = 21$$

$$\dfrac{7d}{7} = \dfrac{21}{7}$$

$$d = 3$$

The solution set is $\{3\}$.

25) $\dfrac{1}{3} + \dfrac{1}{9}(k + 5) - \dfrac{k}{4} = 2$

$$36\left[\dfrac{1}{3} + \dfrac{1}{9}(k + 5) - \dfrac{k}{4}\right] = 36(2)$$

$$36\left(\dfrac{1}{3}\right) + 36 \cdot \dfrac{1}{9}(k + 5) - 36\left(\dfrac{k}{4}\right) = 72$$

$$12 + 4(k + 5) - 9k = 72$$

$$12 + 4k + 20 - 9k = 72$$

$$32 - 5k = 72$$

$$32 - 32 - 5k = 72 - 32$$

$$-5k = 40$$

$$\dfrac{-5k}{-5} = \dfrac{40}{-5}$$

$$k = -8$$

The solution set is $\{-8\}$.

27) $\dfrac{3}{4}(y + 7) + \dfrac{1}{2}(3y - 5) = \dfrac{9}{4}(2y - 1)$

$$4 \cdot \dfrac{3}{4}(y + 7) + 4 \cdot \dfrac{1}{2}(3y - 5) = 4 \cdot \dfrac{9}{4}(2y - 1)$$

$$3(y + 7) + 2(3y - 5) = 9(2y - 1)$$

$$3y + 21 + 6y - 10 = 18y - 9$$

$$9y + 11 = 18y - 9$$

$$9y + 11 - 18y = 18y - 9 - 18y$$

$$-9y + 11 = -9$$

$$-9y + 11 - 11 = -9 - 11$$

$$-9y = -20$$

$$y = \dfrac{20}{9}$$

The solution set is $\left\{\dfrac{20}{9}\right\}$.

29) $\dfrac{2}{3}(3h-5)+1=\dfrac{3}{2}(h-2)+\dfrac{1}{6}h$

$6\cdot\dfrac{2}{3}(3h-5)+6\cdot1=6\cdot\dfrac{3}{2}(h-2)+6\cdot\dfrac{1}{6}h$

$4(3h-5)+6=9(h-2)+h$

$12h-20+6=9h-18+h$

$12h-14=10h-18$

$12h-14-10h=10h-18-10h$

$2h-14+14=-18+14$

$2h=-18+14$

$2h=-4$

$h=\dfrac{-4}{2}$

$h=-2$

The solution set is $\{-2\}$.

31) $0.06d+0.13=0.31$

$100(0.06d+0.13)=100(0.31)$

$6d+13=31$

$6d+13-13=31-13$

$6d=18$

$\dfrac{6d}{6}=\dfrac{18}{6}$

$d=3$

33) $0.04n-0.05(n+2)=0.1$

$100\left[0.04n-0.05(n+2)\right]=100(0.1)$

$4n-5(n+2)=10$

$4n-5n-10=10$

$-n=20$

$\dfrac{-n}{-1}=\dfrac{20}{-1}$

$n=-20$

The solution set is $\{-20\}$.

35) $0.35a-a=0.03(5a+4)$

$0.25a=0.03(5a+4)$

$100(0.25a)=100\left[0.03(5a+4)\right]$

$25a=3(5a+4)$

$25a=15a+12$

$25a-15a=15a=12-15a$

$10a=12$

$\dfrac{10a}{10}=\dfrac{12}{10}$

$a=1.2$

The solution set is $\{1.2\}$

37) $27=0.04y+0.03(y+200)$

$100(27)=100\left[0.04y+0.03(y+200)\right]$

$2700=4y+3(y+200)$

$2700=4y+3y+600$

$2700=7y+600$

$2700-600=7y+600-600$

$2100=7y$

$\dfrac{2100}{7}=\dfrac{7y}{7}$

$300=y$

The solution set is $\{300\}$.

39)
$$0.2(12)+0.08z=0.12(z+12)$$
$$100\left[0.2(12)+0.08z\right]=100\left[0.12(z+12)\right]$$
$$20(12)+8z=12(z+12)$$
$$240+8z=12z+144$$
$$240+8z-8z=12z-8z+144$$
$$240=4z+144$$
$$240-144=4z+144-144$$
$$96=4z$$
$$\frac{96}{4}=\frac{4z}{4}$$
$$24=z$$

The solution set is $\{24\}$.

41) The variable is eliminated, and you get a false statement like 5=12

43)
$$-9r+4r-11+2=3r+7-8r+9$$
$$-5r-9=-5r+16$$
$$-5r+5r-9=-5r+5r+16$$
$$-9\neq16$$
The solution set is $\varnothing$.

45)
$$j-15j+8=-3(4j-3)-2j-1$$
$$j-15j+8=-12j+9-2j-1$$
$$-14j+8=-14j+8$$
$$-14j+14j+8=-14j+14j+8$$
$$8=8$$
The solution set is $\{$all real numbers$\}$.

47)
$$8(3t+4)=10t-3+7(2t+5)$$
$$24t+32=10t-3+14t+35$$
$$24t+32=24t+32$$
$$24t-24t+32=24t-24t+32$$
$$32=32$$
The solution set is $\{$all real numbers$\}$.

49)
$$\frac{5}{6}k-\frac{2}{3}=\frac{1}{6}(5k-4)+\frac{1}{2}$$
$$18\left(\frac{5}{6}k-\frac{2}{3}\right)=18\cdot\frac{1}{6}(5k-4)+18\cdot\frac{1}{2}$$
$$18\cdot\frac{5}{6}k-18\cdot\frac{2}{3}=3\cdot5k-3\cdot4+9$$
$$3\cdot5k-6\cdot2=15k-12+9$$
$$15k-12=15k-3$$
$$15k-12-15k=15k-3-15k$$
$$-12\neq-3$$
The solution set is $\varnothing$.

51)
$$7(2q+3)=6+3(q+5)$$
$$14q+21=6+3q+15$$
$$14q+21=3q+21$$
$$14q+21-3q=3q+21-3q$$
$$11q+21=21$$
$$11q+21-21=21-21$$
$$11q=0$$
$$q=0$$
The solution set is $\{0\}$.

53)
$$016h+0.4(2000)=0.22(2000+h)$$
$$100\cdot0.16h+100\cdot0.4(2000)=100\cdot\left[0.22(2000+h)\right]$$
$$16h+40(2000)=22(2000+h)$$
$$16h+80000=44000+22h$$
$$16h-22h+80000=44000$$
$$-6h+80000-80000=44000-80000$$
$$-6h=-36000$$
$$\frac{-6h}{-6}=\frac{-36000}{-6}$$
$$h=6000$$
The solution set is $\{6000\}$.

55) $t+18=3(5-t)+4t+3$

$t+18=15-3t+4t+3$

$t+18=18+t$

$t+18-t=18+t-t$

$18=18$

The solution set is $\{$all real numbers$\}$.

57) $\dfrac{1}{2}(2r+9)=1+\dfrac{1}{3}(r+12)$

$6\cdot\dfrac{1}{2}(2r+9)=6\cdot1+6\cdot\dfrac{1}{3}(r+12)$

$3\cdot2r+3\cdot9=6+2(r+12)$

$6r+27=6+2r+24$

$6r+27=2r+30$

$6r+27-2r=2r+30-2r$

$4r+27=30$

$4r+27-27=30-27$

$4r=3$

$r=\dfrac{3}{4}$

The solution set is $\left\{\dfrac{3}{4}\right\}$.

59) $2d+7=-4d+3(2d-5)$

$2d+7=-4d+3\cdot2d-3\cdot5$

$2d+7=-4d+6d-15$

$2d+7=2d-15$

$2d+7-2d=2d-15-2d$

$7\neq-15$

The solution set is $\varnothing$.

61) 1) Read the problem until you understand it.

2) Choose a variable to represent an unknown quantity.

3) Translate the problem from English into an equation.

4) Solve the equation.

5) Check the answer in the original problem and interpret the meaning of the solution asit relates to the problem.

63) Let $x=$ the number

$x+12=5$

$x+12-12=5-12$

$x=-7$

The number is -7.

65) Let $x=$ the number

$x-9=12$

$x-9+9=12+9$

$x=21$

The number is 21.

67) Let $x=$ the number

$2x+5=17$

$2x+5-5=17-5$

$2x=12$

$\dfrac{2x}{2}=\dfrac{12}{2}$

$x=6$

The number is 6.

69)   Let x = the number

$$2x + 18 = 8$$

$$2x + 18 - 18 = 8 - 18$$

$$2x = -10$$

$$\frac{2x}{2} = \frac{-10}{2}$$

$$x = -5$$

The number is -5.

71)   Let x = the number

$$3x - 8 = 40$$

$$3x - 8 + 8 = 40 + 8$$

$$3x = 48$$

$$\frac{3x}{3} = \frac{48}{3}$$

$$x = 16$$

The number is 16.

73)   Let x = the number

$$\frac{3}{4}x = 33$$

$$\frac{4}{3} \cdot \frac{3}{4}x = \frac{4}{3} \cdot 33$$

$$1x = 4 \cdot 11$$

$$x = 44$$

The number is 44.

75)   Let x = the number

$$\frac{1}{2}x - 9 = 3$$

$$\frac{1}{2}x - 9 + 9 = 3 + 9$$

$$\frac{1}{2}x = 12$$

$$2 \cdot \frac{1}{2}x = 2 \cdot (12)$$

$$x = 24$$

The number is 24.

77)   Let x = the number

$$x + 6 = 8$$

$$x + 6 - 6 = 8 - 6$$

$$x = 2$$

The number is 2

79)   Let x = the number

$$2x - 3 = x + 8$$

$$2x - 3 - x = x + 8 - x$$

$$x - 3 = 8$$

$$x - 3 + 3 = 8 + 3$$

$$x = 11$$

The number is 11.

81)   Let x = the number

$$\frac{1}{3}x + 10 = x - 2$$

$$3\left(\frac{1}{3}x + 10\right) = 3(x - 2)$$

$$x + 30 = 2x - 6$$

$$x + 30 + 6 = 3x - 6 + 6$$

$$x + 36 = 3x$$

$$x + 36 - x = 3x - x$$

$$36 = 2x$$

$$\frac{36}{2} = \frac{2x}{2}$$

$$18 = x$$

The number is 18.

83)   Let x be the number.

$$x - 45 = \frac{x}{4}$$

$$4(x - 45) = 4\left(\frac{x}{4}\right)$$

$$4x - 180 = x$$

$$4x - 180 - x = x - x$$

$$3x - 180 = 0$$

$$x = 60$$

85)   Let x be the number.

$$x + \frac{2}{3}x = 25$$

$$3\left(x + \frac{2}{3}x\right) = 3(25)$$

$$3x + 2x = 75$$

$$5x = 75$$

$$x = \frac{75}{5}$$

$$x = 15$$

The number is 15.

87)   Let x be the number.

$$x - 2x = 13$$

$$-x = 13$$

$$\frac{-x}{-1} = \frac{13}{-1}$$

$$x = -13$$

The number is -13.

Section 3.3: Exercises

1) $c + 14$

3) $c - 37$

5) $\frac{1}{2}s$

7) $14 - x$

9)   The number of children must be whole number.

11)   It is an even number.

13) Let x=the amount of rain in 2004.
     Then the amount of rain in 1905 = $x+1.2$

$$\left(\begin{array}{c}\text{amount in}\\ \text{2004}\end{array}\right) + \left(\begin{array}{c}\text{amount in}\\ \text{1905}\end{array}\right) = 7.2$$

$$x \quad + \quad x+1.2 \quad\quad = 7.2$$

$$2x + 1.2 = 7.2$$

$$2x = 6.0$$

$$x = 3.0$$

The amount of rain in 1905 was 4.2 inches
and the amount of rain in 2004 was 3.0 inches

15) Let x be the number of titles won
     by Lance Armstrong.
     Then the number of titles won by Miguel
     Indurain is $x-2$

$$\left(\begin{array}{c}\text{Number by}\\ \text{Lance}\end{array}\right) + \left(\begin{array}{c}\text{Number by}\\ \text{Miguel}\end{array}\right) = 12$$

$$x \quad + \quad x-2 \quad = 12$$

$$2x - 2 = 12$$

$$2x = 14$$

$$x = 7$$

$$x - 2 = 5$$

Lance won 7 times and Miguel won 5 times.

17) Let x be the amount of caffeine in decaffeinated coffee.
Then the amount of caffeine regular coffee is $13x$

$$\left(\begin{array}{c}\text{Amount of}\\\text{caffeine in}\\\text{decaffeinated}\\\text{coffee}\end{array}\right)+\left(\begin{array}{c}\text{Amount of}\\\text{caffeine in}\\\text{regular}\\\text{coffee}\end{array}\right)=280$$

$$x\quad+\quad 13x\ =280$$
$$14x=280$$
$$x=20$$

Amount of caffeine in decaffeinated coffee $= 20\,\text{mg}$

Amount of caffeine in regular coffee $= 13x$
$$=13(20)$$
$$=260\ \text{mg}$$

Decaffeinated coffee contains 20 mg and regular coffee contains 260 mg.

19) Let x = number o f students taking the spanish class.
Then the number of students taking

the French class $= \dfrac{2}{3}x.$

$$\left(\begin{array}{c}\#\text{of students}\\\text{in Spanish}\\\text{class}\end{array}\right)+\left(\begin{array}{c}\#\text{of students}\\\text{in French}\\\text{class}\end{array}\right)=310$$

$$x\quad+\quad \frac{2}{3}x\quad=310$$

$$\frac{5}{3}x=310$$

$$x=\frac{3}{5}\cdot 310=186$$

number of students in Spanish class $= 186$

number of students in

French class $= \dfrac{2}{3}\cdot x = \dfrac{2}{3}\cdot 186 = 124$

186 students are studying Spanish and 124 students are studying French.

21) Let x = the length of the shorter piece.

$$(\text{shorter piece})+(\text{longer piece})=36$$
$$x\quad+\quad x+14\quad=36$$
$$2x=22$$
$$x=11$$

shorter piece $= 11$ inches

longer piece $= x+14$
$$=11+14$$
$$=25\,\text{inches}$$

Then the length of the longer piece $= x+14$. The pipe will be cut into an 11 in. piece and a 25 in. piece.

23) Let x = the length of the shorter piece.
Then the length of the longer piece $= 2x$.

$$(\text{longer piece})+(\text{shorter piece})=28.5$$
$$2x\quad+\quad x\quad=28.5$$
$$3x=28.5$$
$$x=9.5\,\text{inches}$$

Shorter piece $= 9.5$ inches

Longer piece $= 2x$
$$=2(9.5)$$
$$=19\,\text{inches}$$

The wire will be cut into a 9.5 in. piece for the bracelet and a 19 in. piece for the necklace.

25) Let x = the length Derek's sandwich.
Then the length of Cory's sandwich $= x+1$

length of Tamara's sandwich $= \dfrac{1}{2}x$

$\left(\begin{array}{c}\text{length of}\\ \text{Derk's sandwich}\end{array}\right) + \left(\begin{array}{c}\text{length of}\\ \text{Cory's sandwich}\end{array}\right)$

$+ \left(\begin{array}{c}\text{length of}\\ \text{Tamara's sandwich}\end{array}\right) = 6$

$x \quad + x+1 \quad + \quad \dfrac{1}{2}x \quad = 6$

$1 + \dfrac{5}{2}x = 6$

$\dfrac{5}{2}x = 5$

$x = \dfrac{2}{5}\cdot 5 = 2\,\text{ft}$

Derek's sandwich $= 2\,\text{ft}$

Cory's sandwich $= x+1 = 3\,ft$

Tamara's sandwich $= \dfrac{1}{2}x = 1\,ft$

Derek's sandwich was 2 ft long,
Cory's sandwich was 3 ft and
Tamara's sandwich was 1 ft.

27) Let x = first integer,
$x+1$ = second integer, and
$x+2$ = third integer

$\left(\begin{array}{c}\text{first}\\ \text{integer}\end{array}\right) + \left(\begin{array}{c}\text{second}\\ \text{integer}\end{array}\right) + \left(\begin{array}{c}\text{third}\\ \text{integer}\end{array}\right) = 126$

$x \quad + \quad x+1 \quad + \quad x+2 \;=126$

$3x+3 = 126$

$3x = 123$

$x = 41$

first integer $= 41$

second integer $= x+1 = 42$

third integer $= x+2 = 43$

The integers are 41, 42 and 43 .

29) Let x = first even integer and
$x+2$ = second even integer

$2\left(\begin{array}{c}\text{first even}\\ \text{integer}\end{array}\right) = 16 + \left(\begin{array}{c}\text{second even}\\ \text{integer}\end{array}\right)$

$2x \quad = 16+ \quad x+2$

$2x = 18 + x$

$x = 18$

first even integer $= 18$

second even integer $= x+2 = 20$

The integers are 18 and 20.

31) Let x = first odd integer,
$x+2$ = second odd integer and
$x+4$ = third odd integer

$\left(\begin{array}{c}\text{first odd}\\ \text{integer}\end{array}\right) + \left(\begin{array}{c}\text{second odd}\\ \text{integer}\end{array}\right)$

$+ \left(\begin{array}{c}\text{third odd}\\ \text{integer}\end{array}\right) = 5 + 4\left(\begin{array}{c}\text{third}\\ \text{odd integer}\end{array}\right)$

$x + x+2 + x+4 \;= 5+4(x+4)$

$3x+6 = 5+4x+16$

$3x+6 = 21+4x$

$6 = 21+x$

$-15 = x$

first odd integer $= -15$

second odd integer $= x+2 = -13$

third odd integer $= x+4 = -11$

The integers are $-15, -13$ and, -11

33) Let x = first page number and
$x+1$ = second page number

$\left(\begin{array}{c}\text{first page}\\ \text{number}\end{array}\right) + \left(\begin{array}{c}\text{second page}\\ \text{number}\end{array}\right) = 215$

$x \quad + \quad x+1 \quad = 215$

$2x = 214$

$x = 107$

$x = $ first page number $= 107$

$x+1 = $ second page number $= 108$

The page numbers are 107 and 108.

35) Let x be the number of trout caught by Kelly
Then the number of trout caught by Jimmy is $x+6$

$$\left(\begin{array}{c}\text{Number of trout}\\\text{caught by Kelly}\end{array}\right)+\left(\begin{array}{c}\text{Number of trout}\\\text{caught by Jimmy}\end{array}\right)=20$$

$x \quad + \quad x+6 \quad = 20$

$2x+6 = 20$

$2x = 14$

$x = 7$

number of trout caught by Kelly $= 7$

number of trout caught by Jimmy $= x+6 = 7+6$

$= 13$

Jimmy: 13. Kelly: 7

37) Let x be the length of the shorter piece
Then the length of the longer piece is $2x+1$

$(\text{shorter piece})+(\text{longer piece})=16$

$x \quad + \quad 2x+1 \quad = 16$

$3x+1 = 16$

$3x = 15$

$x = 5$

shorter piece $= 5\,\text{ft}$

longer piece $= 2x+1$

$= 2\cdot5+1$

$= 11\,\text{ft}$

The pieces are 5 ft and 11 ft.

39) Let x be the attendance at Bomaroo.
Then the attendance at the Lollapalooza is $3x+15000$

$(\text{attendance at Bomaroo})+$

$(\text{attendance at Lollapalooza})=295,000$

$x \quad + \quad 15,000+3x \quad = 295,000$

$4x+15,000 = 295,000$

$4x = 280,000$

$x = 70,000$

attendance at Bomaroo $= 70,000$

attendance at Lollapalooza $= 3x+15,000$

$= 3\cdot70,000+15,000 = 225,000$

Bomaroo: 70,000. Lollapalooza: 225,000.

41) Let x be the first integer,

$x+1 = $ second integer

$x+2 = $ third integer

$$\left(\begin{array}{c}\text{first}\\\text{integer}\end{array}\right)+\left(\begin{array}{c}\text{second}\\\text{integer}\end{array}\right)+\left(\begin{array}{c}\text{third}\\\text{integer}\end{array}\right)=174$$

$x \quad + \quad x+1 \quad + \quad x+2 \quad = 174$

$3x+3 = 174$

$3x = 171$

$x = 57$

first integer $= 57$

second integer $= x+1 = 58$

third integer $= x+2 = 59$

The integers are 57, 58 and 59.

43) Let x be the amount weight lost by Helen.
Then the amount of weight lost by Tara is $x+15$.
The amount of weight lost by Mike is $2x-73$.

$$(\text{Helen's weight loss})+(\text{Tara's weight loss})$$
$$+(\text{Mike's weight loss})=502$$

$$x \quad +x+15 \quad + \quad 2x-73 \quad =502$$
$$4x-58=502$$
$$4x=560$$
$$x=\frac{560}{4}=140\,\text{lb}$$

Helen's weight loss $=140\,\text{lb}$

Tara's weight loss $=x+15=140+15=155\,\text{lb}$

Mike's weight loss $=2x-73=2\cdot140-73$
$$=280-73=207\,\text{lb}$$

Helen: 140 lb. Tara: 155 lb. Mike: 207 lb.

45) Let x be the length of the longest piece.
Then the length of the shortest piece is $\frac{1}{3}x$
Length of the medium-size piece is $x-12$

$$\left(\begin{array}{c}\text{length of}\\\text{shortest piece}\end{array}\right)+\left(\begin{array}{c}\text{length of}\\\text{longest piece}\end{array}\right)$$
$$+\left(\begin{array}{c}\text{length of}\\\text{medium-sized piece}\end{array}\right)=72$$

$$\frac{1}{3}x \quad +x \quad + \quad x-12 \quad =72$$
$$\frac{7}{3}x-12=72$$
$$\frac{7}{3}x=84$$
$$x=\frac{3}{7}\cdot84=36\,\text{in}$$

longest piece $=36\,\text{in}$

shortest piece $=\frac{1}{3}x=\frac{1}{3}\cdot36=12\,\text{in}$

medium-sized piece $=x-12=36-12=24\,\text{in}$

longest: 36 in. medium sized: 24 in. shortest: 12 in.

47) Let x be the number of copies in millions of Coldplay's album.
Then the number of copies of Lil Waynes's album is $x+0.73$
Number of copies of Taylor Swift's album is $x-0.04$

$$(\text{\#of copies of Coldplay's album})+$$
$$(\text{\#of copies of LIl Wayne's album})+$$
$$(\text{\#of copies of Taylor Swift's album})=7.14$$
$$x+x+0.73+x-0.04 \quad =7.14$$
$$3x+0.69=7.14$$
$$3x=7.14-0.69$$
$$3x=6.45$$
$$x=2.15$$

\#of copies of Coldplay's album $=2.15$ million

\#of copies of Lil Wayne's album $=x+0.73$
$$=2.15+0.73=2.88$ million$$

\#of copies of Taylor Swift's album $=x-0.04$
$$=2\cdot15-0.04=2.11$ million$$

49) Let $x=$ first even integer,
$x+2=$ second even integer and
$x+4=$ third even integer

$$\frac{1}{6}\left(\begin{array}{c}\text{first even}\\\text{integer}\end{array}\right)$$
$$=\frac{1}{10}\left[\left(\begin{array}{c}\text{second even}\\\text{integer}\end{array}\right)+\left(\begin{array}{c}\text{third even}\\\text{integer}\end{array}\right)\right]-3$$

$$\frac{1}{6}x=\frac{1}{10}[\quad x+2 \quad + \quad x+4 \quad]-3$$
$$\frac{1}{6}x=\frac{1}{10}[2x+6]-3$$
$$30\left(\frac{1}{6}x\right)=30\left(\frac{1}{10}[2x+6]-3\right)$$
$$5x=3[2x+6]-90$$
$$5x=6x+18-90$$
$$5x=6x-72$$
$$-x=-72$$
$$x=72$$

x = first even integer = 72
$x+2$ = second even integer = 74
$x+4$ = third even integer = 76
The integers are 72, 74 and 76.

Section 3.4: Exercises

1) Amount of Discount
$$= (\text{Rate of Discount})(\text{Original Price})$$
$$= (0.15) \cdot (50.00)$$
$$= 7.50$$
Sale Price
= Original Price − Amount of Discount
$$= \quad 50.00 \quad - \quad 7.50$$
$$= 42.50$$
The sale price is $42.50.

3) Amount of Discount
$$= (\text{Rate of Discount})(\text{Original Price})$$
$$= (0.30) \cdot (29.50)$$
$$= 8.85$$
Sale Price
= Original Price − Amount of Discount
$$= \quad 29.50 \quad - \quad 8.85$$
$$= 20.65$$
The sale price is $20.65.

5) Amount of Discount
$$= (\text{Rate of Discount})(\text{Original Price})$$
$$= \quad (0.60) \quad \cdot \quad (49.00)$$
$$= 29.40$$
Sale Price
= Original Price − Amount of Discount
$$= \quad 49.00 \quad - \quad 29.40$$
$$= 19.60$$
The sale price is $19.60.

7) Let x = the original price of the camera.
Sale Price = Original Price
$\qquad$ − Amount of Discount
$$119 \quad = \quad x \quad - \quad 0.15x$$
$$119 = 0.85x$$
$$140 = x$$
The original price was $140.00.

9) Let x = the original price of the calendar.
Sale Price = Original Price
$\qquad$ − Amount of Discount
$$4.40 \quad = \quad x \quad - \quad 0.75x$$
$$4.40 = 0.25x$$
$$17.60 = x$$
The original price was $17.60

11) Let x = the original price of the coffe maker.
Sale Price = Original Price
$\qquad$ − Discount
$$40.08 \quad = \quad x \quad - \quad 0.40x$$
$$40.08 = 0.60x$$
$$66.80 = x$$
The original price was $66.80.

13) Let x = the number of acres of farmland in 2000.
farmland in 2009
$\qquad$ = farmland in 2000 − Amount of decrease
$$1224 = \quad x \quad - \quad x(0.32)$$
$$1224 = 0.68x$$
$$1800 = x$$
There were 1800 acres of farmland in 2000.

15) Let x = the number of Starbucks in 1996.

Starbucks in 2006

$=$ Starbucks in 1996 + Amount of Increase

$12,440 = \quad x \quad + \quad x(11.26)$

$12,440 = 12.26x$

$10,14.68 \approx x$

17) Let x = the number of employees

at K-mart in 2001.

Number of employees in 2003

$=$ employees in 2001 − Amount of decrease

$2900 = \quad x \quad - \quad x(0.34)$

$2900 = 0.66x$

$4393.93 \approx x$

There were 4400 employees worked in 2001.

19) $I = PRT$

$P = 300, \ R = 0.03, \ T = 1$

$I = (300)(0.03)(1) = 9$

Kristie earned $9 in interest.

21) $I = PRT$

$P = 6500, \ R = 0.07, \ T = 1$

$I = (6500)(0.07)(1) = 455$

Amount in account $= P + I$

$= 6500 + 455$

$= 6955$

There will be $6955 in Jake's account.

23) $I = PRT$

Total Interest Earned $= (3000)(0.065)(1)$

$+ (1500)(0.08)(1)$

$= 195 + 120 = 315$

Rachel earned a total of $315 in interest

from the two accounts.

25) x = amount Amir invested in the 6% account.

$15,000 - x$ = amount Amir invested in

the 7% account.

Total Interest $=$ Interest from 6% account

$+$ Interest from 7% account

$960 = x(0.06)(1) + \ (15,000 - x)(0.07)(1)$

$100(960)$

$= 100[x(0.06)(1) + (15,000 - x)(0.07)(1)]$

$96,000 = 6x + 7(15,000 - x)$

$96,000 = 6x + 105,000 - 7x$

$-9000 = -x$

$9000 = x$

amount invested in 7% $= 15,000 - x$

$= 15,000 - 9000 = 6000$

Amir invested $9000 in the 6% account

and $6000 in the 7% account.

27) x = amount Barney invested in the 6% account.

$x + 450$ = amount barney invested in the

5% account.

Total Interest Earned

$=$ Interest from 6% account

$+$ Interest from 5% account

$204 = x(0.06)(1) + (x + 450)(0.05)(1)$

$100(204)$

$= 100[x(0.06)(1) + (x + 450)(0.05)(1)]$

$20,400 = 6x + 5(x + 450)$

$20,400 = 6x + 5x + 2250$

$18,150 = 11x$

$1650 = x$

amount invested in 5% account $= x + 450$

$= 1650 + 450 = 2100$

Barney invested $1650 in the 6% account

and $2100 in the 5% account.

29) $x = $ amount Taz invested in the 9.5% account.

$7500 - x = $ amount Taz invested in the 6.5% account.

Total Interest Earned = Interest from 9.5% account + Interest from 6.5% account

$577.50 = x(0.095)(1) + (7500 - x)(0.065)(1)$

$1000(577.50)$

$= 1000 \left[x(0.095)(1) + (7500 - x)(0.065)(1) \right]$

$577,500 = 95x + 65(7500 - x)$

$577,500 = 95x + 487,500 - 65x$

$90,000 = 30x$

$3000 = x$

amount invested in 7% $= 7500 - x$

$= 7500 - 3000 = 4500$

Taz invested $3000 in the 9.5% account and $4500 in the 6.5% account.

31) oz of alcohol $= (0.06)(50) = 3$; There are 3 oz of alcohol in the 50 oz solution.

33) mL of acid $= (0.10)(75) + (0.025)(30)$

$\qquad = 7.5 + 0.75 = 8.25$

There are 8.25 mL of acid in the mixture.

35) $x = $ number of oz of 4% acid solution

$24 - x = $ number of oz of 10% acid solution

Solution	Concn	No of oz of soln	No of oz of acid in the soln
4%	0.04	x	$0.04x$
10%	0.10	$24 - x$	$0.10(24 - x)$
6%	0.06	24	$0.06(24)$

$0.04x + 0.10(24 - x) = 0.06(24)$

$100 \left[0.04x + 0.10(24 - x) \right] = 100 \left[0.06(24) \right]$

$4x + 10(24 - x) = 6(24)$

$4x + 240 - 10x = 144$

$-6x = -96$

$x = 16$

# of oz of 10% solution $= 24 - x$

$\qquad = 24 - 16 = 8$

Mix 16 oz of the 4% solution and 8 oz of the 10% solution.

37) $x = $ number of liters of 25% antifreeze solution

$x + 4 = $ number of liters of 45% antifreeze solution

Solution	Concn	No of liters of soln	No of liters of antifreeze in the soln
25%	0.25	x	$0.25x$
60%	0.6	4	$0.6(4)$
45%	0.45	$x + 4$	$0.45(x + 4)$

$0.25x + 0.6(4) = 0.45(x + 4)$

$100 \left[0.25x + 0.6(4) \right] = 100 \left[0.45(x + 4) \right]$

$25x + 60(4) = 45(x + 4)$

$25x + 240 = 45x + 180$

$60 = 20x$

$3 = x$

Add 3 liters of the 25% antifreeze solution.

39) $x = $ number of lbs of cashews

$x + 4 = $ number of lbs mixture

Nuts	Price per Pound	No of lbs of Nuts	Value
cashews	$7.00	x	$7x$
pistachios	$4.00	4	$4(4)$
mixture	$5.00	$x + 4$	$5(x + 4)$

$$7x + 4(4) = 5(x + 4)$$
$$7x + 16 = 5x + 20$$
$$2x = 4$$
$$x = 2$$

Mix 2 lbs of the cashews with the pistachios.

41) $x =$ grams of a 50% silver alloy

$x + 500 =$ grams of a 20% silver alloy

Alloy	Concn	No of grams. of alloy	No of grams. of silver in alloy
50%	0.50	x	$0.50x$
5%	0.05	500	$0.05(500)$
20%	0.20	$x + 500$	$0.20(x + 500)$

$$0.50x + 0.05(500) = 0.20(x + 500)$$
$$100\left[0.50x + 0.05(500)\right] = 100\left[0.20(x + 500)\right]$$
$$50x + 5(500) = 20(x + 500)$$
$$50x + 2500 = 20x + 10000$$
$$30x = 7500$$
$$x = 250$$

Mix 250g of the 50% silver alloy.

43) $x =$ number of gallons of pure acid solution.

$6 + x =$ number of gallons of 20% acid solution

Soln	Concn	No of gallons of soln	No of gallons of acid in the soln
100%	1.00	x	$1.00x$
4%	0.04	6	$0.04(6)$
20%	0.20	$6 + x$	$0.20(6 + x)$

$$1.00x + 0.04(6) = 0.20(6 + x)$$
$$100\left[1.00x + 0.04(6)\right] = 100\left[0.20(6 + x)\right]$$
$$100x + 4(6) = 20(6 + x)$$
$$100x + 24 = 120 + 20x$$
$$80x = 96$$
$$x = 1\frac{1}{5}$$

$1\frac{1}{5}$ gallons of pure acid solution should be added.

45) Let $x =$ Cheryl's Cost.

Price in store = Cheryl's Cost
+Amount of Increase

$$14.00 = x + x(0.60)$$
$$14.00 = 1.60x$$
$$8.75 = x$$

Each stuffed animal cost Cheryl $8.75.

47) Let $x =$ no of people collecting unemployment benefits in September 2005.

Benefits 2006 = Benefits 2005
−Amount of Decrease

$$8330 = x - x(0.02)$$
$$8330 = 0.98x$$
$$8500 = x$$

8500 people were getting unemployment benefits in September 2005.

49) $x =$ amount Erica invested in the 3% CD account.

$2x =$ amount Erica invested in the 4% IRA account.

$x + 1000 =$ amount Erica invested in the 5% mutual fund account.

Total Interest Earned = Interest from CD
+Interest from IRA
+Interest from mutual fund

$$370 = x(0.03)(1) + 2x(0.04)(1)$$
$$+(x+1000)(0.05)(1)$$

$$100(370) = 100 \begin{bmatrix} x(0.03)(1) \\ +2x(0.04)(1) \\ +(x+1000)(0.05)(1) \end{bmatrix}$$

$$37,000 = 3x + 8x + 5(x+1000)$$
$$37,000 = 3x + 8x + 5x + 5000$$
$$37,000 = 16x + 5000$$
$$32,000 = 16x$$
$$2000 = x$$

amount invested CD $= 2x = x(2000)$
$$= 4000$$

Erica invested $2000 in the CD, $4000 in the IRA and $2000 in the mutual fund.

amount invested mutual fund $= x + 1000$
$$= 2000 + 1000 = 3000$$

51) Let x = the original price of the desk lamp.

Sale Price = Original Price − Discount

$$25.60 = \quad x \quad - \quad 0.20x$$
$$25.60 = 0.80x$$
$$32.00 = x$$

The original price was $32.00.

53) Let x = Zoe's previous salary.

Current Salary = Previous Salaray
$$+\text{Amount of Increase}$$

$$40,144 = x + x(0.04)$$
$$40,144 = 1.04x$$
$$38,600 = x$$

Zoe's salary was $38,600 last year.

55) x = number of oz of 9% alcohol solution

$12 - x$ = number of oz of 17% alcohol solution

Soln	Concn	No of oz of solution	No of oz of alcohol in the soln
9%	0.09	x	$0.09x$
17%	0.17	$12-x$	$0.17(12-x)$
15%	0.15	12	$0.15(12)$

$$0.09x + 0.17(12-x) = 0.15(12)$$
$$100[0.09x + 0.17(12-x)] = 100[0.15(12)]$$
$$9x + 17(12-x) = 15(12)$$
$$9x + 204 - 17x = 180$$
$$-8x = -24$$
$$x = 3$$

number of oz of 17% solution $= 12 - x$
$$= 12 - 3 = 9$$

Mix 3 oz of the 9% solution and 9 oz of the 17% solution.

57) x = number of lb of peanuts

$10 - x$ = number of lb of cashews

Nuts	Price per lb	No of lb of Nuts	Value
peanuts	$1.80	x	$1.80x$
cashews	$4.50	$10-x$	$4.50(10-x)$
mixture	$2.61	10	$2.61(10)$

$$1.80x + 4.50(10-x) = 2.61(10)$$
$$100(1.80x) + 100[4.50(10-x)] = 100[2.61(10)]$$
$$180x + 450(10-x) = 261(10)$$
$$180x + 4500 - 450x = 2610$$
$$-270x = -1890$$
$$x = 7$$

cashews $= 10 - x = 10 - 7 = 3$

Mix 7 lb of peanuts and 3 lb of the cashews.

59) x = amount Diego invested in the 4% account.

$20,000 - x$ = amount Diego invested in the 7% account.

Interest Earned = Interest from 4% account
+ Interest from 7% account

$1130 = x(0.04)(1)$
$+ (20,000 - x)(0.07)(1)$

$100(1130) = 100 \begin{bmatrix} x(0.04)(1) \\ +(20,000 - x)(0.07)(1) \end{bmatrix}$

$113,000 = 4x + 7(20,000 - x)$

$113,000 = 4x + 140,000 - 7x$

$-27,000 = -3x$

$9000 = x$

amount invested in 7% $= 20,000 - x$

$= 20,000 - 9000 = 11,000$

61) x = number of ounces of pure orange juice.

$76 - x$ = number of ounces of 20% citrus fruit drink.

Soln	Concn	No of ounces of soln	No of ounces of fruit juice in the soln
100%	1.00	x	$1.00x$
5%	0.05	$76 - x$	$0.05(76 - x)$
25%	0.25	76	$0.25(76)$

$1.00x + 0.05(76 - x) = 0.25(76)$

$100[1.00x + 0.05(76 - x)] = 100[0.25(76)]$

$100x + 5(76 - x) = 25(76)$

$100x + 380 - 5x = 1900$

$95x = 1520$

$x = 16$

number of gallons of 25% solution $= 76 - x$

$76 - 16 = 60$

16 ounces of pure orange juice and 60 ounces of fruit drink mix.

63) Let x = number of procedures performed in 1997.

$2.93x$ = number of procedures performed in 2003

Procedures in 2003 = Procedures in 1997
$+ 2.93$(Procedures in 1997)

$8,253,000 = x + 2.93xx$

$8,253,000 = 3.93x$

$2,100,000 = x$

2,100,000 procedures were performed in 1997.

Section 3.5: Exercises

1) No. The height of the triangle can not be a negative number.

3) Cubic centimeters

5) $A = lw$
$A = 44; \quad l = 16$
$44 = 16w$
$\dfrac{11}{4} = w$

7) $I = PRT$
$240 = P(0.04)(2)$
$240 = 0.08P$
$3000 = P$

9) $d = rt$
$150 = (60)t$
$2.5 = t$

11) $C = 2\pi r$
$C = 2\pi(4.6)$
$C = 9.2\pi$

13) $P = 2l + 2w$

$11 = 2l + 2\left(\dfrac{3}{2}\right)$

$11 = 2l + 3$

$8 = 2l$

$4 = l$

15) $V = lwh$

$52 = (6.5)w(2)$

$52 = 13w$

$4 = w$

17) $V = \dfrac{1}{3}\pi r^2 h$

$48\pi = \dfrac{1}{3}\pi(4)^2 h$

$48\pi = \dfrac{1}{3}\pi(16)h$

$144\pi = 16\pi h$

$9 = h$

19) $S = 2\pi r^2 + 2\pi rh$

$154\pi = 2\pi(7)^2 + 2\pi(7)h$

$154\pi = 2\pi(49) + 2\pi(7)h$

$154\pi = 98\pi + 14\pi h$

$56\pi = 14\pi h$

$4 = h$

21) $A = \dfrac{1}{2}h(b_1 + b_2)$

$136 = \dfrac{1}{2}(16)(7 + b_2)$

$136 = 8(7 + b_2)$

$17 = 7 + b_2$

$10 = b_2$

23) l = length of the tennis court

$A = 2808, \ w = 36$

$A = lw$

$2808 = l \cdot 36$

$78 = l$

The length of the tennis court is 78 ft.

25) h = height of the box

$V = 1232, \ l = 22, \ w = 7$

$V = lwh$

$1232 = (22)(7)h$

$1232 = 154h$

$8 = h$

The height of the flower box is 8 in.

27) A = area of the center circle of a soccer field

$r = 10$

$A = \pi r^2$

$A = \pi(10)^2$

$A = 100\pi$

$A \approx 100(3.14) \approx 314 \text{ yd}^2$

The area is about 314 yd^2.

29) r = average speed of Abbas

$d = rt$

$d = 134, \ t = 2$

$134 = r(2)$

$67 = r$

His average speed was 67 mph.

31) h = the height of the can

$V = 864\pi, \ r = 6$

$V = \pi r^2 h$

$864\pi = \pi(6)^2 h$

$864\pi = 36\pi h$

$24 = h$

The height of the can is 24 in.

33) $A = \dfrac{1}{2}bh$

$A = 6$, $h = 4$

$6 = \dfrac{1}{2}b(4)$

$12 = 4b$

$3 = b$

The length of the base is 3 ft.

35) $I = PRT$

$P = 1500,\ I = 75,\ T = 2$

$75 = (1500)R(2)$

$75 = 3000R$

$0.025 = R$

$R = 2.5\%$

37) Let x be the length of the frame.
Then the width of the frame is $x-10$.

$P = 2l + 2w$

$92 = 2x + 2(x - 10)$

$92 = 2x + 2x - 20$

$112 = 4x$

$28 = x$

length of the frame is 28 in.

width of the frame is $x - 10 = 18$ in.

39) Let x be the width of the basketball court.
Then the length is $2x - 5$

$P = 2l + 2w$

$62 = 2(2x - 5) + 2(x)$

$62 = 4x - 10 + 2x$

$62 = 6x - 10$

$72 = 6x$

$12 = x$

width of the basketball court is 12 ft.

length of the basketball court is

$2x - 5 = 19$ ft.

41) Let x be the length of the other base.
Then the length of the first base is $3x+2$
The height is 5 in.

$A = \dfrac{1}{2}(b_1 + b_2)h$

$25 = \dfrac{1}{2}(x + 3x + 2)5$

$50 = (4x + 2) \cdot 5$

$50 = 20x + 10$

$40 = 20x$

$2 = x$

length of the other base is 2 in.

length of the first base is $3x + 2 = 8$ in.

43) Let x be the side 1.
Then the length of side 2 is x.
Length of side 3 is $x+1$
Perimeter is 5.5 ft.

$P = \text{side } 1 + \text{side } 2 + \text{side } 3$

$5.5 = x + x + x + 1$

$5.5 = 3x + 1$

$4.5 = 3x$

$1.5 = x$

The length of the sides are 1.5 ft, 1.5ft, and 2.5 ft.

45) $m\angle A = (x - 27)^\circ \ \ m\angle B = 83^\circ \ \ m\angle C = x^\circ$

$m\angle A + m\angle B + m\angle C = 180$

$x - 27 \ + \ 83 \ + \ x = 180$

$2x + 56 = 180$

$2x = 124$

$x = 62$

$m\angle A = (x - 27)^\circ = 35^\circ$

$m\angle C = x^\circ = 62^\circ$

47) $m\angle A = x°$ $m\angle B = (2x)°$ $m\angle C = 102°$

$$m\angle A + m\angle B + m\angle C = 180$$
$$x + 2x + 102 = 180$$
$$3x + 102 = 180$$
$$3x = 78$$
$$x = 26$$
$$m\angle A = 26°$$
$$m\angle B = (2x)° = 2(26) = 52°$$

49) $m\angle A = x\left(\dfrac{1}{2}x + 10\right)°$ $m\angle B = x°$ $m\angle C = x°$

$$m\angle A + m\angle B + m\angle C = 180$$
$$\left(\dfrac{1}{2}x + 10\right) + x + x = 180$$
$$\dfrac{5}{2}x + 10 = 180$$
$$\dfrac{5}{2}x = 170$$
$$x = \dfrac{2}{5} \cdot 170$$
$$x = 68$$

$$m\angle B = m\angle C = 68°$$
$$m\angle A = \left(\dfrac{1}{2}x + 10\right)° = 44°$$

51) $(x + 28)° = (3x - 2)°$
$$x + 28 = 3x - 2$$
$$30 = 2x$$
$$15 = x$$
$$(x + 28)° = (15 + 28)° = 43°$$
$$(3x - 2)° = [3(15) - 2]° = 43°$$
The labeled angles measure 43°.

53) $(4x - 12)° = \left(\dfrac{5}{2}x + 57\right)°$

$$4x - 12 = \dfrac{5}{2}x + 57$$
$$2(4x - 12) = 2\left(\dfrac{5}{2}x + 57\right)$$
$$8x - 24 = 5x + 114$$
$$3x = 138$$
$$x = 46$$
$$(4x - 12)° = (184 - 12)° = 172°$$
$$\left(\dfrac{5}{2}x + 57\right)° = \left[\dfrac{5}{2}(46) + 57\right]° = 172°$$
The indicated angles measure 172°.

55) $(3.5x + 3)° = (3x + 8)°$
$$3.5x + 3 = 3x + 8$$
$$0.5x = 5$$
$$x = 10$$
$$(3.5x + 3)° = [3.5(10) + 3]° = 38°$$
$$(3x + 8)° = [3(10) + 8]° = 38°$$
The indicated angles measure 38°.

57) $(4x)° + (x)° = 180$
$$4x° + x° = 180$$
$$5x = 180$$
$$x = 36$$
$$(4x)° = [4(36)]° = 144°$$
$$(x)° = 36°$$
The indicated angles measure 144° and 36°.

59) $$(x)^\circ + \left(\frac{1}{2}x\right)^\circ = 180$$

$$(x)^\circ + \left(\frac{1}{2}x\right)^\circ = 180$$

$$\frac{3}{2}x = 180$$

$$x = 120$$

$$(x)^\circ = 120^\circ$$

$$\left(\frac{1}{2}x\right)^\circ = 60^\circ$$

The labeled angles measure $120^\circ, 60^\circ$.

61) $$(x+30)^\circ + (2x+21)^\circ = 180$$

$$(x+30) + (2x+21) = 180$$

$$3x + 51 = 129$$

$$3x = 36$$

$$x = 43$$

$$(x+30)^\circ = [43+30]^\circ = 73^\circ$$

$$(2x+21)^\circ = [2(43)+21] = 107^\circ$$

The indicated angles measure 73° and 107°.

63)   The supplement is $180-x$.

65)   Let x = the measure of the angle.
$180-x$ = measure of the supplement
$90-x$ = measure of the complement
supplement=2(complement)+63

$$180-x = 2(90-x)+63$$

$$180-x = 180-2x+63$$

$$-x = -2x+63$$

$$x = 63$$

The measure of the angle is 63°.

67)   Let x = the measure of the angle.
$90-x$ = measure of the complement
$180-x$ = measure of the supplement
$6x$ = supplement -12°

$$6x = (180-x)-12$$

$$6x = 168-x$$

$$7x = 168$$

$$24 = x$$

The measure of the angle is 24°.

69)   Let x = the measure of the angle.
$180-x$ = measure of the supplement
$90-x$ = measure of the complement
4(complement)=2(supplement) -40

$$4(90-x) = 2(180-x) \quad -40$$

$$360-4x = 360-2x-40$$

$$360-4x = 320-2x$$

$$360 = 320+2x$$

$$40 = 2x$$

$$20 = x$$

The measure of the angle, complement,
and supplement are 20°, 70°, and 160°.

71)   Let x = the measure of the angle.
$180-x$ = measure of the supplement
$90-x$ = measure of the complement

$$x + \frac{1}{2}(\text{supplement}) = 7(\text{complement})$$

$$x + \frac{1}{2}(180-x) = 7(90-x)$$

$$90 + \frac{1}{2}x = 630-7x$$

$$180 + x = 1260-14x$$

$$15x = 1080$$

$$x = 72$$

The measure of the angle is 72°.

Chapter 3: Linear Equations and Inequalities

73)   Let x = the measure of the angle.
$90 - x$ = measure of the complement

$$4x + 2(\text{complement}) = 270$$
$$4x + 2(90 - x) = 270$$
$$4x + 180 - 2x = 270$$
$$2x = 90$$
$$x = 45$$

The measure of the angle is $45°$.

75)

a) $x + 16 = 37$
$x + 16 - 16 = 37 - 16$
$$x = 21$$

b) $x + h = y$
$x + h - h = y - h$
$$x = y - h$$

c) $x + r = c$
$x + r - r = c - r$
$$x = c - r$$

77)   a) $8c = 56$ b) $ac = d$

$$\frac{8c}{8} = \frac{56}{8} \qquad \frac{ac}{a} = \frac{d}{a}$$
$$c = 7 \qquad\qquad c = \frac{d}{a}$$

c) $mc = v$

$$\frac{mc}{m} = \frac{v}{m}$$
$$c = \frac{v}{m}$$

79) a) $\dfrac{a}{4} = 11$ b) $\dfrac{a}{y} = r$

$$4 \cdot \frac{a}{4} = 11 \cdot 4 \qquad y \cdot \frac{a}{y} = r \cdot y$$
$$a = 44 \qquad\qquad a = ry$$

c) $\dfrac{a}{w} = d$

$$w \cdot \frac{a}{w} = w \cdot d$$
$$a = dw$$

81)   a) $8d - 7 = 17$
$8d - 7 + 7 = 17 + 7$
$$8d = 24$$
$$\frac{8d}{8} = \frac{24}{8}$$
$$d = 3$$

b) $kd - a = z$
$kd - a + a = z + a$
$$kd = z + a$$
$$\frac{kd}{k} = \frac{z + a}{k}$$
$$d = \frac{z + a}{k}$$

83)   a) $9h + 23 = 17$
$9h + 23 - 23 = 17 - 23$
$$9h = -6$$
$$\frac{9h}{9} = \frac{-6}{9}$$
$$h = -\frac{2}{3}$$

b) $qh + v = n$
$qh + v - v = n - v$
$$qh = n - v$$
$$\frac{qh}{q} = \frac{n - v}{q}$$
$$h = \frac{n - v}{q}$$

85) $F = ma$

$$\frac{F}{a} = \frac{ma}{a}$$

$$\frac{F}{a} = m$$

87) $n = \dfrac{c}{v}$

$$n \cdot v = \frac{c}{v} \cdot v$$

$$nv = c$$

89) $E = \sigma T^4$

$$\frac{E}{T^4} = \frac{\sigma T^4}{T^4}$$

$$\frac{E}{T^4} = \sigma$$

91) $V = \dfrac{1}{3}\pi r^2 h$

$$3 \cdot V = 3 \cdot \frac{1}{3}\pi r^2 h$$

$$3V = \pi r^2 h$$

$$\frac{3V}{\pi r^2} = \frac{\pi r^2 h}{\pi r^2}$$

$$\frac{3V}{\pi r^2} = h$$

93) $R = \dfrac{E}{I}$

$$I \cdot R = I \cdot \frac{E}{I}$$

$$IR = E$$

95) $I = PRT$

$$I = PRT$$

$$\frac{I}{PT} = \frac{PRT}{PT}$$

$$\frac{I}{PT} = R$$

97) $P = 2l + 2w$

$$P - 2w = 2l + 2w - 2w$$

$$P - 2w = 2l$$

$$\frac{P - 2w}{2} = \frac{2l}{2}$$

$$\frac{P - 2w}{2} = l \quad \text{or} \quad l = \frac{P}{2} - w$$

99) $H = \dfrac{D^2 N}{2.5}$

$$2.5 \cdot H = 2.5 \cdot \frac{D^2 N}{2.5}$$

$$2.5H = D^2 N$$

$$\frac{2.5H}{D^2} = \frac{D^2 N}{D^2}$$

$$\frac{2.5H}{D^2} = N$$

101) $A = \dfrac{1}{2}h(b_1 + b_2)$

$$2 \cdot A = 2 \cdot \frac{1}{2}h(b_1 + b_2)$$

$$2A = h(b_1 + b_2)$$

$$\frac{2A}{h} = \frac{h(b_1 + b_2)}{h}$$

$$\frac{2A}{h} = b_1 + b_2$$

$$\frac{2A}{h} - b_1 = b_1 - b_1 + b_2$$

$$\frac{2A}{h} - b_1 = b_2 \quad \text{or} \quad b_2 = \frac{2A - hb_1}{h}$$

103) $S = \dfrac{\pi}{4}\left(4h^2 + c^2\right)$

$\dfrac{4}{\pi} \cdot S = \dfrac{4}{\pi} \cdot \dfrac{\pi}{4}\left(4h^2 + c^2\right)$

$\dfrac{4S}{\pi} = 4h^2 + c^2$

$\dfrac{4S}{\pi} - c^2 = 4h^2 + c^2 - c^2$

$\dfrac{4S}{\pi} - c^2 = 4h^2$

$\dfrac{4S}{4\pi} - \dfrac{c^2}{4} = \dfrac{4h^2}{4}$

$\dfrac{S}{\pi} - \dfrac{c^2}{4} = h^2$

105) a) $P = 2l + 2w$

$P - 2l = 2l - 2l + 2w$

$P - 2l = 2w$

$\dfrac{P - 2l}{2} = \dfrac{2w}{2}$

$\dfrac{P - 2l}{2} = w$

b) $w = \dfrac{P - 2l}{2}$

$w = \dfrac{28 - 2(11)}{2}$

$w = \dfrac{28 - 22}{2} = \dfrac{6}{2} = 3 \text{ cm}$

107) a) $C = \dfrac{5}{9}(F - 32)$

$\dfrac{9}{5} \cdot C = \dfrac{9}{5} \cdot \dfrac{5}{9}(F - 32)$

$\dfrac{9}{5}C = F - 32$

$\dfrac{9}{5}C + 32 = F - 32 + 32$

$\dfrac{9}{5}C + 32 = F$

b) $F = \dfrac{9}{5}C + 32$

$F = \dfrac{9}{5}(20) + 32 = 36 + 32 = 68°$

Section 3.6: Exercises

1) Answers may vary, but some possible answers are,

$\dfrac{6}{8}, \dfrac{9}{12}, \dfrac{12}{16}$

3) Yes. A percent can be written as a fraction with a denominator of 100. For example 25% can be written as $\dfrac{5}{20}$.

This can be written as $\dfrac{25}{100}$ or $\dfrac{1}{4}$

5) $\dfrac{16}{12} = \dfrac{4}{3}$

7) $\dfrac{4}{50} = \dfrac{2}{25}$

9) $\dfrac{20}{80} = \dfrac{1}{4}$

11) $\dfrac{2\,\text{ft}}{36\,\text{in}} = \dfrac{2(12)\,\text{in}}{36\,\text{in}}$

$\dfrac{24}{36} = \dfrac{2}{3}$

13) $\dfrac{18\,\text{hours}}{2\,\text{days}} = \dfrac{18\,\text{hours}}{2 \times 24\,\text{hours}}$

$\dfrac{18}{48} = \dfrac{3}{8}$

15)

Unit price for size 8 package $= \dfrac{\$6.29}{8}$

$= 0.78625$

Unit price for size 16 package $= \dfrac{\$12.99}{16}$

$= 0.811875$

Package of 8 : $0.786 per battery is the best buy.

17) Unit price for 8 oz jar$= \dfrac{\$2.69}{8} = 0.336$

Unit price for 15 oz jar$= \dfrac{\$3.59}{15} = 0.239$

Unit price for 48 oz jar$= \dfrac{\$8.49}{48} = 0.177$

48 oz jar : $0.177 per oz is the best buy.

19) Unit price for 11 oz box $= \dfrac{\$4.49}{11}$

$= 0.4082$

Unit price for 16 oz box $= \dfrac{\$5.15}{16} = 0.3219$

Unit price for 24 oz box $= \dfrac{\$6.29}{24} = 0.262$

24 oz box : $0.262 per oz is the best buy.

20) A ratio is a quotient of two quantities. A proportion is a statement that two ratios are equal.

23) True.

$4 \cdot 35 = 20 \cdot 7$

$140 = 140$

25) False.

$72 \cdot 7 = 8 \cdot 54$

$504 \neq 432$

27) True.

$8 \cdot \dfrac{5}{2} = 2 \cdot 10$

$20 = 20$

29) $\dfrac{8}{36} = \dfrac{c}{9}$

$72 = 36c$

$2 = c$

The solution set is $\{2\}$.

31) $\dfrac{w}{15} = \dfrac{32}{12}$

$12w = 480$

$w = 40$

The solution set is $\{40\}$.

33) $\dfrac{40}{24} = \dfrac{30}{a}$

$40a = 720$

$a = 18$

The solution set is $\{18\}$.

35) $\dfrac{2}{k} = \dfrac{9}{12}$

$24 = 9k$

$\dfrac{24}{9} = k$

$\dfrac{8}{3} = k$

The solution set is $\left\{\dfrac{8}{3}\right\}$.

37) $\dfrac{3z+10}{14} = \dfrac{2}{7}$

$7(3z+10) = 2 \cdot 14$

$21z + 70 = 28$

$21z = -42$

$z = -2$

The solution set is $\{-2\}$.

39) $\dfrac{r+7}{9} = \dfrac{r-5}{3}$

$3(r+7) = 9 \cdot (r-5)$

$3r + 21 = 9r - 45$

$66 = 6r$

$11 = r$

The solution set is $\{11\}$.

41) $\dfrac{3h+15}{16} = \dfrac{2h+5}{4}$

$4(3h+15) = 16(2h+5)$

$12h + 60 = 32h + 80$

$-20 = 20h$

$-1 = h$

The solution set is $\{-1\}$.

43) $\dfrac{4m-1}{6} = \dfrac{6m}{10}$

$10(4m-1) = 6(6m)$

$40m - 10 = 36m$

$-10 = -4m$

$\dfrac{-10}{-4} = m$

$\dfrac{5}{2} = m$

The solution set is $\left\{\dfrac{5}{2}\right\}$.

45) Let x = cost of 6 containers of yogurt.

$\dfrac{2.36}{4} = \dfrac{x}{6}$

$6(2.36) = 4x$

$14.16 = 4x$

$3.54 = x$

The cost of 6 contaiers of yogurt is $3.54

47) Let x = amount of orange juice.

$$\frac{2}{3} = \frac{\frac{1}{3}}{x}$$

$$2x = \left(\frac{1}{3}\right)3$$

$$2x = 1$$

$$x = \frac{1}{2}$$

The amount of orange juice is $\frac{1}{2}$ cup.

49) Let x = caffeine in an 18-oz serving

$$\frac{12}{55} = \frac{18}{x}$$

$$12x = 990$$

$$x = 82.5$$

There are 82.5 mg of caffeine in an 18-oz serving of Mountain Dew.

51) Let x = number of smokers who started smoking before they were 21.

$$\frac{9}{10} = \frac{x}{400}$$

$$9(400) = 10x$$

$$3600 = 10x$$

$$360 = x$$

The number of smokers is 360.

53) Let x be the number of pounds of kitchen scraps.

$$\frac{5}{2} = \frac{20}{x}$$

$$5x = 20(2)$$

$$5x = 40$$

$$x = 8$$

The number of pounds of kitchen scraps is 8 lb.

55) Let x be the number of Euros.

$$\frac{20}{14.30} = \frac{50.00}{x}$$

$$20x = 50(14.3)$$

$$20x = 715$$

$$x = \frac{715}{20}$$

$$x = 35.75$$

The number of Euros is 35.75

57) $\frac{7}{5} = \frac{14}{x}$

$$7x = 70$$

$$x = 10$$

59) $\frac{12}{20} = \frac{x}{\frac{65}{3}}$

$$260 = 20x$$

$$13 = x$$

61) $\frac{x}{28} = \frac{45}{20}$

$$20x = 1260$$

$$63 = x$$

63) a) $(\$0.10) \cdot 7 = \$0.70$

    b) $10¢ \cdot 7 = 70¢$

65) a) $(\$0.01) \cdot 422 = \$4.22$

    b) $(1¢) \cdot 422 = 422¢$

67) a) $(\$0.05) \cdot 9 + (\$0.25) \cdot 7$
$$= \$0.45 + \$1.75 = \$2.20$$

    b) $(5¢) \cdot 9 + (25¢) \cdot 7$
$$= 45¢ + 175¢ = 220¢$$

69) a) $0.25 \cdot q = 0.25q$

   b) $25 \cdot q = 25q$

71) a) $0.10 \cdot d = 0.10d$ b) $10 \cdot d = 10d$

73) a) $0.01 \cdot p + 0.05 \cdot n = 0.01p + 0.05n$

   b) $1 \cdot p + 5 \cdot n = p + 5n$

75) x = number of dimes ; $8 + x$ = number of quarters

Value of Dimes + Value of Quarters = Total Value

$$0.10x \quad + \quad 0.25(x+8) \quad = \quad 5.15$$

$$100\left[0.10x + 0.25(x+8)\right] = 100(5.15)$$

$$10x + 25(x+8) = 515$$

$$10x + 25x + 200 = 515$$

$$35x = 315$$

$$x = 9$$

$$\text{quarters} = 9 + 8 = 17$$

There are 9 dimes and 17 quarters.

77) x = number of \$1 bills; $29 - x$ = number of \$5 bills

Value of \$1 bills + Value of \$5 bills = Total Value

$$1x + 5(29 - x) \quad = \quad 73.00$$

$$x + 145 - 5x = 69$$

$$-4x = -72$$

$$x = 18$$

\$5 bills $= 29 - x = 29 - 18 = 11$

There are 11-\$5 bills and 18-\$1 bills.

79) x = number of adult tickets; $2x$ = number of children's ticket

Rev. from adult tickets + Rev. from children's tickets = Total Revenue

$$9x \quad + \quad 7(2x) \quad = \quad 437.00$$

$$9x + 14x = 437$$

$$23x = 437$$

$$x = 19$$

children's tickets $= 2x = 2(19) = 38$

There were 19 adult tickets and

38 children's tickets sold.

81) $x = $ cost of ticket for Marc Anthony concert

$x - 19.50 = $ cost of ticket for Santana concert

cost of Marc Anthony concert + cost of Santana concert = Total Amount Paid

$$5x + 2(x - 19.5) = 563$$
$$5x + 2x - 39 = 563$$
$$7x = 602$$
$$x = 86$$

cost of Santana concert ticket $= x - 19.5 = \$66.50$

Marc Anthony: \$86; Santana: \$66.50

83) miles

85) $r = $ the rate of the northbound plane. Then the rate of the southbound plane is $r + 50$.

	d	=	r	$\cdot$	t
northbound	$2r$		r		2
southbound	$2(r + 50)$		$r + 50$		2

northbound distance + southbound distance = 900

$$2r + 2(r + 50) = 900$$
$$2r + 2r + 100 = 900$$
$$4r + 100 = 900$$
$$4r = 800$$
$$r = 200$$
$$\text{speed of southbound plane} = r + 50 = 250$$

northbound: 200 mph; southbound: 250 mph

87) $t = $ the amount of time traveling until they are 200 miles apart

	d	=	r	$\cdot$	t
Lance	22t		22		t
Danica	18t		18		t

Lance's distance + Danica's distance = 200

$$22t \qquad + \qquad 18t = 200$$
$$40t = 200$$
$$t = 5 \text{ hours}$$

89) t = the time Ahmad traveling.

$t - \dfrac{20}{60} = t - \dfrac{1}{3}$ is the time Davood has been traveling

	d	$=$ r	$\cdot$ t
Ahmad	$30t$	30	t
Davood	$36\left(t - \dfrac{1}{3}\right)$	36	$\left(t - \dfrac{1}{3}\right)$

Ahmad's distance = Davood's distance

$$30t \quad = \quad 36\left(t - \frac{1}{3}\right)$$

$$30t = 36t - \frac{36}{3}$$

$$30t = 36t - 12$$

$$-6t = -12$$

$$t = 2 \, \text{hours}$$

Ahmad has been traveling for 2 hours.

Davood has been traveling for $2 - \dfrac{1}{3} = 1\dfrac{2}{3}$ hours.

So Davood catches up with Ahmad in $1\dfrac{2}{3}$ hours.

91) t = the amount of time traveling until they are 6 miles apart

	d	$=$ r	$\cdot$ t
Truck	$35t$	35	t
Car	$45t$	45	t

Truck's distance $+ 6 =$ Car's distance

$$35t + 6 = 45t$$

$$6 = 10t$$

$$\frac{3}{5} = t$$

$$\frac{3}{5} \cdot 60 = 36 \, \text{min}$$

They will be 6 miles apart in 36 minutes.

93) r = Nick's speed

Then Scott's speed is $x - 2$

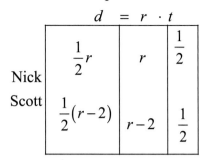

Nick's distance + Scott's distance = 13

$$\frac{1}{2}r + \frac{1}{2}(r - 2) = 13$$

$$2 \cdot \frac{1}{2}r + 2 \cdot \frac{1}{2}(r - 2) = 2 \cdot 13$$

$$r + r - 2 = 26$$
$$2r = 28$$
$$r = 14$$

Nick's speed= 14 mph

Scott's speed= 12 mph

95)　　r = Speed of Freight train

Then speed of Passenger train is $r + 20$

	d	= r ·	t
Freight train	$5r$	r	5
Passenger train	$(r + 20)$	$r + 20$	5

Freight train's distance + Passenger train's distance = 400

$$5r + 5(r + 20) = 400$$
$$5r + 5r + 100 = 400$$
$$10r + 100 = 400$$
$$10r = 300$$
$$r = 30$$

Freight train's speed= 30 mph

Passenger train's speed= 50 mph

97) Let x be the number of Yen.

$$\frac{4.00}{442} = \frac{70.00}{x}$$

$$4x = 442(70)$$

$$4x = 30940$$

$$x = \frac{30940}{4}$$

$$x = 7735 \text{ Yen}$$

The number of Yen is 7735.

99) t = the amount of time traveling until they are 6 miles apart

	d	= r	$\cdot$ t
Bill	$14t$	14	t
Sherri	$10t$	10	t

Bill's distance+Sherri's distance = 6

$$14t + 10t = 6$$

$$24t = 6$$

$$t = \frac{6}{24}$$

$$t = \frac{1}{4} \text{ hour} = 15 \text{ minutes}$$

So they will be 6 miles apart in 15 minutes.

101) x = number of quarters; $11 + x$ = number of dimes

Value of Quarters + Value of Dimes = Total Value

$$0.25x + 0.10(11 + x) = 6.70$$

$$100\left[0.25x + 0.10(11 + x)\right] = 100(6.70)$$

$$25x + 10(11 + x) = 670$$

$$25x + 110 + 10x = 670$$

$$35x = 560$$

$$x = 16$$

dimes = $x + 11 = 16 + 11 = 27$

There are 27 dimes and 16 quarters.

103) r = the speed of the small plane

$2r$ = the speed of the jet

d	=	r	·	t
small Plane	$\dfrac{3}{4}r$	r	$\dfrac{45}{60} = \dfrac{3}{4}$
jet	$\dfrac{3}{4}(2r)$	$2r$	$\dfrac{3}{4}$

small Plane's distance $+150 =$ jet's distance

$$\frac{3}{4}r \ + \ 150 = \frac{3}{4}(2r)$$

$$3r + 600 = 6r$$

$$600 = 3r$$

$$200 = r$$

the speed of the jet $= 2r = 2(200) = 400$

The jet is traveling at 400 mph, and the small plane is traveling at 200 mph.

105) Let x = the number of girls that are babysitters.

$$\frac{3}{5} = \frac{x}{400}$$

$$400 \cdot 3 = 5x$$

$$1200 = 5x$$

$$240 = x$$

There are 240 girls that are babysitters.

Section 3.7: Exercises

1) You use brackets when there is a $\leq$ or $\geq$ symbol.

3) $(-\infty, 4)$

5) $[-3, \infty)$

7)

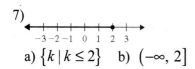

a) $\{k \mid k \leq 2\}$ b) $(-\infty, 2]$

9)

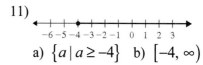

a) $\left\{c \,\middle|\, c < \dfrac{5}{2}\right\}$ b) $\left(-\infty, \dfrac{5}{2}\right)$

11)

a) $\{a \mid a \geq -4\}$ b) $[-4, \infty)$

13) when you multiply or divide an inequality by a negative number

Chapter 3: Linear Equations and Inequalities

15) $k+9\ge 7$
$k+9-9\ge 7-9$

$k\ge -2$

a) $\{k\,|\,k\ge -2\}$ b) $[-2,\infty)$

17) $c-10\le -6$
$c-10+10\le -6+10$

$c\le 4$

a) $\{c\,|\,c\le 4\}$ b) $(-\infty,4]$

19) $-3+d<-4$
$-3+d+3<-4+3$

$d<-1$

a) $\{d\,|\,d<-1\}$ b) $(-\infty,-1)$

21) $16<z+11$
$16-11<z+11-11$

$5<z$

$z>5$

a) $\{z\,|\,z>5\}$ b) $(5,\infty)$

23) $5m>15$
$\dfrac{5m}{5}>\dfrac{15}{5}$

$m>3$

a) $\{m\,|\,m>3\}$ b) $(3,\infty)$

25) $12x<-21$
$\dfrac{12x}{12}<\dfrac{-21}{12}$

$x<-\dfrac{7}{4}$

a) $\left\{x\,\middle|\,x<-\dfrac{7}{4}\right\}$ b) $\left(-\infty,-\dfrac{7}{4}\right)$

27) $-4b\le 32$
$\dfrac{-4b}{-4}\ge \dfrac{32}{-4}$

$b\ge -8$

a) $\{b\,|\,b\ge -8\}$ b) $[-8,\infty)$

29) $-24a<-40$
$\dfrac{-24a}{-24}>\dfrac{-40}{-24}$

$a>\dfrac{5}{3}$

a) $\left\{a\,\middle|\,a>\dfrac{5}{3}\right\}$ b) $\left(\dfrac{5}{3},\infty\right)$

31) $\dfrac{1}{3}k\ge -5$

$3\cdot\dfrac{1}{3}k\ge 3\cdot(-5)$

$k\ge -15$

a) $\{k\,|\,k\ge -15\}$ b) $[-15,\infty)$

33) $-\dfrac{3}{8}c<-3$

$-\dfrac{8}{3}\cdot\left(-\dfrac{3}{8}c\right)>-\dfrac{8}{3}\cdot(-3)$

$c>8$

a) $\{c\,|\,c>8\}$ b) $(8,\infty)$

35) $4p-11\le 17$
$4p-11+11\le 17+11$

$4p\le 28$

$p\le 7$

$(-\infty,7]$

74

37) $9 - 2w \leq 11$

$\qquad 9 - 2w - 9 \leq 11 - 8$

$\qquad\qquad -2w \leq 2$

$\qquad\qquad\qquad w \geq -1$

$\qquad [-1, \infty)$

39) $\qquad -\dfrac{3}{4}m + 10 > 1$

$\qquad 4 \cdot \left(-\dfrac{3}{4}m\right) + 4 \cdot 10 > 4 \cdot 1$

$\qquad\qquad -3m + 40 > 4$

$\qquad\qquad\qquad -3m > -36$

$\qquad\qquad\qquad\qquad d < 12$

$\qquad (-\infty, 12)$

41) $\qquad 3c + 10 > 5c + 13$

$\qquad 3c + 10 - 10 > 5c + 13 - 10$

$\qquad\qquad 3c > 5c + 3$

$\qquad\qquad 3c - 5c > 3$

$\qquad\qquad\qquad -2c > 3$

$\qquad\qquad\qquad\qquad c < -\dfrac{3}{2}$

$\qquad \left(-\infty, -\dfrac{3}{2}\right)$

43) $3(n+1) - 16 \leq 2(6-n)$

$\qquad 3n + 3 - 16 \leq 12 - 2n$

$\qquad\qquad 3n - 13 \leq 12 - 2n$

$\qquad 3n - 13 + 13 \leq 12 + 13 - 2n$

$\qquad\qquad\qquad 3n \leq 25 - 2n$

$\qquad\qquad 3n + 2n \leq 25 - 2n + 2n$

$\qquad\qquad\qquad 5n \leq 25$

$\qquad\qquad\qquad\qquad n \leq 5$

$\qquad (-\infty, 5]$

45) $\qquad \dfrac{8}{3}(2k+1) > \dfrac{1}{6}k + \dfrac{8}{3}$

$\qquad 6\left[\dfrac{8}{3}(2k+1)\right] > 6\left(\dfrac{1}{6}k + \dfrac{8}{3}\right)$

$\qquad\qquad 16(2k+1) > k + 16$

$\qquad\qquad 32k + 16 > k + 16$

$\qquad\qquad\qquad 31k > 0$

$\qquad\qquad\qquad\qquad k > 0$

$\qquad (0, \infty)$

47) $\qquad 0.05x + 0.09(40-x) > 0.07(40)$

$100(0.05x) + 100[0.09(40-x)] > 100[0.07(40)]$

$\qquad\qquad 5x + 9(40-x) > 7(40)$

$\qquad\qquad 5x + 360 - 9x > 280$

$\qquad\qquad\qquad -4x + 360 > 280$

$\qquad\qquad\qquad\qquad -4x > -80$

$\qquad\qquad\qquad\qquad\qquad x < 20$

$\qquad (-\infty, 20)$

49) $[-5, 3]$

51) $(-3, 0]$

53)

a) $\{y \mid -4 < y < 0\}$　b) $(-4, 0)$

55)

a) $\{k \mid -3 \leq k \leq 2\}$　b) $[-3, 2]$

57)

a) $\left\{n \mid \dfrac{1}{2} < n \leq 3\right\}$　b) $\left(\dfrac{1}{2}, 3\right]$

59) $\qquad -11 \leq b - 8 \leq -7$

$\qquad -11 + 8 \leq b - 8 + 8 \leq -7 + 8$

$\qquad\qquad\qquad -3 \leq b \leq 1$

$[-3, 1]$

75

61) $-10 < 2a < 7$

$$\dfrac{-10}{2} < \dfrac{2a}{2} < \dfrac{7}{2}$$

$$-5 < a < \dfrac{7}{2}$$

$\left(-5, \dfrac{7}{2}\right)$

63)

$$-5 \le 4k - 13 \le 7$$

$$-5 + 13 \le 4k - 13 + 13 \le 7 + 13$$

$$8 \le 4k \le 20$$

$$\dfrac{8}{4} \le \dfrac{4k}{4} \le \dfrac{20}{4}$$

$$2 \le k \le 5$$

$[2, 5]$

65) $-17 < \dfrac{3}{2}c - 5 < 1$

$$-17 + 5 < \dfrac{3}{2}c - 5 + 5 < 1 + 5$$

$$-12 < \dfrac{3}{2}c < 6$$

$$\dfrac{2}{3} \cdot (-12) < \dfrac{2}{3} \cdot \dfrac{3}{2}c < \dfrac{2}{3} \cdot 6$$

$$-8 < c < 4$$

$(-8, 4)$

67)

$$-6 \le 4c - 13 < -1$$

$$-6 + 13 \le 4c - 13 + 13 < -1 + 13$$

$$7 \le 4c < 12$$

$$\dfrac{7}{4} \le \dfrac{4c}{4} < \dfrac{12}{4}$$

$$\dfrac{7}{4} \le c < 3$$

$\left[\dfrac{7}{4}, 3\right)$

69) $4 \le \dfrac{k+11}{4} \le 5$

$$4 \cdot 4 \le 4\left(\dfrac{k+11}{4}\right) \le 4 \cdot 5$$

$$16 \le k + 11 \le 20$$

$$16 - 11 \le k + 11 - 11 \le 20 - 11$$

$$5 \le k \le 9$$

$[5, 9]$

71) $-7 \le 8 - 5y < 3$

$$-7 - 8 \le 8 - 8 - 5y < 3 - 8$$

$$-15 \le -5y < -5$$

$$\dfrac{-15}{-5} \ge \dfrac{-5y}{-5} > \dfrac{-5}{-5}$$

$$3 \ge y > 1$$

$$1 < y \le 3$$

$(1, 3]$

73)

$$2 < 10 - p \le 5$$

$$2 - 10 < 10 - 10 - p \le 5 - 10$$

$$-8 < -p \le -5$$

$$\dfrac{-8}{-1} > \dfrac{-p}{-1} \ge \dfrac{-5}{-1}$$

$$8 > p \ge 5$$

$$5 \le p < 8$$

$[5, 8)$

75) Let x = the number of additional chidren after the first 10 children.
$x + 10$ = Total number of chidren
$(\text{Total Cost}) = (\text{Cost of first 10 chidren}) + (\text{Cost for each additonal child})$
$$= \qquad 180 \qquad + \qquad 16x$$
$$180 + 16x \le 300$$
$$16x \le 120$$
$$x \le \frac{120}{16} = \frac{15}{2}$$
$$x \le 7\frac{1}{2}$$
Leslie can afford at most 7 additional children.
Total number of children $= x + 10 = 7 + 10 = 17$

77) Let x = the number of people he can invite for the party.
$(\text{Total Cost}) = (\text{Cost to rent the tent }) + (\text{Cost of food per person})$
$$= \qquad 150 \qquad + \qquad 11.50x$$
$$150 + 11.50x \le 450$$
$$11.5x \le 300$$
$$x \le 26.09$$
The greatest number of people Heinrich can invite is 26.

79) Let x = the number of $\dfrac{1}{5}$ mi. $\dfrac{1}{5}x$ = Total Mileage
$(\text{Total Cost}) = (\text{Initial Cost}) + (\text{Cost per Mile})$
$$= \quad 2.50 \quad + \quad 0.40x$$
$$2.50 + 0.40x \le 14.50$$
$$0.40x \le 12$$
$$x \le 30$$
Total Mileage $= \dfrac{1}{5}x = \dfrac{1}{5}(30) = 6$. You can go at most 6 miles.

81) Let x = the grade she needs to make . $87 + 94 + x$ = Total of three scores
$$\frac{87 + 94 + x}{3} \ge 90$$
$$3 \cdot \frac{181 + x}{3} \ge 3 \cdot 90$$
$$181 + x \ge 270$$
$$x \ge 89$$
Melinda must make an 89 or higher.

Chapter 3: Linear Equations and Inequalities

Section 3.8: Exercises

1) $A \cap B$ means "A intersect B." $A \cap B$ is the set of all numbers which are in set A and set B.

3) $\{8\}$

5) $\{2, 4, 5, 6, 7, 8, 9, 10\}$

7) $\varnothing$

9) $\{1, 2, 3, 4, 5, 6, 8, 10\}$

11) $\{$Liliane Bettancourt, Alice Walton$\}$

13) $\left\{\begin{array}{l}\text{Liliane Bettancourt, J.K.Rowling,} \\ \text{Oprah Winfrey}\end{array}\right\}$

15)

$[-3, 2]$

17)

$(-1, 3)$

19)

$[3, \infty)$

21)

$\varnothing$

23)

$[2, 5]$

25) $b - 7 > -9$ and $8b < 24$

$b > -2$ and $b < 3$

$(-2, 3)$

27) $5w + 9 \le 29$ and $\dfrac{1}{3}w - 8 > -9$

$5w \le 20$ and $\dfrac{1}{3}w > -1$

$w \le 4$ and $w > -3$

$(-3, 4]$

29) $2m + 15 \ge 19$ and $m + 6 < 5$

$2m \ge 4$ and $m < -1$

$m \ge 2$ and $m < -1$

$\varnothing$

31) $r - 10 > -10$ and $3r - 1 > 8$

$r > 0$ and $3r > 9$

$r > 0$ and $r > 3$

$(3, \infty)$

33) $9 - n \le 13$ and $n - 8 \le -7$

$-n \le 4$ and $n \le 1$

$n \ge -4$ and $n \le 1$

$[-4, 1]$

35)

$(-\infty, -1) \cup (5, \infty)$

37)

$\left(-\infty, \dfrac{5}{3}\right] \cup (4, \infty)$

39)

$(1, \infty)$

41)

$(-\infty, \infty)$

43)

$(-\infty, -1) \cup (3, \infty)$

45) $6m \leq 21$ or $m - 5 > 1$

$m \leq \dfrac{7}{2}$ or $m > 6$

$\left(-\infty, \dfrac{7}{2}\right] \cup (6, \infty)$

47) $3t + 4 > -11$ or $t + 19 > 17$

$3t > -15$ or $t > -2$

$t > -5$ or $t > -2$

$(-\infty, -6) \cup [-3, \infty)$

49) $-2v - 5 \leq 1$ or $\dfrac{7}{3}v < -14$

$-2v \leq 6$ or $v < -6$

$v \geq -3$ or $v < -6$

$(-\infty, -6) \cup [-3, \infty)$

51) $c + 3 \geq 6$ or $\dfrac{4}{5}c \leq 10$

$c \geq 3$ or $c \leq \dfrac{25}{2}$

$(-\infty, \infty)$

54) $7 - 6n \geq 19$ or $n + 14 < 11$

$-6n \geq 12$ or $n < -3$

$n \leq -2$ or $n < -3$

$(-\infty, -2]$

Chapter 3 Review

1) No.

$$\dfrac{3}{2}k - 5 = 1$$

$$\dfrac{3}{2}(-4) - 5 = 1$$

$$-6 - 5 = 1$$

$$-11 \neq 1$$

3) The variables are eliminated and you get a false statement like 5=13.

5) $h + 14 = -5$

$h + 14 - 14 = -5 - 14$

$h = -19$

The solution set is $\{-19\}$.

7) $-7g = 56$

$$\dfrac{-7g}{-7} = \dfrac{56}{-7}$$

$g = -8$

The solution set is $\{-8\}$.

9) $4 = \dfrac{c}{9}$

$9 \cdot 4 = 9 \cdot \dfrac{c}{9}$

$36 = c$

The solution set is $\{36\}$.

11)
$$23 = 4m - 7$$
$$23 + 7 = 4m + 2 - 2$$
$$30 = 4m$$
$$\frac{30}{4} = \frac{4m}{4}$$
$$\frac{15}{2} = m$$

The solution set is $\left\{\dfrac{15}{2}\right\}$.

13) $4c + 9 + 2(c - 12) = 15$
$$4c + 9 + 2c - 24 = 15$$
$$6c - 15 = 15$$
$$6c - 15 + 15 = 15 + 15$$
$$6c = 30$$
$$c = 5$$
The solution set is $\{5\}$.

15)
$$2z + 11 = 8z - 15$$
$$2z - 8z + 11 = 8z - 8z + 15$$
$$-6z + 11 = 15$$
$$-6z + 11 - 11 = 15 - 11$$
$$-6z = 4$$
$$z = \frac{4}{-6}$$
$$z = -\frac{2}{3}$$

The solution set is $\left\{-\dfrac{2}{3}\right\}$.

17) $k + 3(2k - 5) = 4(k - 2) - 7$
$$k + 6k - 15 = 4k - 8 - 7$$
$$7k - 15 = 4k - 15$$
$$7k - 15 + 15 = 4k + 15 - 15$$
$$7k = 4k$$
$$7k - 4k = 4k - 4k$$
$$3k = 0$$
$$k = 0$$
The solution set is $\{0\}$.

19)
$$0.18a + 0.1(20 - a) = 0.14(20)$$
$$100\left[0.18a + 0.1(20 - a)\right] = 100\left[0.14(20)\right]$$
$$18a + 10(20 - a) = 14(20)$$
$$18a + 200 - 10a = 280$$
$$8a + 200 = 280$$
$$8a = 80$$
$$a = 10$$
The solution set is $\{10\}$.

21) $3(r + 4) - r = 2(r + 6)$
$$3r + 12 - r = 2r + 12$$
$$2r + 12 = 2r + 12$$
$$12 = 12$$
The solution set is $\{$all real numbers$\}$.

23) Let $x =$ a number
$$2x - 9 = 25$$
$$2x - 9 + 9 = 25 + 9$$
$$2x = 34$$
$$x = 17$$
The number is 17.

25) Let x be the number of e-mails received on Thursday.
Then the number of e-mails received on Friday $= x - 24$

$$\left(\begin{array}{c}\text{number of e-mails}\\ \text{received}\\ \text{on Thursday}\end{array}\right) + \left(\begin{array}{c}\text{number of e-mail}\\ \text{received}\\ \text{on Friday}\end{array}\right) = 126$$

$$x \quad + x - 24 = 126$$
$$2x - 24 = 126$$
$$2x = 150$$
$$x = 75$$
$$\text{e-mails received on Friday} = x - 24$$
$$= 75 - 24$$
$$= 51$$

Kendrick received 75 e-mails on Thursday, and 51 on Friday.

27) Let x be the length of the shorter pipe.

Then the length of the longer pipe is x+8
$$(\text{length of shorter pipe}) + (\text{length of longer pipe}) = 36$$

$$x \quad + x + 8 = 36$$
$$2x + 8 = 36$$
$$2x = 28$$
$$x = 14$$
$$\text{length of the longer pipe} = x + 8$$
$$= 14 + 8$$
$$= 22$$

The pipes are 14 in., and 22 in.

29) Let x be the weight of implant 20 years ago.
Then the weight of today's implant is $0.5x$.

$$\left(\begin{array}{c}\text{weight of today's}\\ \text{implant}\end{array}\right) = 3$$

$$0.5x = 3$$
$$2(0.5x) = 2 \cdot 3$$
$$x = 6$$

The weight of implant 20 years ago was 6 lb.

31) $x =$ amount Jerome invested in the 2% account.
$6000 - x =$ amount Jerome invested in the 4% account.
Total Interest Earned = Interest from 2% account + Interest from 4% account

$$210 = x(0.02)(1) + (6000 - x)(0.04)(1)$$

$$100(210) = 100\left[x(0.02)(1) + (6000 - x)(0.04)(1)\right]$$

$$21,000 = 2x + 4(3000 - x)$$

$$21,000 = 2x + 12,000 - 4x$$

$$9000 = 6x$$

$$1500 = x$$

amount invested in 4% $= 6000 - x = 6000 + 1500 = 4500$
Jose invested \$1500 in the 2% account and \$4500 in the 4% account.

33)
$$P = 2l + 2w$$
$$32 = 2(9) + 2w$$
$$32 = 18 + 2w$$
$$14 = 2w$$
$$7 = w$$

35) $A = \dfrac{1}{2}bh$

$$42 = \dfrac{1}{2}(12)h$$
$$42 = 6h$$
$$7 = h$$
The height is 7 in.

37) $m\angle A = (x)^\circ \quad m\angle B = (x)^\circ$

$$m\angle C = (x + 15)^\circ$$
$$m\angle A + m\angle B + m\angle C = 180$$
$$x + x + (x + 15) = 180$$
$$3x + 15 = 180$$
$$3x = 165$$
$$x = 55$$

$$m\angle A = (x)^\circ = 55^\circ$$
$$m\angle B = (x)^\circ = 55^\circ$$
$$m\angle C = (x + 15)^\circ = (55 + 15)^\circ = 70^\circ$$

39) $(6x + 7)^\circ = (9x - 20)^\circ$

$$6x + 7 = 9x - 20$$
$$27 = 3x$$
$$9 = x$$

$$(6x + 7)^\circ = (6(9) + 7)^\circ = 61^\circ$$

$$(9x - 20)^\circ = (9(9) - 20)^\circ = 61^\circ$$

41) $p - n = z$
$$p - n + n = z + n$$
$$p = z + n$$

43) $A = \dfrac{1}{2}bh$

$$2A = bh$$
$$\dfrac{2A}{b} = h$$

45) Yes, it can be written as $\dfrac{15}{100}$ or $\dfrac{3}{20}$.

47) $\dfrac{12}{15} = \dfrac{4}{5}$

49) $\dfrac{x}{15} = \dfrac{8}{10}$
$$10x = 120$$
$$x = 12$$
$$\{12\}$$

51) Let x be the number of students that have used alcohol.

$$\frac{x}{2500} = \frac{9}{20}$$

$$20x = 2500 \cdot 9$$

$$20x = 22500$$

$$x = 1125$$

1125 students would be expected to have used alcohol.

53) $x =$ number of $10 bills;

$2 + x =$ number of $20 bills

Value of $10 bills + Value of $20 bills = Total Value

$$10x + 20(2+x) = 340.00$$

$$10x + 40 + 20x = 340$$

$$30x = 300$$

$$x = 10$$

$20 bills $= 2 + x = 2 + 10 = 12$. There are 12-$20 bills and 10-$10 bills.

55) $x =$ Meg's speed. Then Jared's speed is $x - 1$

	d	$=$	r	$\cdot$	t
Meg	$1(x)$		x		1
Jared	$1(x-1)$		$x-1$		1

Meg's distance + Jared's distance $= 11$

$$x + x - 1 = 11$$

$$2x - 1 = 11$$

$$2x = 12$$

$$x = 6 \, \text{mph}$$

Meg's speed: 6 mph

Jared's speed: $x - 1 = 5$ mph

57) $w + 8 > 5$

$w > -3$

$(-3, \infty)$

59) $5x - 2 \leq 18$

$5x - 2 + 2 \leq 18 + 2$

$5x \leq 20$

$$\frac{5x}{5} \leq \frac{20}{5}$$

$x \leq 4$

$(-\infty, 4]$

61) $-19 \le 7p + 9 \le 2$

$-19 - 9 \le 7p + 9 - 9 \le 2 - 9$

$-28 \le 7p \le -7$

$-4 \le p \le -1$

$[-4, -1]$

63) $\dfrac{1}{2} < \dfrac{1-4t}{6} < \dfrac{3}{2}$

$6\left(\dfrac{1}{2}\right) < 6 \cdot \dfrac{1-4t}{6} < 6 \cdot \dfrac{3}{2}$

$3 < 1 - 4t < 9$

$3 - 1 < 1 - 4t - 1 < 9 - 1$

$2 < -4t < 8$

$\dfrac{2}{-4} > t > \dfrac{8}{-4}$

$-\dfrac{1}{2} > t > -2$

$-2 < t < -\dfrac{1}{2}$

$\left(-2, -\dfrac{1}{2}\right)$

65) {Toyota}

67) $\{10, 20, 25, 30, 35, 40, 50\}$

69) $a + 6 \le 9$ and $7a - 2 \ge 5$

$a \le 3$ and $7a \ge 7$

$a \le 3$ and $a \ge 1$

$[1, 3]$

71) $8 - y < 9$ or $\dfrac{1}{10}y > \dfrac{3}{5}$

$-y < 1$ or $y > 6$

$y > -1$ or $y > 6$

$(-1, \infty)$

73) $-8k + 13 = -7$

$-8k + 13 - 13 = -7 - 13$

$-8k = -20$

$\dfrac{-8k}{-8} = \dfrac{-20}{-8}$

$k = \dfrac{5}{2}$

The solution set is $\left\{\dfrac{5}{2}\right\}$.

75) $29 = -\dfrac{4}{7}m + 5$

$7(29) = 7\left(-\dfrac{4}{7}\right)m + 7(5)$

$203 = -4m + 35$

$203 - 35 = -4m$

$\dfrac{168}{-4} = m$

$-42 = m$

The solution set is $\{-42\}$.

77) $10p + 11 = 5(2p + 3) - 1$

$10p + 11 = 10p + 15 - 1$

$10p + 11 = 10p + 14$

$10p + 11 - 10p = 10p + 14 - 10p$

$11 \ne 14$

The solution set is $\{\varnothing\}$

Chapter 3 Review

79) $\dfrac{2x+9}{5}=\dfrac{x+1}{2}$

$2(2x+9)=5(x+1)$
$4x+18=5x+5$
$4x+18-5=5x+5-5$
$4x+13=5x+5-5$
$4x+13=5x$
$4x-4x+13=5x-4x$
$13=x$

81) $\dfrac{5}{6}-\dfrac{3}{4}(r+2)=\dfrac{1}{2}r+\dfrac{7}{12}$

$12\left(\dfrac{5}{6}\right)-12\left(\dfrac{3}{4}(r+2)\right)=12\left(\dfrac{1}{2}r\right)+12\left(\dfrac{7}{12}\right)$
$2(5)-3(3(r+2))=6r+7$
$10-9r-18=6r+7$
$-9r-8=6r+7$
$-9r-8+9r=6r+9r+7$
$-8=15r+7$
$-8-7=15r$
$-15=15r$
$-1=r$

The solution set is $\{-1\}$.

83) $x=$ number of oz of 5% alcohol solution
$60+x=$ number of oz of 9% alcohol solution

Soln	Concn	No of oz of soln	No of oz of alcohol in the soln
5%	0.05	x	$0.05x$
17%	0.17	60	$0.17(60)$
9%	0.09	$60+x$	$0.09(60+x)$

$0.05x+0.17(60)=0.09(60+x)$
$100[0.05x+0.17(60)]=100[0.09(60+x)]$
$5x+17(60)=9(60+x)$
$5x+1020=540+9x$
$-4x=-480$
$x=120$
120 oz of the 5% solution must be mixed.

85

85)   Let x be the first odd integer.

$x + 2 =$ second odd integer

$$\left(\begin{array}{c}\text{first odd}\\\text{integer}\end{array}\right) + \left(\begin{array}{c}\text{second odd}\\\text{integer}\end{array}\right) = 3\left(\begin{array}{c}\text{larger}\\\text{integer}\end{array}\right) - 21$$

$$x \quad + \quad x+2 \quad = 3(x+2) - 21$$
$$2x + 2 = 3x + 6 - 21$$
$$2x + 2 = 3x - 15$$
$$2x + 2 - 2 = 3x - 15 - 2$$
$$2x = 3x - 17$$
$$2x - 3x = 3x - 17 - 3x$$
$$-x = -17$$
$$x = 17$$

$x = $ first odd integer $= 17$

$x + 2 =$ second odd integer $= 17 + 2 = 19$

The integers are 17 and 19.

87)   Let x be the length of the shortest side.

Then the length of the other side is $x + 3$ and length of the longest side is $2x$.

shortest side $+$ other side $+$ longest side $=$ Perimeter

$$x + (x + 3) + 2x = 35$$
$$4x + 3 = 35$$
$$4x = 32$$
$$x = 8$$

$$\text{shortest side is } 8\,\text{cm}$$
$$\text{other side is } x + 3 = 8 + 3 = 11\,\text{cm}$$
$$\text{longest side is } 2x = 2(8) = 16\,\text{cm}$$

89)   Let $x =$ number of residents who want to secede.

$$\frac{9}{25} = \frac{x}{1000}$$
$$9(1000) = 25x$$
$$9000 = 25x$$
$$\frac{9000}{25} = x$$
$$360 = x$$

360 residents out of 1000 want to secede.

Chapter 3 Test

1) $-18y = 14$

$$\frac{-18y}{-18} = \frac{14}{-18}$$

$$y = -\frac{7}{9}$$

$$\left\{-\frac{7}{9}\right\}$$

3) $$\frac{8}{3}n - 11 = 5$$

$$3\left(\frac{8}{3}n\right) - 3 \cdot 11 = 3 \cdot 5$$

$$8n - 33 = 15$$

$$8n = 48$$

$$n = 6$$

$$\{6\}$$

5) $$\frac{1}{2} - \frac{1}{6}(x-5) = \frac{1}{3}(x+1) + \frac{2}{3}$$

$$6 \cdot \frac{1}{2} - 6\left(\frac{1}{6}(x-5)\right) = 6\left(\frac{1}{3}(x+1)\right) + 6 \cdot \frac{2}{3}$$

$$3 - 1(x-5) = 2(x+1) + 4$$

$$3 - x + 5 = 2x + 2 + 4$$

$$8 - x = 2x + 6$$

$$2 = 3x$$

$$\frac{2}{3} = x$$

$$\left\{\frac{2}{3}\right\}$$

7) $$\frac{9-w}{4} = \frac{3w+1}{2}$$

$$2(9-w) = 4(3w+1)$$

$$18 - 2w = 12w + 4$$

$$14 = 14w$$

$$1 = w$$

$$\{1\}$$

9) Let x = first even integer.

Then the second even integer is $x+2$

Third even integer is $x+4$

$$x + x + 2 + x + 4 = 114$$

$$3x + 6 = 114$$

$$3x = 108$$

$$x = 36$$

Second even intger=38

Third even integer=40

The integers are 36, 38 and 40.

11) Let x = amount spent by Debra.

$$\frac{14}{40.60} = \frac{11}{x}$$

$$14x = 11(40.60)$$

$$14x = 446.60$$

$$x = 31.90$$

Debra spent \$31.90.

13) x = speed of eastbound car. Then

speed of westbound car is x+6.

	d	= r	· t
westbound	$2.5(x)$	x	2.5
eastbound	$2.5(x+6)$	$x+6$	2.5

westbound car's distance + eastbound car's distance = 345

$$2.5x + 2.5(x+6) = 345$$

$$2.5x + 2.5x + 15 = 345$$

$$5x + 15 = 345$$

$$5x = 330$$

$$x = 66$$

westbound car's speed: 66 mph

eastbound car's speed: $x + 6$= 7 mph

15)
$$S - 2\pi r^2 = 2\pi r^2 - 2\pi r^2 + 2\pi rh$$
$$S - 2\pi r^2 = 2\pi rh$$
$$\frac{S}{2\pi r} - \frac{2\pi r^2}{2\pi r} = \frac{2\pi rh}{2\pi r}$$
$$\frac{S - 2\pi r^2}{2\pi r} = h$$
$$\text{or } h = \frac{S}{2\pi r} - r$$

17)
$$6m + 19 \le 7$$
$$6m \le -12$$
$$\frac{6m}{6} \le \frac{-12}{6}$$
$$m \le -2$$

$$(-\infty, -2]$$

19)
$$-\frac{5}{6} < \frac{4c-1}{6} \le \frac{3}{2}$$
$$6\left(-\frac{5}{6}\right) < 6\left(\frac{4c-1}{6}\right) < 6\left(\frac{3}{2}\right)$$
$$-5 < 4c - 1 \le 9$$
$$-4 < 4c \le 10$$
$$-1 < c \le \frac{5}{2}$$

$$\left(-1, \frac{5}{2}\right]$$

21) $3n + 5 > 12$ or $\frac{1}{4}n < -2$
$$3n > 7 \text{ or } n < -8$$
$$n > \frac{7}{3} \text{ or } n < -8$$
$$(-\infty, -8) \cup \left(\frac{7}{3}, \infty\right)$$

23) $6 - p < 10$ or $p - 7 < 2$
$$-p < 4 \text{ or } p < 9$$
$$p > -4 \text{ or } p < 9$$
$$(-\infty, \infty)$$

Cumulative Review: Chapters 1-3.

1) $\dfrac{3}{8} - \dfrac{5}{6} = \dfrac{9}{24} - \dfrac{20}{24} = -\dfrac{11}{24}$

3) $26 - 14 \div 2 + 5 \cdot 7 = 26 - 7 + 35 = 54$

5)
$$-39 - |7 - 15| = -39 - |-8| = -39 - 8 = -47$$

7) $\{-5, 0, 9\}$

9) $\{0, 9\}$

11) No. For example, $10 - 3 \ne 3 - 10$

13) $\dfrac{35r^{16}}{28r^4} = \dfrac{5}{4}r^{16-4} = \dfrac{5}{4}r^{12}$

15) $\left(-12z^{10}\right)\left(\dfrac{3}{8}z^{-16}\right) = -\dfrac{9}{2}z^{10-16} = -\dfrac{9}{2}z^{-6} = -\dfrac{9}{2z^6}$

17) $0.00000895 : 0.00000895$
$$= 8.95 \times 10^{-6}$$

19)
$$\frac{3}{2}n + 14 = 20$$
$$2 \cdot \frac{3}{2}n + 2 \cdot 14 = 2 \cdot 20$$
$$3n + 28 = 40$$
$$3n = 12$$
$$n = 4$$
$$\{4\}$$

21) $\dfrac{x+3}{10} = \dfrac{2x-1}{4}$

$4(x+3) = 10(2x-1)$

$4x+12 = 20x-10$

$22 = 16x$

$\dfrac{22}{16} = x$

$\dfrac{11}{8} = x$

$\left\{\dfrac{11}{8}\right\}$

23) $x =$ speed of the train

speed of the car $= x-10$

	d	$=$	r	$\cdot$	t
train	140		x		$\dfrac{140}{x}$
car	120		$x-10$		$\dfrac{120}{x-10}$

time by train = time by car

$\dfrac{140}{x} = \dfrac{120}{x-10}$

$140(x-10) = 120x$

$140x-1400 = 120x$

$-1400 = -20x$

$70 = x$

speed of train=70 mph

speed of car=70−10= 60 mph

25) $8x \le -24$ or $4x-5 \ge 6$

$x \le -3$ or $4x \ge 11$

$x \le -3$ or $x \ge \dfrac{11}{4}$

$(-\infty, -3] \cup \left[\dfrac{11}{4}, \infty\right)$

Chapter 4: Linear Equations in Two Variables

Section 4.1 Exercises

1) 16.1 gallons

3) 2004 and 2006: 15.9 gallons

5) Consumption was increasing

7) New Jersey: 86.3%

9) Florida's graduation rate is about 32.4% less than New Jersey's.

11) Answers may vary

13) Yes
$$2x+5y=1$$
$$2(-2)+5(1)=1$$
$$-4+5=1$$
$$1=1$$

15) Yes
$$-3x-2y=-15$$
$$-3(7)-2(-3)=-15$$
$$-21+6=-15$$
$$-15=-15$$

17) No
$$y=-\frac{3}{2}x-7$$
$$(5)=-\frac{3}{2}(8)-7$$
$$5=-12-7$$
$$5\neq-19$$

19) Yes
$$y=-7$$
$$-7=-7$$

21) $y=3x-7$
$$y=3(4)-7$$
$$y=12-7$$
$$y=5$$

23) $2x-15y=13$
$$2x-15\left(-\frac{4}{3}\right)=13$$
$$2x+20=13$$
$$2x=-7$$
$$x=-\frac{7}{2}$$

25) $x=5$

27) $y=2x-4$
$$y=2(0)-4 \qquad y=2(-1)-4$$
$$y=0-4 \qquad y=-2-4$$
$$y=-4 \qquad y=-6$$

$$y=2(1)-4 \qquad y=2(-2)-4$$
$$y=2-4 \qquad y=-4-4$$
$$y=-2 \qquad y=-8$$

x	y
0	−4
1	−2
−1	−6
−2	−8

29) $y = 4x$

$$y = 4(0) \qquad y = 4\left(\frac{1}{2}\right)$$
$$y = 0$$
$$\qquad\qquad\qquad y = 2$$

$$-20 = 4x \qquad 12 = 4x$$
$$\frac{-20}{4} = x \qquad \frac{12}{4} = x$$
$$-5 = x \qquad\quad 3 = x$$

x	y
0	0
$\frac{1}{2}$	2
3	12
−5	−20

31) $5x + 4y = -8$

$$5x + 4(0) = -8$$
$$5(0) + 4y = -8 \qquad 5x + 0 = -8$$
$$4y = -8 \qquad\qquad 5x = -8$$
$$y = \frac{-8}{4} = -2 \qquad x = -\frac{8}{5}$$

$$5x + 4y = -8$$
$$5(1) + 4y = -8 \qquad 5\left(-\frac{12}{5}\right) + 4y = -8$$
$$5 + 4y = -8 \qquad -12 + 4y = -8$$
$$4y = -13 \qquad\qquad 4y = 4$$
$$y = -\frac{13}{4} \qquad\qquad y = \frac{4}{4} = 1$$

x	y
0	−2
$-\frac{8}{5}$	0
1	$-\frac{13}{4}$
$-\frac{12}{5}$	1

33) $y = -2$

x	y
0	−2
−3	−2
8	−2
17	−2

35) Answers may vary.

37) A: $(-2, 1)$ quadrant II

   B: $(5, 0)$ no quadrant

   C: $(-2, -1)$ quadrant III

   D: $(0, -1)$ no quadrant

   E: $(2, -2)$ quadrant IV

   F: $(3, 4)$ quadrant I

39–42)

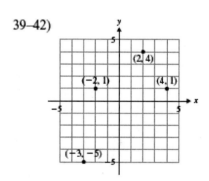

43–46)

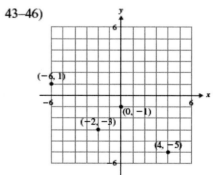

47–50)

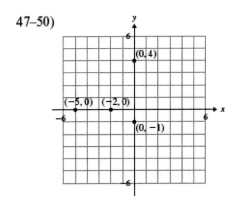

x	y
0	3
$\dfrac{3}{4}$	0
2	−5
−1	7

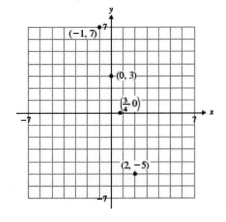

51–54)

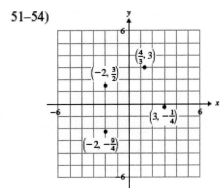

55–56)

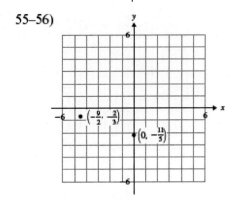

59) $\qquad y = x$

$y = (0) \qquad (3) = x$

$y = 0 \qquad\qquad 3 = x$

$y = (-1) \qquad (-5) = x$

$y = -1 \qquad\qquad -5 = x$

x	y
0	0
−1	−1
3	3
−5	−5

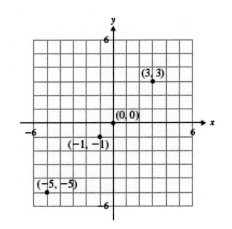

57) $\qquad y = -4x + 3$

$y = -4(0) + 3 \qquad\qquad (0) = -4x + 3$

$y = 0 + 3 \qquad\qquad\quad -3 = -4x$

$y = 3 \qquad\qquad\qquad \dfrac{3}{4} = x$

$y = -4(2) + 3 \qquad\qquad (7) = -4x + 3$

$y = -8 + 3 \qquad\qquad\quad 4 = -4x$

$y = -5 \qquad\qquad\qquad -1 = x$

61) $3x + 4y = 12$

$3(0) + 4y = 12$ $3x + 4(0) = 12$

$4y = 12$ $3x = 12$

$y = 3$ $x = 4$

$3(1) + 4y = 12$ $3x + 4(6) = 12$

$3 + 4y = 12$ $3x + 24 = 12$

$4y = 9$ $3x = -12$

$y = \dfrac{9}{4}$ $x = -4$

x	y
0	3
4	0
1	$\dfrac{9}{4}$
-4	6

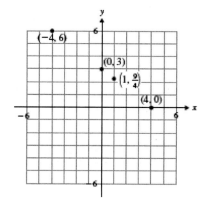

63) $y + 1 = 0$

$y = -1$

x	y
0	-1
1	-1
-3	-1
-1	-1

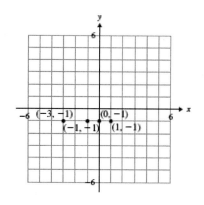

65) $y = \dfrac{1}{4}x + 2$

$y = \dfrac{1}{4}(0) + 2$ $y = \dfrac{1}{4}(4) + 2$

$y = 0 + 2$ $y = 1 + 2$

$y = 2$ $y = 3$

$y = \dfrac{1}{4}(-2) + 2$ $y = \dfrac{1}{4}(-1) + 2$

$y = -\dfrac{1}{2} + 2$ $y = -\dfrac{1}{4} + \dfrac{8}{4}$

$y = \dfrac{3}{2}$ $y = \dfrac{7}{4}$

x	y
0	2
-2	$\dfrac{3}{2}$
4	3
-1	$\dfrac{7}{4}$

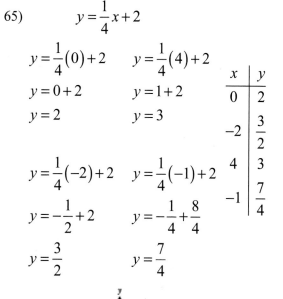

67) $y = \dfrac{2}{3}x - 7$

a) $x = 3$ $y = \dfrac{2}{3}(3) - 7$

$y = 2 - 7$

$y = -5$

$$x = 6 \qquad y = \frac{2}{3}(6) - 7$$
$$y = 4 - 7$$
$$y = -3$$

$$x = -3 \qquad y = \frac{2}{3}(-3) - 7$$
$$y = -2 - 7$$
$$y = -9$$

$$(3, -5), (6, -3), (-3, -9)$$

b) $\quad x = 1 \qquad y = \frac{2}{3}(1) - 7$

$$y = \frac{2}{3} - \frac{21}{3}$$
$$y = -\frac{19}{3}$$

$$x = 5 \qquad y = \frac{2}{3}(5) - 7$$
$$y = \frac{10}{3} - \frac{21}{3}$$
$$y = -\frac{11}{3}$$

$$x = -2 \qquad y = \frac{2}{3}(-2) - 7$$
$$y = -\frac{4}{3} - \frac{21}{3}$$
$$y = -\frac{25}{3}$$

$$\left(1, -\frac{19}{3}\right), \left(5, -\frac{11}{3}\right), \left(-2, -\frac{25}{3}\right)$$

c) The x-values in part a) are multiples of the denominator of $\frac{2}{3}$. When you multiply $\frac{2}{3}$ by a multiple of 3, the fraction is eliminated.

69) negative

71) negative

73) positive

75) zero

77) a) x represents the year
   y represents the number of visitors in millions

b) In 2004 there were 37.4 million visitors to Las Vegas.

c) 38.9 million

d) 2005

e) 2 million

f) (2007, 39.2)

79) a) (1985, 52.9), (1990, 50.6), (1995, 42.4), (2000, 41.4), (2005, 40.0)

b)

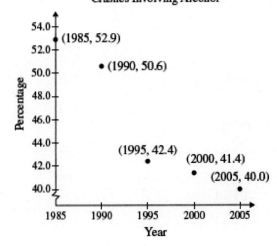

94

c) In 2000, 41.4 % of all fatal
accidents involved alcohol.

81) a) $y = 0.095x$

$y = 0.095(100.00)$ $y = 0.095(210.72)$
$y = 9.50$ $y = 20.0184$

$y = 0.095(140.00)$ $y = 0.095(250.00)$
$y = 13.30$ $y = 23.75$

x	y
100.00	9.50
140.00	13.30
210.72	20.0184
250.00	23.75

$(100, 9.50), (140.00, 13.30),$

$(210.72, 20.0184), (250.00, 23.75)$

b)

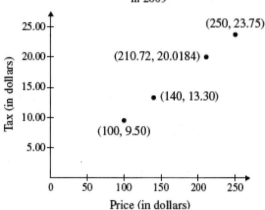

c) If the bill totals $40.00, then
sales tax will be $13.30.

d) $20.02

e) Yes, they lie on a straight line.

f) $y = 0.095x$
$19.00 = 0.095x$
$100(19.00) = 100(0.095x)$
$1900 = 9.5x$
$\dfrac{1900}{9.5} = x$
$\$20.00 = x$
The cost of the item purchased is $20.00.

Section 4.2 Exercises

1) line

3) $y = -2x + 4$

$y = -2(0) + 4$ $y = -2(2) + 4$
$y = 0 + 4$ $y = -4 + 4$
$y = 4$ $y = 0$

$y = -2(-1) + 4$ $y = -2(3) + 4$
$y = 2 + 4$ $y = -6 + 4$
$y = 6$ $y = -2$

x	y
0	4
−1	6
2	0
3	−2

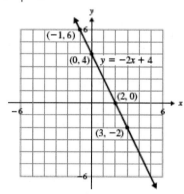

5) $y = \frac{3}{2}x + 7$

$y = \frac{3}{2}(0) + 7$ $y = \frac{3}{2}(-2) + 7$

$y = 0 + 7$ $y = -3 + 7$

$y = 7$ $y = 4$

$y = \frac{3}{2}(2) + 7$ $y = \frac{3}{2}(-4) + 7$

$y = 3 + 7$ $y = -6 + 7$

$y = 10$ $y = 1$

x	y
0	7
2	10
−2	4
−4	1

x	y
$\frac{3}{2}$	0
0	3
$\frac{1}{2}$	2
−1	5

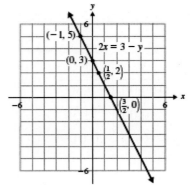

9) $x = -\frac{4}{9}$

x	y
$-\frac{4}{9}$	5
$-\frac{4}{9}$	0
$-\frac{4}{9}$	−1
$-\frac{4}{9}$	−2

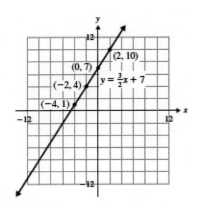

7) $2x = 3 - y$

$2x = 3 - 0$ $2(\frac{1}{2}) = 3 - y$

$x = \frac{3}{2}$ $1 = 3 - y$

    $2 = y$

$2(0) = 3 - y$ $2x = 3 - 5$

$0 = 3 - y$ $2x = -2$

$3 = y$ $x = -1$

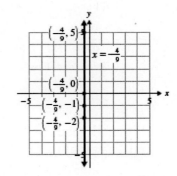

11) It is the point where the graph
intersects the y-axis.
Let $x = 0$, and solve for y.

13) $y = x - 1$
x-int: Let $y = 0$, and solve for x.
$(0) = x - 1$
$0 = x - 1$
$1 = x$ $\qquad$ $(1, 0)$
y-int: Let $x = 0$, and solve for y.
$y = (0) - 1$
$y = -1$ $\qquad$ $(0, -1)$
Let $y = 1$.
$(1) = x - 1$
$2 = x$ $\qquad$ $(2, 1)$

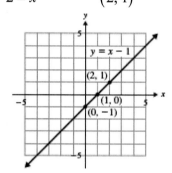

15) $3x - 4y = 12$
x-int: Let $y = 0$, and solve for x.
$3x - 4(0) = 12$
$3x - 0 = 12$
$3x = 12$
$x = 4$ $\qquad$ $(4, 0)$
y-int: Let $x = 0$, and solve for y.
$3(0) - 4y = 12$
$0 - 4y = 12$
$-4y = 12$
$y = -3$ $\qquad$ $(0, -3)$
Let $x = 2$.

$3(2) - 4y = 12$
$6 - 4y = 12$
$-4y = 6$
$y = -\dfrac{3}{2}$ $\qquad$ $\left(2, -\dfrac{3}{2}\right)$

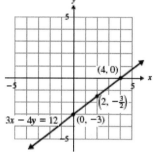

17) $x = -\dfrac{4}{3}y - 2$
x-int: Let $y = 0$, and solve for x.
$x = -\dfrac{4}{3}(0) - 2$
$x = 0 - 2$
$x = -2$ $\qquad$ $(-2, 0)$
y-int: Let $x = 0$, and solve for y.
$(0) = -\dfrac{4}{3}y - 2$
$2 = -\dfrac{4}{3}y$
$-\dfrac{3}{2} = y$ $\qquad$ $\left(0, -\dfrac{3}{2}\right)$
Let $y = -3$.
$x = -\dfrac{4}{3}(-3) - 2$
$x = 4 - 2$
$x = 2$ $\qquad$ $(2, -3)$

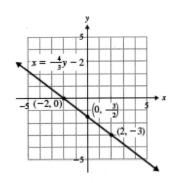

19) $2x - y = 8$

x-int: Let $y = 0$, and solve for x.

$2x - (0) = 8$

$2x = 8$

$x = 4 \qquad (4, 0)$

y-int: Let $x = 0$, and solve for y.

$2(0) - y = 8$

$0 - y = 8$

$-y = 8$

$y = -8 \qquad (0, -8)$

Let $x = 2$.

$2(2) - y = 8$

$4 - y = 8$

$-y = 4$

$y = -4 \qquad (2, -4)$

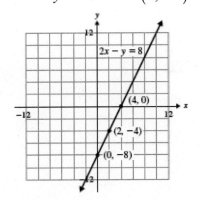

21) $y = -x$

x-int: Let $y = 0$, and solve for x.

$(0) = -x$

$0 = x \qquad (0, 0)$

y-int: Let $x = 0$, and solve for y.

$y = -(0)$

$y = 0 \qquad (0, 0)$

Let $x = 1$.

$y = -(1)$

$y = -1 \qquad (1, -1)$

Let $x = -1$.

$y = -(-1)$

$y = 1 \qquad (-1, 1)$

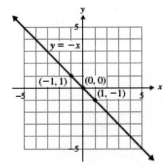

23) $4x - 3y = 0$

x-int: Let $y = 0$, and solve for x.

$4x - 3(0) = 0$

$4x - 0 = 0$

$4x = 0$

$x = 0 \qquad (0, 0)$

y-int: Let $x = 0$, and solve for y.

$4(0) - 3y = 0$

$0 - 3y = 0$

$3y = 0$

$y = 0 \qquad (0, 0)$

Let $x = 3$.

$4(3) - 3y = 0$

$12 - 3y = 0$

$12 = 3y$

$4 = y \qquad (3, 4)$

Let $x = -3$.

$4(-3) - 3y = 0$

$-12 - 3y = 0$

$-12 = 3y$

$-4 = y \qquad (-3, -4)$

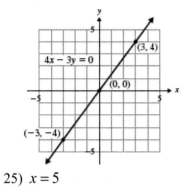

25) $x = 5$

$(5, 0), (5, 2), (5, -1)$

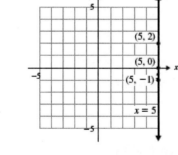

27) $y = 0$

$(0, 0), (1, 0), (-2, 0)$

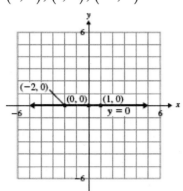

29) $x - \dfrac{4}{3} = 0$

$x = \dfrac{4}{3}$

$\left(\dfrac{4}{3}, 0\right), \left(\dfrac{4}{3}, 1\right), \left(\dfrac{4}{3}, -2\right)$

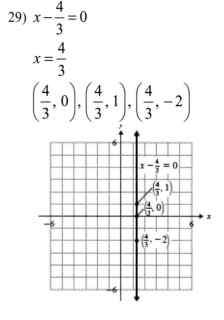

31) $4x - y = 9$

x-int: Let $y = 0$, and solve for x.

$4x - (0) = 9$

$4x = 9$

$x = \dfrac{9}{4} \qquad \left(\dfrac{9}{4}, 0\right)$

y-int: Let $x = 0$, and solve for y.

$4(0) - y = 9$

$0 - y = 9$

$y = -9 \qquad (0, -9)$

Let $y = 3$.

$4x - (3) = 9$

$4x = 12$

$x = 3 \qquad (3, 3)$

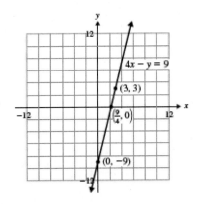

33) (0, 0)

35) $y = 1.29x$

   a) $x = 0$ $y = 1.29(0)$
                     $y = 0$

        $x = 4$ $y = 1.29(4)$
                     $y = 5.16$

        $x = 7$ $y = 1.29(7)$
                     $y = 9.03$

       $x = 12$ $y = 1.29(12)$
                     $y = 15.48$

x	y
0	0
4	5.16
7	9.03
12	15.48

    $(0, 0), (4, 5.16),$

    $(7, 9.03), (12, 15.48)$

  b) (0, 0): If no songs are purchased, the cost is $0. (4, 5.16): The cost of downloading 4 songs is $5.16. (7, 9.03): The cost of downloading 7 songs is $9.03. (12, 15.48): The cost of downloading 12 songs is $15.48.

c)

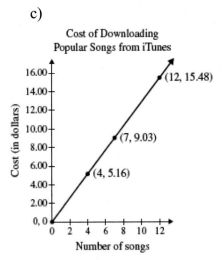

Cost of Downloading Popular Songs from iTunes

d) $y = 1.29x$

       $11.61 = 1.29x$

       $\dfrac{11.61}{1.29} = x$

       $9 = x$

9 songs can be downloaded for $11.61.

37) a) $2004 : 26,275;\ 2007 : 31,801$

    b) $y = 1662x + 24,916$

        $x = 1$ for 2004

        $y = 1662(1) + 24,916$

        $y = 1662 + 24,916$

        $y = 26,578$

        $x = 4$ for 2007

        $y = 1662(4) + 24916$

        $y = 6648 + 24,916$

        $y = 31,564$

       Yes, they are close.

c)

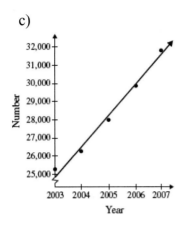

d) The *y*-intercept is 24,916. In 2003, approximately 24,916 science and engineering doctorates were awarded. It looks like it is within about 300 units of the plotted point.

e) $y = 1662x + 24,916$

$x = 9$ for 2012

$y = 1662(9) + 24,916$

$y = 14,958 + 24,916$

$y = 39,874$

Section 4.3 Exercises

1) The slope of a line is the ratio of vertical change to horizontal change.

It is $\dfrac{\text{change in } y}{\text{change in } x}$ or $\dfrac{\text{rise}}{\text{run}}$ or $\dfrac{y_2 - y_1}{x_2 - x_1}$ where (x_1, y_1) and (x_2, y_2) are points on the line.

3) It slants upward from left to right.

5) undefined

7) a) Vertical change: 3 units
Horizontal change: 4 units

$\text{Slope} = \dfrac{3}{4}$

b) $(x_1, y_1) = (1, -1)$
$(x_2, y_2) = (5, 2)$

$m = \dfrac{y_2 - y_1}{x_2 - x_1} = \dfrac{2 - (-1)}{5 - 1} = \dfrac{3}{4}$

9) a) Vertical change: -2 units
Horizontal change: 3 units

$\text{Slope} = \dfrac{-2}{3} = -\dfrac{2}{3}$

b) $(x_1, y_1) = (1, 5)$
$(x_2, y_2) = (4, 3)$

$m = \dfrac{y_2 - y_1}{x_2 - x_1} = \dfrac{3 - 5}{4 - 1} = \dfrac{-2}{3}$

$= -\dfrac{2}{3}$

11) a) Vertical change: -3 units
Horizontal change: 1 unit

$\text{Slope} = \dfrac{-3}{1} = -3$

b) $(x_1, y_1) = (2, 3)$
$(x_2, y_2) = (3, 0)$

$m = \dfrac{y_2 - y_1}{x_2 - x_1} = \dfrac{0 - 3}{3 - 2} = \dfrac{-3}{1} = -3$

13) a) Vertical change: 6 units
Horizontal change: 0 units

$\text{Slope} = \dfrac{6}{0}$ is undefined.

b) $(x_1, y_1) = (-3, -4)$
$(x_2, y_2) = (-3, 2)$

$m = \dfrac{y_2 - y_1}{x_2 - x_1} = \dfrac{2 - (-4)}{-3 - (-3)} = \dfrac{6}{0}$

Slope is undefined.

15)

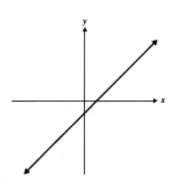

17) $(x_1, y_1) = (2,1) \quad (x_2, y_2) = (0, -3)$

$$m = \frac{y_2 - y_1}{x_2 - x_1} = \frac{-3-1}{0-2} = \frac{-4}{-2} = 2$$

19) $(x_1, y_1) = (2, -6) \quad (x_2, y_2) = (-1, 6)$

$$m = \frac{y_2 - y_1}{x_2 - x_1} = \frac{6-(-6)}{-1-2} = \frac{12}{-3} = -4$$

21) $(x_1, y_1) = (-4, 3) \quad (x_2, y_2) = (1, -8)$

$$m = \frac{y_2 - y_1}{x_2 - x_1} = \frac{-8-3}{1-(-4)} = \frac{-11}{5} = -\frac{11}{5}$$

23) $(x_1, y_1) = (-2, -2) \quad (x_2, y_2) = (-2, 7)$

$$m = \frac{y_2 - y_1}{x_2 - x_1} = \frac{7-(-2)}{(-2)-(-2)} = \frac{9}{0}$$

Slope is undefined.

25) $(x_1, y_1) = (3, 5) \quad (x_2, y_2) = (-1, 5)$

$$m = \frac{y_2 - y_1}{x_2 - x_1} = \frac{5-5}{-1-3} = \frac{0}{-4} = 0$$

27) $(x_1, y_1) = \left(\frac{2}{3}, \frac{5}{2}\right)$

$(x_2, y_2) = \left(-\frac{1}{2}, 2\right)$

$$m = \frac{y_2 - y_1}{x_2 - x_1} = \frac{2 - \left(\dfrac{5}{2}\right)}{-\dfrac{1}{2} - \dfrac{2}{3}}$$

$$= \frac{\dfrac{4}{2} - \dfrac{5}{2}}{-\dfrac{3}{6} - \dfrac{4}{6}} = \frac{-\dfrac{1}{2}}{-\dfrac{7}{6}}$$

$$= -\frac{1}{2} \div \left(-\frac{7}{6}\right) = -\frac{1}{2} \cdot \left(-\frac{6}{7}\right) = \frac{3}{7}$$

29) $(x_1, y_1) = (3.5, -1.4)$

$(x_2, y_2) = (7.5, 1.6)$

$$m = \frac{y_2 - y_1}{x_2 - x_1} = \frac{1.6 - (-1.4)}{7.5 - 3.5}$$

$$= \frac{3.0}{4.0} = 0.75$$

31) $\quad m = \dfrac{\text{rise}}{\text{run}} = \dfrac{-60}{395} = -\dfrac{12}{79}$

33) $m = \dfrac{\text{rise}}{\text{run}} = \dfrac{6}{9} = \dfrac{2}{3} = 0.\overline{6}$

No. The slope of the slide is $0.\overline{6}$.
This is more than the recommended
slope of 0.577.

35) $m = \dfrac{\text{rise}}{\text{run}} = \dfrac{0.75}{20} = 0.0375$

Yes. The slope of the driveway is 0.0375.
This is less than the maximum
allowed slope of 0.05.

37) $m = \dfrac{\text{rise}}{\text{run}} = \dfrac{12}{26} = \dfrac{6}{13}$. Slope is $\dfrac{6}{13}$.

39) a) 2003:2.89 million; 2005: 2.70
million

b) The line slants downward from
left to right; therefore it has a
negative slope.

c) The number of injuries is decreasing over time.

d) $m = \dfrac{\text{rise}}{\text{run}} = \dfrac{2.49 - 2.89}{4} = \dfrac{-0.4}{4} = -0.1$

The number of injuries is decreasing by 0.1 million or 100,000 per year.

41)

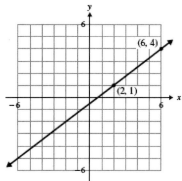

43)

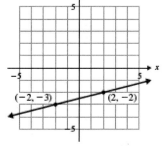

45)

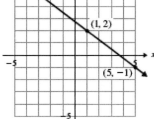

47)

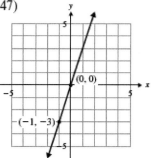

49)

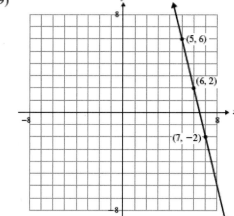

51)

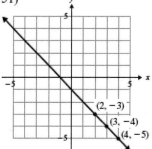

53)

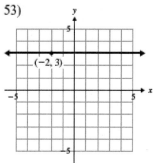

55)

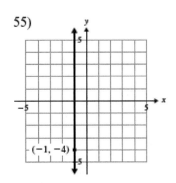

57)

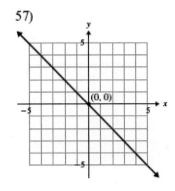

Section 4.4 Exercises

1) The slope is m, and the y-intercept is $(0, b)$.

3) $m = \dfrac{2}{5}$, y-int: $(0, -6)$

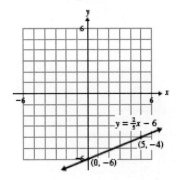

5) $m = -\dfrac{3}{2}$, y-int: $(0, 3)$

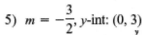
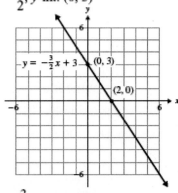

7) $m = \dfrac{3}{4}$, y-int: $(0, 2)$

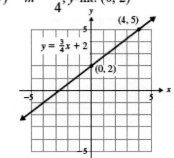

9) $m = -2$, y-int: $(0, -3)$

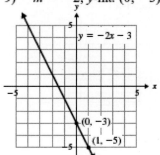

11) $m = 5$, y-int: $(0, 0)$

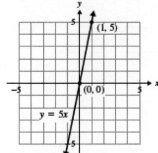

13) $m = -\frac{3}{2}$, y-int: $\left(0, -\frac{7}{2}\right)$

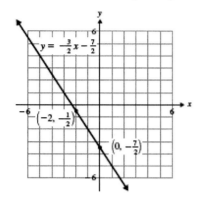

15) $m = 0$, y-int: $(0, 6)$

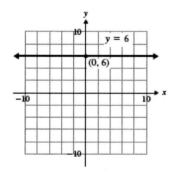

17) $x + 3y = -6$
$3y = -x - 6$
$y = -\frac{1}{3}x - 2$
$m = -\frac{1}{3}$, y-int : $(0, -2)$

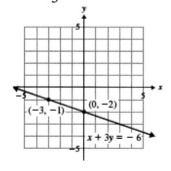

19) $4x + 3y = 21$
$3y = -4x + 21$
$y = -\frac{4}{3}x + 7$
$m = -\frac{4}{3}$, y-int : $(0, 7)$

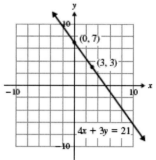

21) $2 = x + 3$
$-1 = x$
The slope is undefined,
and no y-intercept

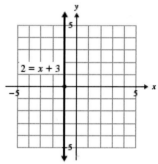

23) $2x = 18 - 3y$
$3y = -2x + 18$
$y = -\frac{2}{3}x + 6$
$m = -\frac{2}{3}$, $y-$int : $(0,6)$

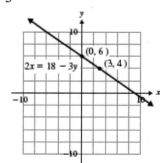

25) $y + 2 = -3$
      $y = -5$

$m = 0, \ y\text{-int} : (0, -5)$

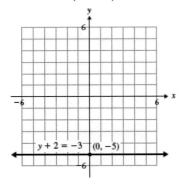

27)   a) (0, 0). If Kolya works 0 hours, he earns $0.

      b) $m = 8.5$; Kolya earns $8.50 per hour.

      c) $102.00
          $P = 8.50h$
          $P = 8.50(12)$
          $P = 102.00$

29)   a) (0, 18); when the joey comes out of the pouch it weighs 18 oz

      b) $y = 2x + 18$
          $y = 2(3) + 18 = 24 \, oz$

      c) A joey gains 2 oz per week after coming out of its mother's pouch.

      d) $y = 2x + 18$
          $y = 2x + 18$
          $32 = 2x + 18$
          $14 = 2x$
          $7 = x$

31)   a) $(0, 0)$; $0 = 0$ rupees

b) $m = 48.2$: each American dollar is worth 48.2 rupees.

c) $r = 48.2d$
    $r = 48.2(80.00)$
    $r = 3856$

    She gets 3856 rupees.

d) $r = 48.2d$
        $r = 48.2d$
    $2410 = 48.2d$
    $\dfrac{2410}{48.2} = d$
    $50 = d$

    One gets $50 for 2410 rupees.

33) $y = mx + b \quad m = -4; \, b = 7$
      $y = -4x + 7$

35) $y = mx + b \quad m = \dfrac{9}{5}; \, b = -3$
      $y = \dfrac{9}{5}x - 3$

37) $y = mx + b \quad m = -\dfrac{5}{2}; \, b = -1$
      $y = -\dfrac{5}{2}x - 1$

39) $(x_1, y_1) = (0, 2); \ m = 1$
      $y - y_1 = m(x - x_1)$
      $y - 2 = 1(x - 0)$
      $y - 2 = x$
      $y = x + 2$

41) $y = mx + b \quad m = 0; \, b = 0$
      $y = 0$

43)   Their slopes are negative reciprocals.

45) $y = -x - 5 \qquad y = x + 8$
      $m = -1 \qquad\qquad m = 1$
      perpendicular

47) $y = \dfrac{2}{9}x + 4 \qquad 4x - 18y = 9$

$-18y = -4x + 9$

$y = \dfrac{2}{9}x - \dfrac{1}{2}$

$m = \dfrac{2}{9} \qquad m = \dfrac{2}{9}$

parallel

49) $3x - y = 4 \qquad 2x - 5y = -9$

$-y = -3x + 4 \qquad -5y = -2x - 9$

$y = 3x - 4 \qquad y = \dfrac{2}{5}x + \dfrac{9}{5}$

$m = 3$

$\qquad\qquad\qquad m = \dfrac{2}{5}$

neither

51) $-x + y = -21 \qquad y = 2x + 5$

$y = x - 21$

$\qquad\qquad\qquad m = 2$

$m = 1$

neither

53) $x + 7y = 4 \qquad y - 7x = 4$

$7y = -x + 4$

$\qquad\qquad\qquad y = 7x + 4$

$y = -\dfrac{1}{7}x + \dfrac{4}{7}$

$m = -\dfrac{1}{7} \qquad m = 7$

perpendicular

55) $y = -\dfrac{1}{2}x \qquad x + 2y = 4$

$\qquad\qquad\qquad 2y = -x + 4$

$\qquad\qquad\qquad y = -\dfrac{1}{2}x + 2$

$m = -\dfrac{1}{2} \qquad m = -\dfrac{1}{2}$

parallel

57) $x = -1 \qquad\qquad y = 6$

$m = $ undefined $\qquad m = 0$

perpendicular

59) $x = -4.3 \qquad\qquad x = 0$

$m = $ undefined $\qquad m = $ undefined

parallel

61) $L_1: m = \dfrac{8 - (-7)}{2 - (-1)} = \dfrac{15}{3} = 5$

$L_2: m = \dfrac{4 - 2}{0 - 10} = \dfrac{2}{-10} = -\dfrac{1}{5}$

perpendicular

63) $L_1: m = \dfrac{8 - 10}{3 - 1} = \dfrac{-2}{2} = -1$

$L_2: m = \dfrac{-17 - 4}{-5 - 2} = \dfrac{-21}{-7} = 3$

neither

65) $L_1: m = \dfrac{-1 - 6}{4 - (-3)} = \dfrac{-7}{7} = -1$

$L_2: m = \dfrac{-1 - (-5)}{-10 - (-6)} = \dfrac{4}{-4} = -1$

parallel

107

Chapter 4: Linear Equations in Two Variables

67) $L_1: m = \dfrac{1-2}{-6-(-6)} = \dfrac{-1}{0} =$ undefined

$L_2: m = \dfrac{-5-0}{4-4} = \dfrac{-5}{0} =$ undefined

parallel

69) $L_1: m = \dfrac{5-2}{7-7} = \dfrac{3}{0} =$ undefined

$L_2: m = \dfrac{0-0}{1-(-2)} = \dfrac{0}{3} = 0$

perpendicular

Section 4.5 Exercises

1) $y = -2x - 4$
$2x + y = -4$

3) $x = y + 1$
$x - y = 1$

5) $y = \dfrac{4}{5}x + 1$
$5 \cdot y = 5\left(\dfrac{4}{5}x + 1\right)$
$5y = 4x + 5$
$-4x + 5y = 5$
$4x - 5y = -5$

7) $y = -\dfrac{1}{3}x - \dfrac{5}{4}$
$12 \cdot y = 12\left(-\dfrac{1}{3}x - \dfrac{5}{4}\right)$
$12y = -4x - 15$
$4x + 12y = -15$

9) Use $y = mx + b$ and substitute the slope and y-intercept values into the equation.

11) $y = mx + b$ $m = -7; b = 2$
$y = -7x + 2$

13) $y = mx + b$ $m = -4; b = 6$
$y = -4x + 6$
$4x + y = 6$

15) $y = mx + b$ $m = \dfrac{2}{7}; b = -3$
$y = \dfrac{2}{7}x - 3$
$7 \cdot y = 7\left(\dfrac{2}{7}x - 3\right)$
$7y = 2x - 21$
$-2x + 7y = -21$
$2x - 7y = 21$

17) $y = mx + b$ $m = -1; b = 0$
$y = -x$

19) a) $y - y_1 = m(x - x_1)$

b) Use the point-slope formula and substitute the point on the line and the slope into the equation.

21) $(x_1, y_1) = (5, 7); m = 1$
$y - y_1 = m(x - x_1)$
$y - 7 = 1(x - 5)$
$y - 7 = x - 5$
$y = x + 2$

108

23) $(x_1,\ y_1)=(4,\ -1);\ \ m=-5$

$$y-y_1=m(x-x_1)$$
$$y-(-1)=-5(x-4)$$
$$y+1=-5(x-4)$$
$$y+1=-5x+20$$
$$y=-5x+19$$

25) $(x_1,\ y_1)=(-2,\ -1);\ \ m=4$

$$y-y_1=m(x-x_1)$$
$$y-(-1)=4(x-(-2))$$
$$y+1=4x+8$$
$$-4x+y=7$$
$$4x-y=-7$$

27) $(x_1,\ y_1)=(-5,\ 8);\ \ m=\dfrac{2}{5}$

$$y-y_1=m(x-x_1)$$
$$y-8=\dfrac{2}{5}(x-(-5))$$
$$y-8=\dfrac{2}{5}x+2$$
$$5(y-8)=5\left(\dfrac{2}{5}x+2\right)$$
$$5y-40=2x+10$$
$$-2x+5y=50$$
$$2x-5y=-50$$

29) $(x_1,\ y_1)=(5,\ 1);\ \ m=-\dfrac{5}{4}$

$$y-y_1=m(x-x_1)$$
$$y-1=-\dfrac{5}{4}(x-5)$$
$$y-1=-\dfrac{5}{4}x+\dfrac{25}{4}$$
$$y=-\dfrac{5}{4}x+\dfrac{25}{4}+1$$
$$y=-\dfrac{5}{4}x+\dfrac{29}{4}$$

31) $(x_1,\ y_1)=(-3,0);\ \ m=\dfrac{5}{6}$

$$y-y_1=m(x-x_1)$$
$$y-0=\dfrac{5}{6}(x-(-3))$$
$$y=\dfrac{5}{6}x+\dfrac{5}{2}$$
$$6\cdot y=6\left(\dfrac{5}{6}x+\dfrac{5}{2}\right)$$
$$6y=5x+15$$
$$-5x+6y=15$$
$$5x-6y=-15$$

33) Use the points to find the slope of the line, and then use the slope and either one of the points in the point-slope formula.

35) $m = \dfrac{-5-7}{3-(-1)} = \dfrac{-12}{4} = -3$

$(x_1, y_1) = (-1, 7)$

$y - y_1 = m(x - x_1)$

$y - 7 = -3(x - (-1))$

$y - 7 = (-3)(x + 1)$

$y = -3x - 3 + 7$

$y = -3x + 4$

37) $m = \dfrac{11-5}{7-4} = \dfrac{6}{3} = 2$

$(x_1, y_1) = (4, 5)$

$y - y_1 = m(x - x_1)$

$y - 5 = 2(x - 4)$

$y - 5 = 2x - 8$

$y = 2x - 3$

39) $m = \dfrac{3-4}{1-(-2)} = \dfrac{-1}{3} = -\dfrac{1}{3}$

$(x_1, y_1) = (-2, 4)$

$y - y_1 = m(x - x_1)$

$y - 4 = -\dfrac{1}{3}(x - (-2))$

$y - 4 = -\dfrac{1}{3}(x + 2)$

$y - 4 = -\dfrac{1}{3}x - \dfrac{2}{3}$

$y = -\dfrac{1}{3}x + \dfrac{10}{3}$

41) $m = \dfrac{-2-1}{4-(-5)} = \dfrac{-3}{9} = -\dfrac{1}{3}$

$(x_1, y_1) = (-5, 1)$

$y - y_1 = m(x - x_1)$

$y - 1 = -\dfrac{1}{3}(x - (-5))$

$y - 1 = -\dfrac{1}{3}x - \dfrac{5}{3}$

$3(y - 1) = 3\left(-\dfrac{1}{3}x - \dfrac{5}{3}\right)$

$3y - 3 = -x - 5$

$x + 3y = -2$

43) $m = \dfrac{-1-(-11)}{3-(-3)} = \dfrac{10}{6} = \dfrac{5}{3}$

$(x_1, y_1) = (-3, -1)$

$y - y_1 = m(x - x_1)$

$y - (-11) = \dfrac{5}{3}(x - (-3))$

$y + 11 = \dfrac{5}{3}x + 5$

$3(y + 11) = 3\left(\dfrac{5}{3}x + 5\right)$

$3y + 33 = 5x + 15$

$3y - 5x = -18$

$5x - 3y = 18$

45) $m = \dfrac{-13.9-8.3}{5.1-(-2.3)} = \dfrac{-22.2}{7.4} = -3.0$

$(x_1, y_1) = (-2.3, 8.3)$

$y - y_1 = m(x - x_1)$

$y - 8.3 = -3.0(x - (-2.3))$

$y - 8.3 = -3.0x - 6.9$

$y = -3.0x + 1.4$

47) $m = \dfrac{-4-(-1)}{(-4)-0} = \dfrac{-3}{-4} = \dfrac{3}{4}$

$y = mx + b \quad m = \dfrac{3}{4}; b = -1$

$y = \dfrac{3}{4}x - 1$

49) $m = \dfrac{-7-2}{1-(-2)} = \dfrac{-9}{3} = -3$

$(x_1, y_1) = (-2, 2)$

$y - y_1 = m(x - x_1)$

$y - 2 = -3(x - (-2))$

$y - 2 = -3x - 6$

$y = -3x - 4$

51) $y = 3$

53) $m = \dfrac{-1-7}{2-(-4)} = \dfrac{-8}{6} = -\dfrac{4}{3}$

$(x_1, y_1) = (-4, 7)$

$y - y_1 = m(x - x_1)$

$y - 7 = -\dfrac{4}{3}(x - (-4))$

$y - 7 = -\dfrac{4}{3}(x + 4)$

$y - 7 = -\dfrac{4}{3}x - \dfrac{16}{3}$

$y = -\dfrac{4}{3}x - \dfrac{16}{3} + 7$

$y = -\dfrac{4}{3}x - \dfrac{16}{3} + \dfrac{21}{3}$

$y = -\dfrac{4}{3}x + \dfrac{5}{3}$

55) $m = 1$

$(x_1, y_1) = (3, 5)$

$y - y_1 = m(x - x_1)$

$y - 5 = 1(x - 3)$

$y - 5 = x - 3$

$y = x + 2$

57) $m = 7; b = 6$

$y = mx + b$

$y = 7x + 6$

59) Vertical Line

$(c, d) = (3, 5)$

$x = c$

$x = 3$

61) Horizontal Line

$(c, d) = (2, 3)$

$y = d$

$y = 3$

63) $m = -4; b = -4$

$y = mx + b$

$y = -4x - 4$

65) $m = -3$

$(x_1, y_1) = (10, -10)$

$y - y_1 = m(x - x_1)$

$y - (-10) = -3(x - 10)$

$y + 10 = -3x + 30$

$y = -3x + 20$

67) $m = \dfrac{-1-(-4)}{2-(-4)} = \dfrac{3}{6} = \dfrac{1}{2}$

$(x_1, y_1) = (-4, -4)$

$y - y_1 = m(x - x_1)$

$y - (-4) = \dfrac{1}{2}(x - (-4))$

$y + 4 = \dfrac{1}{2}x + 2$

$y = \dfrac{1}{2}x - 2$

69) They have the same slopes and different y-intercepts.

71) $y = mx + b \qquad m = 4; \; b = 2$

$y = 4x + 2$

73) $(x_1, y_1) = (-1, -4); \; m = 4$

$y - y_1 = m(x - x_1)$

$y - (-4) = 4(x - (-1))$

$y + 4 = 4x + 4$

$y = 4x$

$4x - y = 0$

75) Determine the slope.

$x + 2y = 22$

$2y = -x + 22$

$y = -\dfrac{1}{2}x + 11$

$(x_1, y_1) = (-4, 7); \; m = -\dfrac{1}{2}$

$y - y_1 = m(x - x_1)$

$y - 7 = -\dfrac{1}{2}(x - (-4))$

$y - 7 = -\dfrac{1}{2}(x + 4)$

$2(y - 7) = 2\left(-\dfrac{1}{2}x - 2\right)$

$2y - 14 = -x - 4$

$x + 2y = 10$

77) Determine the slope.

$15x - 3y = 1$

$-3y = -15x + 1$

$y = 5x - \dfrac{1}{3}$

$(x_1, y_1) = (-2, -12); \; m = 5$

$y - y_1 = m(x - x_1)$

$y - (-12) = 5(x - (-2))$

$y + 12 = 5x + 10$

$y = 5x - 2$

79) $(x_1, y_1) = (4, 2); \; m_{\text{perp}} = \dfrac{3}{2}$

$y - y_1 = m(x - x_1)$

$y - 2 = \dfrac{3}{2}(x - 4)$

$y = \dfrac{3}{2}x - 6$

$y = \dfrac{3}{2}x - 4$

81) $(x_1, y_1) = (10, 0)$; $m_{perp} = \dfrac{1}{5}$

$$y - y_1 = m(x - x_1)$$

$$y - 0 = \dfrac{1}{5}(x - 10)$$

$$5 \cdot y = 5\left(\dfrac{1}{5}x - 2\right)$$

$$5y = x - 10$$

$$-x + 5y = -10$$

$$x - 5y = 10$$

83) $(x_1, y_1) = (4, -9)$; $m_{perp} = -1$

$$y - y_1 = m(x - x_1)$$

$$y - (-9) = -1(x - 4)$$

$$y + 9 = -x + 4$$

$$y = -x - 5$$

85) Determine the slope.

$$x + 3y = 18$$

$$3y = -x + 18;$$

$$y = -\dfrac{1}{3}x + 6$$

$$m = -\dfrac{1}{3}$$

$(x_1, y_1) = (4, 2)$; $m_{perp} = 3$

$$y - y_1 = m(x - x_1)$$

$$y - 2 = 3(x - 4)$$

$$y - 2 = 3x - 12$$

$$y = 3x - 10$$

$$y - 3x = -10$$

$$3x - y = 10$$

87) Determine the slope.

$$3x + y = 8$$

$$y = -3x + 8$$

$$m = -3$$

$(x_1, y_1) = (-4, 0)$; $m = -3$

$$y - y_1 = m(x - x_1)$$

$$y - 0 = -3(x - (-4))$$

$$y = -3x - 12$$

89) Determine the slope.

$$y = x - 2$$

$$m = 1$$

$(x_1, y_1) = (2, 9)$; $m_{perp} = -1$

$$y - y_1 = m(x - x_1)$$

$$y - 9 = -1(x - 2)$$

$$y - 9 = -x + 2$$

$$y = -x + 11$$

91) Determine the slope.

$$y = 1$$

$$m = 0$$

$(x_1, y_1) = (-3, 4)$; $m = 0$

$$y - y_1 = m(x - x_1)$$

$$y - 4 = 0(x - (-3))$$

$$y - 4 = 0$$

$$y = 4$$

93) Determine the slope.

$$x = 0$$

$$m = \text{undefined}$$

$(x_1, y_1) = (9, 2)$; $m_{perp} = 0$

$$y - y_1 = m(x - x_1)$$

$$y - 2 = 0(x - 9)$$

$$y - 2 = 0$$

$$y = 2$$

Chapter 4: Linear Equations in Two Variables

95) Determine the slope.

$$21x - 6y = 2$$
$$-6y = -21x + 2$$
$$\frac{-6y}{-6} = \frac{-21}{-6}x + \frac{2}{-6}$$
$$y = \frac{7}{2}x - \frac{1}{3} \qquad m = \frac{7}{2}$$

$$(x_1, y_1) = (4, -1); \quad m_{\text{perp}} = -\frac{2}{7}$$

$$y - y_1 = m(x - x_1)$$
$$y - (-1) = -\frac{2}{7}(x - 4)$$
$$y + 1 = -\frac{2}{7}x + \frac{8}{7}$$
$$y = -\frac{2}{7}x + \frac{1}{7}$$

97) Determine the slope.

$$y = 0$$
$$m = 0$$

$$(x_1, y_1) = \left(4, -\frac{3}{2}\right); \quad m = 0$$

$$y - y_1 = m(x - x_1)$$
$$y - \left(-\frac{3}{2}\right) = 0(x - 4)$$
$$y + \frac{3}{2} = 0$$
$$y = -\frac{3}{2}$$

99) a) $2005 : (0, 81150);$
$2008 : (3, 94960)$

$$m = \frac{94{,}960 - 81{,}150}{3 - 0}$$
$$= \frac{13{,}810}{3} = 4603.\overline{3}$$

$$b = 81{,}150$$
$$y = mx + b$$
$$y = 4603.\overline{3}x + 81{,}150$$

b) The average salary of a mathematician is increasing by \$4603.$\overline{3}$ per year.

c) $y = 4603.3x + 81{,}150$

In 2014, $x = 15$

$$y = 4603.\overline{3}(9) + 81{,}150$$
$$y = 41{,}430 + 81{,}150$$
$$y = \$122{,}580$$

101) a) $m = -15000$

The year 2007 corresponds to $x = 0$.

Then 500,000 corresponds to $y = 500{,}000$.

A point on the line is $(0, 500000) = (x_1, y_1)$.

$$y - y_1 = m(x - x_1)$$
$$y - 500{,}000 = -15000(x - 0)$$
$$y = -15000x + 500{,}000$$

b) $y = mx + b$
$$y = -1.5x + 211$$

The budget is cut by \$15,000 per year.

c) $y = -15000x + 500,000$

In 2010, $x = 3$

$y = -15000(3) + 500,000$

$y = -45,000 + 500,000$

$= \$455,000$

The budget in 2010 \$455,000.

d) $y = -15000x + 500,000$

$365,000 = -15000(x) + 500,000$

$365,000 - 500,000 = -15,000x$

$-135,000 = -15,000x$

$\dfrac{-135000}{-15000} = x$

$9 = x$

The year is 2007+9=2016.

103) a) $(0, 100)$; $m = 8$ g; $b = 100$

$y = mx + b$

$y = 8x + 100$

b) A kitten gains about 8 g per day.

c) $y = 8x + 100$

$x = 5$

$y = 8(5) + 100$

$y = 40 + 100$

$y = 140$ g

x=2 weeks= 14 days

$y = 8(14) + 100$

$y = 112 + 100$

$y = 212$ g

d) $y = mx + b$

$284 = 8x + 100$

$184 = 8x$

$23 = x$

105) a) $(6, 38)$; $(8.5, 42)$

$m = \dfrac{42 - 38}{8.5 - 6} = \dfrac{4}{2.5} = 1.6;$

$y - y = m(x - x_1)$

$y = E;\ x = A$

$E - 38 = 1.6(x - 6)$

$E - 38 = 1.6A - 9.6$

$E = 1.6x + 28.4$

Section 4.6 Exercises

1) a) any set of ordered pairs

b) Answers may vary.

c) Answers may vary.

3) Domain: $\{-8, -2, 1, 5\}$

Range: $\{-3, 4, 6, 13\}$

Function

5) Domain: $\{1, 9, 25\}$

Range: $\{-3, -1, 1, 5, 7\}$

Not a function

7) Domain: $\{-1, 2, 5, 8\}$

Range: $\{-7, -3, 12, 19\}$

Not a function

9) Domain: $(-\infty, \infty)$

Range: $(-\infty, \infty)$

Function

11) Domain: $(-\infty, 4]$

Range: $(-\infty, \infty)$

Not a function

13) Domain: $(-\infty, \infty)$

    Range: $(-\infty, 6]$

    Function

15) yes

17) yes

19) no

21) no

23) Domain: $(-\infty, \infty)$; Function

25) Domain: $(-\infty, \infty)$; Function

27) Domain: $[0, \infty)$; Not a function

29) $x \neq 0$

    Domain: $(-\infty, 0) \cup (0, \infty)$;

    Function

31) $x + 4 = 0$

    $x = -4$

    Domain: $(-\infty, -4) \cup (-4, \infty)$;

    Function

33) $x - 5 = 0$

    $x = 5$

    Domain: $(-\infty, 5) \cup (5, \infty)$;

    Function

35) $5x - 3 = 0$

    $5x = 3$

    $x = \dfrac{3}{5}$

    Domain: $\left(-\infty, \dfrac{3}{5}\right) \cup \left(\dfrac{3}{5}, \infty\right)$;

    Function

37) $3x + 4 = 0$

    $3x = -4$

    $x = -\dfrac{4}{3}$

    Domain: $\left(-\infty, -\dfrac{4}{3}\right) \cup \left(-\dfrac{4}{3}, \infty\right)$;

    Function

39) $9 - 3x = 0$

    $-3x = -9$

    $x = 3$

    Domain: $(-\infty, 3) \cup (3, \infty)$;

    Function

41) Domain: $(-\infty, \infty)$; Function

43) y is a function, and y is a

    function of x.

45) a) $y = 5(3) - 8$ b) $f(3) = 5(3) - 8$

        $y = 15 - 8$ $f(3) = 15 - 8$

        $y = 7$ $f(3) = 7$

47) $f(5) = -4(5) + 7$

    $f(5) = -20 + 7$

    $f(5) = -13$

49) $f(0) = -4(0) + 7$

    $f(0) = 0 + 7$

    $f(0) = 7$

51) $g(4) = (4)^2 + 9(4) - 2$

    $g(4) = 16 + 36 - 2$

    $g(4) = 50$

53) $g(-1)=(-1)^2+9(-1)-2$

$g(-1)=1-9-2$

$g(-1)=-10$

55) $g\left(-\dfrac{1}{2}\right)=\left(-\dfrac{1}{2}\right)^2+9\left(-\dfrac{1}{2}\right)-2$

$g\left(-\dfrac{1}{2}\right)=\dfrac{1}{4}-\dfrac{9}{2}-2=\dfrac{1}{4}-\dfrac{18}{4}-\dfrac{8}{4}$

$g\left(-\dfrac{1}{2}\right)=-\dfrac{25}{4}$

57) $f(6)=-4(6)+7$

$f(6)=-24+7$

$f(6)=-17$

$g(6)=(6)^2+9(6)-2$

$g(6)=36+54-2$

$g(6)=90-2$

$g(6)=88$

$f(6)-g(6)=-17-88=-105$

59) $f(-1)=10,\ f(4)=-5$

61) $f(-1)=6,\ f(4)=2$

63) $f(-1)=7,\ f(4)=3$

65) $10=-3x-2$

$12=-3x$

$-4=x$

67) $5=\dfrac{2}{3}x+1$

$4=\dfrac{2}{3}x$

$6=x$

69) $f(x)=4x-5$

$f(k+6)=4(k+6)-5$ Substitute

$k+6$ for x.

$=4k+24-5$ Distribute.

$=4k+19$ Simplify.

71) a) $f(c)=-7(c)+2=-7c+2$

b) $f(t)=-7(t)+2=-7t+2$

c) $f(a+4)=-7(a+4)+2$

$f(a+4)=-7a-28+2$

$f(a+4)=-7a-26$

d) $f(z-9)=-7(z-9)+2$

$f(z-9)=-7z+63+2$

$f(z-9)=-7z+65$

e) $g(k)=(k)^2-5(k)+12$

$g(k)=k^2-5k+12$

f) $g(m)=(m)^2-5(m)+12$

$g(m)=m^2-5m+12$

g) $f(x+h)=-7(x+h)+2$

$=-7x-7h+2$

h)

$f(x+h)-f(x)=-7(x+h)+2-(-7x+2)$

$=-7x-7h+2+7x-2$

$=-7h$

73) $f(x) = x - 4$

$f(0) = (0) - 4$

$f(0) = -4$

x	$f(x)$
0	-4
1	-3
3	-1

$f(1) = (1) - 4$

$f(1) = -3$

$f(3) = (3) - 4$

$f(3) = -1$

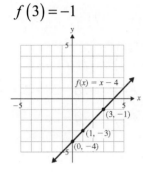

75) $f(x) = \dfrac{2}{3}x + 2$

$f(-3) = \dfrac{2}{3}(-3) + 2$

$f(-3) = -2 + 2 = 0$

x	$f(x)$
-3	0
0	2
3	4

$f(0) = \dfrac{2}{3}(0) + 2$

$f(0) = 0 + 2 = 2$

$f(3) = \dfrac{2}{3}(3) + 2$

$f(3) = 2 + 2 = 4$

77) $h(x) = -3$

$h(-2) = -3$

x	$h(x)$
-2	-3
0	-3
1	-3

$h(0) = -3$

$h(1) = -3$

79) $g(x) = 3x + 3$

x-int: Let $g(x) = 0$, and solve for x.

$0 = 3x + 3$

$-3 = 3x$

$-1 = x$ $\qquad$ $(-1, 0)$

y-int: Let $x = 0$, and find $g(0)$.

$g(0) = 3(0) + 3$

$g(0) = 3$ $\qquad$ $(0, 3)$

Let $x = -2$.

$g(-2) = 3(-2) + 3$

$g(-2) = -6 + 3 = -3$ $\qquad$ $(-2, -3)$

118

81) $f(x) = -\frac{1}{2}x + 2$

x-int: Let $f(x) = 0$, and solve for x.

$$0 = -\frac{1}{2}x + 2$$

$$-2 = -\frac{1}{2}x$$

$$4 = x \qquad (4, 0)$$

y-int: Let $x = 0$, and find $f(0)$.

$$f(0) = -\frac{1}{2}(0) + 2$$

$$f(0) = 2 \qquad (0, 2)$$

Let $x = 2$.

$$f(2) = -\frac{1}{2}(2) + 2$$

$$f(2) = -1 + 2 = 1 \qquad (2, 1)$$

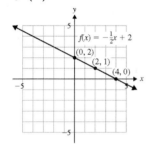

83) $h(x) = x$

x-int: Let $h(x) = 0$, and solve for x.

$$0 = x \qquad (0, 0)\text{-also } y\text{-int}$$

Let $x = -2$.

$$h(-2) = (-2) = -2 \qquad (-2, -2)$$

Let $x = 1$.

$$h(1) = (1) = 1 \qquad (1, 1)$$

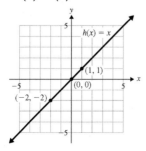

85) $m = -4$, y-int: $(0, -1)$

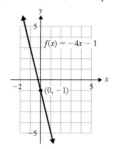

87) $m = \frac{3}{5}$, y-int: $(0, -2)$

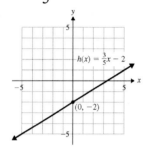

89) $m = 2$, y-int: $\left(0, \frac{1}{2}\right)$

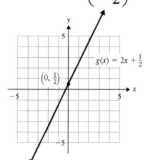

91)

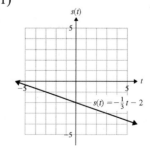

93)

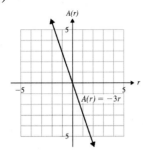

95) a) $D(2) = 54(2)$

$D(2) = 108$

The truck will travel 108 miles in 2 hours.

b) $135 = 54t$

$2.5 = t$

In 2.5 hours the truck will travel 135 miles.

c)

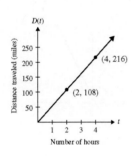

97) a) $E(10) = 7.50(10)$

$E(10) = 75.0$

When Jenelle works for 10 hr, she earns $75.00.

b) $210 = 7.50t$

$28 = t$

For Jenelle to earn $210.00, she must work 28 hr.

99) a) $D(12) = 21.13(12)$

$D(12) = 253.56$

253.56 MB can be recorded in 12 seconds.

b) $D(60) = 21.13(60)$

$D(60) = 1267.8$

1267.8 MB can be recorded in 1 minute.

c) $422.6 = 21.13t$

$20 = t$

In 20 seconds 422.6 MB can be recorded.

d)

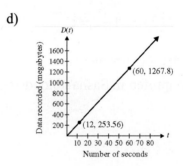

101) a) $F(2) = 30(2)$

$F(2) = 60$

60,000 fingerprints can be compared in 2 seconds.

b) $105 = 30s$

$3.5 = s$

It would take AFIS 3.5 seconds.

103) a) $S(50) = 44,100(50)$

$S(50) = 2,205,000$

After 50 sec the CD player reads
2,205,000 samples of sound.

b) $2,646,000 = 44,100t$

$60 = t$

The CD player reads 2,646,000
samples of sound in 60 sec
(or 1 minute).

104) a) 2 hours; 400 mg
  b) after about 30 min and
    after 6 hours
  c) 200 mg
  d) $A(8) = 50$

After 8 hours there are 50 mg of
ibuprofen in Sasha's bloodstream

Chapter 4 Review

1) Yes

$5x - y = 13$

$5(2) - (-3) = 13$

$10 + 3 = 13$

$13 = 13$

3) Yes

$y = -\dfrac{4}{3}x + \dfrac{7}{3}$

$(-3) = -\dfrac{4}{3}(4) + \dfrac{7}{3}$

$-3 = -\dfrac{16}{3} + \dfrac{7}{3}$

$-3 = \dfrac{-9}{3}$

$-3 = -3$

5) $y = -2x + 4$

$y = -2(-5) + 4$

$y = 10 + 4$

$y = 14$

$(-5,\ 14)$

7) 7) $y = -9$

$(7,\ -9)$

$(6,\ 12)$

9) $y = x - 14$

$y = (0) - 14$ \quad $y = (-3) - 14$

$y = -14$ \quad\quad $y = -17$

$y = (6) - 14$ \quad $y = (-8) - 14$

$y = -8$ \quad\quad $y = -22$

x	y
0	−14
6	−8
−3	−17
−8	−22

11)

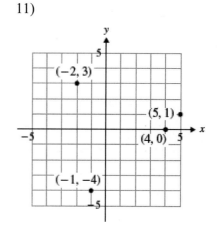

13) a) $y = 0.50x + 45.00$

$y = 0.50(10) + 45.00$ $y = 0.50(29) + 45.00$

$y = 5.00 + 45.00$ $y = 14.5 + 45.00$

$y = 50$ $y = 59.5$

$y = 0.50(18) + 45.00$ $y = 0.50(36) + 45.00$

$y = 9.00 + 45.00$ $y = 18.00 + 45.00$

$y = 54$ $y = 63$

x	y
10	50
18	54
29	59.50
36	63

$(10, 50), (18, 54),$
$(29, 59.50), (36, 63)$

b)

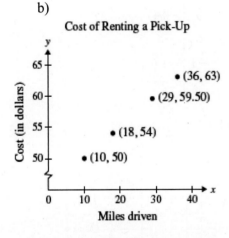

Cost of Renting a Pick-Up

c) The cost of renting the pick-up is
$74.00 if it is driven 58 miles.

15) $y = -2x + 4$

$y = -2(0) + 4$ $y = -2(2) + 4$

$y = 0 + 4$ $y = -4 + 4$

$y = 4$ $y = 0$

$y = -2(1) + 4$ $y = -2(3) + 4$

$y = -2 + 4$ $y = -6 + 4$

$y = 2$ $y = -2$

x	y
0	4
1	2
2	0
3	-2

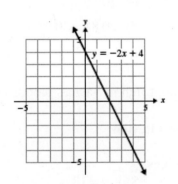

17) $x - 2y = 2$

x-int: Let $y = 0$, and solve for x.

$x - 2(0) = 2$

$x - 0 = 2$

$x = 2$ $(2, 0)$

y-int: Let $x = 0$, and solve for y.

$(0) - 2y = 2$

$-2y = 2$

$y = -1$ $(0, -1)$

Let $x = 4$.

$(4) - 2y = 2$

$-2y = -2$

$y = 1$ $(4, 1)$

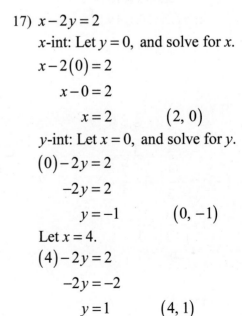

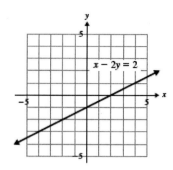

19) $y = -\dfrac{1}{2}x + 1$

x-int: Let $y = 0$, and solve for *x*.

$(0) = -\dfrac{1}{2}x + 1$

$-1 = -\dfrac{1}{2}x$

$2 = x$ $(2, 0)$

y-int: Let $x = 0$, and solve for *y*.

$y = -\dfrac{1}{2}(0) + 1$

$y = 0 + 1$

$y = 1$ $(0, 1)$

Let $x = 4$.

$y = -\dfrac{1}{2}(4) + 1$

$y = -2 + 1$

$y = -1$ $(4, -1)$

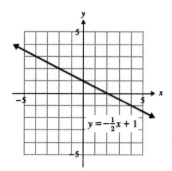

21) $y = 4$

     $(0, 4), (2, 4), (-1, 4)$

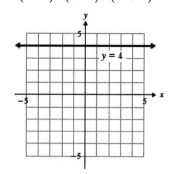

23) $m = \dfrac{\text{rise}}{\text{run}} = \dfrac{3}{2}$

25) $(x_1, y_1) = (5, 8)$

     $(x_2, y_2) = (1, -12)$

     $m = \dfrac{y_2 - y_1}{x_2 - x_1} = \dfrac{-12 - 8}{1 - 5} = \dfrac{-20}{-4} = 5$

27) $(x_1, y_1) = (-7, -2)$

     $(x_2, y_2) = (2, 4)$

     $m = \dfrac{y_2 - y_1}{x_2 - x_1} = \dfrac{4 - (-2)}{2 - (-7)} = \dfrac{6}{9} = \dfrac{2}{3}$

29) $(x_1, y_1) = \left(-\dfrac{1}{4}, 1\right)$

     $(x_2, y_2) = \left(\dfrac{3}{4}, -6\right)$

     $m = \dfrac{y_2 - y_1}{x_2 - x_1} = \dfrac{-6 - 1}{\dfrac{3}{4} - \left(-\dfrac{1}{4}\right)} = \dfrac{-7}{1} = -7$

31) $(x_1, y_1) = (-2, 5)$ $(x_2, y_2) = (4, 5)$

     $m = \dfrac{y_2 - y_1}{x_2 - x_1} = \dfrac{5 - 5}{4 - (-2)} = \dfrac{0}{6} = 0$

     Slope $= 0$

33) a) In 1975, Christine paid $4.00
        for the album.

    b) The slope is positive, so the
        value of the album is increasing
        over time.

    c) $m = \dfrac{34-4}{2005-1975} = \dfrac{30}{30} = 1.$
        The value of the

    album is increasing by $1.00 per year.

35)

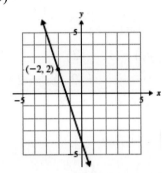

37)

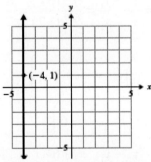

39) $m = -1$, y-int $= (0, 5)$

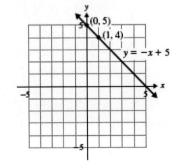

41) $m = \dfrac{2}{5}$, y-int $= (0, -6)$

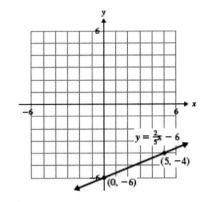

43) $x + 3y = -6$
     $3y = -x - 6$

     $y = -\dfrac{1}{3}x - 2$

     $m = -\dfrac{1}{3}$, y-int $= (0, -2)$

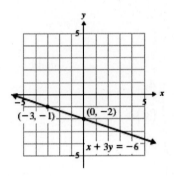

45) $x + y = 0$
     $y = -x$

     $m = -1$, y-int :$(0, 0)$

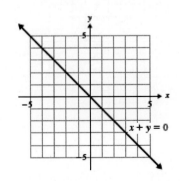

47) a) $(0, 197.6)$; In 2003 the squash crop was worth about \$197.6 million.

b) It has been increasing by \$7.9 million per year.

c) Estimate: \$213 million.
$$y = 7.9x + 197.6$$
$$y = 7.9(2) + 197.6$$
$$y = 15.8 + 197.6$$
$$y = 213.4$$
\$213.4 million

49) $x - 4y = 20$ $\qquad$ $-x + 4y = 6$
$$-4y = -x + 20 \qquad 4y = x + 6$$
$$y = \frac{1}{4}x - 5 \qquad y = \frac{1}{4}x + \frac{3}{2}$$
$$m = \frac{1}{4} \qquad\qquad m = \frac{1}{4}$$
parallel

51) $x = 7$ $\qquad\qquad$ $y = -3$
$m = $ undefined $\qquad$ $m = 0$
perpendicular

53) $(x_1, y_1) = (-1, 4)$; $m = 6$
$$y - y_1 = m(x - x_1)$$
$$y - 4 = 6(x - (-1))$$
$$y - 4 = 6x + 6$$
$$y = 6x + 10$$

55) $y = mx + b \qquad m = -\frac{3}{4}; b = 7$
$$y = -\frac{3}{4}x + 7$$

57) $m = \dfrac{-3-1}{6-4} = \dfrac{-4}{2} = -2$
$$(x_1, y_1) = (4, 1)$$
$$y - y_1 = m(x - x_1)$$
$$y - 1 = -2(x - 4)$$
$$y - 1 = -2x + 8$$
$$y = -2x + 9$$

59) Horizontal Line
$$(c, d) = (3, 7)$$
$$y = d$$
$$y = 7$$

61) $m = \dfrac{-10-5}{-1-4} = \dfrac{-15}{-5} = 3$
$$(x_1, y_1) = (4, 5)$$
$$y - y_1 = m(x - x_1)$$
$$y - 5 = 3(x - 4)$$
$$y - 5 = 3x - 12$$
$$y = 3x - 7$$
$$-3x + y = -7$$
$$3x - y = 7$$

63) $(x_1, y_1) = \left(1, -\dfrac{3}{2}\right)$; $m = \dfrac{5}{2}$
$$y - y_1 = m(x - x_1)$$
$$y - \left(-\frac{3}{2}\right) = \frac{5}{2}(x - 1)$$
$$y + \frac{3}{2} = \frac{5}{2}x - \frac{5}{2}$$
$$2y + 3 = 5x - 5$$
$$2y = 5x - 8$$
$$2y - 5x = -8$$
$$5x - 2y = 8$$

65) $y = mx + b \quad m = -4; \; b = 0$

$\quad y = (-4)x - 0$

$4x + y = 0$

67) $m = \dfrac{5-1}{2-6} = \dfrac{4}{-4} = -1$

$(x_1, y_1) = (6, 1)$

$y - y_1 = m(x - x_1)$

$y - 1 = -1(x - 6)$

$y - 1 = -x + 6$

$\quad y = -x + 7$

$x + y = 7$

69) a) $2005 : (62,000); \; 2010 : (79,500)$

$m = \dfrac{79,500 - 62,000}{2010 - 2005} = \dfrac{17,500}{5}$

$m = 3500; \; b = 62,000$

$y = mx + b$

$y = 3500x + 62,000$

b) Mr. Romanki's salary is increasing by $3500 per year.

c) $y = 3500x + 62000$

In 2008, $x = 3$

$y = 3500(3) + 62000$

$y = 10,500 + 62,000$

$y = 72,500$

He earned $72,500.

d) $y = 3500x + 62000$

$93,500 = 3500x + 62,000$

$31,500 = 3500x$

$x = 9$

He expects to earn $93,500 in 2014.

71) $(x_1, y_1) = (-1, 14); \; m = -8$

$y - y_1 = m(x - x_1)$

$y - 14 = -8(x - (-1))$

$y - 14 = -8x - 8$

$\quad y = -8x + 6$

73) Determine the slope.

$x - 2y = 6$

$-2y = -x + 6$

$y = \dfrac{1}{2}x - 3$

$(x_1, y_1) = (4, 11); \; m = \dfrac{1}{2}$

$y - y_1 = m(x - x_1)$

$y - 11 = \dfrac{1}{2}(x - 4)$

$2(y - 11) = 2\left(\dfrac{1}{2}x - 2\right)$

$2y - 22 = x - 4$

$-x + 2y = 18$

$x - 2y = -18$

75) Determine the slope.

$x + 5y = 10$

$5y = -x + 10$

$y = -\dfrac{1}{5}x + 2$

$(x_1, y_1) = (15, 7); \; m = -\dfrac{1}{5}$

$y - y_1 = m(x - x_1)$

$y - 7 = -\dfrac{1}{5}(x - 15)$

$y - 7 = -\dfrac{1}{5}x + 3$

$y = -\dfrac{1}{5}x + 10$

77) $(x_1,\, y_1)=(3,\,-9);\ m_{perp}=1$

$\quad y-y_1=m(x-x_1)$

$\quad y-(-9)=1(x-3)$

$\quad y+9=x-3$

$\quad y=x-12$

79) Determine the slope.

$\quad 2x+3y=-3$

$\quad 3y=-2x-3$

$\quad y=-\dfrac{2}{3}x-1$

$\quad (x_1,\, y_1)=(-4,\,-4);\ m_{perp}=\dfrac{3}{2}$

$\quad y-y_1=m(x-x_1)$

$\quad y-(-4)=\dfrac{3}{2}(x-(-4))$

$\quad y+4=\dfrac{3}{2}x+6$

$\quad y=\dfrac{3}{2}x+2$

81) $y=4$

83) Domain: $\{-3,5,\,12\}$

Range: $\{-3,1,\,3,4\}$

Not a function

85) Domain: $\{$Beagle, Siamese, Parrot$\}$

Range: $\{$Dog, Cat, Bird$\}$

Function

87) Domain: $[0,4]$

Range: $[0,2]$

Not a function

89) $x+3=0$

$\quad x=-3$

Domain: $(-\infty,\,-3)\cup(-3,\,\infty)$

Function

91) Domain: $[0,\infty)$ not a function

93) $7x-2=0$

$\quad 7x=2$

$\quad x=\dfrac{7}{2}$

Domain: $\left(-\infty,\,\dfrac{7}{2}\right)\cup\left(\dfrac{7}{2},\,\infty\right)$

Function

95) $f(3)=27;\ f(-2)=-8$

97) a) $f(4)=5(4)-12=20-12=8$

b) $f(-3)=5(-3)-12$

$\quad f(-3)=-15-12=-27$

c) $g(3)=(3)^2+6(3)+5$

$\quad g(3)=9+18+5=32$

d) $g(0)=(0)^2+6(0)+5$

$\quad g(0)=0+0+5=5$

e) $f(a)=5(a)-12=5a-12$

f) $g(t)=(t)^2+6(t)+5$

$\quad g(t)=t^2+6t+5$

g) $f(k+8)=5(k+8)-12$

$\quad f(k+8)=5k+40-12$

$\quad f(k+8)=5k+28$

h) $f(c-2)=5(c-2)-12$

$\quad f(c-2)=5c-10-12$

$\quad f(c-2)=5c-22$

i) $f(x+h)=5(x+h)-12$

$\quad f(x+h)=5x+5h-12$

j)

$f(x+h)-f(x)=5(x+h)-12-(5x-12)$

$\qquad\qquad\qquad = 5x+5h-12-5x+12$

$\qquad\qquad\qquad = 5h$

99) $\dfrac{11}{2}=\dfrac{3}{2}x+5$

$\quad \dfrac{1}{2}=\dfrac{3}{2}x$

$\quad \dfrac{1}{3}=x$

101) a) $f(x)=\dfrac{2}{3}x-1$

x-int: Let $f(x)=0,$ and solve for $x.$

$\quad 0=\dfrac{2}{3}x-1$

$\quad 1=\dfrac{2}{3}x$

$\quad \dfrac{3}{2}=x \qquad \left(\dfrac{3}{2},0\right)$

y-int: Let $x=0,$ and find $f(0).$

$\quad f(0)=\dfrac{2}{3}(0)-1$

$\quad f(0)=0-1$

$\quad f(0)=-1 \qquad (0,-1)$

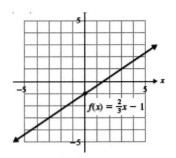

b) $f(x)=-3x+2$

x-int: Let $f(x)=0,$ and solve for $x.$

$\quad 0=-3x+2$

$\quad -2=-3x$

$\quad \dfrac{2}{3}=x \qquad \left(\dfrac{2}{3},0\right)$

y-int: Let $x=0,$ and find $f(0).$

$\quad f(0)=-3(0)+2$

$\quad f(0)=2 \qquad (0,2)$

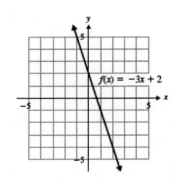

103)

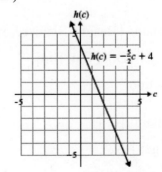

128

105) a) $f(2) = 480(2)$

   $f(2) = 960$ MB

   $f(6) = 480(6)$

   $f(6) = 2880$ MB

   b) $1200 = 480t$

   $2.5 = t$ 2.5 sec

Chapter 4 Test

1) Yes

$$2x - 7y = 8$$
$$2(-3) - 7(-2) = 8$$
$$-6 + 14 = 8$$
$$8 = 8$$

3) Positive; Negative.

5)

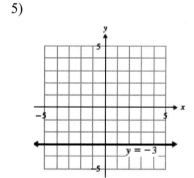

7) $m = \dfrac{5}{4}, b = 0$

   a) $(x_1, y_1) = (3, -1)$

   $(x_2, y_2) = (-5, 9)$

   $m = \dfrac{y_2 - y_1}{x_2 - x_1} = \dfrac{9 - (-1)}{-5 - 3}$

   $= \dfrac{10}{-8} = -\dfrac{5}{4}$

b) $(x_1, y_1) = (8, 6)$

   $(x_2, y_2) = (11, 6)$

   $m = \dfrac{y_2 - y_1}{x_2 - x_1} = \dfrac{6 - 6}{11 - 8}$

   $= \dfrac{0}{8} = 0$

9)

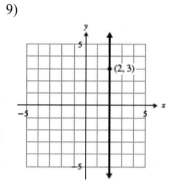

11) $y = mx + b$ $m = 7; b = -10$

   $y = 7x - 10$

13) $4x + 18y = 9$

   $9x - 2y = -6$

   $18y = -4x + 9$ $-2y = -9x - 6$

   $y = -\dfrac{4}{18}x + \dfrac{9}{18}$ $y = \dfrac{-9}{-2}x - \dfrac{6}{-2}$

   $y = -\dfrac{2}{9}x + \dfrac{1}{2}$ $y = \dfrac{9}{2}x + 3$

   $m = -\dfrac{2}{9}$ $m = \dfrac{9}{2}$

perpendicular

15) a) $2007 : (2, 399)$

   The number in 2007 was 399.

b) $2005:(0,\ 419);\quad 2010:(5,\ 374)$

$$m=\frac{374-419}{5-0}=\frac{-45}{5}=-9;\ \ b=419$$

$$y=mx+b$$

$$y=-9x+419$$

c) $y=-9x+419$

$$y=-9(2)+419$$

$$=-18+419$$

$$=401$$

According to the equation, 401 students attended the school in 2007.

The actual number was 399.

d) $m=-9$

The school is losing 9 students per year.

e) y-int:$(0,\ 419)$

In 2005, 419 students attended this school.

f) $y=-9x+419;\ 2013:x=8$

$$y=-9(8)+419$$

$$=-72+419$$

$$=347$$

347 children are expected to attend in 2013.

17) Domain: $\{-2,\ 1,\ 3,8\}$

Range: $\{-5,-1,\ 1,4\}$

Function

19) a) $(-\infty,\infty)$ b) yes

21) $f(2)=-3$

23) $f(6)=-4(6)+2=-24+2=-22$

25) $g(t)=(t)^{2}-3(t)+7=t^{2}-3t+7$

27)

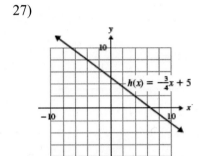

Cumulative Review: Chapters 1 - 4

1) $\dfrac{336}{792}=\dfrac{336\div24}{792\div24}=\dfrac{14}{33}$

3) $-3^{4}=-81$

5) $\dfrac{3}{8}-2=\dfrac{3}{8}-\dfrac{16}{8}=-\dfrac{13}{8}$

7) $2(17)-9;\ 34-9;\ 25$

9) $\left(\dfrac{30w^{5}}{15w^{-3}}\right)^{-4}=\left(\dfrac{15w^{-3}}{30w^{5}}\right)^{4}=\left(\dfrac{1}{2w^{8}}\right)^{4}=\dfrac{1}{16w^{32}}$

11) $\frac{3}{2}(7c-5)-1=\frac{2}{3}(2c+1)$

$6\cdot\frac{3}{2}(7c-5)-6\cdot1=6\cdot\frac{2}{3}(2c+1)$

$9(7c-5)-6=4(2c+1)$

$63c-45-6=8c+4$

$63c-51=8c+4$

$55c=55$

$c=1$

The solution set is $\{1\}$.

13) $3x+14\le7x+4$

$\quad\quad 3x\le7x-10$

$\quad -4x\le-10$

$\quad\quad x\ge\frac{-10}{-4}$

$\quad\quad x\ge\frac{5}{2}$

$\quad\quad \left[\frac{5}{2},\infty\right)$

15) $m\angle A=x°\quad m\angle B=(5x-14)°$

$m\angle C=20°$

$m\angle A\ +\ m\angle B\ +\ m\angle C\ =180$

$\quad x\ +\ 5x-14\ +\ \ 20\ =180$

$\quad\quad\quad\quad\quad\quad 6x+6=180$

$\quad\quad\quad\quad\quad\quad 6x=174$

$\quad\quad\quad\quad\quad\quad x=29$

$m\angle A=x°=29°$

$m\angle B=(5x-14)°=\left[5(29)-14\right]°=131°$

17) $(x_1,y_1)=(-7,8)$

$(x_2,y_2)=(2,17)$

$m=\frac{y_2-y_1}{x_2-x_1}=\frac{17-8}{2-(-7)}=\frac{9}{9}=1$

19) $(x_1,y_1)=(-8,1);\ \ m=-\frac{5}{4}$

$y-y_1=m(x-x_1)$

$y-1=-\frac{5}{4}(x-(-8))$

$y-1=-\frac{5}{4}(x+8)$

$4(y-1)=4\left(-\frac{5}{4}x-10\right)$

$4y-4=-5x-40$

$5x+4y=-36$

21) $x+7=0$

$\quad x=-7$

Domain: $(-\infty,-7)\cup(7,\infty)$

23) $f(a)=8(a)+3=8a+3$

25)

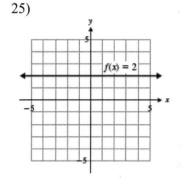

Section 5.1 Exercises

1) Yes

$$x+2y=-6 \qquad -x-3y=13$$
$$8+2(-7)\overset{?}{=}-6 \quad -8-3(-7)\overset{?}{=}13$$
$$8-14\overset{?}{=}-6 \qquad -8+21\overset{?}{=}13$$
$$-6=-6 \qquad 13=13$$

3) No

$$5x+y=21 \qquad 2x-3y=11$$
$$5(4)+1\overset{?}{=}21 \quad 2(4)-3(1)\overset{?}{=}21$$
$$20+1\overset{?}{=}21 \qquad 8-3\overset{?}{=}21$$
$$21=21 \qquad 5\neq21$$

5) Yes

$$5y-4x=-5$$
$$5(-3)-4\left(-\frac{5}{2}\right)\overset{?}{=}-5$$
$$-15+10\overset{?}{=}-5$$
$$-5=-5$$

$$6x+2y=-21$$
$$6\left(-\frac{5}{2}\right)+2(-3)\overset{?}{=}-21$$
$$-15-6\overset{?}{=}-21$$
$$-21=-21$$

7) No

$$y=-x+11$$
$$9\overset{?}{=}-0+11$$
$$9\neq11$$

$$x=5y-2$$
$$0\overset{?}{=}5(9)-2$$
$$0\overset{?}{=}45-2$$
$$0\neq43$$

9) The lines are parallel.

11) $(3,1)$

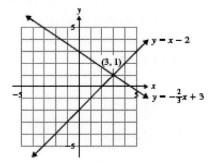

13) $(2,3)$

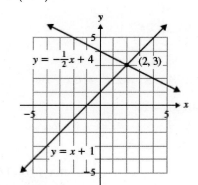

15) $(4,-5)$

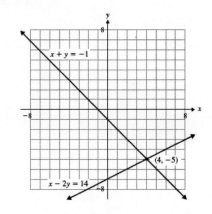

17) $(-1, -4)$

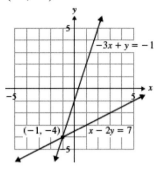

19) $\varnothing$; incosistent system

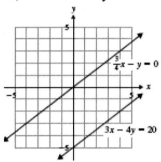

21) Infinite number of solutions of the form

$$\left\{ (x, y) \mid y = \frac{1}{3}x - 2 \right\}$$

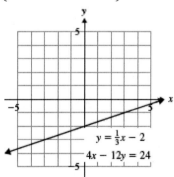

23) $(0, 2)$

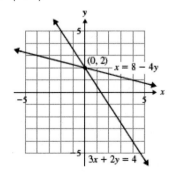

25) infinite number of solutions of the form $\left\{ (x, y) \mid y = -3x + 1 \right\}$;

dependent system

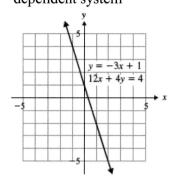

27) $(-2, 2)$

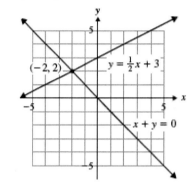

29) $(1, -1)$

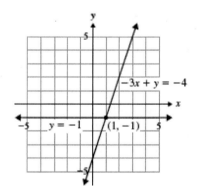

31) $\varnothing$; inconsistent system

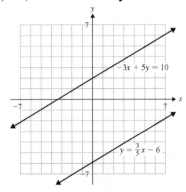

33) Answers may vary.

35) Answers may vary.

37) Answers may vary.

39) C. $(-3, 4)$ is in quadrant II.

41) B; $(4.1, 0)$ is the only point on the positive x-axis.

43) The slopes are different.

45) One solution; the slopes are different

47) No solution; the slopes are same; the lines are parallel.
$$6x + 16y = -9$$
$$16y = -6x - 9$$
$$y = -\frac{6}{16}x - \frac{9}{16} \qquad y = -\frac{3}{8}x + 1$$
$$y = -\frac{3}{8}x - \frac{9}{16}$$

49) Infinite number of solutions
$$-15x + 9y = 27$$
$$9y = 15x + 27$$
$$y = \frac{15}{9}x + \frac{27}{9}$$
$$y = \frac{5}{3}x + 3$$
$$10x - 6y = -18$$
$$6y = 10x + 18$$
$$y = \frac{10}{6}x + \frac{18}{3}$$
$$y = \frac{5}{3}x + 6$$

51) One solution;
the lines have different slopes.
$$3x + 12y = 9$$
$$12y = -3x + 9$$
$$y = -\frac{3}{12}x + \frac{9}{12}$$
$$y = -\frac{1}{4}x + \frac{3}{4}$$
$$x - 4y = 3$$
$$4y = x - 3$$
$$y = \frac{1}{4}x - \frac{3}{4}$$

53) No solution
The lines are parallel

55) a) 1985-2000

b) $(2000, 1.4)$; in the year 2000, 1.4% of foreign students were from Hong Kong and 1.4% were from Malaysia

c) 1985-1990 and 2000-2005

d) 1985-1990; this line segment has the most negative slope.

57) $(3,-4)$

59) $(4,1)$

61) $(-2.25,-1.6)$

Section 5.2 Exercises

1) It is the only variable with a coefficient of 1.

3) The variables are eliminated, and you get a false statement.

5) Substitute $y = 4x - 3$ into
$$5x + y = 15$$
$$5x + (4x - 3) = 15$$
$$9x - 3 = 15$$
$$9x = 18$$
$$x = 2$$
Substitute $x = 2$ into
$$y = 4x - 3$$
$$y = 4(2) - 3$$
$$y = 8 - 3$$
$$y = 5 \qquad (2,5)$$

7) Substitute $x = 7y + 11$ into
$$4x - 5y = -2$$
$$4(7y + 11) - 5y = -2$$
$$28y + 44 - 5y = -2$$
$$23y + 44 = -2$$
$$23y = -2 - 44$$
$$23y = -46$$
$$y = -2$$
Substitute $y = -2$ into
$$x = 7y + 11$$
$$x = 7(-2) + 11$$
$$x = -14 + 11$$
$$x = -3 \qquad (-3,-2)$$

9) Substitute $x = -2y - 3$ into
$$4x + 5y = -6$$
$$4(-2y - 3) + 5y = -6$$
$$-8y - 12 + 5y = -6$$
$$-3y - 12 = -6$$
$$-3y = -6 + 12$$
$$-3y = 6$$
$$y = -2$$
Substitute $y = -2$ into
$$x + 2y = -3$$
$$x + 2(-2) = -3$$
$$x - 4 = -3$$
$$x = -3 + 4$$
$$x = 1 \qquad (1,-2)$$

11) Substitute $y = 4x - 7$ into
$$2y - 7x = -14$$
$$2(4x - 7) - 7x = -14$$
$$8x - 14 - 7x = -14$$
$$x - 14 = -14$$
$$x = -14 + 14$$
$$x = 0$$
Substitute $x = 0$ into
$$4x - y = 7$$
$$4(0) - y = 7$$
$$0 - y = 7$$
$$-y = 7$$
$$y = -7 \qquad (0,-7)$$

13) Substitute $y = 2x - 3$ into
$$9y - 18x = 5$$
$$9(2x - 3) - 18x = 5$$
$$18x - 27 - 18x = 5$$
$$-27 \neq 5$$
There is no solution. $\varnothing$

15) Substitute $x = 2y + 10$ into

$$3x - 6y = 30$$
$$3(2y + 10) - 6y = 30$$
$$6y + 30 - 6y = 30$$
$$30 = 30$$

Infinite number of solutions of the form

$$\{(x,y) \mid x - 2y = 10\}$$

Substitute $y = 5$ into

$$x = -\frac{3}{5}y + 7$$
$$x = \left(-\frac{3}{5}\right)5 + 7$$
$$x = -3 + 7$$
$$x = 4 \qquad\qquad (4,5)$$

17) Substitute $y = -10x - 5$ into

$$-5x + 2y = 10$$
$$-5x + 2(-10x - 5) = 10$$
$$-5x - 20x - 10 = 10$$
$$-25x - 10 = 10$$
$$-25x = 20$$
$$x = -\frac{4}{5}$$

Substitute $x = -\frac{4}{5}$ into

$$10x + y = -5$$
$$10\left(-\frac{4}{5}\right) + y = -5$$
$$-8 + y = -5$$
$$y = -5 + 8$$
$$y = 3 \qquad\qquad \left(-\frac{4}{5}, 3\right)$$

19) Substitute $x = -\frac{3}{5}y + 7$ into

$$x + 4y = 24$$
$$\left(-\frac{3}{5}y + 7\right) + 4y = 24$$
$$-3y + 35 + 20y = 120$$
$$17y + 35 = 120$$
$$17y = 120 - 35$$
$$17y = 85$$
$$y = 5$$

21) Substitute $x = 2y - 2$ into

$$4y = 2x + 4$$
$$4y = 2(2y - 2) + 4$$
$$4y = 4y - 4 + 4$$
$$4y = 4y$$

Infinite number of solutions of the form

$$\{(x,y) \mid -x + 2y = 2\}$$

23) $\qquad 2x + 3y = 6$

$$2x = 6 - 3y$$
$$x = \frac{1}{2}(6 - 3y)$$

Substitute $x = \frac{1}{2}(6 - 3y)$ into

$$5x + 2y = -7$$
$$5\left[\frac{1}{2}(6 - 3y)\right] + 2y = -7$$
$$15 - \frac{15}{2}y + 2y = -7$$
$$30 - 15y + 4y = -14$$
$$-11y + 30 = -14$$
$$-11y = -44$$
$$y = 4$$

Substitute $y = 4$ into

$$2x + 3y = 6$$
$$2x + 3(4) = 6$$
$$2x + 12 = 6$$
$$2x = 6 - 12$$
$$2x = -6$$
$$x = -3 \qquad (-3, 4)$$

25) $\qquad 9x - 2y = 11$
$$-2y = 11 - 9x$$
$$y = \frac{9}{2}x - \frac{11}{2}$$

Substitute $y = \frac{9}{2}x - \frac{11}{2}$ into

$$6x - 7y = -4$$
$$6x - 7\left(\frac{9}{2}x - \frac{11}{2}\right) = -4$$
$$6x - \frac{63}{2}x + \frac{77}{2} = -4$$
$$12x - 63x + 77 = -8$$
$$-51x + 77 = -8$$
$$-51x = -8 - 77$$
$$-51x = -85$$
$$x = \frac{85}{51}$$
$$x = \frac{5}{3}$$

Substitute $x = \frac{5}{3}$ into

$$6x - 7y = -4$$
$$6\left(\frac{5}{3}\right) - 7y = -4$$
$$10 - 7y = -4$$
$$-7y = -4 - 10$$
$$-7y = -14$$
$$y = 2 \qquad\qquad \left(\frac{5}{3}, 2\right)$$

27) $18x + 6y = -66$
$$6y = -66 - 18x$$
$$y = -11 - 3x$$

Substitute $y = -11 - 3x$ into

$$12x + 4y = -19$$
$$12x + 4(-11 - 3x) = -19$$
$$12x - 44 - 12x = -19$$
$$-44 \neq -19$$

There is no solution $\quad \varnothing$

29) Multiply the equation by the LCD of the fractions to eliminate the fractions.

31) $\qquad \frac{1}{4}x - \frac{1}{2}y = 1$
$$4\left(\frac{1}{4}x - \frac{1}{2}y\right) = 4 \cdot 1$$
$$x - 2y = 4$$
$$x = 2y + 4$$

$$\frac{2}{3}x + \frac{1}{6}y = \frac{25}{6}$$
$$6\left(\frac{2}{3}x + \frac{1}{6}y\right) = 6 \cdot \frac{25}{6}$$
$$4x + y = 25$$

Substitute $x = 2y + 4$ into

$$4x + y = 25$$
$$4(2y + 4) + y = 25$$
$$8y + 16 + y = 25$$
$$9y + 16 = 25$$
$$9y = 9$$
$$y = 1$$

Substitute $y = 1$ into

$$x = 2y + 4$$
$$x = 2(1) + 4$$
$$x = 2 + 4$$
$$x = 6 \qquad\qquad (6, 1)$$

33) $\dfrac{1}{6}x + \dfrac{4}{3}y = \dfrac{13}{3}$

$x + 8y = 26$

$x = -8y + 26$

Substitute $x = -8y + 26$ into

$\dfrac{2}{5}x + \dfrac{3}{2}y = \dfrac{18}{5}$

$4x + 15y = 36$

$4(-8y + 26) + 15y = 36$

$-32y + 104 + 15y = 36$

$-17y = 36 - 104$

$-17y = -68$

$y = 4$

Substitute $y = 4$ into

$x = -8y + 26$

$x = -8(4) + 26$

$x = -32 + 26$

$x = -6$ $\qquad (-6,\ 4)$

35) $\dfrac{x}{10} - \dfrac{y}{2} = \dfrac{13}{10}$

$x - 5y = 13$

$x = 5y + 13$

$\dfrac{x}{3} + \dfrac{5}{4}y = -\dfrac{3}{2}$

$4x + 15y = -18$

Substitute $x = 5y + 13$ into

$4x + 15y = -18$

$4(5y + 13) + 15y = -18$

$20y + 52 + 15y = -18$

$35y + 52 = -18$

$35y = -70$

$y = -2$

Substitute $y = -2$ into

$x = 5y + 13$

$x = 5(-2) + 13$

$x = -10 + 13$

$x = 3$ $\qquad (3,\ -2)$

37) $y - \dfrac{5}{2}x = -2$

$y = \dfrac{5}{2}x - 2$

Substitute $y = \dfrac{5}{2}x - 2$ into

$\dfrac{3}{4}x - \dfrac{3}{10}y = \dfrac{3}{5}$

$\dfrac{3}{4}x - \dfrac{3}{10}\left(\dfrac{5}{2}x - 2\right) = \dfrac{3}{5}$

$\dfrac{3}{4}x - \dfrac{3}{4}x + \dfrac{3}{5} = \dfrac{3}{5}$

$\dfrac{3}{5} = \dfrac{3}{5}$

There are infinite number of solutions

of the form $\left\{(x, y)\,\middle|\, y - \dfrac{5}{2}x = -2\right\}$

39) $\dfrac{3}{4}x + \dfrac{1}{2}y = 6$

$3x + 2y = 24$

Substitute $x = 3y + 8$ into

$3x + 2y = 24$

$3(3y + 8) + 2y = 24$

$9y + 24 + 2y = 24$

$11y = 0$

$y = 0$

Substitute $y = 0$ into

$x = 3y + 8$

$x = 3(0) + 8$

$x = 0 + 8$

$x = 8$ $\qquad (8,\ 0)$

41) $\quad 0.2x - 0.1y = 0.1$

$10(0.2x - 0.1y) = 10(0.1)$

$2x - y = 1$

$y = 2x - 1$

$0.01x + 0.04y = 0.23$

$100(0.01x + 0.04y) = 100(0.23)$

$x + 4y = 23$

Substitute $y = 2x - 1$ into

$x + 4y = 23$

$x + 4(2x - 1) = 23$

$x + 8x - 4 = 23$

$9x - 4 = 23$

$9x = 27$

$x = 3$

Substitute $x = 3$ into

$y = 2x - 1$

$y = 2(3) - 1$

$y = 6 - 1$

$x = 5$ $\qquad (3, 5)$

43) $\quad 0.6x - 0.1y = 1$

$10(0.6x - 0.1y) = 10(1)$

$6x - y = 10$

$y = 6x - 10$

$-0.4x + 0.5y = -1.1$

$10(-0.4x + 0.5y) = 10(-1.1)$

$-4x + 5y = -11$

Substitute $y = 6x - 10$ into

$-4x + 5y = -11$

$-4x + 5(6x - 10) = -11$

$-4x + 30x - 50 = -11$

$26x = -11 + 50$

$26x = 39$

$x = 1.5$

Substitute $x = 1.5$ into

$y = 6x - 10$

$y = 6(1.5) - 10$

$y = 9 - 10$

$y = -1$ $\qquad (1.5, -1)$

45) $\quad 0.02x + 0.01y = -0.44$

$100(0.02x + 0.01y) = 100(-0.44)$

$2x + y = -44$

$y = -2x - 44$

$-0.1x - 0.2y = 4$

$10(-0.1x - 0.2y) = 10(4)$

$-x - 2y = 40$

Substitute $y = -2x - 44$ into

$-x - 2y = 40$

$-x - 2(-2x - 44) = 40$

$-x + 4x + 88 = 40$

$3x + 88 = 40$

$3x = -48$

$x = -16$

Substitute $x = -16$ into

$y = -2x - 44$

$y = -2(-16) - 44$

$y = 32 - 44$

$y = -12$ $\qquad (-16, -12)$

47) $\quad 2.8x + 0.7y = 0.1$

$10(2.8x + 0.7y) = 10(0.1)$

$28x + 7y = 1$

$0.04x + 0.01y = -0.06$

$100(0.04x + 0.01y) = 100(-0.06)$

$4x + y = -6$

$y = -6 - 4x$

Substitute $y = -6 - 4x$ into
$$28x + 7y = 1$$
$$28x + 7(-6 - 4x) = 1$$
$$28x - 42 - 28x = 1$$
$$-42 \neq 1$$
There is no solution $\varnothing$

49) $8 + 2(3x - 5) - 7x + 6y = 16$
$$8 + 6x - 10 - 7x + 6y = 16$$
$$-x - 2 + 6y = 16$$
$$-x + 6y = 18$$
$$x = 6y - 18$$
$$9(y - 2) + 5x - 13y = -4$$
$$9y - 18 + 5x - 13y = -4$$
$$5x - 4y - 18 = -4$$
$$5x - 4y = 14$$
Substitute $x = 6y - 18$ into
$$5x - 4y = 14$$
$$5(6y - 18) - 4y = 14$$
$$30y - 90 - 4y = 14$$
$$26y = 14 + 90$$
$$26y = 104$$
$$y = 4$$
Substitute $y = 4$ into
$$x = 6y - 18$$
$$x = 6(4) - 18$$
$$x = 24 - 18$$
$$x = 6 \qquad (6, 4)$$

51) $10(x + 3) - 7(y + 4) = 2(4x - 3y) + 3$
$$10x + 30 - 7y - 28 = 8x - 6y + 3$$
$$10x - 7y + 2 = 8x - 6y + 3$$
$$10x - 7y - 8x + 6y = 3 - 2$$
$$2x - y = 1$$
$$10 - 3(2x - 1) + 5y = 3y - 7x - 9$$

$$10 - 6x + 3 + 5y = 3y - 7x - 9$$
$$13 - 6x + 5y = 3y - 7x - 9$$
$$-6x + 5y - 3y + 7x = -9 - 13$$
$$x + 2y = -22$$
$$x = -22 - 2y$$
Substitute $x = -22 - 2y$ into

$$2x - y = 1$$
$$2(-22 - 2y) - y = 1$$
$$-44 - 4y - y = 1$$
$$-44 - 5y = 1$$
$$-5y = 1 + 44$$
$$-5y = 45$$
$$y = -9$$
Substitute $y = -9$ into
$$2x - y = 1$$
$$2x - (-9) = 1$$
$$2x + 9 = 1$$
$$2x = 1 - 9$$
$$2x = -8$$
$$x = -4 \qquad (-4, -9)$$

53) $-(y + 3) = 5(2x + 1) - 7x$
$$-y - 3 = 10x + 5 - 7x$$
$$-y - 3 = 3x + 5$$
$$-y - 3x = 5 + 3$$
$$-y - 3x = 8$$
$$y = -3x - 8$$
$$x + 12 - 8(y + 2) = 6(2 - y)$$
$$x + 12 - 8y - 16 = 12 - 6y$$
$$x - 8y - 4 = 12 - 6y$$
$$x - 8y + 6y = 12 + 4$$
$$x - 2y = 16$$

Substitute $y = -3x - 8$ into

$$x - 2y = 16$$
$$x - 2(-3x - 8) = 16$$
$$x + 4x + 16 = 16$$
$$5x + 16 = 16$$
$$5x = 16 - 16$$
$$5x = 0$$
$$x = 0$$

Substitute $x = 0$ into

$$y = -3x - 8$$
$$y = -3(0) - 8$$
$$y = 0 - 8$$
$$y = -8 \qquad (0, -8)$$

55) A+ Rental: $y = 0.60x$
Rock Bottom Rental: $y = 0.25x + 70$

a) How much would it cost to drive 160 miles?

A+: $y = 0.60(160) = 96.00$ $96.00

Rock: $y = 0.25x + 70$
$$y = 0.25(160) + 70$$
$$y = 40 + 70 = 110 \qquad \$110.00$$

b) How much would it cost to drive 300 miles?

A+: $y = 0.60x = 0.60(300)$
$$= 180 \qquad \$180.00$$

Rock Bottom:
$$y = 0.25x + 70$$
$$y = 0.25(300) + 70$$
$$y = 75 + 70 = 145 \qquad \$145.00$$

c) Requirement: A+Rental and Rock Bottom Rental to be the same Break Even Condition

Substitute $y = 0.60x$ into $y = 0.25x + 70$

$$0.60x = 0.25x + 70$$
$$100(0.60x) = 100(0.25x + 70)$$
$$60x = 25x + 7000$$
$$60x - 25x = 7000$$
$$35x = 7000$$
$$x = 200$$

If the cargo trailer is driven 200 miles, the cost would be the same from each company: $120.00

$$(200, 120)$$

d) If the cargo trailer is driven less than 200 miles, it is cheaper to rent from A+ if the distance driven is more than 200 miles, then, it is cheaper to rent from Rock Bottom Rentals. If the trailer is driven exactly 200 miles, then the cost would be the same from each company.

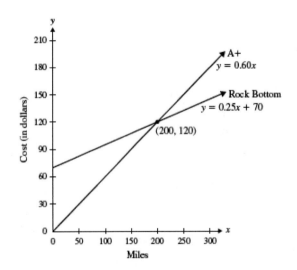

Section 5.3 Exercises

1)   Add the equations

Chapter 5: Solving Systems of Linear Equations

3) Add the equations.

$$x - y = -3$$
$$\underline{2x + y = 18}$$
$$3x = 15$$
$$x = 5$$

Substitute $x = 5$ into
$$x - y = -3$$
$$5 - y = -3$$
$$-y = -3 - 5$$
$$y = 8 \qquad (5,8)$$

5) Add the equations
$$-x + 2y = 2$$
$$\underline{x - 7y = 8}$$
$$-5y = 10$$
$$y = -2$$

Substitute y = -2 into
$$-x + 2y = 2$$
$$-x + 2(-2) = 2$$
$$-x - 4 = 2$$
$$-x = 2 + 4$$
$$x = -6 \qquad (-6,-2)$$

7) Add the equations
$$x + 4y = 1$$
$$\underline{3x - 4y = -29}$$
$$4x = -28$$
$$x = -7$$

Substitute $x = -7$ into
$$x + 4y = 1$$
$$-7 + 4y = 1$$
$$4y = 1 + 7$$
$$4y = 8$$
$$y = 2 \qquad (-7,2)$$

9) Multiply the second equation by 2.
$$4x - 7y = 8$$
$$8x - 14y = 16$$
Add the equations
$$8x - 14y = 16$$
$$\underline{-8x + 5y = -16}$$
$$-11y = 0$$
$$y = 0$$
Substitute $y = 0$ into
$$4x - 7y = 8$$
$$4x - 7(0) = 8$$
$$4x - 0 = 8$$
$$4x = 8$$
$$x = 2 \qquad (2,0)$$

11) Multiply the equation by -3
$$3x + 5y = 16$$
$$-9x - 15y = 48$$
Add the equations
$$4x + 15y = 13$$
$$\underline{-9x - 15y = -48}$$
$$-5x = -35$$
$$x = 7$$
Substitute $x = 7$ into
$$3x + 5y = 16$$
$$3(7) + 5y = 16$$
$$21 + 5y = 16$$
$$5y = -5$$
$$y = -1 \qquad (7,-1)$$

13) Multiply the first equation by 3
$$9x - 7y = -14$$
$$27x - 21y = -42$$
Multiply the second equation by 7
$$4x + 3y = 6$$
$$28x + 21y = 42$$
Add the equations
$$28x + 21y = \ \ 42$$
$$\underline{27x - 21y = -42}$$
$$x \qquad\quad = 0$$
Substitute $x = 0$ into
$$9x - 7y = -14$$
$$9(0) - 7y = -14$$
$$0 - 7y = -14$$
$$-7y = -14$$
$$y = 2 \qquad\qquad (0, 2)$$

15) Multiply the first equation by 3
$$-9x + 2y = -4$$
$$-27x + 6y = -12$$
Multiply the second equation by 2
$$6x - 3y = 11$$
$$12x - 6y = 22$$
Add the equations
$$-27x + 6y = -12$$
$$\underline{12x - 6y = \ \ 22}$$
$$-15x \qquad = 10$$
$$x = -\frac{2}{3}$$
Substitute $x = -\dfrac{2}{3}$ into

$$-9x + 2y = -4$$
$$-9\left(-\frac{2}{3}\right) + 2y = -4$$
$$6 + 2y = -4$$
$$2y = -10$$
$$y = -5 \qquad\qquad \left(-\frac{2}{3}, -5\right)$$

17) Multiply the equation by -2
$$9x - y = 2$$
$$-18x + 2y = -4$$

Add the equations.
$$18x - 2y = \ \ 4$$
$$\underline{-18x + 2y = -4}$$
$$0 = 0$$
There are infinite number of
solutions of the form $\{(x, y) \mid 9x - y = 2\}$

19) Multiply the first equation by -2
$$x = 12 - 4y$$
$$-2x = -24 + 8y$$
$$-2x - 8y = -24$$
Write the second equation in standard form.
$$2x - 7 = 9y$$
$$2x - 9y = 7$$
Add the equations.
$$-2x - 8y = -24$$
$$\underline{2x - 9y = \ \ \ 7}$$
$$-17y = -17$$
$$y = 1$$
Substitute $y = 1$ into
$$x = 12 - 4y$$
$$x = 12 - 4(1)$$
$$x = 12 - 4$$
$$x = 8 \qquad\qquad (8, 1)$$

21) Multiply the first equation by -3, then write it in standard form.

$$4y = 9 - 3x$$
$$-12y = -27 + 9x$$
$$-9x - 12y = -27$$

Multiply the second equation by 2, then write it in the standard form.

$$5x - 16 = -6y$$
$$10x - 32 = -12y$$
$$10x + 12y = 32$$

Add the equations.

$$-9x - 12y = -27$$
$$\underline{10x + 12y = 32}$$
$$x = 5$$

Substitute $x = 5$ into

$$4y = 9 - 3x$$
$$4y = 9 - 3(5)$$
$$4y = 9 - 15$$
$$4y = -6$$
$$y = -\frac{3}{2} \qquad \left(5, -\frac{3}{2}\right)$$

23) Multiply the first equation by 5, then write it in standard form.

$$2x - 9 = 8y$$
$$10x - 45 = 40y$$
$$10x - 40y = 45$$

Multiply the second equation by 2.

$$20y - 5x = 6$$
$$40y - 10x = 12$$
$$-10x + 40y = 12$$

Add the equations.

$$10x - 40y = 45$$
$$\underline{-10x + 40y = 12}$$
$$0 \neq 57$$

There is no solution. $\qquad \varnothing$

25) Multiply the first equation by 7.

$$6x - 11y = -1$$
$$42x - 77y = -7$$

Multiply the second equation by 6.

$$-7x + 13y = 2$$
$$-42x + 78y = 12$$

Add the equations.

$$42x - 77y = -7$$
$$\underline{-42x + 78y = 12}$$
$$y = 5$$

Substitute $y = 5$ into

$$6x - 11y = -1$$
$$6x - 11(5) = -1$$
$$6x - 55 = -1$$
$$6x = -1 + 55$$
$$6x = 54$$
$$x = 9 \qquad\qquad (9, 5)$$

27) Multiply the first equation by 2.

$$9x + 6y = -2$$
$$18x + 12y = -4$$

Multiply the second equation by 3.

$$-6x - 4y = 11$$
$$-18x - 12y = 33$$

Add the equations.

$$-18x - 12y = 33$$
$$\underline{18x + 12y = -4}$$
$$0 \neq 29$$

There is no solution. $\qquad \varnothing$

29) Eliminate the fractions. Multiply the first equation by 4 and multiply the second equation by 24.

31) Multiply the first equation by -10.
$$\frac{4}{5}x - \frac{1}{2}y = -\frac{3}{2}$$
$$-8x + 5y = 15$$
Multiply the second equation by 4

$$2x - \frac{1}{4}y = \frac{1}{4}$$
$$8x - y = 1$$

Add the equations.
$$-8x + 5y = 15$$
$$\underline{8x - y = 1}$$
$$4y = 16$$
$$y = 4$$
Substitute $y = 4$ into
$$8x - y = 1$$
$$8x - 4 = 1$$
$$8x = 1 + 4$$
$$8x = 5$$
$$x = \frac{5}{8} \qquad \left(\frac{5}{8}, 4\right)$$

33) Multiply the first equation by -16.
$$\frac{5}{4}x - \frac{1}{2}y = \frac{7}{8}$$
$$-20x + 8y = -14$$
Multiply the second equation by 50.

$$\frac{2}{5}x - \frac{1}{10}y = -\frac{1}{2}$$
$$20x - 5y = -25$$
Add the equations.
$$-20x + 8y = -14$$
$$\underline{20x - 5y = -25}$$
$$3y = -39$$
$$y = -13$$
Substitute $y = -13$ into

$$20x - 5y = -25$$
$$20x - 5(-13) = -25$$
$$20x + 65 = -25$$
$$20x = -25 - 65$$
$$20x = -90$$
$$x = -\frac{9}{2} \qquad \left(-\frac{9}{2}, -13\right)$$

35) Multiply the first equation by 36.
$$\frac{x}{4} + \frac{y}{2} = -1$$
$$9x + 18y = -36$$
Multiply the second equation by -24.
$$\frac{3}{8}x + \frac{5}{3}y = -\frac{7}{12}$$
$$-9x - 40y = 14$$
Add the equations.
$$9x + 18y = -36$$
$$\underline{-9x - 40y = 14}$$
$$-22y = -22$$
$$y = 1$$
Substitute $y = 1$ into
$$9x + 18y = -36$$
$$9x + 18(1) = -36$$
$$9x + 18 = -36$$
$$9x = -36 - 18$$
$$9x = -54$$
$$x = -6 \qquad (-6, 1)$$

Chapter 5: Solving Systems of Linear Equations

37) Multiply the first equation by 24.
$$\frac{x}{12} - \frac{y}{8} = \frac{7}{8}$$
$$2x - 3y = 21$$
Multiply the second equation by 3
and write it in standard form.
$$y = \frac{2}{3}x - 7$$
$$3y = 2x - 21$$
$$-2x + 3y = -21$$

Add the equations.
$$2x - 3y = 21$$
$$\underline{-2x + 3y = -21}$$
$$0 = 0$$

There are infinite solutions of the form
$$\left\{ (x, y) \mid y = \frac{2}{3}x - 7 \right\}.$$

39) Multiply the first equation by 8.
$$-\frac{1}{2}x + \frac{5}{4}y = \frac{3}{4}$$
$$-4x + 10y = 6$$
Multiply the second equation by 10.
$$\frac{2}{5}x - \frac{1}{2}y = -\frac{1}{10}$$
$$4x - 5y = -1$$

Add the equations.
$$-4x + 10y = 6$$
$$\underline{4x - 5y = -1}$$
$$5y = 5$$
$$y = 1$$
Substitute $y = 1$ into
$$4x - 5y = -1$$
$$4x - 5(1) = -1$$
$$4x - 5 = -1$$
$$4x = -1 + 5$$
$$4x = 4$$
$$x = 1 \qquad\qquad (1,1)$$

41) Multiply the first equation by 400.
$$0.08x + 0.07y = -0.84$$
$$32x + 28y = -336$$
Multiply the second equation by -100.
$$0.32x - 0.06y = -2$$
$$-32x + 6y = 200$$
Add the equations
$$32x + 28y = -336$$
$$\underline{-32x + 6y = 200}$$
$$34y = -136$$
$$y = -4$$
Substitute $y = -4$ into $-32x + 6y = 200$
$$-32x + 6(-4) = 200$$
$$-32x - 24 = 200$$
$$-32x = 200 + 24$$
$$-32x = 224$$
$$x = -7 \qquad\qquad (-7,-4)$$

43) Multiply the first equation by 30.
$$0.1x + 2y = -0.8$$
$$3x + 60y = -24$$

146

Multiply the second equation by -100.

$0.03x + 0.10y = 0.26$

$-3x - 10y = -26$

Add the equations

$3x + 60y = -24$

$\underline{-3x - 10y = -26}$

$50y = -50$

$y = -1$

Substitute $y = -1$ into

$3x + 60y = -24$

$3x + 60(-1) = -24$

$3x - 60 = -24$

$3x = -24 + 60$

$3x = 36$

$x = 12 \qquad (12, -1)$

45) Multiply the first equation by 30.

$-0.4x + 0.2y = 0.1$

$-12x + 6y = 3$

Multiply the second equation by 20.

$0.6x - 0.3y = 1.5$

$12x - 6y = 30$

Add the equations

$-12x + 6y = \ \ 3$

$\underline{12x - 6y = 30}$

$0 \neq 33$

There is no solution. $\varnothing$

47) Multiply the first equation by 300.

$0.04x + 0.03y = 0.16$

$12x + 9y = 48$

Multiply the second equation by -20.

$0.6x + 0.2y = 1.15$

$-12x - 4y = -23$

Add the equations

$-12x - 4y = -23$

$\underline{12x + 9y = 48}$

$5y = 25$

$y = 5$

Substitute $y = 5$ into

$12x + 9y = 48$

$12x + 9(5) = 48$

$12x + 45 = 48$

$12x = 48 - 45$

$12x = 3$

$x = 0.25 \qquad (0.25, 5)$

49) $17x - 16(y+1) = 4(x-y)$

$17x - 16y - 16 = 4x - 4y$

$13x - 12y = 16$

$19 - 10(x+2) = -4(x+6) - y + 2$

$19 - 10x - 20 = -4x - 24 - y + 2$

$-6x + y = -21$

Multiply the second equation by 12.

$-6x + y = -21$

$-72x + 12y = -252$

Add the new equations

$-72x + 12y = -252$

$\underline{13x - 12y = \ \ 16}$

$-59x \qquad = -236$

$x = 4$

Substitute $x = 4$ into

$$13x - 12y = 16$$
$$13(4) - 12y = 16$$
$$52 - 12y = 16$$
$$-12y = -36$$
$$y = 3 \qquad (4,3)$$

51) $5 - 3y = 6(3x + 4) - 8(x + 2)$
$$5 - 3y = 18x + 24 - 8x - 16$$
$$-10x - 3y = 3$$

Multiply the above equation by 4.
$$-40x - 12y = 12$$
$$6x - 2(5y + 2) = -7(2y - 1) - 4$$
$$6x - 10y - 4 = -14y + 7 - 4$$
$$6x + 4y = 7$$

Multiply the above equation by 3.
$$18x + 12y = 21$$

Add the new equations
$$18x + 12y = 21$$
$$\underline{-40x - 12y = 12}$$
$$-22x \qquad = 33$$
$$x = -\frac{3}{2}$$

Substitute $x = -\frac{3}{2}$ into

$$6x + 4y = 7$$
$$6\left(-\frac{3}{2}\right) + 4y = 7$$
$$-9 + 4y = 7$$
$$4y = 16$$
$$y = 4 \qquad \left(-\frac{3}{2}, 4\right)$$

53) $6(x - 3) + x - 4y = 1 + 2(x - 9)$
$$6x - 18 + x - 4y = 1 + 2x - 18$$
$$5x - 4y = 1$$

Multiply the above equation by 2.
$$10x - 8y = 2$$
$$4(2y - 3) + 10x = 5(x + 1) - 4$$
$$8y - 12 + 10x = 5x + 5 - 4$$
$$5x + 8y = 13$$

Add the new equations.
$$10x - 8y = 2$$
$$\underline{5x + 8y = 13}$$
$$15x \qquad = 15$$
$$x = 1$$

Substitute $x = 1$ into
$$5x + 8y = 13$$
$$5(1) + 8y = 13$$
$$5 + 8y = 13$$
$$8y = 8$$
$$y = 1 \qquad (1,1)$$

55) Multiply the first equation by 3.
$$4x + 5y = -6$$
$$12x + 15y = -18$$

Multiply the second equation by -4.
$$3x + 8y = 15$$
$$-12x - 32y = -60$$

Add the equations.
$$12x + 15y = -18$$
$$-12x - 32y = -60$$
$$-17y = -78$$
$$y = \frac{78}{17}$$

Multiply the original first equation by 8.
$$4x + 5y = -6$$
$$32x + 40y = -48$$

Multiply the original second equation by -5.

$$3x + 8y = 15$$
$$-15x - 40y = -75$$

Add the equations.

$$-15x - 40y = -75$$
$$32x + 40y = -48$$
$$17x = -123$$
$$x = -\frac{123}{17}$$

$$\left(-\frac{123}{17}, \frac{78}{17}\right)$$

57) Multiply the first equation by 3

$$4x + 9y = 7$$
$$12x + 27y = 21$$

Multiply the second equation by -2.

$$6x + 11y = -14$$
$$-12x - 22y = 28$$

Add the equations.

$$12x + 27y = 21$$
$$-12x - 22y = 28$$
$$5y = 49$$
$$y = \frac{49}{5}$$

Multiply the original first equation by 11

$$4x + 9y = 7$$
$$44x + 99y = 77$$

Multiply the original second equation by -9

$$6x + 11y = -14$$
$$-54x - 99y = 126$$

Add the equations.

$$-54x - 99y = 126$$
$$44x + 99y = 77$$
$$-10x = 203$$
$$x = -\frac{203}{10}$$

$$\left(-\frac{203}{10}, \frac{49}{5}\right)$$

59) Ensure $(5, 4)$ is a solution; substitute for (x, y) in

$$2x - 3y = -2$$
$$2(5) - 3(4) = -2$$
$$10 - 12 = -2$$
$$-2 = -2$$

To obtain the value of k, substitute $(5, 4)$ in

$$x + ky = 17$$
$$5 + k(4) = 17$$
$$4k = 17 - 5$$
$$4k = 12$$
$$k = 3$$

61) Ensure that $(-7, 3)$ is a solution; substitute for (x, y) in

$$3x + 4y = -9$$
$$3(-7) + 4(3) = -9$$
$$-21 + 12 = -9$$
$$-9 = -9$$

To calculate k, substitute $(-7, 3)$ in

$$kx - 5y = 41$$
$$k(-7) - 5(3) = 41$$
$$-7k - 15 = 41$$
$$-7k = 41 + 15$$
$$-7k = 56$$
$$k = -8$$

63) a) To have infinite solutions,

the variables need to be eliminated
Multiply the equation by -1

$$x - y = 5$$
$$-x + y = -5$$

Add to the second equation

$$x - y = c$$
$$\underline{-x + y = -5}$$
$$0 = c - 5$$

To satisfy the above equation,

c needs to be 5.

b) Any value for c other than 5
will produce no solution.

65) a) To have infinite solutions,

the variables need to be eliminated
Multiply the equation by -3

$$ax + 4y = -5$$
$$-3ax - 12y = 15$$

Add to the first equation

$$9x + 12y = -15$$
$$\underline{-3ax - 12y = 15}$$
$$x(9 - 3a) = 0$$
$$9 - 3a = 0$$
$$a = 3$$

To satisfy the above equation,

a needs to be 3

b) Any value for a other than 3 will
produce exactly one solution

67) Add the equations

$$-5x + 4by = 6$$
$$\underline{5x + 3by = 8}$$
$$7by = 14$$

$$y = \frac{14}{7b} = \frac{2}{b}$$

Substitute $y = \frac{2}{b}$ in

$$5x + 3by = 8$$

$$5x + 3b(\frac{2}{b}) = 8$$

$$5x + 6 = 8$$

$$5x = 8 - 6$$

$$x = \frac{2}{5} \qquad \left(\frac{2}{5}, \frac{2}{b}\right)$$

69) Add the equations.

$$3ax + by = 4$$
$$\underline{ax - by = -5}$$
$$4ax \quad\;\; = -1$$

$$x = -\frac{1}{4a}$$

Substitute $x = -\frac{1}{4a}$ in

$$ax - by = -5$$

$$a\left(-\frac{1}{4a}\right) - by = -5$$

$$-\frac{1}{4} - by = -5$$

$$-by = -5 + \frac{1}{4}$$

$$-by = -\frac{19}{4}$$

$$y = \frac{19}{4b} \qquad \left(-\frac{1}{4a}, \frac{19}{4b}\right)$$

Chapter 5: Putting It All Together

1) Elimination method; none of the coefficients is 1 or -1

$$2x - 3y = -8$$
$$-4(2x - 3y) = -4(-8)$$
$$-8x + 12y = 32$$

Add the equations.

$$8x - 5y = 10$$
$$+ \ -8x + 12y = 32$$
$$\overline{\quad 7y = 42 \quad}$$
$$y = 6$$

Substitute $y = 6$ into

$$2x - 3y = -8$$
$$2x - 3(6) = -8$$
$$2x - 18 = -8$$
$$2x = 10$$
$$x = 5 \qquad (5, 6)$$

3) Since the coefficient of y in the second equation is 1, you can solve for y and use substitution. Or, multiply the second equation by -5 and use the elimination method. Either method will work well.

$$8x + y = -1$$
$$y = -1 - 8x$$

Substitute $y = -8x - 1$ into

$$12x - 5y = 18$$
$$12x - 5(-8x - 1) = 18$$
$$12x + 40x + 5 = 18$$
$$52x + 5 = 18$$
$$52x = 13$$
$$x = \frac{1}{4}$$

Substitute $x = \frac{1}{4}$ into

$$8x + y = -1$$
$$8\left(\frac{1}{4}\right) + y = -1$$
$$2 = y = -1$$
$$y = -3 \qquad \left(\frac{1}{4}, -3\right)$$

5) Substitution; the second equation is solved for x and does not contain any fractions.

Substitute $x = y + 8$ into

$$y - 4x = -11$$
$$y - 4(y + 8) = -11$$
$$y - 4y - 32 = -11$$
$$-3y - 32 = -11$$
$$-3y = 21$$
$$y = -7$$

Substitute $y = -7$ into

$$x = y + 8$$
$$x = (-7) + 8$$
$$x = 1 \qquad (1, -7)$$

7) Substitute $x = 3y + 6$ into

$$4x + 5y = 24$$
$$4(3y + 6) + 5y = 24$$
$$12y + 24 + 5y = 24$$
$$17y + 24 = 24$$
$$17y = 0$$
$$y = 0$$

Substitute $y = 0$ into

$x = 3y + 6$

$x = 3(0) + 6$

$x = 0 + 6$

$x = 6$ $\qquad$ (6, 0)

9) $\quad 6x + 15y = -1$

$\quad 2(6x + 15y) = 2(-1)$

$\quad 12x + 30y = -2$

$\qquad 9x = 10y - 8$

$\qquad 3(9x) = 3(10y - 8)$

$\qquad 27x = 30y - 24$

$27x - 30y = -24$

Add the equations.

$27x - 30y = -24$

$\underline{12x + 30y = -2}$

$39x \qquad = -26$

$\qquad x = -\dfrac{2}{3}$

Substitute $x = -\dfrac{2}{3}$ into

$\quad 6x + 15y = -1$

$6\left(-\dfrac{2}{3}\right) + 15y = -1$

$\quad -4 + 15y = -1$

$\qquad 15y = 3$

$\qquad y = \dfrac{1}{5}$ $\qquad \left(-\dfrac{2}{3}, \dfrac{1}{5}\right)$

11) $\quad 10x + 4y = 7$

$\quad 3(10x + 4y) = 3(7)$

$\quad 30x + 12y = 21$

$\quad 15x + 6y = -2$

$\quad -2(15x + 6y) = -2(-2)$

$\quad -30x - 12y = 4$

Add the equations.

$\qquad 30x + 12y = 21$

$\qquad \underline{-30x - 12y = 4}$

$\qquad \qquad \quad 0 \ne 25$

$\varnothing$

13) $\quad$ Substitute $x = -\dfrac{1}{2}$ into

$\quad 10x + 9y = 4$

$10\left(-\dfrac{1}{2}\right) + 9y = 4$

$\quad -5 + 9y = 4$

$\qquad 9y = 9$

$\qquad y = 1$ $\qquad \left(-\dfrac{1}{2}, 1\right)$

15) $\quad 7y - 2x = 13$

$\qquad 15x + 6y = -2$

$\quad -2(15x + 6y) = -2(-2)$

$\qquad -30x - 12y = 4$

Add the equations.

$\qquad -6x + 21y = 39$

$\qquad \underline{6x - 4y = 12}$

$\qquad \qquad 17y = 51$

$\qquad \qquad y = 3$

Substitute $y = 3$ into

$\quad 3x - 2y = 6$

$\quad 3x - 2(3) = 6$

$\quad 3x - 6 = 6$

$\qquad 3x = 12$

$\qquad x = 4$ $\qquad (4, 3)$

17) $\frac{2}{5}x + \frac{4}{5}y = -2$

$-5\left(\frac{2}{5}x + \frac{4}{5}y\right) = -5(-2)$

$-2x - 4y = 10$

$\frac{1}{6}x + \frac{1}{6}y = \frac{1}{3}$

$12\left(\frac{1}{6}x + \frac{1}{6}y\right) = 12\left(\frac{1}{3}\right)$

$2x + 2y = 4$

Add the equations.

$-2x - 4y = 10$

$\underline{2x + 2y = 4}$

$-2y = 14$

$y = -7$

Substitute $y = -7$ into

$2x + 2y = 4$

$2x + 2(-7) = 4$

$2x - 14 = 4$

$2x = 18$

$x = 9 \qquad (9, -7)$

19) $-0.3x + 0.1y = 0.4$

$10(-0.3x + 0.1y) = 10(0.4)$

$-3x + y = 4$

$0.01x + 0.05y = 0.2$

$300(0.01x + 0.05y) = 300(0.2)$

$3x + 15y = 60$

Add the equations.

$-3x + y = 4$

$\underline{3x + 15y = 60}$

$16y = 64$

$y = 4$

Substitute $y = 4$ into

$-3x + y = 4$

$-3x + 4 = 4$

$-3x = 0$

$x = 0 \quad (0, 4)$

21) $-6x + 2y = -10$

$7(-6x + 2y) = 7(-10)$

$-42x + 14y = -70$

$21x - 7y = 35$

$2(21x - 7y) = 2(35)$

$42x - 14y = 70$

Add the equations.

$-42x + 14y = -70$

$\underline{42x - 14y = 70}$

$0 = 0$

There are infinite solutions of the form

$\{(x, y) \mid -6x + 2y = -10\}$.

23) $y = \frac{3}{2}x - \frac{1}{2}$

$10(y) = 10\left(\frac{3}{2}x - \frac{1}{2}\right)$

$10y = 15x - 5$

$-15x + 10y = -5$

$2 = 5y - 8x$

$-2(2) = -2(5y - 8x)$

$-4 = -10y + 16x$

$16x - 10y = -4$

Add the equations.

$-15x+10y=-5$

$\underline{16x-10y=-4}$

$x=-9$

Substitute $x=-9$ into

$16x-10y=-4$

$16(-9)-10y=-4$

$-144-10y=-4$

$-10y=140$

$y=-14$ $\qquad(-9,-14)$

25) $\quad 2x-3y=-8$

$10(2x-3y)=10(-8)$

$20x-30y=-80$

$7x+10y=4$

$3(7x+10y)=3(4)$

$21x+30y=12$

Add the equations.

$21x+30y=\ \ \ 12$

$\underline{20x-30y=-80}$

$41x\qquad=-68$

$x=-\dfrac{68}{41}$

$7(2x-3y)=7(-8)$

$14x-21y=-56$

$-2(7x+10y)=-2(4)$

$-14x-20y=-8$

Add the equations.

$14x-21y=-56$

$-14x-20y=-8$

$-41y=-64$

$y=\dfrac{64}{41}$ $\quad\left(-\dfrac{68}{41},\dfrac{64}{41}\right)$

27) $\quad 6(2x-3)=y+4(x-3)$

$12x-18=y+4x-12$

$8x-y=6$

$7(8x-y)=7(6)$

$56x-7y=42$

$5(3x+4)+4y=11-3y+27x$

$15x+20+4y=11-3y+27x$

$-12x+7y=-9$

Add the equations

$56x-7y=42$

$\underline{-12x+7y=-9}$

$44x\qquad=33$

$x=\dfrac{3}{4}$

Substitute $x=\dfrac{3}{4}$ into

$8x-y=6$

$8\left(\dfrac{3}{4}\right)-y=6$

$6-y=6$

$y=0$ $\qquad\left(\dfrac{3}{4},0\right)$

29) $\quad 2y-2(3x+4)=-5(y-2)-17$

$2y-6x-8=-5y+10-17$

$-6x+7y=1$

$4(-6x+7y)=4(1)$

$-24x+28y=4$

$4(2x+3)=10+5(y+1)$

$8x+12=10+5y+5$

$8x-5y=3$

$3(8x-5y)=3(3)$

$24x-15y=9$

Add the equations

$$-24x + 28y = 4$$
$$\underline{24x - 15y = 9}$$
$$13y = 13$$
$$y = 1$$

Substitute $y = 1$ into
$$8x - 5y = 3$$
$$8x - 5(1) = 3$$
$$8x - 5 = 3$$
$$8x = 8$$
$$x = 1 \qquad (1,1)$$

31) Substitute $y = -4x$ into
$$10x + 2y = -5$$
$$10x + 2(-4x) = -5$$
$$10x - 8x = -5$$
$$2x = -5$$
$$x = -\frac{5}{2}$$

Substitute $x = -\frac{5}{2}$ into
$$y = -4x$$
$$y = -4\left(-\frac{5}{2}\right)$$
$$y = 10 \qquad \left(-\frac{5}{2}, 10\right)$$

33) $(2,2)$

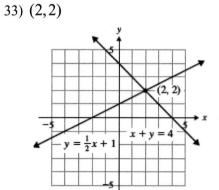

35) $(-4, -4)$

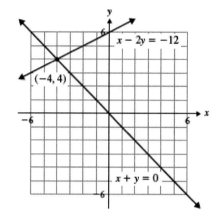

37) $\varnothing$

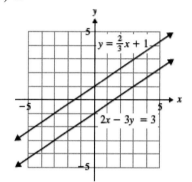

39) Using a graphing calculator, the solution set is $\{(-1.25, \ -0.5)\}$

Section 5.4 Exercises

1) Let $x =$ one number

   $y =$ other number, $x + 11$

   The sum of the numbers is 87
   $$x + y = 87$$
   $$x + x + 11 = 87$$
   $$2x + 11 = 87$$
   $$2x = 76$$
   $$x = 38$$
   The numbers are 38 and 49.

3) Let $x = $ Money made by Dark Knight
let $y = $ Money made by Transformers

Dark Knight $= 6.6 + $ Transformers
$$x = 6.6 + y$$
Dark Knight $+$ Transformers $= 127.8$
$$x + y = 127.8$$

Use substitution.
$$(6.6 + y) + y = 127.8$$
$$6.6 + 2y = 127.8$$
$$2y = 121.2$$
$$y = 60.6$$
$$x = 6.6 + y; \ y = 60.6$$
$$x = 60.6 + 6.6$$
$$x = 67.2$$
Dark Knight: $67.2 million;

Transformer: $60.6 million

5) Let $x = $ BET Awards nominated - Beyonce
Let $y = $ BET awards nominated - T.I.

Beyonce $+$ T.I. $= 27$
$$x + y = 27$$
Beyonce $= T.I. + 5$
$$x = y + 5$$
Use substitution.
$$(y + 5) + y = 27$$
$$2y + 5 = 27$$
$$2y = 22$$
$$y = 11$$
$$x = y + 5; \ y = 11$$
$$x = 11 + 5$$
$$x = 16$$
Beyonce: -16; T.I.: -11

7) Let $x = $ number who speak Urdu
Let $y = $ number who speak Polish

Who speak Urdu $= \dfrac{1}{2}$(who speak Polish)
$$x = \dfrac{1}{2}y$$
who speak Urdu $+$ who speak Polish $= 975,000$
$$x + y = 975,000$$
Use substitution.
$$y + \dfrac{1}{2}(y) = 975,000$$
$$2y + y = 1,950,000$$
$$3y = 1,950,000$$
$$y = 650,000$$
$$x = \dfrac{1}{2}y; \ y = 650,000$$
$$x = \dfrac{1}{2}(650,000)$$
$$x = 325,000$$
Polish Speaking: 650,000;

Urdu Speaking: 325,000

9) Let $x = $ minutes by Yuri
let $y = $ minutes by Shepard.
Mins. by Yuri $= $ Mins. by Shepard $+ 93$
$$x = y + 93$$
Mins. by Yuri $+$ Mins. by Shepard $= 123$
$$x + y = 123$$
Use substitution.
$$y + (y + 93) = 123$$
$$2y + 93 = 123$$
$$2y = 30$$
$$y = 15$$
$$x = y + 93; \ y = 15$$
$$x = 93 + 15$$
$$x = 108$$

Minutes in space by Yuri = 108;

by Shepard = 15.

11) Let w = the width, and

let h = the height.

height = width − 50

$h = w - 50$

$2h + 2w = 220$

Use substitution.

$2(h - 50) + 2h = 220$

$4h - 100 = 220$

$4h = 320$

$h = 80$

$w = h - 50;\ h = 80$

$w = h - 50$

$w = 30$

Height: 80 in. width: 30 in.

13) Let w = the width, and

let l = the length

$2l + 2w = 343.6$

Length = width + 48.2

$l = w + 48.2$

Use substitution.

$2(w + 48.2) + 2w = 343.6$

$2w + 96.4 + 2w = 343.6$

$4w + 96.4 = 343.6$

$4w = 247.2$

$w = 61.8$

$l = w + 48.2;\ w = 61.8$

$l = 61.8 + 48.2$

$l = 110.0$

Length: 110 mm; width: 61.8 mm

15) Let w = the width, and

let l = the length

$l + 2w = 119$

length = 1.5(width)

$l = 1.5(w)$

Use substitution.

$1.5w + 2w = 119$

$3.5w = 119$

$7w = 238$

$w = 34$

$l = 1.5w;\ w = 34$

$l = 1.5(34)$

$l = 51$

length: 51 ft; width: 34 ft.

17) $x° = \frac{3}{5}y°$

supplementary angles

$x° + y° = 180°$

Use substitution.

$\frac{3}{5}y + y = 180$

$\frac{8}{5}y = 180$

$y = 112.5$

$x = \frac{3}{5}y;\ y = 112.5$

$x = \frac{3}{5}(112.5)$

$x = 67.5$

$m\angle x = 67.5°;\ m\angle y = 112.5°$

19) Let x = cost of a t-shirt

and let y = cost of a souvenir puck.

Kenny's Purchase:

t-shirt+2 puck = 36

$x + 2y = 36$

Kyle's Purchase:

2 t-shirts+3 pucks = 64

$2x + 3y = 64$

$x + 2y = 36$

$-2(x + 2y) = -2(36)$

$-2x - 4y = -72$

Add the equations.

$-2x - 4y = -72$

$\underline{2x + 3y = 64}$

$-y = -8$

$y = 8$

Substitute $y = 8$ into

$x + 2y = 36$

$x + 2(8) = 36$

$x + 16 = 36$

$x = 20$

t-shirt: \$20; souvenir puck: \$8

21) Let x = cost of a Bobblehead

let y = cost of a mug

3(bobblehead)+4(mug) $= 105$

$3x + 4y = 105$

2(bobblehead)+3(mug) $= 74$

$2x + 3y = 74$

$2(3x + 4y) = 2(105)$

$6x + 8y = 210$

$-3(2x + 3y) = -3(74)$

$-6x - 9y = -222$

Add the equations

$-6x - 9y = -222$

$\underline{6x + 8y = 210}$

$-y = -12$

$y = 12$

Substitute $y = 12$ into

$2x + 3y = 74$

$2x + 3(12) = 74$

$2x + 36 = 74$

$2x = 38$

$x = 19$

Bobblehead:\$19.00; Mug:\$12.00

23) Let x = cost of hamburger

let y = cost of fries.

$5 \cdot$ hamburgers $+ 1 \cdot$ fries $= 4.44$

$5x + y = 4.44$

$4 \cdot$ hamburgers $+ 2 \cdot$ fries $= 5.22$

$4x + 2y = 5.22$

$5x + y = 4.44$

$-2(5x + y) = -2(4.44)$

$-10x - 2y = -8.88$

Add the equations.

$-10x - 2y = -8.88$

$\underline{4x + 2y = 5.22}$

$-6x = -3.66$

$x = 0.61$

Substitute $x = 0.61$ into

$5x + y = 4.44$

$5(0.61) + y = 4.44$

$3.05 + y = 4.44$

$y = 1.39$

hamburger : \$0.61; fries: \$1.39

25) Let x = wrapping paper

let y = gift bag

$4 \cdot$ wrapping paper $+ 3 \cdot$ gift bag $= 52$

$4x + 3y = 52$

$3 \cdot$ wrapping paper $+ 1 \cdot$ gift bag $= 29$

$3x + y = 29$

$3x + y = 29$

$-3(3x + y) = -3(29)$

$-9x - 3y = -87$

Add the equations

$$-9x - 3y = -87$$
$$\underline{4x + 3y = 52}$$
$$-5x = -35$$
$$x = 7$$

Substitute $x = 7$ into

$$4x + 3y = 52$$
$$4(7) + 3y = 52$$
$$28 + 3y = 52$$
$$3y = 24$$
$$y = 8$$

wrapping paper: $7.00;
gift bag: $8.00

27) x = number of ounces of 9% solution

y = number of ounces of 17% solution

Soln	Concn	No of Oz of soln	No of Oz of alcohol in the soln
9%	0.09	x	$0.09x$
17%	0.17	y	$0.17y$
15%	0.15	12	$0.15(12)$

$$x + y = 12$$
$$y = 12 - x$$
$$0.09x + 0.17y = 0.15(12)$$
$$100(0.09x + 0.17y) = 100\big[0.15(12)\big]$$
$$9x + 17y = 15(12)$$

Use substitution.

$$9x + 17(12 - x) = 15(12)$$
$$9x + 204 - 17x = 180$$
$$-8x + 204 = 180$$
$$-8x = -24$$
$$x = 3$$

$$y = 12 - x;\ x = 3$$
$$y = 12 - 3$$
$$y = 9$$
$$9\% : 3\,oz;\ 17\% : 9\,oz$$

29) x = amount of 25% acid solution
y = amount of 100% acid solution

Solution	Amount	Concn.	Amount in Solution
25%	x	0.25	$0.25x$
100%	y	1.00	y
40%	10	0.40	$0.40(10)$

$$x + y = 10 \qquad\qquad 0.25x + y = 0.4(10)$$
$$y = 10 - x \quad 100(0.25x + y) = 100\big[0.4(10)\big]$$
$$25x + 100y = 400$$

Use substitution.

$$25x + 100(10 - x) = 400$$
$$25x + 1000 - 100x = 400$$
$$-75x + 1000 = 400$$
$$-75x = -600$$
$$x = 8$$
$$y = 10 - x;\ x = 8$$
$$y = 10 - 8$$
$$y = 2$$

Pure acid: 2 Liters; 25% Solution: 8 Liters

31) x = amount of Asian Treasure tea
y = amount of Pearadise tea

$$x = 60 - y$$
$$x + y = 60$$
$$x(7.5) + y(5.0) = 60(6.0)$$
$$7.5x + 5.0y = 360$$

Use substitution.

$$x = 60 - y$$
$$7.5(60 - y) + 5y = 360$$
$$450 - 7.5y + 5y = 360$$
$$-2.5y = -90$$
$$y = 36$$
$$x = 60 - y; \; y = 36$$
$$x = 60 - 36$$
$$x = 24$$

Asian Treasure: 24 oz; Pearadise: 36 oz.

33) x = sodium in chalupa;

y = sodium in crunchy taco

$$x + 3y = 1640$$
$$2x + 2y = 1960$$

Use substituion.
$$x = 1640 - 3y$$
$$2(1640 - 3y) + 2y = 1960$$
$$3280 - 6y + 2y = 1960$$
$$3280 - 4y = 1960$$
$$-4y = -1320$$
$$y = 330$$
$$x = 1640 - 3y; \; y = 330$$
$$x = 1640 - 3(330)$$
$$x = 1640 - 990$$
$$x = 650$$

sodium in chicken chalupa: 650 mg;

sodium in crunchy taco: 330 mg

35) x = Amount invested at 2%

y = Amount invested at 4%

$$x + y = 6000 \qquad x(0.02) + y(0.04) = 190$$
$$y = 6000 - x \quad 100(0.02x + 0.04y) = 100(190)$$
$$2x + 4y = 19000$$

Use substitution.
$$2x + 4(6000 - x) = 19,000$$
$$2x + 24,000 - 4x = 19,000$$
$$24,000 - 2x = 19,000$$
$$-2x = -5000$$
$$x = 2500$$
$$y = 6000 - x; \; x = 2500$$
$$y = 6000 - 2500$$
$$y = 3500$$

Amount invested at 2%: $2500;

Amount invested at 4%: $3500

37) x = Number of $0.44 stamps;

y = Number of $0.28 stamps

$$x + y = 16 \qquad 0.44x + 0.28y = 6.40$$
$$x = 16 - y \quad 100(0.44x + 0.28y) = 100(6.40)$$
$$44x + 28y = 640$$

Use substitution.
$$44(16 - y) + 28y = 640$$
$$704 - 44y + 28y = 640$$
$$704 - 16y = 640$$
$$-16y = -64$$
$$y = 4$$
$$x = 16 - y; \; y = 4$$
$$x = 16 - 4$$
$$x = 12$$

$0.44 Stamps: 12; $0.28 Stamps: 4

39) $x =$ Michael's Speed; $y =$ Jan's Speed

$x = y + 1$

Distance Traveled in 3 Hrs.

Michael: $3x = 3(y+1)$

Jan: $3y$

Since Jan and Michael traveled in opposite direction,

$3(y+1) + 3y = 51$

$3y + 3 + 3y = 51$

$6y + 3 = 51$

$6y = 48$

$y = 8$

$x = y + 1$

$x = 8 + 1$

$x = 9$

Michael: 9 mph; Jan: 8 mph

41) $x =$ Speed of small plane;

$y =$ Speed of jet

$x = y - 160$

$2x + 2y = 1280$

Use substitution.

$2(y-160) + 2y = 1280$

$2y - 320 + 2y = 1280$

$4y - 320 = 1280$

$4y = 1600$

$y = 400$

$x = y - 160; y = 400$

$x = 400 - 160$

$x = 240$

Speed of small plane: 240 mph;

Speed of jet: 400 mph

43) $x =$ Pam's Speed; $y =$ Jim's Speed

$x = y - 2$ $\qquad$ $\dfrac{1}{2}x + \dfrac{1}{2}y = 9$

$\qquad\qquad\qquad 2\left(\dfrac{1}{2}x + \dfrac{1}{2}y\right) = 2(9)$

$\qquad\qquad\qquad\qquad x + y = 18$

Use Substitution

$(y-2) + y = 18$ $\qquad x = y - 2; y = 10$

$y - 2 + y = 18$ $\qquad\quad x = 10 - 2$

$2y - 2 = 18$ $\qquad\qquad x = 8$

$2y = 20$

$y = 10$

Pam's speed: 8 mph;

Jim's speed: 10 mph

45) $2(x+y) = 14$

$2x + 2y = 14$

$2(x-y) = 10$

$2x - 2y = 10$

Add the equations

$\begin{array}{r} 2x + 2y = 14 \\ 2x - 2y = 10 \\ \hline 4x \quad\quad = 24 \end{array}$

$x = 6$

Substitute $x = 6$ into

$2x + 2y = 14$

$2(6) + 2y = 14$

$12 + 2y = 14$

$2y = 2$

$y = 1$

Speed of boat in still water: 6 mph;

Speed of current: 1 mph

47) $x =$ Speed of boat;

$y =$ Speed of current

$$5x + 5y = 80$$
$$8(5x + 5y) = 8(80)$$
$$40x + 40y = 640$$
$$8x - 8y = 80$$
$$5(8x - 8y) = 5(80)$$
$$40x - 40y = 400$$

Add the equations.

$$40x - 40y = 400$$
$$\underline{40x + 40y = 640}$$
$$80x = 1040$$
$$x = 13$$

Substitute $x = 13$ into

$$5x + 5y = 80$$
$$5(13) + 5y = 80$$
$$65 + 5y = 80$$
$$5y = 15$$
$$y = 3$$

Speed of boat: 13 mph;

Speed of current: 3 mph

49) $x =$ Speed of jet;

$y =$ Speed of wind

Add the equations

$$2.5x - 2.5y = 1000$$
$$\underline{2.5x + 2.5y = 1250}$$
$$5x = 2250$$
$$x = 450$$

Substitute $x = 450$ into

$$2.5x + 2.5y = 1250$$
$$2.5(450) + 2.5y = 1250$$
$$1125 + 2.5y = 1250$$
$$2.5y = 125$$
$$y = 50$$

Speed of jet: 450 mph;

Speed of wind: 50 mph

Section 5.5 Exercises

1) $4x + 3y - 7z = -6$

$$4(-2) + 3(3) - 7(1) \overset{?}{=} -6$$
$$-8 + 9 - 7 \overset{?}{=} -6$$
$$-6 = -6$$

$$x - 2y + 5z = -3$$
$$(-2) - 2(3) + 5(1) \overset{?}{=} -3$$
$$-2 - 6 + 5 \overset{?}{=} -3$$
$$-3 = -3$$

$$-x + y + 2z = 7$$
$$-(-2) + (3) + 2(1) \overset{?}{=} 7$$
$$2 + 3 + 2 \overset{?}{=} 7$$
$$7 = 7$$

The ordered triple is a solution of the system.

3) $-x + y - 2z = 2$

$$-(0) + (6) - 2(2) \overset{?}{=} 2$$
$$6 - 4 \overset{?}{=} 2$$
$$2 = 2$$

$$3x - y + 5z = 4$$
$$3(0) - (6) + 5(2) \overset{?}{=} 4$$
$$0 - 6 + 10 \overset{?}{=} 4$$
$$4 = 4$$

$2x + 3y - z = 7$

$2(0) + 3(6) - (2) \overset{?}{=} 7$

$0 + 18 - 2 \overset{?}{=} 7$

$16 \neq 7$

The ordered triple is not a solution of the system.

5) Answers may vary

7) $\boxed{\text{I}}$ $x + 3y + z = 3$

$\boxed{\text{II}}$ $4x - 2y + 3z = 7$

$\boxed{\text{III}}$ $-2x + y - z = -1$

Add $\boxed{\text{I}}$ and $\boxed{\text{III}}$

$x + 3y + z = 3$

$\underline{-2x + y - z = -1}$

$\boxed{\text{A}}$ $-x + 4y = 2$

Add $\boxed{\text{I}} \cdot (-3)$ and $\boxed{\text{II}}$

$-3x - 9y - 3z = -9$

$\underline{4x - 2y + 3z = 7}$

$\boxed{\text{B}}$ $x - 11y = -2$

Add $\boxed{\text{A}}$ and $\boxed{\text{B}}$

$-x + 4y = 2$

$\underline{x - 11y = -2}$

$-7y = 0$

$y = 0$

Substitute $y = 0$ into $\boxed{\text{A}}$

$-x + 4(0) = 2$

$-x + 0 = 2$

$x = -2$

Substitute $x = -2$ and $y = 0$ into $\boxed{\text{I}}$

$(-2) + 6(0) + z = 3$

$-2 + 0 + z = 3$

$z = 5$

$(-2, 0, 5)$

9) $\boxed{\text{I}}$ $5x + 3y - z = -2$

$\boxed{\text{II}}$ $-2x + 3y + 2z = 3$

$\boxed{\text{III}}$ $-x + 6y + z = -1$

Add $\boxed{\text{II}}$ and $\boxed{\text{I}} \cdot (2)$

$-2x + 3y + 2z = 3$

$\underline{10x + 6y - 2z = -4}$

$\boxed{\text{A}}$ $8x + 9y = -1$

Add $\boxed{\text{III}} \cdot (-2)$ and $\boxed{\text{II}}$

$-2x - 12y - 2z = 2$

$\underline{-2x + 3y + 2z = 3}$

$\boxed{\text{B}}$ $-4x - 9y = 5$

Add $\boxed{\text{A}}$ and $\boxed{\text{B}}$

$8x + 9y = -1$

$\underline{-4x - 9y = 5}$

$4x = 4$

$x = 1$

Substitute $x = 1$ into $\boxed{\text{B}}$

$-4(1) - 9y = 5$

$-9y = 9$

$y = -1$

Substitute $y = -1$ and $x = 1$ into $\boxed{\text{III}}$

$1 + 6(-1) + z = -1$

$-5 + z = -1$

$z = 4$

$(1, -1, 4)$

11) $\boxed{\text{I}}$ $-3a + 5b - 3c = -4$

$\boxed{\text{II}}$ $a - 3b + c = 6$

$\boxed{\text{III}}$ $-4a + 6b + 2c = -6$

Add $\boxed{\text{II}} \cdot (3)$ and $\boxed{\text{I}}$

$3a - 9b + 3c = 18$

$\underline{3a + 5b - 3c = -4}$

$\boxed{A}\ 6a - 4b = 14$

Add $\boxed{II}\cdot(-2)$ and $\boxed{III}$

$-2a + 6b - 2c = -12$

$\underline{-4a + 6b + 2c = -6}$

$\boxed{B} -6a + 12b = -18$

Add $\boxed{A}$ and $\boxed{B}$

$6a - 4b\ =\ 14$

$\underline{-6a + 12b = -18}$

$\qquad\quad 8b = -4$

$\qquad\quad b = -\dfrac{1}{2}$

Substitute $b = -\dfrac{1}{2}$ into $\boxed{A}$

$6a - 4\left(-\dfrac{1}{2}\right) = 14$

$6a + 2 = 14$

$6a = 12$

$a = 2$

Substitute $a = 2$ and $b = -\dfrac{1}{2}$ into $\boxed{III}$

$-4(2) + 6\left(-\dfrac{1}{2}\right) + 2c = -6$

$-8 - 3 + 2c = -6$

$2c = 5$

$c = \dfrac{5}{2}$

$\left(2, -\dfrac{1}{2}, \dfrac{5}{2}\right)$

13) $\boxed{I}\qquad a - 5b + c = -4$

$\boxed{II}\quad 3a + 2b - 4c = -3$

$\boxed{III}\quad 6a + 4b - 8c = 9$

Add $\boxed{II}$ and $\boxed{I}\cdot(4)$

$3a +\ 2b - 4c = -\ 3$

$\underline{4a - 20b + 4c = -16}$

$\boxed{A}\ 7a - 18b = -19$

Add $\boxed{III}$ and $\boxed{II}\cdot(-2)$

$6a + 4b - 8c = 9$

$\underline{-6a - 4b + 8c = 6}$

$\boxed{B}\qquad\quad 0 \neq 15$

$\varnothing$; Inconsistent

15) $\boxed{I}\quad -15x - 3y + 9z = 3$

$\boxed{II}\qquad 5x + y - 3z = -1$

$\boxed{III}\quad 10x + 2y - 6z = -2$

Equation $\boxed{II} = \dfrac{1}{2}\boxed{III}$

Equation $\boxed{II} = -\dfrac{1}{3}\boxed{I}$

$\{(x, y, z) \mid 5x + y - 3z = -1\}$; Dependent

17) $\boxed{I}\quad -3a + 12b - 9c = -3$

$\boxed{II}\quad 5a - 20b + 15c = 5$

$\boxed{III}\quad -a + 4b - 3c = -1$

Equation $\boxed{III}\ =\ -\dfrac{1}{3}\boxed{I}$

Equation $\boxed{II}\ =\ -5\boxed{III}$

$\{(a, b, c) \mid -a + 4b - 3c = -1\}$; Dependent

Section 5.5: Solving Systems Three Equations and Applications

19) $\boxed{\text{I}}$ $5x-2y+z=-5$

$\boxed{\text{II}}$ $x-y-2z=7$

$\boxed{\text{III}}$ $4y+3z=5$

Add $\boxed{\text{I}}$ and $(-5)\cdot\boxed{\text{II}}$

$5x-2y+\ z=-\ 5$

$-5x+5y+10z=-35$

$\boxed{\text{A}}$ $3y+11z=-40$

Add $\boxed{\text{III}}\cdot(-3)$ and $\boxed{\text{A}}\cdot(4)$

$-12y-9z=-15$

$12y+44z=-160$

$35z=-175$

$z=-5$

Substitute $z=-5$ into $\boxed{\text{A}}$

$3y+11(-5)=-40$

$3y-55=-40$

$3y=15$

$y=5$

Substitute $z=-5$ and $y=5$ into $\boxed{\text{II}}$

$x-(5)-2(-5)=7$

$x+10=7$

$x=-3$

$(-3,5,-5)$

21) $\boxed{\text{I}}$ $a+15b=5$

$\boxed{\text{II}}$ $4a+10b+c=-6$

$\boxed{\text{III}}$ $-2a-5b-2c=-3$

Add $\boxed{\text{II}}$ and $\boxed{\text{III}}\cdot(2)$

$4a+10b+\ c=-6$

$-4a-10b-4c=-6$

$-3c=-12$

$c=4$

Substitute $c=4$ into $\boxed{\text{III}}$

$-2a-5b-2(4)=-3$

$-2a-5b-8=-3$

$\boxed{\text{A}}$ $-2a-5b=5$

Add $\boxed{\text{I}}$ and $\boxed{\text{A}}\cdot(3)$

$a+15b=5$

$-6a-15b=15$

$-5a=20$

$a=-4$

Substitute $a=-4$ into $\boxed{\text{I}}$

$-4+15b=5$

$15b=9$

$b=\dfrac{3}{5}$

$\left(-4,\dfrac{3}{5},4\right)$

23) $\boxed{\text{I}}$ $x+2y+3z=4$

$\boxed{\text{II}}$ $-3x+y=-7$

$\boxed{\text{III}}$ $4y+3z=-10$

Add $\boxed{\text{I}}\cdot(3)$ and $\boxed{\text{II}}$

$3x+6y+9z=12$

$-3x+y\ \ \ \ \ =-7$

$\boxed{\text{A}}$ $7y+9z=5$

Add $\boxed{\text{III}}\cdot(-3)$ and $\boxed{\text{A}}$

$-12y-9z=30$

$7y+9z=\ 5$

$-5y=35$

$y=-7$

Substitute $y=-7$ into $\boxed{\text{III}}$

$4(-7)+3z=-10$

$-28+3z=-10$

$3z=18$

$z=6$

Substitute $y=-7$ into $\boxed{\text{II}}$

$-3x+(-7)=-7$

$-3x=0$

$x=0$

$(0,-7,6)$

25) $\boxed{\text{I}}\quad -5x+z=-3$

$\boxed{\text{II}}\quad 4x-y=-1$

$\boxed{\text{III}}\quad 3y-7z=1$

Add $\boxed{\text{III}}$ and $\boxed{\text{I}}\cdot(7)$

$\quad 3y-7z=\quad 1$

$\underline{-35x\quad\quad +7z=-21}$

$\boxed{\text{A}}\;-35x+3y=-20$

Add $\boxed{\text{II}}\cdot(3)$ and $\boxed{\text{A}}$

$12x-3y=-3$

$\underline{-35x+3y=-20}$

$-23x=-23$

$x=1$

Substitute $x=1$ into $\boxed{\text{II}}$

$4(1)-y=-1$

$-y=-5$

$y=5$

Substitute $y=5$ into $\boxed{\text{III}}$

$3y-7z=1$

$3(5)-7z=1$

$15-7z=1$

$-7z=-14$

$z=2$

$(1,5,2)$

27) $\boxed{\text{I}}\quad 4a+2b=-11$

$\boxed{\text{II}}\;-8a-3c=-7$

$\boxed{\text{III}}\quad b+2c=1$

Add $\boxed{\text{III}}\cdot(-2)$ and $\boxed{\text{I}}$

$-2b-4c=-2$

$\underline{4a+2b\quad\quad=-11}$

$\boxed{\text{A}}\;4a-4c=-13$

Add $\boxed{\text{A}}\cdot(2)$ and $\boxed{\text{II}}$

$8a-8c=-26$

$\underline{-8a-3c=-7}$

$-11c=-33$

$c=3$

Substitute $c=3$ into $\boxed{\text{III}}$

$b+2(3)=1$

$b+6=1$

$b=-5$

Substitute $b=-5$ into $\boxed{\text{I}}$

$4a+2(-5)=-11$

$4a-10=-11$

$4a=-1$

$a=-\dfrac{1}{4}$

$\left(-\dfrac{1}{4},-5,3\right)$

29) $\boxed{\text{I}}\quad 6x+3y-3z=-1$

$\boxed{\text{II}}\quad 10x+5y-5z=4$

$\boxed{\text{III}}\quad x-3y+4z=6$

Add $\boxed{\text{II}}\cdot(-3)$ and $\boxed{\text{I}}\cdot(5)$

$-30x-15y+15z=-12$

$\underline{30x+15y-15z=-5}$

$0\neq-17$

$\varnothing$; Inconsistent

31) $\boxed{\text{I}}$ $7x+8y-z=16$

$\boxed{\text{II}}$ $-\dfrac{1}{2}x-2y+\dfrac{3}{2}z=1$

$\boxed{\text{III}}$ $\dfrac{4}{3}x+4y-3z=-\dfrac{2}{3}$

Add equations $\boxed{\text{I}}\cdot(3)$ and $\boxed{\text{II}}\cdot(2)$

$21x+24y-3z=48$

$\underline{-x-\ \ 4y+3z=2}$

$\boxed{\text{A}}\ 20x+20y=50$

Add equations $\boxed{\text{I}}\cdot(9)$ and $\boxed{\text{III}}\cdot(-3)$

$63x+72y-9z=144$

$\underline{-4x-12y+9z=2}$

$\boxed{\text{B}}\ \ 59x+60y=146$

Add equations $\boxed{\text{A}}\cdot(-3)$ and $\boxed{\text{B}}$

$-60x-60y=-150$

$\underline{\ \ 59x+60y=146}$

$-x=-4$

$x=4$

Substitute $x=4$ into $\boxed{\text{A}}$

$20(4)+20y=50$

$80+20y=50$

$20y=-30$

$y=-\dfrac{3}{2}$

Substitute $x=4$ and $y=-\dfrac{3}{2}$ into $\boxed{\text{I}}$

$7(4)+8\left(-\dfrac{3}{2}\right)-z=16$

$28-12-z=16$

$z=0$

$\left(4,-\dfrac{3}{2},0\right)$

33) $\boxed{\text{I}}$ $2a-3b=-4$

$\boxed{\text{II}}$ $3b-c=8$

$\boxed{\text{III}}\ -5a+4c=-4$

Add Equations $\boxed{\text{I}}$ and $\boxed{\text{II}}$.

$2a-3b\ \ \ \ =-4$

$\underline{\ \ \ \ \ 3b-c=8}$

$\boxed{\text{A}}\ 2a\ \ \ \ -c=4$

Add equations $4\cdot\boxed{\text{A}}$ and $\boxed{\text{III}}$

$8a-4c=16$

$\underline{-5a+4c=-4}$

$3a=12$

$a=4$

Substitute $a=4$ into $\boxed{\text{I}}$

$2(4)-3b=-4$

$8-3b=-4$

$-3b=-12$

$b=4$

Substitute $b=4$ into $\boxed{\text{II}}$

$3(4)-c=8$

$12-c=8$

$c=4$

$(4,4,4)$

35) $\boxed{\text{I}}$ $-4x+6y+3z=3$

$\boxed{\text{II}}$ $-\dfrac{2}{3}x+y+\dfrac{1}{2}z=\dfrac{1}{2}$

$\boxed{\text{III}}$ $12x-18y-9z=-9$

Equations $\boxed{\text{I}}\cdot(-3)$ and $\boxed{\text{III}}$ are equivalent.

Equations $\boxed{\text{I}}$ and $\boxed{\text{II}}\cdot(6)$ are equivalent.

$\{(x,y,z)\,|-4x+6y+3z=3\}$; dependent

37) $\boxed{\text{I}}$ $\quad a+b+9c=-3$

$\boxed{\text{II}}$ $\quad -5a-2b+3c=10$

$\boxed{\text{III}}$ $\quad 4a+3b+6c=-15$

Add equations $\boxed{\text{I}}$ and $\boxed{\text{II}}\cdot(-3)$

$a+\ b+9c=-3$

$\underline{15a+6b-9c=-30}$

A $\quad 16a+7b=-33$

Add equations $\boxed{\text{II}}\cdot(-2)$ and $\boxed{\text{III}}$

$10a+4b-6c=-20$

$\underline{4a+3b+6c=-15}$

$\boxed{\text{B}}\ 14a+7b=-35$

Add equations $A\cdot(-1)$ and B

$-16a-7b=33$

$\underline{14a+7b=-35}$

$\quad -2a=-2$

$a=1$

Substitute $a=-1$ into A

$16(1)+7b=-33$

$16+7b=-33$

$7b=-49$

$b=-7$

Substitute $a=1$ and $b=-7$ into I

$1+(-7)+9c=-3$

$-6+9c=-3$

$9c=3$

$c=\dfrac{1}{3}$

$\left(1,-7,\dfrac{1}{3}\right)$

39) $\boxed{\text{I}}$ $\quad x+5z=10$

$\boxed{\text{II}}$ $\quad 4y+z=-2$

$\boxed{\text{III}}$ $\quad 3x-2y=2$

Add equations $\boxed{\text{I}}\cdot$ and $\boxed{\text{II}}\cdot(-5)$

$x\quad\ +5z=10$

$\underline{-20y-5z=10}$

$\boxed{\text{A}}\ x-20y=20$

Add equations $\boxed{\text{A}}$ and $\boxed{\text{III}}\cdot(-10)$

$x-20y=20$

$\underline{-30x+20y=-20}$

$\quad -29x=0$

$\quad x=0$

Substitute $x=0$ into $\boxed{\text{III}}$

$3(0)-2y=2$

$-2y=2$

$y=-1$

Substitute $y=-1$ into $\boxed{\text{II}}$

$4(-1)+z=-2$

$-4+z=-2$

$z=2$

$(0,-1,2)$

41) $\boxed{\text{I}}$ $\quad 2x-y+4z=-1$

$\boxed{\text{II}}$ $\quad x+3y+z=-5$

$\boxed{\text{III}}$ $\quad -3x+2y=7$

Add equations $\boxed{\text{I}}$ and $\boxed{\text{II}}\cdot(-4)$

$2x-\ y+4z=-1$

$\underline{-4x-12y-4z=20}$

$\boxed{\text{A}}\ -2x-13y=19$

Add equations $\boxed{A} \cdot (-3)$ and $\boxed{III} \cdot (2)$

$6x + 39y = -57$

$\underline{-6x + 4y = 14}$

$43y = -43$

$y = -1$

Substitute $y = -1$ into $\boxed{III}$

$-3x + 2(-1) = 7$

$-3x - 2 = 7$

$-3x = 9$

$x = -3$

Substitute $x = -3$ and $y = -1$ into $\boxed{II}$

$-3 + 3(-1) + z = -5$

$-6 + z = -5$

$z = 1$

$(-3, -1, 1)$

43) Answers may vary.

45) $x =$ Hot Dogs; $y =$ Fries; $z =$ Soda

$\boxed{I} \quad 2x + 2y + z = 9.0$

$\boxed{II} \quad 2x + y + 2z = 9.50$

$\boxed{III} \quad 3x + 2y + z = 11.00$

Add equations $\boxed{II}$ and $\boxed{I} \cdot (-2)$

$2x + y + 2z = 9.5$

$\underline{-4x - 4y - 2z = -18}$

$\boxed{A} \quad -2x - 3y = -8.5$

Add equations $\boxed{II}$ and $\boxed{III} \cdot (-2)$

$2x + y + 2z = 9.50$

$\underline{-6x - 4y - 2z = -22}$

$\boxed{B} \quad -4x - 3y = -12.5$

Add equations $\boxed{A}$ and $\boxed{B} \cdot (-1)$

$-2x - 3y = -8.5$

$\underline{4x + 3y = 12.5}$

$2x = 4.0$

$x = 2.0$

Substitute $x = 2$ into $\boxed{A}$

$-2(2) - 3y = -8.5$

$-4 - 3y = -8.5$

$-3y = -4.5$

$y = 1.5$

Substitute $x = 2$ and $y = 1.5$ into $\boxed{I}$

$2(2) + 2(1.5) + z = 9$

$4 + 3 + z = 9$

$z = 2$

Hot Dog: \$2.00; Fries: \$1.50 and Soda: \$2.00

47) $x =$ Protein in Clif Bar;

$y =$ Protein in Balance Bar;

$z =$ Protein in PowerBar

$\boxed{I} \qquad x = y - 4$

$\boxed{II} \qquad z = y + 9$

$\boxed{III} \quad x + y + z = 50$

Substitute $\boxed{I}$ and $\boxed{II}$ into $\boxed{III}$

$(y - 4) + y + (y + 9) = 50$

$3y + 5 = 50$

$3y = 45$

$y = 15$

Substitute $y = 15$ into $\boxed{\text{I}}$

$x = 15 - 4$

$x = 11$

Substitute $y = 15$ into $\boxed{\text{II}}$

$z = 15 + 9$

$z = 24$

Protein in Clif Bar: 11g;

Protein in PowerBar: 24g;

Protein in Balance Bar: 15g

49) $x =$ Knicks Revenue;

$y =$ Lakers Revenue;

$z =$ Bulls Revenue

$\boxed{\text{I}} \quad x + y + z = 428$

$\boxed{\text{II}} \qquad\quad y = z + 30$

$\boxed{\text{III}} \qquad\quad x = y + 11$

Substitute $\boxed{\text{II}}$ and $\boxed{\text{III}}$ in $\boxed{\text{I}}$

$(z + 30) + 11 + (z + 30) + z = 428$

$3z + 71 = 428$

$3z = 357$

$z = 119$

Substitute $z = 119$ into $\boxed{\text{II}}$

$y = 119 + 30$

$y = 149$

Substitute $y = 149$ into $\boxed{\text{III}}$

$x = 149 + 11$

$x = 160$

Knicks Revenue: 160 million;

Lakers Revenue: 149 Million and

Bulls Revenue : 119 Million

51) $x =$ Value Date;

$y =$ Regular Date;

$z =$ Prime Date

$\boxed{\text{I}} \qquad 4x + 4y + 3z = 367$

$\boxed{\text{II}} \qquad 3x + 3y + z = 219$

$\boxed{\text{III}} \qquad 4x + 3y + 2z = 286$

Add equations $\boxed{\text{I}}$ and $\boxed{\text{II}} \cdot (-3)$

$4x + 4y + 3z = 367$

$\underline{-9x - 9y - 3z = -657}$

$\boxed{\text{A}} \ -5x - 5y = -290$

Add equations $\boxed{\text{III}}$ and $\boxed{\text{II}} \cdot (-2)$

$4x + 3y + 2z = 286$

$\underline{-6x - 6y - 2z = -438}$

$\boxed{\text{B}} \ -2x - 3y = -152$

Add equations $\boxed{\text{A}} \cdot (2)$ and $\boxed{\text{B}} \cdot (-5)$

$-10x - 10y = -580$

$\underline{10x + 15y = 760}$

$5y = 180$

$y = 36$

Substitute $y = 36$ into $\boxed{\text{B}}$

$-2x - 3(36) = -152$

$-2x - 108 = -152$

$-2x = -44$

$x = 22$

Substitute $x = 22$ and $y = 36$ into $\boxed{\text{II}}$

$3(22) + 3(36) + z = 219$

$66 + 108 + z = 219$

$z = 219 - 174$

$z = 45$

Value Tickets: \$22;

Regular: \$36;

Prime: \$45

53) $x =$ Largest Angle;

$y =$ Middle Angle;

$z =$ Smallest Angle

I $\quad\quad\quad x = 2y$

II $\quad\quad\quad z = y - 28$

III $\quad\quad\quad x + y + z = 180$

Substitute I and II into III

$(2y) + y + (y - 28) = 180$

$4y = 208$

$y = 52$

Substitute $y = 52$ into I

$x = 2(52)$

$x = 104$

Substitute $y = 52$ into II

$z = 52 - 28$

$z = 24$

Largest Angle: 104°;

Middle Angle: 52°;

Smallest Angle: 24°

55) $x =$ Smallest Angle;

$y =$ Largest Angle;

$z =$ Middle Angle

I $\quad\quad\quad x = y - 44$

II $\quad\quad\quad x + z = y + 20$

III $\quad\quad\quad x + y + z = 180$

Substitute II into III

$y + (y + 20) = 180$

$2y = 160$

$y = 80$

Substitute $y = 80$ into I

$x = 80 - 44$

$x = 36$

Substitute $x = 36$ and $y = 80$ into III

$36 + 80 + z = 180$

$z = 64$

Smallest Angle: 36°;

Largest Angle: 80°;

Middle Angle: 64°

57) $x =$ Longest Side;

$y =$ Shortest Side;

$z =$ Middle side

I $\quad\quad\quad x + y + z = 29$

II $\quad\quad\quad x = y + 5$

III $\quad\quad\quad y + z = x + 5$

Substitute III into I

$x + (x + 5) = 29$

$2x = 24$

$x = 12$

Substitute $x = 12$ into II

$12 = y + 5$

$y = 7$

Substitute $x = 12$ and $y = 7$ into III

$7 + z = 12 + 5$

$z = 10$

Largset Side: 12 cm;

Smallest Side: 7 cm;

Middle Side: 10 cm

59) $\boxed{\text{I}}$ $\quad a-2b-c+d=0$

$\boxed{\text{II}}$ $\quad -a+2b+3c+d=6$

$\boxed{\text{III}}$ $\quad 2a+b+c-d=8$

$\boxed{\text{IV}}$ $\quad a-b+2c+3d=7$

Add Equations $\boxed{\text{I}}$ and $\boxed{\text{III}}$.

$a-2b-c+d=0$

$\underline{2a+b+c-d=0}$

$\boxed{\text{A}} \quad 3a-b \quad\quad =8$

Add Equations $\boxed{\text{II}}$ and $\boxed{\text{III}}$.

$-a+2b+3c+d=6$

$\underline{2a+b+c-d=8}$

$\boxed{\text{B}} \quad a+3b+4c \quad =14$

Add Equations $-3\cdot(\boxed{\text{I}})$ and $\boxed{\text{IV}}$.

$-3a+6b+3c-3d=0$

$\underline{a-b+2c+3d=7}$

$\boxed{\text{C}} \quad -2a+5b+5c \quad =7$

We now have the system of three equations:

$\boxed{\text{A}} \quad 3a-b \quad\quad = 8$

$\boxed{\text{B}} \quad a+3b+4c \quad =14$

$\boxed{\text{C}} \quad -2a+5b+5c = 7$

Add Equations $5\cdot(\boxed{\text{B}})$ and $-4(\boxed{\text{C}})$.

$5a+15b+20c = 70$

$\underline{8a-20b-20c=-28}$

$\boxed{\text{D}} \; 13a-5b \quad = 42$

Add Equations $\boxed{\text{D}}$ and $-5(\boxed{\text{A}})$.

$13a-5b= 42$

$\underline{-15a+5b=-40}$

$-2a \quad\quad = 2$

$\quad a=-1$

Substitute $a=-1$ into Equation $\boxed{\text{D}}$.

$13(-1)-5b=42$

$-13-5b=42$

$-5b=55$

$b=-11$

Substitute $a=-1$ and $b=-11$ into $\boxed{\text{B}}$.

$-1+3(-11)+4c=14$

$-1-33+4c=14$

$4c=48$

$c=12$

Substitute $a=-1$, $b=-11$ and $c=12$ into $\boxed{\text{I}}$.

$(-1)-2(-11)-(12)+d=0$

$9+d=0$

$d=-9$

The solution set is $(-1,-11,12,-9)$.

61) $\boxed{\text{I}}$ $\quad 3a+4b+c-d = -7$

$\boxed{\text{II}}$ $\quad -3a-2b-c+d=1$

$\boxed{\text{III}}$ $\quad a+2b+3c-2d=5$

$\boxed{\text{IV}}$ $\quad 2a+b+c-d = 2$

Add Equations $\boxed{\text{I}}$ and $\boxed{\text{II}}$

$3a+4b+c-d = -7$

$\underline{-3a-2b-c+d=1}$

$2b \quad\quad =-6$

$\quad b=-3$

Add Equations $\boxed{\text{II}}$ and $\boxed{\text{IV}}$.

$-3a-2b-c+d=1$

$\underline{2a+b+c-d = 2}$

$\boxed{\text{A}} \quad -a-b \quad\quad =3$

Substitute $b=-3$ into $\boxed{\text{A}}$

$$-a-(-3)=3$$
$$-a+3=3$$
$$a=0$$

Add Equations $2\cdot(\boxed{\text{II}})$ and $\boxed{\text{III}}$.
$$-6a-4b-2c+2d=2$$
$$\underline{a+2b+3c-2d\quad=5}$$
$$\boxed{\text{B}}\quad -5a-2b+c\quad\quad=7$$

Substitute $a=0$ and $b=-3$ into $\boxed{\text{B}}$
$$-5(0)-2(-3)+c=7$$
$$6+c=7$$
$$c=1$$

Substitute $a=0$, $b=-3$, and $c=1$ into Equation $\boxed{\text{II}}$.
$$-3(0)-2(-3)-(1)+d=1$$
$$5+d=1$$
$$d=-4$$

The solution set is $(0,-3,1,-4)$.

Chapter 5 Review

1) No
$$x-5y=13$$
$$(-4)-5(-5)=13$$
$$-4+25=13$$
$$21\neq13$$

3) The Lines are Parallel

5) $\varnothing$

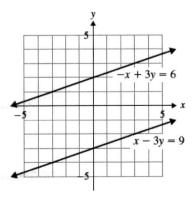

7) $(-3,-1)$

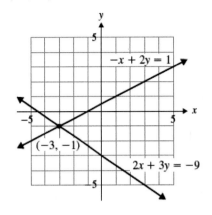

9) Infinite number of solutions: slopes and y-intercepts are the same

11)
$$9x-2y=8$$
$$y=2x+1$$
$$9x-2(2x+1)=8$$
$$9x-4x-2=8$$
$$5x=10$$
$$x=2$$
$$y=2x+1$$
$$y=2(2)+1$$
$$y=5\qquad(2,5)$$

13)
$$-x+8y=19$$
$$x=8y-19$$
$$4x-3y=11$$
$$4(8y-19)-3y=11$$
$$32y-3y-76=11$$
$$29y=87$$
$$y=3$$
$$x=8(3)-19$$
$$x=5\qquad(5,3)$$

Chapter 5: Solving Systems of Linear Equations

15) Add the equations

$$x - 7y = 3$$
$$-x + 5y = -1$$
$$-2y = 2$$
$$y = -1$$

Substitute $y = -1$ into

$$x - 7y = 3$$
$$x - 7(-1) = 3$$
$$x + 7 = 3$$
$$x = -4 \qquad (-4, -1)$$

17) Multiply the second equation by 2
$$5x - 2y = 4$$
$$10x - 4y = 8$$

Add the equations

$$-10x + 4y = -8$$
$$10x - 8y = 8$$
$$0 = 0$$

There are infinite solutions of the form
$$\{(x, y) \mid 5x - 2y = 4\}$$

19) $\boxed{\text{I}}$ $\quad 2x + 9y = -6$

$\boxed{\text{II}}$ $\quad 5x + y = 3$

Add equations $\boxed{\text{I}}$ and $\boxed{\text{II}} \cdot (-9)$

$$-45x - 9y = -27$$
$$2x + 9y = -6$$
$$-43x = -33$$
$$x = \frac{33}{43}$$

Add the equations $\boxed{\text{I}} \cdot (5)$ and $\boxed{\text{II}} \cdot (-2)$.

$$10x + 45y = -30$$
$$-10x - 2y = -6$$
$$43y = -36$$

$$y = -\frac{36}{43} \qquad \left(\frac{33}{43}, -\frac{36}{43}\right)$$

21) when one of the variables has a coefficiant of 1 or -1.

23) $6x + y = -8$
$$y = -6x - 8$$

Substitute $y = -6x - 8$ into

$$9x + 7y = -1$$
$$9x + 7(-6x - 8) = -1$$
$$9x - 42x - 56 = -1$$
$$-33x = 55$$
$$x = -\frac{5}{3}$$

Substitute $x = -\frac{5}{3}$ into

$$9x + 7y = -1$$
$$9\left(-\frac{5}{3}\right) + 7y = -1$$
$$-15 + 7y = -1$$
$$7y = 14$$
$$y = 2 \qquad \left(-\frac{5}{3}, 2\right)$$

25) $\boxed{\text{I}}$ $\quad \dfrac{1}{3}x - \dfrac{2}{9}y = -\dfrac{2}{3}$

$\boxed{\text{II}}$ $\quad \dfrac{5}{12}x + \dfrac{1}{3}y = 1$

Add equations $\boxed{\text{I}} \cdot (36)$ and $\boxed{\text{II}} \cdot (24)$

$12x - 8y = -24$

$\underline{10x + 8y = 24}$

$\qquad 22x = 0$

$\qquad\quad x = 0$

Substitute $x = 0$ into

$\quad 10x + 8y = 24$

$\qquad 0 + 8y = 24$

$\qquad\quad 8y = 24$

$\qquad\quad\; y = 3 \qquad\qquad (0, 3)$

27) $\quad 6(2x - 3) = y + 4(x - 3)$

$\qquad 12x - 18 = y + 4x - 12$

$\boxed{\text{I}} \qquad 8x - y = 6$

$\quad 5(3x + 4) + 4y = 11 - 3y + 27x$

$\quad 15x + 20 + 4y = 11 - 3y + 27x$

$\boxed{\text{II}} \quad -12x + y = -9$

Add equations $\boxed{\text{I}}$ and $\boxed{\text{II}}$

$\quad 8x - y = 6$

$\underline{-12x + y = -9}$

$\quad -4x = -3$

$\qquad x = \dfrac{3}{4}$

Substitute $x = \dfrac{3}{4}$ into $\boxed{\text{I}}$

$8\left(\dfrac{3}{4}\right) - y = 6$

$\qquad 6 - y = 6$

$\qquad\quad y = 0 \qquad \left(\dfrac{3}{4}, 0\right)$

29) $\boxed{\text{I}}$ $\quad \dfrac{3}{4}x - \dfrac{5}{4}y = \dfrac{7}{8}$

$\quad 4 - 2(x + 5) - y = 3(1 - 2y) + x$

$\quad 4 - 2x - 10 - y = 3 - 6y + x$

$\boxed{\text{II}} \quad -3x + 5y = 9$

Add equations $\boxed{\text{I}} \cdot (8)$ and $\boxed{\text{II}} \cdot (2)$

$\quad 6x - 10y = 7$

$\underline{-6x + 10y = 18}$

$\qquad\quad 0 \neq 25$

$\varnothing$; There is no solution

31) $x =$ white milk; $y =$ chocolate milk

$\boxed{\text{I}} \qquad\quad x = 2y$

$\boxed{\text{II}} \qquad x + y = 141$

Substitute $x = 2y$ into $\boxed{\text{II}}$

$\quad (2y) + y = 141$

$\qquad\quad 3y = 141$

$\qquad\quad\; y = 47$

Substitute $y = 47$ into $\boxed{\text{I}}$

$x = 2(47)$

$x = 94$

white milk: 94; chocolate milk: 47

33) x = Edwin's speed; y = Camille's speed

$\boxed{\text{I}}$ $\qquad x = y + 2$

$\boxed{\text{II}}$ $\qquad 0.5x + 0.5y = 7$

Substitute $x = y + 2$ into $\boxed{\text{II}}$

$0.5(y + 2) + 0.5y = 7$

$y + 1 = 7$

$y = 6$

Substitute $y = 6$ into $\boxed{\text{I}}$

$x = 6 + 2$

$x = 8$

Edwin's speed: 8 mph;

Camille's speed: 6 mph

35) x = width; y = length

$\boxed{\text{I}}$ $\qquad x = y - 5$

$\boxed{\text{II}}$ $\qquad 2x + 2y = 38$

Substitute $x = y - 5$ into $\boxed{\text{II}}$

$2(y - 5) + 2y = 38$

$4y - 10 = 38$

$4y = 48$

$y = 12$

Substitute $y = 12$ into $\boxed{\text{I}}$

$x = 12 - 5$

$x = 7$

width: 7 cm; length: 12 cm

37) x = quarters; y = dimes

$\boxed{\text{I}}$ $\qquad x + y = 63$

$0.25x + 0.10y = 11.55$

$100(0.25x + 0.10y) = 100(11.55)$

$\boxed{\text{II}}$ $\qquad 25x + 10y = 1155$

Add equations $\boxed{\text{I}} \cdot (-10)$ and $\boxed{\text{II}}$

$-10x - 10y = -630$

$\underline{25x + 10y = 1155}$

$15x = 525$

$x = 35$

Substitute $x = 35$ into $\boxed{\text{I}}$

$35 + y = 63$

$y = 28$

quarters: 35; dimes: 28

39) x = Hand Warmers; y = Socks

$\boxed{\text{I}}$ $\qquad 2x + y = 27.50$

$\boxed{\text{II}}$ $\qquad 5x + 3y = 78.00$

Add equations $\boxed{\text{I}} \cdot (-3)$ and $\boxed{\text{II}}$

$-6x - 3y = -82.50$

$\underline{5x + 3y = \quad 78.00}$

$-x = -4.50$

$x = 4.5$

Substitute $x = 4.5$ into $\boxed{\text{I}}$

$2(4.5) + y = 27.50$

$9 + y = 27.50$

$y = 18.50$

Hand Warmers: $ 4.50;

Socks: $18.50

41) $x - 6y + 4z = 13$

$5x + y + 7z = 8$

$2x + 3y - z = -5$

$(-3, -2, 1)$

$2(-3) + 3(-2) - 1 = -5$

$-6 - 6 - 1 = -5$

$-13 \neq -5$

No

43) $\boxed{\text{I}}$ $2x - 5y - 2z = 3$

$\boxed{\text{II}}$ $x + 2y + z = 5$

$\boxed{\text{III}}$ $-3x - y + 2z = 0$

Add equations $\boxed{\text{I}}$ and $\boxed{\text{II}} \cdot (2)$

$2x - 5y - 2z = 3$

$\underline{2x + 4y + 2z = 10}$

$\boxed{\text{A}}$ $4x - y = 13$

Add equations $\boxed{\text{III}}$ and $\boxed{\text{II}} \cdot (-2)$

$-3x - y + 2z = 0$

$\underline{-2x - 4y - 2z = -10}$

$\boxed{\text{B}}$ $-5x - 5y = -10$

Add equations $\boxed{\text{A}} \cdot (-5)$ and $\boxed{\text{B}}$

$-20x + 5y = -65$

$\underline{-5x - 5y = -10}$

$-25x = -75$

$x = 3$

Substitute $x = 3$ into $\boxed{\text{A}}$

$4(3) - y = 13$

$12 - y = 13$

$y = -1$

Substitute $x = 3$ and $y = -1$ into $\boxed{\text{II}}$

$3 + 2(-1) + z = 5$

$1 + z = 5$

$z = 4$ $(3, -1, 4)$

45) $\boxed{\text{I}}$ $5a - b + 2c = -6$

$\boxed{\text{II}}$ $-2a - 3b + 4c = -2$

$\boxed{\text{III}}$ $a + 6b - 2c = 10$

Add equations $\boxed{\text{I}}$ and $\boxed{\text{III}}$

$5a - b + 2c = -6$

$\underline{a + 6b - 2c = 10}$

$\boxed{\text{A}}$ $6a + 5b = 4$

Add equations $\boxed{\text{II}}$ and $\boxed{\text{III}} \cdot (2)$

$-2a - 3b + 4c = -2$

$\underline{2a + 12b - 4c = 20}$

$9b = 18$

$b = 2$

Substitute $b = 2$ into $\boxed{\text{A}}$

$6a + 5(2) = 4$

$6a + 10 = 4$

$6a = -6$

$a = -1$

Substitute $a = -1$ and $b = 2$ into $\boxed{\text{I}}$

$-5 - 2 + 2c = -6$

$2c = 1$

$c = \dfrac{1}{2}$ $\left(-1, 2, \dfrac{1}{2}\right)$

47) $\boxed{\text{I}}$ $4x - 9y + 8z = 2$

$\boxed{\text{II}}$ $x + 3y = 5$

$\boxed{\text{III}}$ $6y + 10z = -1$

Rewrite $\boxed{\text{II}}$ as $x = 5 - 3y$

and substitute into $\boxed{\text{I}}$.

$4(5 - 3y) - 9y + 8z = 2$

$20 - 21y + 8z = 2$

$\boxed{\text{A}}$ $-21y + 8z = -18$

Add equations $\boxed{\text{A}} \cdot (2)$ and $\boxed{\text{III}} \cdot (7)$

$-42y + 16z = -36$

$\underline{42y + 70z = -7}$

$86z = -43$

$$z = -\frac{1}{2}$$

Substitute $z = -\frac{1}{2}$ into $\boxed{\text{III}}$

$$6y + 10\left(-\frac{1}{2}\right) = -1$$

$$6y - 5 = -1$$

$$6y = 4$$

$$y = \frac{2}{3}$$

Substitute $y = \frac{2}{3}$ into $\boxed{\text{II}}$

$$x + 3\left(\frac{2}{3}\right) = 5$$

$$x + 2 = 5$$

$$x = 3 \qquad \left(3, \frac{2}{3}, -\frac{1}{2}\right)$$

49) $\boxed{\text{I}} \qquad x + 3y - z = 0$

$\boxed{\text{II}} \qquad 11x - 4y + 3z = 8$

$\boxed{\text{III}} \qquad 5x + 15y - 5z = 1$

Add equations $\boxed{\text{I}} \cdot (-5)$ and $\boxed{\text{III}}$.

$$-5x - 15y + 5z = 0$$
$$\underline{5x + 15y - 5z = 1}$$
$$0 \neq 1$$

$\varnothing$; inconsistent

51) $\boxed{\text{I}} \qquad 12a - 8b + 4c = 8$

$\boxed{\text{II}} \qquad 3a - 2b + c = 2$

$\boxed{\text{III}} \qquad -6a + 4b - 2c = -4$

Equations $\boxed{\text{I}} \cdot (3)$ and $\boxed{\text{II}}$ are equivalent.

Equations $\boxed{\text{II}} \cdot (-2)$ and $\boxed{\text{III}}$ are equivalent.

Dependent; $\{(a, b, c) \mid 3a - 2b + c = 2\}$

53) $\boxed{\text{I}} \qquad 5y + 2z = 6$

$\boxed{\text{II}} \qquad -x + 2y = -1$

$\boxed{\text{III}} \qquad 4x - z = 1$

Add Equations $\boxed{\text{I}}$ and $\boxed{\text{III}} \cdot (2)$

$$5y + 2z = 6$$
$$\underline{8x \qquad -2z = 2}$$
$$\boxed{\text{A}} \quad 8x + 5y = 8$$

Add Equations $\boxed{\text{A}}$ and $\boxed{\text{II}} \cdot (8)$

$$8x + 5y = 8$$
$$\underline{-8x + 16y = -8}$$
$$21y = 0$$
$$y = 0$$

Substitute $y = 0$ into $\boxed{\text{II}}$

$$-x + 2(0) = -1$$
$$x = 1$$

Substitute $x = 1$ into $\boxed{\text{III}}$

$$4(1) - z = 1$$
$$4 - z = 1$$
$$z = 3 \qquad (1, 0, 3)$$

55) $\boxed{\text{I}} \qquad 8x + z = 7$

$\boxed{\text{II}} \qquad 3y + 2z = -4$

$\boxed{\text{III}} \qquad 4x - y = 5$

Add Equations $\boxed{\text{II}}$ and $\boxed{\text{I}} \cdot (-2)$

$$3y + 2z = -4$$
$$\underline{-16x \qquad -2z = -14}$$
$$\boxed{\text{A}} -16x + 3y = -18$$

Add Equations $\boxed{\text{A}}$ and $\boxed{\text{III}} \cdot (3)$

$$-16x + 3y = -18$$
$$\underline{12x - 3y = 15}$$
$$-4x = -3$$

$$x = \frac{3}{4}$$

Substitute $x = \dfrac{3}{4}$ into $\boxed{\text{I}}$

$8\left(\dfrac{3}{4}\right) + z = 7$

$6 + z = 7$

$z = 1$

Substitute $z = 1$ into $\boxed{\text{II}}$

$3y + 2(1) = -4$

$3y + 2 = -4$

$3y = -6$

$y = -2 \qquad \left(\dfrac{3}{4}, -2, 1\right)$

57) $x = $ Powerade; $y = $ Propel; $z = $ Gatorade

$\boxed{\text{I}} \qquad\qquad x = y + 17$

$\boxed{\text{II}} \qquad\qquad z = x + 58$

$\boxed{\text{III}} \qquad x + y + z = 197$

Substitute $\boxed{\text{I}}$ into $\boxed{\text{II}}$

$z = x + 58 = (y + 17) + 58 = y + 75$

$\boxed{\text{A}} \qquad\qquad z = y + 75$

Substitute $\boxed{\text{I}}$ and $\boxed{\text{A}}$ into $\boxed{\text{III}}$.

$x + y + z = 197$

$(y + 17) + y + (y + 75) = 197$

$3y + 92 = 105$

$3y = 105$

$y = 35$

$x = y + 17 = 35 + 17 = 52$

$z = x + 58 = 52 + 58 = 110$

Propel: 35 mg; Powerade: 52 mg;
Gatorade: 110 mg

59) $x = $ Serena; $y = $ Blair; $z = $ Chuck

$\boxed{\text{I}} \qquad x + y + z = 140$

$\boxed{\text{II}} \qquad\qquad y = x + 15$

$\boxed{\text{III}} \qquad\qquad z = 0.5x$

Substitute $\boxed{\text{II}}$ and $\boxed{\text{III}}$ into $\boxed{\text{I}}$

$x + (x + 15) + 0.5x = 140$

$2.5x = 125$

$x = 50$

$y = x + 15 = 50 + 15 = 65$

$z = 0.5x = 0.5(50) = 25$

Serena; 50; Blair: 65; Chuck: 25

61) $x = $ Ice Cream Cone; $y = $ Shake; $z = $ Sundae

$\boxed{\text{I}} \qquad 2x + 3y + z = 13.5$

$\boxed{\text{II}} \qquad 3x + y + 2z = 13.0$

$\boxed{\text{III}} \qquad 4x + y + z = 11.50$

Add Equations $\boxed{\text{I}}$ and $\boxed{\text{III}} \cdot (-1)$

$2x + 3y + z = 13.5$

$\underline{-4x - y - z = -11.5}$

$\boxed{\text{A}} \quad -2x + 2y = 2.0$

Add Equations $\boxed{\text{II}}$ and $\boxed{\text{III}} \cdot (-2)$

$3x + y + 2z = 13$

$\underline{-8x - 2y - 2z = -23}$

$\boxed{\text{B}} \quad -5x - y = -10$

Add Equations $\boxed{\text{A}}$ and $\boxed{\text{B}} \cdot (2)$

$-2x + 2y = 2$

$\underline{-10x - 2y = -20}$

$-12x = -18$

$x = 1.5$

Substitute $x = 1.5$ into $\boxed{\text{A}}$

$-2(1.5) + 2y = 2.0$

$-3 + 2y = 2$

$2y = 5$

$y = 2.5$

Substitute $x = 1.5$ and $y = 2.5$ into $\boxed{\text{III}}$

$4(1.5) + 2.5 + z = 11.50$

$6 + 2.5 + z = 11.50$

$z = 3.00$

Ice Cream Cone: $1.50;

Shake: $2.50;

Sundae: $3.00

63) $x = $ Smallest Angle;

$y = $ Middle Angle;

$z = $ Largest Angle

$$x = \frac{1}{3}y$$

$\boxed{\text{I}}$ $\qquad y = 3x$

$\boxed{\text{II}}$ $\qquad z = x + 70$

$\boxed{\text{III}}$ $\qquad x + y + z = 180$

Substitute $\boxed{\text{I}}$ and $\boxed{\text{II}}$ into $\boxed{\text{III}}$

$x + 3x + (x + 70) = 180$

$5x = 110$

$x = 22$

$y = 3(22)$

$y = 66$

$z = 22 + 70$

$z = 92$

Smallest angle : $22°$; Middle angle: $66°$;

Largest angle: $92°$

Chapter 5 Test

1) $\qquad\qquad 9x + 5y = 14$

$$9\left(-\frac{2}{3}\right) + 5(4) \overset{?}{=} 14$$

$$-6 + 20 \overset{?}{=} 14$$

$$14 = 14$$

$$-6x - y = 0$$

$$-6\left(-\frac{2}{3}\right) - 4 \overset{?}{=} 0$$

$$4 - 4 \overset{?}{=} 0$$

$$0 = 0$$

Yes

3) $\varnothing$

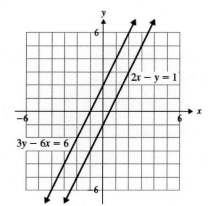

5) Substitute $x = -9 - 8y$ into
$$3x - 10y = -10$$
$$3(-9 - 8y) - 10y = -10$$
$$-27 - 24y - 10y = -10$$
$$-34y = 17$$
$$y = -\frac{1}{2}$$

Substitute $y = -\frac{1}{2}$ into

$$x + 8y = -9$$
$$x + 8(-\frac{1}{2}) = -9$$
$$x - 4 = -9$$
$$x = -5 \qquad \left(-5, -\frac{1}{2}\right)$$

7) Add the equations
$$2x + 5y = 11$$
$$\underline{7x - 5y = 16}$$
$$9x = 27$$
$$x = 3$$

Substitute $x = 3$ into
$$2x + 5y = 11$$
$$2(3) + 5y = 11$$
$$6 + 5y = 11$$
$$5y = 5$$
$$y = 1 \qquad (3,1)$$

9) $\boxed{\text{I}} \qquad -6x + 9y = 14$

$\boxed{\text{II}} \qquad 4x - 6y = 5$

Add Equations $\boxed{\text{I}} \cdot (2)$ and $\boxed{\text{II}} \cdot (3)$
$$-12x + 18y = 28$$
$$\underline{12x - 18y = 15}$$
$$0 \neq 43$$
$$\varnothing$$

11) $\boxed{\text{I}} \qquad \frac{5}{8}x + \frac{1}{4}y = \frac{1}{4}$

$\boxed{\text{II}} \qquad \frac{1}{3}x + \frac{1}{2}y = -\frac{4}{3}$

Add Equations $\boxed{\text{I}} \cdot (24)$ and $\boxed{\text{II}} \cdot (-12)$
$$15x + 6y = 6$$
$$\underline{-4x - 6y = 16}$$
$$11x = 22$$
$$x = 2$$

Substitute $x = 2$ into $\boxed{\text{II}} \cdot (-12)$
$$-4(2) - 6y = 16$$
$$-8 - 6y = 16$$
$$-6y = 24$$
$$y = -4 \qquad (2,-4)$$

13) $\boxed{\text{I}} \qquad -x + 4y + 3z = 6$

$\boxed{\text{II}} \qquad 3x - 2y + 6z = -18$

$\boxed{\text{III}} \qquad x + y + 2z = -1$

Add Equations $\boxed{\text{I}} \cdot (2)$ and $\boxed{\text{II}} \cdot (-1)$
$$-2x + 8y + 6z = 12$$
$$\underline{-3x + 2y - 6z = 18}$$
$$\boxed{\text{A}} -5x + 10y = 30$$

Add Equations $\boxed{\text{II}}$ and $\boxed{\text{III}} \cdot (-3)$
$$3x - 2y + 6z = -18$$
$$\underline{-3x - 3y - 6z = 3}$$
$$-5y = -15$$
$$y = 3$$

Substitute $y = 3$ into $\boxed{\text{A}}$
$$-5x + 10(3) = 30$$
$$-5x = 0$$
$$x = 0$$

Substitute $x = 0$ and $y = 3$ into $\boxed{\text{III}}$
$$0 + 3 + 2z = -1$$
$$2z = -4$$
$$z = -2 \qquad (0,3,-2)$$

15) x = Area of Yellowstone;

   y = Area of Death Valley

   $\boxed{I}$ $x + y = 5.5$

   $\boxed{II}$ $x = y - 1.1$

   Substitute $\boxed{II}$ into $\boxed{I}$

   $(y - 1.1) + y = 5.5$

   $\qquad 2y = 6.6$

   $\qquad y = 3.3$

   $\qquad x = 3.3 - 1.1$

   $\qquad x = 2.2$

   Yellowstone: 2.2 mil acres;

   Death Valley: 3.3 mil acres

17) x = Width; y = Length

   $\boxed{I}$ $2x + 2y = 114$

   $\boxed{II}$ $x = \dfrac{1}{2}y$

   Substitute $\boxed{II}$ into $\boxed{I}$

   $2\left(\dfrac{1}{2}\right)y + 2y = 114$

   $\qquad 3y = 114$

   $\qquad y = 38$

   $\qquad x = \dfrac{1}{2}(38)$

   $\qquad x = 19$

   Width: 19 cm; Length: 38 cm

19) x = Rory's Speed; y = Loreli's Speed

   $\boxed{I}$ $x = y + 4$

   $\boxed{II}$ $1.5x + 1.5y = 120$

   Substitute $\boxed{I}$ into $\boxed{II}$

   $1.5(y + 4) + 1.5y = 120$

   $\qquad 3y + 6 = 120$

$\qquad 3y = 114$

$\qquad y = 38$

$\qquad x = 38 + 4$

$\qquad x = 42$

Rory's Speed: 42 mph; Loreli's Speed: 38 mph

Cumulative Review: Chapters 1-5

1) $\dfrac{7}{15} + \dfrac{9}{10} = \dfrac{14}{30} + \dfrac{27}{30} = \dfrac{41}{30}$

3) $3(5 - 7)^3 + 18 \div 6 - 8$

   $= 3(-2)^3 + 18 \div 6 - 8$

   $= 3(-8) + 18 \div 6 - 8$

   $= -24 + 3 - 8$

   $= -29$

5) $-3(4x^2 + 5x - 1)$

   $= -12x^2 - 15x + 3$

7) $9x^2 \cdot 7x^{-6}$

   $= 63x^{-4} = \dfrac{63}{x^4}$

9) $0.0007319 = 7.319 \cdot 10^{-4}$

11) $11 - 3(2k - 1) = 2(6 - k)$

   $11 - 6k + 3 = 12 - 2k$

   $\qquad -4k = -2$

   $\qquad k = \dfrac{1}{2}$

   The solution set is $\left\{\dfrac{1}{2}\right\}$.

13) $x =$ Gas Milege for the old car

$y =$ Gas Milege for the new car

$y = 1.61x$

Substitute the value for y as 25.4

$x = \dfrac{25.4}{1.61}$

$x = 15.8$

The mileage was 15.8 mpg.

15)

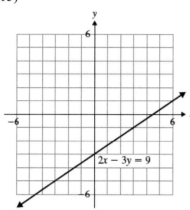

$2x - 3y = 9$

17) Slope, $m = \dfrac{-1-2}{-9-3}$

$m = \dfrac{-3}{-12}$

$m = \dfrac{1}{4}$

Use the point $(3, 2)$ with $m = \dfrac{1}{4}$ in the

point slope form.

$y - y_1 = m(x - x_1)$

$y - 2 = \dfrac{1}{4}(x - 3)$

$y = \dfrac{1}{4}x - \dfrac{3}{4} + 2$

$y = \dfrac{1}{4}x + \dfrac{5}{4}$

19) $9x - 3y = 6$

$\boxed{\text{I}}$ $y = 3x - 2$

$\boxed{\text{II}}$ $3x - 2y = -8$

Substitute $\boxed{\text{I}}$ into $\boxed{\text{II}}$

$3x - 2(3x - 2) = -8$

$3x - 6x + 4 = -8$

$-3x = -12$

$x = 4$

$y = 3(4) - 2$

$y = 12 - 2$

$y = 10$ $(4, 10)$

21) $\boxed{\text{I}}$ $x + 2y = 4$

$\boxed{\text{II}}$ $-3x - 6y = 6$

Add Equations $\boxed{\text{I}} \cdot (3)$ and $\boxed{\text{II}}$

$3x + 6y = 12$

$\underline{-3x - 6y = 6}$

$0 \neq 18$

$\varnothing$

23) $\boxed{\text{I}}$ $4a - 3b = -5$

$\boxed{\text{II}}$ $-a + 5c = 2$

$\boxed{\text{III}}$ $2b + c = -2$

Add Equations $\boxed{\text{I}}$ and $\boxed{\text{II}} \cdot (4)$

$4a - 3b \quad\quad = -5$

$\underline{-4a \quad\quad + 20c = 8}$

$\boxed{\text{A}} -3b + 20c = 3$

Add Equations $\boxed{\text{A}} \cdot (2)$ and $\boxed{\text{III}} \cdot (3)$

$-6b + 40c = 6$

$\underline{-6b + 3c = -6}$

$43c = 0$

$c = 0$

Substitute $c = 0$ into $\boxed{\text{II}}$

$$-a + 5(0) = 2$$
$$a = -2$$

Substitute $a = -2$ into $\boxed{\text{I}}$

$$4(-2) - 3b = -5$$
$$-8 - 3b = -5$$
$$-3b = 3$$
$$b = -1 \qquad (-2, -1, 0)$$

25) $x =$ Awards won by Joanes;

$\quad y =$ Awards won by Sanz;

$\quad z =$ Awards won by Shakira

I $\qquad x = y + 3$

II $\qquad y = 2z$

III $\qquad x + y + z = 38$

Substitute I and II into III

$$(y + 3) + y + z = 38$$
$$(2z) + 3 + (2z) + z = 38$$
$$5z = 35$$
$$z = 7$$
$$y = 2(7)$$
$$y = 14$$
$$x = 14 + 3$$
$$x = 17$$

Joanes: 17; Sanz: 14; Shakira: 7

Section 6.1 Exercises

1) quotient rule; k^6

3) power rule for a product; $16h^4$

5) $2^2 \cdot 2^4 = 2^{2+4} = 2^6 = 64$

7) $\dfrac{(-4)^8}{(-4)^5} = (-4)^{8-5} = (-4)^3 = -64$

9) $6^{-1} = \dfrac{1}{6}$

11) $\left(\dfrac{1}{9}\right)^{-2} = 9^2 = 81$

13) $\left(\dfrac{3}{2}\right)^{-4} = \left(\dfrac{2}{3}\right)^4 = \dfrac{2^4}{3^4} = \dfrac{16}{81}$

15) $6^0 + \left(-\dfrac{1}{2}\right)^{-5} = 1 + (-2)^5$
$= 1 - 32$
$= -31$

17) $\dfrac{8^5}{8^7} = 8^{5-7} = 8^{-2} = \left(\dfrac{1}{8}\right)^2 = \dfrac{1}{64}$

19) $t^5 \cdot t^8 = t^{5+8} = t^{13}$

21) $(-8c^4)(2c^5) = -16c^{4+5} = -16c^9$

23) $(z^6)^4 = z^{6\cdot4} = z^{24}$

25) $(5p^{10})^3 = 5^3 p^{10\cdot3} = 125p^{30}$

27) $\left(-\dfrac{2}{3}a^7 b\right)^3 = \left(-\dfrac{2}{3}\right)^3 a^{7\cdot3}b^3$
$= -\dfrac{8}{27}a^{21}b^3$

29) $\dfrac{f^{11}}{f^7} = f^{11-7} = f^4$

31) $\dfrac{35v^9}{5v^8} = 7v^{9-8} = 7v$

33) $\dfrac{9d^{10}}{54d^6} = \dfrac{d^{10-6}}{6} = \dfrac{d^4}{6}$

35) $\dfrac{x^3}{x^9} = x^{3-9} = x^{-6} = \dfrac{1}{x^6}$

37) $\dfrac{m^2}{m^3} = m^{2-3} = m^{-1} = \dfrac{1}{m}$

39) $\dfrac{45k^{-2}}{30k^2} = \dfrac{3}{2}k^{-2-2} = \dfrac{3}{2}k^{-4} = \dfrac{3}{2k^4}$

41) $5(2m^4n^7)^2 = 5(4m^8n^{14}) = 20m^8n^{14}$

43) $(6y^2)(2y^3)^2 = 6y^2 \cdot 4y^6 = 24y^8$

45) $\left(\dfrac{7a^4}{b^{-1}}\right)^{-2} = \left(\dfrac{b^{-1}}{7a^4}\right)^2 = \dfrac{b^{-2}}{49a^8} = \dfrac{1}{49a^8b^2}$

47) $\dfrac{a^{-12}b^7}{a^{-9}b^2} = a^{-12-(-9)}b^{7-2} = a^{-3}b^5 = \dfrac{b^5}{a^3}$

49) $\dfrac{(x^2y^{-3})^4}{x^5y^8} = \dfrac{x^8y^{-12}}{x^5y^8} = x^3y^{-20} = \dfrac{x^3}{y^{20}}$

51) $\dfrac{12a^6bc^{-9}}{\left(3a^2b^{-7}c^4\right)^2}=\dfrac{12a^6bc^{-9}}{9a^4b^{-14}c^8}=\dfrac{4a^2b^{15}c^{-17}}{3}=\dfrac{4a^2b^{15}}{3c^{17}}$

53) $\left(xy^{-3}\right)^{-5}=x^{-5}y^{15}=\dfrac{y^{15}}{x^5}$

55) $\left(\dfrac{a^2b}{4c^2}\right)^{-3}=\left(\dfrac{4c^2}{a^2b}\right)^3=\dfrac{64c^6}{a^6b^3}$

57) $\left(\dfrac{7h^{-1}k^9}{21h^{-5}k^5}\right)^{-2}=\left(\dfrac{21h^{-5}k^5}{7h^{-1}k^9}\right)^2$

$=\left(3h^{-5-(-1)}k^{5-9}\right)^2$

$=\left(3h^{-4}k^{-4}\right)^2$

$=9h^{-8}k^{-8}$

$=\dfrac{9}{h^8k^8}$

59) $\left(\dfrac{15cd^{-4}}{5c^3d^{-10}}\right)^{-3}=\left(\dfrac{5c^3d^{-10}}{15cd^{-4}}\right)^3$

$=\left(\dfrac{1}{3}c^{3-1}d^{-10-(-4)}\right)^3$

$=\left(\dfrac{1}{3}c^2d^{-6}\right)^3$

$=\dfrac{1}{27}c^6d^{-18}=\dfrac{c^6}{27d^{18}}$

61) $\dfrac{\left(2u^{-5}v^2w^4\right)^{-5}}{\left(u^6v^{-7}w^{-10}\right)^2}=\dfrac{1}{\left(u^6v^{-7}w^{-10}\right)^2\left(2u^{-5}v^2w^4\right)^5}$

$=\dfrac{1}{u^{12}v^{-14}w^{-20}2^5u^{-25}v^{10}w^{20}}$

$=\dfrac{1}{2^5u^{-13}v^{-4}w^0}$

$=\dfrac{u^{13}v^4}{32}$

63) $A=5x\cdot2x=10x^2$ sq. units

$P=2(5x)+2(2x)=10x+4x$

$=14x$ units

65) $A=\dfrac{1}{4}p\cdot\dfrac{3}{4}p=\dfrac{3}{16}p^2$ sq. units

$P=2\left(\dfrac{1}{4}p\right)+2\left(\dfrac{3}{4}p\right)$

$=\dfrac{1}{2}p+\dfrac{3}{2}p$

$=2p$ units

67) $k^{4a}\cdot k^{2a}=k^{4a+2a}=k^{6a}$

69) $\left(g^{2x}\right)^4=g^{2x\cdot4}=g^{8x}$

71) $\dfrac{x^{7b}}{x^{4b}}=x^{7b-4b}=x^{3b}$

73) $\left(2r^{6m}\right)^{-3}=\dfrac{1}{\left(2r^{6m}\right)^3}=\dfrac{1}{8r^{18m}}$

Section 6.2 Exercises

1) Yes; The coefficients are real numbers and the exponents are whole numbers.

3) No; One of the exponents is a negative number.

5) No; Two of the exponents are fractions.

7) binomial

9) trinomial

11) monomial

13) It is the same as the degree of the term in the polynomial with the highest degree.

15) Add the exponents on the variables.

17) Degree of polynomial $= 4$

Term	Coeff	Degree
$3y^4$	3	4
$7y^3$	7	3
$-2y$	-2	1
8	8	0

19) Degree of polynomial $= 5$

Term	Coeff	Degree
$-4x^2y^3$	-4	5
$-x^2y^2$	-1	4
$\frac{2}{3}xy$	$\frac{2}{3}$	2
$5y$	5	1

21) a) $2r^2 - 7r + 4,\ r = 3$

$2(3)^2 - 7(3) + 4 = 18 - 21 + 4 = 1$

b) $2r^2 - 7r + 4,\ r = -1$

$2(-1)^2 - 7(-1) + 4 = 2 + 7 + 4 = 13$

23) $9x + 4y,\ x = 5; y = -2$

$9(5) + 4(-2) = 45 - 8 = 37$

25) $x^2y^2 - 5xy + 2y,\ x = 5; y = -2$

$(5)^2(-2)^2 - 5(5)(-2) + 2(-2)$

$= 25 \cdot 4 + 50 - 4$

$= 100 + 50 - 4$

$= 146$

27) $\frac{1}{2}xy - 4x - y,\ x = 5; y = -2$

$\frac{1}{2}(5)(-2) - 4(5) - (-2)$

$= -5 - 20 + 2$

$= -23$

29) a) $y = 60x + 380,\ x = 5$

$y = 60(5) + 380 = 300 + 380 = 680$

If he rents the equipment for 5 hours, the cost of building will be $680.00

b) $y = 60x + 380,\ x = 9$

$y = 60(9) + 380 = 540 + 380 = 920$

If he keeps the equipment for 9 hours, the cost of building will be $920.00

c) $y = 60x + 380,\ y = 860.00$

$860 = 60x + 380$

$480 = 60x$

$8 = x$

If the cost of road is $860.00, then he needed to rent the equipment for 8 hours.

31) $-6z + 8z + 11z = 13z$

33) $5c^2 + 9c - 16c^2 + c - 3c = -11c^2 + 7c$

35) $6.7t^2 - 9.1t^6 - 2.5t^2 + 4.8t^6 = -4.3t^6 + 4.2t^2$

37) $7a^4b^4 + 4ab^2 - 16ab^2 - a^2b^2 + 5ab^2 = 6a^2b^2 - 7ab^2$

39) $\quad 5n - 8$
$+\ \underline{4n + 3}$
$\quad 9n - 5$

41) $\quad -7a^3 + 11a$
$+\ \underline{2a^3 - 4a}$
$\quad -5a^3 + 7a$

43)
$$9r^2 + 16r + 2$$
$$+\ \underline{3r^2 - 10r + 9}$$
$$12r^2 + 6r + 11$$

45)
$$b^2 - 8b - 14$$
$$+\ \underline{3b^2 + 8b + 11}$$
$$4b^2 - 3$$

47)
$$\frac{5}{6}w^4 - \frac{2}{3}w^2 \qquad\quad +\frac{1}{2}$$
$$+\ \underline{-\frac{4}{9}w^4 + \frac{1}{6}w^2 - \frac{3}{8}w - 2}$$
$$\frac{7}{18}w^4 - \frac{1}{2}w^2 - \frac{3}{8}w - \frac{3}{2}$$

49) $\left(6m^2 - 5m + 10\right) + \left(-4m^2 + 8m + 9\right)$
$$= 2m^2 + 3m + 19$$

51) $\left(-2c^4 - \dfrac{7}{10}c^3 + \dfrac{3}{4}c - \dfrac{2}{9}\right) +$

$\left(12c^4 + \dfrac{1}{2}c^3 - c + 3\right)$

$$= 10c^4 - \frac{1}{5}c^3 - \frac{1}{4}c + \frac{25}{9}$$

53) $\left(2.7d^3 + 5.6d^2 - 7d + 3.1\right)$
$$+\left(-1.5d^3 + 2.1d^2 - 4.3d - 2.5\right)$$
$$= 1.2d^3 + 7.7d^2 - 11.3d + 0.6$$

55)
$$15w + 7 \qquad\qquad 15w + 7$$
$$-\ \underline{\ 3w + 11\ } \qquad +\ \underline{-3w - 11}$$
$$\qquad\qquad\qquad\qquad 12w - 4$$

57)
$$y - 6 \qquad\qquad y - 6$$
$$-\ \underline{\ 2y - 8\ } \qquad +\ \underline{-2y + 8}$$
$$\qquad\qquad\qquad\qquad -y + 2$$

59)
$$3b^2 - 8b + 12$$
$$-\ \underline{\ 5b^2 + 2b - 7\ }$$

$$3b^2 - 8b + 12$$
$$+\ \underline{-5b^2 - 2b +\ \ 7}$$
$$-2b^2 - 10b + 19$$

61)
$$f^4 - 6f^3 + 5f^2 - 8f + 13$$
$$-\ \underline{\left(-3f^4 + 8f^3 - f^2 \qquad\ \ +14\right)}$$

$$f^4 - 6f^3 + 5f^2 - 8f + 13$$
$$+\ \underline{\ 3f^4 - 8f^3 + f^2 \qquad\quad -4}$$
$$4f^4 - 14f^3 + 6f^2 - 8f + 9$$

63)
$$10.7r^2 + 1.2r + 9$$
$$-\ \underline{\ 4.9r^2 - 5.3r - 2.8}$$

$$10.7r^2 + 1.2r + 9$$
$$+\ \underline{-4.9r^2 + 5.3r + 2.8}$$
$$5.8r^2 + 6.5r + 11.8$$

65) $\left(j^2 + 16j\right) - \left(-6j^2 + 7j + 5\right)$
$$= \left(j^2 + 16j\right) + \left(6j^2 - 7j - 5\right)$$
$$= 7j^2 + 9j - 5$$

67) $\left(17s^5 - 12s^2\right) - \left(9s^5 + 4s^4 - 8s^2 - 1\right)$
$$= \left(17s^5 - 12s^2\right) + \left(-9s^5 - 4s^4 + 8s^2 + 1\right)$$
$$= 8s^5 - 4s^4 - 4s^2 + 1$$

69) $\left(-\dfrac{3}{8}r^2+\dfrac{2}{9}r+\dfrac{1}{3}\right)-\left(-\dfrac{7}{16}r^2-\dfrac{5}{9}r+\dfrac{7}{6}\right)$

$=\left(-\dfrac{6}{16}r^2+\dfrac{2}{9}r+\dfrac{2}{6}\right)+\left(\dfrac{7}{16}r^2+\dfrac{5}{9}r-\dfrac{7}{6}\right)$

$=\dfrac{1}{16}r^2+\dfrac{7}{9}r-\dfrac{5}{6}$

71) Answers may vary.

73) No. If the coefficients of the like terms are opposite in sign, their sum will be zero. Example:

$\left(3x^2+4x+5\right)+\left(2x^2-4x+1\right)=5x^2+6$

75) $\left(8a^4-9a^2+17\right)-\left(15a^4+3a^2+3\right)$

$=\left(8a^4-9a^2+17\right)+\left(-15a^4-3a^2-3\right)$

$=-7a^4-12a^2+14$

77) $\left(-11n^2-8n+21\right)+\left(4n^2+15n-3\right)$

$+\left(7n^2-10\right)=7n+8$

79) $\left(w^3+5w^2+3\right)-\left(6w^3-2w^2+w+12\right)$

$+\left(9w^3+7\right)$

$=\left(w^3+5w^2+3\right)+\left(-6w^3+2w^2-w-12\right)$

$+\left(9w^3+7\right)=4w^3+7w^2-w-2$

81) $\left(y^3-\dfrac{3}{4}y^2-5y+\dfrac{3}{7}\right)+\left(\dfrac{1}{3}y^3-y^2+8y-\dfrac{1}{2}\right)$

$=\left(\dfrac{3}{3}y^3-\dfrac{3}{4}y^2-5y+\dfrac{6}{14}\right)+\left(\dfrac{1}{3}y^3-\dfrac{4}{4}y^2+8y-\dfrac{7}{14}\right)$

$=\dfrac{4}{3}y^3-\dfrac{7}{4}y^2+3y-\dfrac{1}{14}$

83) $\left(3m^3-5m^2+m+12\right)-\left(7m^3+4m^2-m+11\right)$

$+\left(-5m^3-2m^2+6m+8\right)$

$=\left(3m^3-5m^2+m+12\right)-\left(2m^3+2m^2+5m+19\right)$

$=\left(3m^3-5m^2+m+12\right)+\left(-2m^3-2m^2-5m-19\right)$

$=m^3-7m^2-4m-7$

85) $\left(p^2-7\right)+\left(8p^2+2p-1\right)=9p^2+2p-8$

87) $\left(6z^6+z^2+9\right)-\left(z^6-8z^2+13\right)$

$=\left(6z^6+z^2+9\right)+\left(-z^6+8z^2-13\right)$

$=5z^6+9z^2-4$

89) $\left(2p^2+p+5\right)-\left[\left(6p^2+1\right)+\left(3p^2-8p+5\right)\right]$

$=\left(2p^2+p+5\right)-\left[9p^2-8p+5\right]$

$=\left(2p^2+p+5\right)+\left(-9p^2+8p-5\right)$

$=-7p^2+9p$

91) $\left(5w+17z\right)-\left(w+3z\right)=\left(5w+17z\right)+\left(-w-3z\right)$

$=4w+14z$

93) $\left(ac+8a+6c\right)+\left(-6ac+4a-c\right)=-5ac+12a+5c$

95) $\left(-6u^2v^2+11luv+14\right)-\left(-10u^2v^2-20uv+18\right)$

$=\left(-6u^2v^2+11luv+14\right)+\left(10u^2v^2+20uv-18\right)$

$=4u^2v^2+31luv-4$

Chapter 6: Polynomials

97) $\left(12x^3y^2-5x^2y^2+9x^2y-17\right)+\left(5x^3y^2+x^2y-1\right)$

$-\left(6x^2y^2+10x^2y+2\right)$

$=\left(17x^3y^2-5x^2y^2+10x^2y-18\right)$

$+\left(-6x^2y^2-10x^2y-2\right)$

$=17x^3y^2-11x^2y^2-20$

99) $P=2(2x+7)+2(x-4)$

$=4x+14+2x-8$

$=6x+6$ units

101) $P=2\left(5p^2-2p+3\right)+2(p-6)$

$=10p^2-4p+6+2p-12$

$=10p^2-2p-6$ units

103) a) $f(-3)=5(-3)^2+7(-3)-8$

$f(-3)=5(9)+7(-3)-8$

$f(-3)=45-21-8$

$f(-3)=16$

b) $f(1)=5(1)^2+7(1)-8$

$f(1)=5(1)+7(1)-8$

$f(1)=5+7-8$

$f(1)=4$

105) a) $P(3)=(3)^3-2(3)^2+5(3)+8$

$P(3)=27-2(9)+5(3)+8$

$P(3)=27-18+15+8$

$P(3)=32$

b) $P(0)=(0)^3-2(0)^2+5(0)+8$

$P(0)=0-2(0)+5(0)+8$

$P(0)=0-0+0+8$

$P(0)=8$

107) $13=-3z+11$

$2=-3z$

$-\dfrac{2}{3}=z$

109) $14=\dfrac{3}{5}k-4$ 110)

$18=\dfrac{3}{5}k$

$\dfrac{90}{3}=k$

$30=k$

Section 6.3 Exercises

1) Answers may vary.

3) $\left(3m^5\right)\left(8m^3\right)=24m^8$

5) $(-8c)\left(4c^5\right)=-32c^6$

7) $5a(2a-7)=(5a)(2a)+(5a)(-7)$

$=10a^2-35a$

9) $-6c(7c+2)=(-6c)(7c)+(-6c)(2)$

$=-42c^2-12c$

190

11) $6v^3\left(v^2-4v-2\right)$

$=\left(6v^3\right)\left(v^2\right)+\left(6v^3\right)\left(-4v\right)+\left(6v^3\right)\left(-2\right)$

$=6v^5-24v^4-12v^3$

13) $-9b^2\left(4b^3-2b^2-6b-9\right)$

$=\left(-9b^2\right)\left(4b^3\right)+\left(-9b^2\right)\left(-2b^2\right)$

$\quad+\left(-9b^2\right)\left(-6b\right)+\left(-9b^2\right)\left(-9\right)$

$=-36b^5+18b^4+54b^3+81b^2$

15) $3a^2b\left(ab^2+6ab-13b+7\right)$

$=\left(3a^2b\right)\left(ab^2\right)+\left(3a^2b\right)\left(6ab\right)$

$\quad+\left(3a^2b\right)\left(-13b\right)+\left(3a^2b\right)\left(7\right)$

$=3a^3b^3+18a^3b^2-39a^2b^2+21a^2b$

17) $-\dfrac{3}{5}k^4\left(15k^2+20k-3\right)$

$=\left(-\dfrac{3}{5}k^4\right)\left(15k^2\right)+\left(-\dfrac{3}{5}k^4\right)\left(20k\right)$

$+\left(-\dfrac{3}{5}k^4\right)\left(-3\right)=-9k^6-12k^5+\dfrac{9}{5}k^4$

19) $\left(c+4\right)\left(6c^2-13c+7\right)$

$=\left(c\right)\left(6c^2\right)+\left(c\right)\left(-13c\right)+\left(c\right)\left(7\right)$

$+\left(4\right)\left(6c^2\right)+\left(4\right)\left(-13c\right)+\left(4\right)\left(7\right)$

$=6c^3-13c^2+7c+24c^2-52c+28$

$=6c^3+11c^2-45c+28$

21) $\left(f-5\right)\left(3f^2+2f-4\right)$

$=\left(f\right)\left(3f^2\right)+\left(f\right)\left(2f\right)+\left(f\right)\left(-4\right)$

$+\left(-5\right)\left(3f^2\right)+\left(-5\right)\left(2f\right)+\left(-5\right)\left(-4\right)$

$=3f^3+2f^2-4f-15f^2-10f+20$

$=3f^3-13f^2-14f+20$

23) $\left(4x^3-x^2+6x+2\right)\left(2x-5\right)$

$=\left(4x^3\right)\left(2x\right)+\left(4x^3\right)\left(-5\right)+\left(-x^2\right)\left(2x\right)$

$\quad+\left(-x^2\right)\left(-5\right)+\left(6x\right)\left(2x\right)$

$\quad+\left(6x\right)\left(-5\right)+\left(2\right)\left(2x\right)+\left(2\right)\left(-5\right)$

$=8x^4-20x^3-2x^3+5x^2+12x^2-30x+4x-10$

$=8x^4-22x^3+17x^2-26x-10$

25) $\left(\dfrac{1}{3}y^2+4\right)\left(12y^2+7y-9\right)$

$=\left(\dfrac{1}{3}y^2\right)\left(12y^2\right)+\left(\dfrac{1}{3}y^2\right)\left(7y\right)+\left(\dfrac{1}{3}y^2\right)\left(-9\right)$

$\quad+\left(4\right)\left(12y^2\right)+\left(4\right)\left(7y\right)+\left(4\right)\left(-9\right)$

$=4y^4+\dfrac{7}{3}y^3-3y^2+48y^2+28y-36$

$=4y^4+\dfrac{7}{3}y^3+45y^2+28y-36$

27) $\left(s^2-s+2\right)\left(s^2+4s-3\right)$

$=\left(s^2\right)\left(s^2\right)+\left(s^2\right)\left(4s\right)+\left(s^2\right)\left(-3\right)$

$\quad+\left(-s\right)\left(s^2\right)+\left(-s\right)\left(4s\right)$

$\quad+\left(-s\right)\left(-3\right)+\left(2\right)\left(s^2\right)+\left(2\right)\left(4s\right)$

$\quad+\left(2\right)\left(-3\right)$

$=s^4+4s^3-3s^2-s^3-4s^2+3s$

$\quad+2s^2+8s-6$

$=s^4+3s^3-5s^2+11s-6$

29) $\left(4h^2 - h + 2\right)\left(-6h^3 + 5h^2 - 9h\right)$

$= \left(4h^2\right)\left(-6h^3\right) + \left(4h^2\right)\left(5h^2\right)$

$\quad + \left(4h^2\right)\left(-9h\right) + \left(-h\right)\left(-6h^3\right)$

$\quad + \left(-h\right)\left(5h^2\right) + \left(-h\right)\left(-9h\right) + \left(2\right)\left(-6h^3\right)$

$\quad + \left(2\right)\left(5h^2\right) + \left(2\right)\left(-9h\right)$

$= -24h^5 + 20h^4 - 36h^3 + 6h^4 - 5h^3$

$\quad + 9h^2 - 12h^3 + 10h^2 - 18h$

$= -24h^5 + 26h^4 - 53h^3 + 19h^2 - 18h$

31) $\left(3y - 2\right)\left(5y^2 - 4y + 3\right)$

$= \left(3y\right)\left(5y^2\right) + \left(3y\right)\left(-4y\right)$

$\quad + \left(3y\right)\left(3\right) + \left(-2\right)\left(5y^2\right)$

$\quad + \left(-2\right)\left(-4y\right) + \left(-2\right)\left(3\right)$

$= 15y^3 - 12y^2 + 9y - 10y^2 + 8y - 6$

$= 15y^3 - 22y^2 + 17y - 6$

$$
\begin{array}{r}
5y^2 - 4y + 3 \\
\times \qquad 3y - 2 \\
\hline
-10y^2 + 8y - 6 \\
+ \ 15y^3 - 12y^2 + 9y \qquad \\
\hline
15y^3 - 22y^2 + 17y - 6
\end{array}
$$

33) First, Outer, Inner, Last

35) $\left(w + 5\right)\left(w + 7\right) = w^2 + 7w + 5w + 35$

$\qquad\qquad = w^2 + 12w + 35$

37) $\left(r - 3\right)\left(r + 9\right) = r^2 + 9r - 3r - 27$

$\qquad\qquad = r^2 + 6r - 27$

39) $\left(y - 7\right)\left(y - 1\right) = y^2 - y - 7y + 7$

$\qquad\qquad = y^2 - 8y + 7$

41) $\left(3p + 7\right)\left(p - 2\right) = 3p^2 - 6p + 7p - 14$

$\qquad\qquad = 3p^2 + p - 14$

43) $\left(7n + 4\right)\left(3n + 1\right) = 21n^2 + 7n + 12n + 4$

$\qquad\qquad = 21n^2 + 19n + 4$

45) $\left(5 - 4w\right)\left(3 - w\right) = 15 - 5w - 12w + 4w^2$

$\qquad\qquad = 4w^2 - 17w + 15$

47) $\left(4a - 5b\right)\left(3a + 4b\right)$

$= 12a^2 + 16ab - 15ab - 20b^2$

$= 12a^2 + ab - 20b^2$

49) $\left(6x + 7y\right)\left(8x + 3y\right)$

$= 48x^2 + 18xy + 56xy + 21y^2$

$= 48x^2 + 74xy + 21y^2$

51) $\left(v + \dfrac{1}{3}\right)\left(v + \dfrac{3}{4}\right)$

$= v^2 + \dfrac{3}{4}v + \dfrac{1}{3}v + \dfrac{3}{12}$

$= v^2 + \dfrac{9}{12}v + \dfrac{4}{12}v + \dfrac{1}{4}$

$= v^2 + \dfrac{13}{12}v + \dfrac{1}{4}$

53) $\left(\dfrac{1}{2}a + 5b\right)\left(\dfrac{2}{3}a - b\right)$

$= \dfrac{2}{6}a^2 - \dfrac{1}{2}ab + \dfrac{10}{3}ab - 5b^2$

$= \dfrac{1}{3}a^2 - \dfrac{3}{6}ab + \dfrac{20}{6}ab - 5b^2$

$= \dfrac{1}{3}a^2 + \dfrac{17}{6}ab - 5b^2$

55) a) $P = 2(y+5) + 2(y-3)$

$\quad = 2y + 10 + 2y - 6$

$\quad = 4y + 4$ units

b) $A = (y+5)(y-3)$

$\quad = y^2 - 3y + 5y - 15$

$\quad = y^2 + 2y - 15$ sq. units

57) a) $P = 2(3m) + 2(m^2 - 2m + 7)$

$\quad = 6m + 2m^2 - 4m + 14$

$\quad = 2m^2 + 2m + 14$ units

b) $A = (3m)(m^2 - 2m + 7)$

$\quad = 3m^3 - 6m^2 + 21m$ sq. units

59) $A = \dfrac{1}{2}(6n-5)n$

$\quad = \dfrac{1}{2}(6n^2 - 5n)$

$\quad = 3n^2 - \dfrac{5}{2}n$ sq. units

61) Both are correct.

63) $2(n+3)(4n-5)$

$\quad = (2n+6)(4n-5)$

$\quad = 8n^2 - 10n + 24n - 30$

$\quad = 8n^2 + 14n - 30$

65) $-5z^2(z-8)(z-2)$

$\quad = -5z^2(z^2 - 2z - 8z + 16)$

$\quad = -5z^2(z^2 - 10z + 16)$

$\quad = -5z^4 + 50z^3 - 80z^2$

67) $(c+3)(c+4)(c-1)$

$\quad = (c^2 + 4c + 3c + 12)(c-1)$

$\quad = (c^2 + 7c + 12)(c-1)$

$\quad = c^3 + 7c^2 + 12c - c^2 - 7c - 12$

$\quad = c^3 + 6c^2 + 5c - 12$

69) $(3x-1)(x-2)(x-6)$

$\quad = (3x-1)(x^2 - 6x - 2x + 12)$

$\quad = (3x-1)(x^2 - 8x + 12)$

$\quad = 3x^3 - 24x^2 + 36x - x^2 + 8x - 12$

$\quad = 3x^3 - 25x^2 + 44x - 12$

71) $8p\left(\dfrac{1}{4}p^2 + 3\right)(p^2 + 5)$

$\quad = (2p^3 + 24p)(p^2 + 5)$

$\quad = 2p^5 + 10p^3 + 24p^3 + 120p$

$\quad = 2p^5 + 34p^3 + 120p$

73) $(y+5)(y-5) = y^2 - 5^2 = y^2 - 25$

75) $(a-7)(a+7) = (a+7)(a-7)$

$\quad = a^2 - 7^2$

$\quad = a^2 - 49$

77) $(3-p)(3+p) = (3+p)(3-p)$

$\quad = 3^2 - p^2$

$\quad = 9 - p^2$

79) $\left(u + \dfrac{1}{5}\right)\left(u - \dfrac{1}{5}\right) = u^2 - \left(\dfrac{1}{5}\right)^2 = u^2 - \dfrac{1}{25}$

Chapter 6: Polynomials

81) $\left(\dfrac{2}{3}-k\right)\left(\dfrac{2}{3}+k\right)=\left(\dfrac{2}{3}+k\right)\left(\dfrac{2}{3}-k\right)$

$=\left(\dfrac{2}{3}\right)^2-k^2$

$=\dfrac{4}{9}-k^2$

83) $(2r+7)(2r-7)=(2r)^2-7^2$

$=4r^2-49$

85) $-(8j-k)(8j+k)$

$=-(8j+k)(8j-k)$

$=-\left[(8j)^2-k^2\right]$

$=-\left(64j^2-k^2\right)$

$=k^2-64j^2$

87) $(d+4)^2=d^2+2(d)(4)+4^2$

$=d^2+8d+16$

89) $(n-13)^2=n^2-2(n)(13)+(13)^2$

$=n^2-26n+169$

91) $(h-0.6)^2=h^2-2(h)(0.6)+(0.6)^2$

$=h^2-1.2h+0.36$

93) $(3u+1)^2=(3u)^2+2(3u)(1)+1^2$

$=9u^2+6u+1$

95) $(2d-5)^2=(2d)^2-2(2d)(5)+(5)^2$

$=4d^2-20d+25$

97) $(3c+2d)^2=(3c)^2+2(3c)(2d)+(2d)^2$

$=9c^2+12cd+4d^2$

99) $\left(\dfrac{3}{2}k+8m\right)^2$

$=\left(\dfrac{3}{2}k\right)^2+2\left(\dfrac{3}{2}k\right)(8m)+(8m)^2$

$=\dfrac{9}{4}k^2+24km+64m^2$

101) $\left[(2a+b)+3\right]^2$

$=(2a+b)^2+2(2a+b)(3)+(3)^2$

$=(2a)^2+2(2a)(b)+b^2+12a+6b+9$

$=4a^2+4ab+b^2+12a+6b+9$

103) $\left[(f-3g)+4\right]\left[(f-3g)-4\right]$

$=(f-3g)^2-4^2$

$=f^2-2(f)(3g)+(-3g)^2-16$

$=f^2-6fg+9g^2-16$

105) No. The order of operations tell us to perform exponents, $(r+2)^2$, before multiplying by 3.

107) $7(y+2)^2=7\left[y^2+2(y)(2)+2^2\right]$

$=7\left[y^2+4y+4\right]$

$=7y^2+28y+28$

109) $4c(c+3)^2=4c\left[c^2+2(c)(3)+3^2\right]$

$=4c\left[c^2+6c+9\right]$

$=4c^3+24c^2+36c$

111) $(r+5)^3 = (r+5)^2(r+5)$

$= (r^2 + 10r + 25)(r+5)$

$= r^3 + 10r^2 + 25r + 5r^2$

$\qquad + 50r + 125$

$= r^3 + 15r^2 + 75r + 125$

113) $(g-4)^3 = (g-4)^2(g-4)$

$= (g^2 - 8g + 16)(g-4)$

$= g^3 - 4g^2 - 8g^2 + 32g$

$\qquad + 16g - 64$

$= g^3 - 12g^2 + 48g - 64$

115) $(2a-1)^3$

$= (2a-1)^2(2a-1)$

$= ((2a)^2 - 2(2a)(1) + (-1)^2)(2a-1)$

$= (4a^2 - 4a + 1)(2a-1)$

$= 8a^3 - 4a^2 - 8a^2 + 4a + 2a - 1$

$= 8a^3 - 12a^2 + 6a - 1$

117) $(h+3)^4 = (h+3)^2(h+3)^2$

$= (h^2 + 6h + 9)(h^2 + 6h + 9)$

$= h^4 + 6h^3 + 9h^2 + 6h^3 + 36h^2$

$\qquad + 54h + 9h^2 + 54h + 81$

$= h^4 + 12h^3 + 54h^2 + 108h + 81$

119) $(5t-2)^4 = (5t-2)^2(5t-2)^2$

$= (25t^2 - 20t + 4)(25t^2 - 20t + 4)$

$= 625t^4 - 500t^3 + 100t^2$

$\qquad -500t^3 + 400t^2 - 80t$

$\qquad +100t^2 - 80t + 16$

$= 625t^4 - 1000t^3 + 600t^2$

$\qquad -160t + 16$

121) No; $(x+2)^2 = x^2 + 4x + 4$

123) $(c-12)(c+7) = c^2 + 7c - 12c - 84$

$= c^2 - 5c - 84$

125) $4(6-5a)(2a-1)$

$= (24 - 20a)(2a-1)$

$= 48a - 24 - 40a^2 + 20a$

$= -40a^2 + 68a - 24$

127) $(2k-9)(5k^2 + 4k - 1)$

$= (2k)(5k^2) + (2k)(4k) + (2k)(-1)$

$\qquad + (-9)(5k^2) + (-9)(4k) + (-9)(-1)$

$= 10k^3 + 8k^2 - 2k - 45k^2 - 36k + 9$

$= 10k^3 - 37k^2 - 38k + 9$

129) $\left(\dfrac{1}{6} - h\right)\left(\dfrac{1}{6} + h\right) = \left(\dfrac{1}{6} + h\right)\left(\dfrac{1}{6} - h\right)$

$= \left(\dfrac{1}{6}\right)^2 - h^2$

$= \dfrac{1}{36} - h^2$

131) $(3c+1)^3 = (3c+1)^2(3c+1)$

$= \left((3c)^2 + 2(3c)(1) + (1)^2\right)(3c+1)$

$= \left(9c^2 + 6c + 1\right)(3c+1)$

$= 27c^3 + 9c^2 + 18c^2 + 6c + 3c + 1$

$= 27c^3 + 27c^2 + 9c + 1$

133) $\left(\dfrac{3}{8}p^7\right)\left(\dfrac{3}{4}p^4\right) = \dfrac{9}{32}p^{11}$

135) $\left(a^2 + 7b^2\right)^2$

$= \left(a^2\right)^2 + 2\left(a^2\right)\left(7b^2\right) + \left(7b^2\right)^2$

$= a^4 + 14a^2b^2 + 49b^4$

137) $-5z(z-3)^2 = -5z\left[z^2 - 2(z)(3) + (3)^2\right]$

$= -5z\left[z^2 - 6z + 9\right]$

$= -5z^3 + 30z^2 - 45z$

139) $\left[(x-4y)+5\right]\left[(x-4y)-5\right]$

$= (x-4y)^2 - 5^2$

$= x^2 - 2(x)(4y) + (4y)^2 - 25$

$= x^2 - 8xy + 16y^2 - 25$

141) $V = (a+4)^3$

$= (a+4)^2(a+4)$

$= \left(a^2 + 8a + 16\right)(a+4)$

$= a^3 + 4a^2 + 8a^2 + 32a + 16a + 64$

$= a^3 + 12a^2 + 48a + 64$ cubic units

143) $A = \pi(k+5)^2$

$= \pi\left(k^2 + 10k + 25\right)$

$= \pi k^2 + 10\pi k + 25\pi$ sq. units

145) $A = (3c-2)(3c-2) - \dfrac{1}{2}(6)(c)$

$= (3c)^2 - 2(3c)(2) + (-2)^2 - 3c$

$= 9c^2 - 12c + 4 - 3c$

$= 9c^2 - 15c + 4$ sq. units

Section 6.4: Exercises

1) dividend: $6c^3 + 15c^2 - 9c$;

   divisor: $3c$;

   quotient: $2c^2 + 5c - 3$

3) Answers may vary.

   Divide each term in the polynomial

   by the monomial and simplify.

5) $\dfrac{49p^4 + 21p^3 + 28p^2}{7} = \dfrac{49p^4}{7} + \dfrac{21p^3}{7} + \dfrac{28p^2}{7}$

$= 7p^4 + 3p^3 + 4p^2$

7) $\dfrac{12w^3 - 40w^2 - 36w}{4w} = \dfrac{12w^3}{4w} - \dfrac{40w^2}{4w} - \dfrac{36w}{4w}$

$= 3w^2 - 10w - 9$

9) $\dfrac{22z^6 + 14z^5 - 38z^3 + 2z}{2z}$

$= \dfrac{22z^6}{2z} + \dfrac{14z^5}{2z} - \dfrac{38z^3}{2z} + \dfrac{2z}{2z}$

$= 11z^5 + 7z^4 - 19z^2 + 1$

11) $\dfrac{9h^8+54h^6-108h^3}{9h^2}$

$=\dfrac{9h^8}{9h^2}+\dfrac{54h^6}{9h^2}-\dfrac{108h^3}{9h^2}$

$=h^6+6h^4-12h$

13) $\dfrac{36r^7-12r^4+6}{12r}=\dfrac{36r^7}{12r}-\dfrac{12r^4}{12r}+\dfrac{6}{12r}$

$=3r^6-r^3+\dfrac{1}{2r}$

15) $\dfrac{8d^6-12d^5+18d^4}{2d^4}=\dfrac{8d^6}{2d^4}-\dfrac{12d^5}{2d^4}+\dfrac{18d^4}{2d^4}$

$=4d^2-6d+9$

17) $\dfrac{28k^7+8k^5-44k^4-36k^2}{4k^2}$

$=\dfrac{28k^7}{4k^2}+\dfrac{8k^5}{4k^2}-\dfrac{44k^4}{4k^2}-\dfrac{36k^2}{4k^2}$

$=7k^5+2k^3-11k^2-9$

19) $\left(35d^5-7d^2\right)\div\left(-7d^2\right)$

$=\dfrac{35d^5-7d^2}{-7d^2}$

$=\dfrac{35d^5}{-7d^2}-\dfrac{7d^2}{-7d^2}$

$=-5d^3+1$

21) $\dfrac{10w^5+12w^3-6w^2+2w}{6w^2}$

$=\dfrac{10w^5}{6w^2}+\dfrac{12w^3}{6w^2}-\dfrac{6w^2}{6w^2}+\dfrac{2w}{6w^2}$

$=\dfrac{5}{3}w^3+2w-1+\dfrac{1}{3w}$

23) $\left(12k^8-4k^6-15k^5-3k^4+1\right)\div\left(2k^5\right)$

$=\dfrac{12k^8-4k^6-15k^5-3k^4+1}{2k^5}$

$=\dfrac{12k^8}{2k^5}-\dfrac{4k^6}{2k^5}-\dfrac{15k^5}{2k^5}-\dfrac{3k^4}{2k^5}+\dfrac{1}{2k^5}$

$=6k^3-2k-\dfrac{15}{2}-\dfrac{3}{2k}+\dfrac{1}{2k^5}$

25) $\dfrac{48p^5q^3+60p^4q^2-54p^3q+18p^2q}{6p^2q}$

$=\dfrac{48p^5q^3}{6p^2q}+\dfrac{60p^4q^2}{6p^2q}$

$-\dfrac{54p^3q}{6p^2q}+\dfrac{18p^2q}{6p^2q}$

$=8p^3q^2+10p^2q-9p+3$

27) $\dfrac{14s^6t^6-28s^5t^4-s^3t^3+21st}{7s^2t}$

$=\dfrac{14s^6t^6}{7s^2t}-\dfrac{28s^5t^4}{7s^2t}-\dfrac{s^3t^3}{7s^2t}+\dfrac{21st}{7s^2t}$

$=2s^4t^5-4s^3t^3-\dfrac{1}{7}st^2+\dfrac{3}{s}$

29) The answer is incorrect. When you divide 5p by 5p, you get 1. The quotient should be $8p^2-2p+1$.

31) dividend: $12w^3-2w^2-23w-7$

divisor: $3w+1$

quotient: $4w^2-2w-7$

33) 2

35)

$$
\begin{array}{r}
158 \\
6)\overline{949} \\
\underline{6} \\
34 \\
\underline{30} \\
49 \\
\underline{48} \\
1
\end{array}
$$

Answer: $158\frac{1}{6}$

37)

$$
\begin{array}{r}
437 \\
9)\overline{3937} \\
\underline{36} \\
33 \\
\underline{27} \\
67 \\
\underline{63} \\
4
\end{array}
$$

Answer: $437\frac{4}{9}$

39) $g+5\overline{)\,g^2+9g+20}$ $g+4$
$$\underline{-(g^2+5g)}$$
$$4g+20$$
$$\underline{-(4g+20)}$$
$$0$$

41) $a+7\overline{)\,a^2+13a+42}$ $a+6$
$$\underline{-(a^2+7a)}$$
$$6a+42$$
$$\underline{-(6a+42)}$$
$$0$$

43) $k+5\overline{)\,k^2-k-30}$ $k-6$
$$\underline{-(k^2+5k)}$$
$$-6k-30$$
$$\underline{-(-6k-30)}$$
$$0$$

45) $3h-4\overline{)\,6h^3+7h^2-17h-4}$ $2h^2+5h+1$
$$\underline{-(6h^3-8h^2)}$$
$$15h^2-17h$$
$$\underline{-(15h^2-20h)}$$
$$3h-4$$
$$\underline{-(3h-4)}$$
$$0$$

47) $4p+1\overline{)\,12p^3+23p^2+p-1}$ $3p^2+5p-1$
$$\underline{-(12p^3+3p^2)}$$
$$20p^2+p$$
$$\underline{-(20p^2+5p)}$$
$$-4p-1$$
$$\underline{-(-4p-1)}$$
$$0$$

49)

$$
\begin{array}{r}
7m+12 \\
m-4{\overline{\smash{\big)}\,7m^2-16m-41}} \\
\underline{-(7m^2-28m)} \\
12m-41 \\
\underline{-(12m-48)} \\
7
\end{array}
$$

$$\left(7m^2-16m-41\right)\div(m-4)$$

$$=7m+12+\frac{7}{m-4}$$

51)

$$
\begin{array}{r}
4a^2-7a\ +2 \\
5a-2{\overline{\smash{\big)}\,20a^3-43a^2+24a-12}} \\
\underline{-(20a^3-8a^2)} \\
-35a^2+24a \\
\underline{-(-35a^2+14a)} \\
10a-12 \\
\underline{-(10a-4)} \\
-8
\end{array}
$$

$$\left(24a+20a^3-12-43a^2\right)\div(5a-2)$$

$$=4a^2-7a+2-\frac{8}{5a-2}$$

53)

$$
\begin{array}{r}
n^2-3n+9 \\
n+3{\overline{\smash{\big)}\,n^3+0n^2+0n+27}} \\
\underline{-(n^3+3n^2)} \\
-3n^2+0n \\
\underline{-(-3n^2-9n)} \\
9n+27 \\
\underline{-(9n+27)} \\
0
\end{array}
$$

55)

$$
\begin{array}{r}
2r^2+4r+5 \\
4r-5{\overline{\smash{\big)}\,8r^3+6r^2+0r-25}} \\
\underline{-(8r^3-10r^2)} \\
16r^2+0r \\
\underline{-(16r^2-20r)} \\
20r-25 \\
\underline{-(20r-25)} \\
0
\end{array}
$$

57)

$$
\begin{array}{r}
6x^2-9x+5 \\
2x+3{\overline{\smash{\big)}\,12x^3+0x^2-17x+4}} \\
\underline{-(12x^3+18x^2)} \\
-18x^2-17x \\
\underline{-(-18x^2-27x)} \\
10x+4 \\
\underline{-(10x+15)} \\
-11
\end{array}
$$

$$\left(12x^3-17x+4\right)\div(2x+3)$$

$$=6x^2-9x+5-\frac{11}{2x+3}$$

59)

$$
\begin{array}{r}
k^2+k+5 \\
k^2+4{\overline{\smash{\big)}\,k^4+k^3+9k^2+4k+20}} \\
\underline{-(k^4\qquad 4k^2)} \\
k^3+5k^2+4k \\
\underline{-(k^3\qquad +4k)} \\
5k^2\qquad +20 \\
\underline{-(5k^2\qquad +20)} \\
0
\end{array}
$$

61)

$$5t^2 - 1 \overline{\smash{\big)}\ 15t^4 - 40t^3 - 33t^2 + 10t + 2}$$

quotient: $3t^2 - 8t - 6$

$-(15t^4 \qquad\quad -3t^2)$

$\qquad -40t^3 - 30t^2 + 10t$

$\qquad -(-40t^3 \qquad\quad +8t)$

$\qquad\qquad -30t^2 + 2t + 2$

$\qquad\qquad -(-30t^2 \qquad +6)$

$\qquad\qquad\qquad 2t - 4$

$$\frac{15t^4 - 40t^3 - 33t^2 + 10t + 2}{5t^2 - 1}$$

$$= 3t^2 - 8t - 6 + \frac{2t - 4}{5t^2 - 1}$$

63) No. For example, $\dfrac{12x + 8}{3x} = 4x + \dfrac{8}{3x}$. The

quotient is not a polynomial because one term has a variable in the denominator.

65) Use synthetic division when the divisor is in the form $x - c$.

67)

$$4 \,\rfloor\ 1 \quad 5 \quad -36$$
$$\qquad\quad 4 \quad 36$$
$$\overline{\quad\ 1 \quad 9 \quad\ 0}$$

$$\left(t^2 + 5t - 36\right) \div (t - 4) = t + 9$$

69)

$$-3 \,\rfloor\ 5 \quad 21 \quad 20$$
$$\qquad\quad\ -15 \quad -18$$
$$\overline{\quad\ 5 \quad\ 6 \quad\ 2}$$

$$\frac{5n^2 + 21n + 20}{n + 3} = 5n + 6 + \frac{2}{n + 3}$$

71)

$$-5 \,\rfloor\ 2 \quad 7 \quad -10 \quad 21$$
$$\qquad\quad\ -10 \quad 15 \quad -25$$
$$\overline{\quad\ 2 \quad -3 \quad\ 5 \quad -4}$$

$$\left(2y^3 + 7y^2 - 10y + 21\right) \div (y + 5)$$

$$= 2y^2 - 3y + 5 - \frac{4}{y + 5}$$

73)

$$-4 \,\rfloor\ 2 \quad 10 \quad 3 \quad -20$$
$$\qquad\quad\ -8 \quad -8 \quad 20$$
$$\overline{\quad\ 2 \quad\ 2 \quad -5 \quad\ 0}$$

$$\left(2c^3 + 10c^2 + 3c - 20\right) \div (c + 4)$$

$$= 2c^2 + 2c - 5$$

75)

$$2 \,\rfloor\ 1 \quad -4 \quad 7 \quad 1 \quad -8$$
$$\qquad\quad\ 2 \quad -4 \quad 6 \quad 14$$
$$\overline{\quad\ 1 \quad -2 \quad 3 \quad 7 \quad 6}$$

$$\left(w^4 - 4w^3 + 7w^2 + w - 8\right) \div (w - 2)$$

$$= w^3 - 2w^2 + 3w + 7 + \frac{6}{w - 2}$$

77)

$$3 \,\rfloor\ 1 \quad 0 \quad 0 \quad 0 \quad -81$$
$$\qquad\quad\ 3 \quad 9 \quad 27 \quad 81$$
$$\overline{\quad\ 1 \quad 3 \quad 9 \quad 27 \quad 0}$$

$$\frac{m^4 - 81}{m - 3} = m^3 + 3m^2 + 9m + 27$$

79)

$$\frac{1}{2} \,\rfloor\ 2 \quad 7 \quad -16 \quad 6$$
$$\qquad\qquad\ 1 \quad 4 \quad -6$$
$$\overline{\quad\ 2 \quad 8 \quad -12 \quad 0}$$

$$\left(2x^3 + 7x^2 - 16x + 6\right) \div \left(x - \frac{1}{2}\right)$$

$$= 2x^2 + 8x - 12$$

81) $\dfrac{50a^4b^4+30a^4b^3-a^2b^2+2ab}{10a^2b^2}$

$=\dfrac{50a^4b^4}{10a^2b^2}+\dfrac{30a^4b^3}{10a^2b^2}-\dfrac{a^2b^2}{10a^2b^2}+\dfrac{2ab}{10a^2b^2}$

$=5a^2b^2+3a^2b-\dfrac{1}{10}+\dfrac{1}{5ab}$

83)
$$\begin{array}{r}
-3f^3+6f^2-2f+9 \\
5f-2{\overline{\smash{\big)}\,-15f^4+36f^3-22f^2+49f+5}}
\end{array}$$

$\underline{-(-15f^4+6f^3)}$

$\qquad 30f^3-22f^2$

$\qquad \underline{-\ (30f^3-12f^2)}$

$\qquad\qquad -10f^2+49f$

$\qquad\qquad \underline{-\ (-10f^2+4f)}$

$\qquad\qquad\qquad 45f+5$

$\qquad\qquad\qquad \underline{-(45f-18)}$

$\qquad\qquad\qquad\qquad 23$

$\left(-15f^4+36f^3-22f^2+49f+5\right)$

$\div\left(5f-2\right)=-3f^3+6f^2-2f+9+\dfrac{23}{5f-2}$

85)
$$\begin{array}{r|rrr}
3 & 8 & -19 & -4 \\
  &   & 24 & 15 \\
\hline
  & 8 & 5 & 11
\end{array}$$

$\dfrac{8t^2-19t-4}{t-3}=8t+5+\dfrac{11}{t-3}$

87)
$$\begin{array}{r}
16p^2+12p+9 \\
4p-3{\overline{\smash{\big)}\,64p^3+0p^2+0p-27}}
\end{array}$$

$\underline{-(64p^3-48p^2)}$

$\qquad 48p^2+0p$

$\qquad \underline{-\ (48p^2-36p)}$

$\qquad\qquad 36p-27$

$\qquad\qquad \underline{-(36p-27)}$

$\qquad\qquad\qquad 0$

89)
$$\begin{array}{r}
6x^2+x-7 \\
x^2+3{\overline{\smash{\big)}\,6x^4+x^3+11x^2+3x-21}}
\end{array}$$

$\underline{-(6x^4\qquad 18x^2)}$

$\qquad x^3-7x^2+3x$

$\qquad \underline{-\ (x^3\qquad +3x)}$

$\qquad\qquad -7x^2\qquad -21$

$\qquad\qquad \underline{-(-7x^2\qquad -21)}$

$\qquad\qquad\qquad 0$

91)
$$\begin{array}{r}
5h^2-3h-2 \\
2h^2-9{\overline{\smash{\big)}\,10h^4-6h^3-49h^2+27h+19}}
\end{array}$$

$\underline{-(10h^4\qquad -45h^2)}$

$\qquad -6h^3-4h^2+27h$

$\qquad \underline{-\ (-6h^3\qquad +27h)}$

$\qquad\qquad -4h^2\qquad +19$

$\qquad\qquad \underline{-(-4h^2\qquad +18)}$

$\qquad\qquad\qquad 1$

$\dfrac{10h^4-6h^3-49h^2+27h+19}{2h^2-9}$

$=5h^2-3h-2+\dfrac{1}{2h^2-9}$

93)

$$\begin{array}{r} j^2 +1 \\ j^2-1\overline{)j^4+0j^3+0j^2+0j-1} \\ \underline{-(j^4-j^2)} \\ j^2+0j-1 \\ \underline{-(j^2-1)} \\ 0 \end{array}$$

95) $\dfrac{9q^2+42q^4-9+6q-8q^3}{3q^2}$

$= \dfrac{42q^4-8q^3+9q^2+6q-9}{3q^2}$

$= \dfrac{42q^4}{3q^2}-\dfrac{8q^3}{3q^2}+\dfrac{9q^2}{3q^2}+\dfrac{6q}{3q^2}-\dfrac{9}{3q^2}$

$= 14q^2-\dfrac{8}{3}q+3+\dfrac{2}{q}-\dfrac{3}{q^2}$

97)

$$\begin{array}{r} 3p^2-5p+1 \\ 7p^2+2p-4\overline{)21p^4-29p^3-15p^2+28p+16} \\ \underline{-(21p^4+6p^3-12p^2)} \\ -35p^3-3p^2+28p \\ \underline{-(-35p^3-10p^2+20p)} \\ 7p^2+8p+16 \\ \underline{-(7p^2+2p-4)} \\ 6p+20 \end{array}$$

$\dfrac{21p^4-29p^3-15p^2+28p+16}{7p^2+2p-4}$

$= 3p^2-5p+1+\dfrac{6p+20}{7p^2+2p-4}$

99) $w =$ width of the rectangle.

$6x^2+23x+21 = (2x+3)w$

$\dfrac{6x^2+23x+21}{2x+3} = w$

$$\begin{array}{r} 3x+7 \\ 2x+3\overline{)6x^2+23x+21} \\ \underline{-(6x^2+9x)} \\ 14x+21 \\ \underline{-(14x+21)} \\ 0 \end{array}$$

$width = 3x+7$

101) $b =$ base of the triangle

$15n^3-18n^2+6n = \dfrac{1}{2}(\text{base})(\text{height})$

$15n^3-18n^2+6n = \dfrac{1}{2}(b)(n)$

$2\left(15n^3-18n^2+6n\right) = (b)(n)$

$\dfrac{30n^3-36n^2+12n}{n} = b$

$\dfrac{30n^3}{n}+\dfrac{-36n^2}{n}+\dfrac{12n}{n} = b$

$30n^2-36n+12 = b$

$base = 30n^2-36n+12$

103) distance $=$ (rate)(time)

Let r be the rate.

$$\left(3x^3 + 5x^2 - 26x + 8\right) = (r)(x+4)$$

$$\frac{3x^3 + 5x^2 - 26x + 8}{x+4} = r$$

$$\begin{array}{r} 3x^2 - 7x + 2 \\ x+4\overline{)3x^3 + 5x^2 - 26x + 8} \\ \underline{-(3x^3 + 12x^2)} \\ -7x - 26x \\ \underline{-(-7x - 28x)} \\ 2x + 8 \\ \underline{-(2x+8)} \\ 0 \end{array}$$

rate $= 3x^2 - 7x + 2$ mph

Chapter 6 Review

1) $\dfrac{2^{11}}{2^6} = 2^{11-6} = 2^5 = 32$

3) $\left(\dfrac{2}{5}\right)^{-3} = \left(\dfrac{5}{2}\right)^3 = \dfrac{125}{8}$

5) $\left(p^7\right)^4 = p^{7 \cdot 4} = p^{28}$

7) $\dfrac{60t^9}{12t^3} = 5t^{9-3} = 5t^6$

9) $(-7c)\left(6c^8\right) = -42c^{1+8} = -42c^9$

11) $\dfrac{k^7}{k^{12}} = k^{7-12} = k^{-5} = \dfrac{1}{k^5}$

13) $\left(-2r^2 s\right)^3 \left(6r^{-9} s\right)$

$= \left(-8r^6 s^3\right)\left(6r^{-9} s\right)$

$= -48r^{-3} s^4$

$= -\dfrac{48s^4}{r^3}$

15) $\left(\dfrac{2xy^{-8}}{3x^{-2} y^6}\right)^{-2} = \left(\dfrac{3x^{-2} y^6}{2xy^{-8}}\right)^2$

$= \left(\dfrac{3y^{14}}{2x^3}\right)^2$

$= \dfrac{9y^{28}}{4x^6}$

17) $\dfrac{m^{-1} n^8}{mn^{14}} = m^{-1-1} n^{8-14} = m^{-2} n^{-6} = \dfrac{1}{m^2 n^6}$

19) $A = (3f)(4f)$

$= 12f^2$ sq. units

$P = 2(3f) + 2(4f)$

$= 6f + 8f$

$= 14f$ units

21) $y^{4a} \cdot y^{3a} = y^{4a+3a} = y^{7a}$

23) $\dfrac{r^{11x}}{r^{2x}} = r^{11x-2x} = r^{9x}$

25) Degree of polynomial $= 3$

Term	Coeff	Degree
$7s^3$	7	3
$-9s^2$	-9	2
s	1	1
6	6	0

27) $2r^2 - 8r - 11$

$\quad 2(-3)^2 - 8(-3) - 11$

$\quad = 2(9) + 24 - 11$

$\quad = 42 - 11 = 31$

29) a) $\quad h(x) = 5x^2 - 3x - 6$

$\quad\quad h(-2) = 5(-2)^2 - 3(-2) - 6$

$\quad\quad h(-2) = 5(4) + 6 - 6$

$\quad\quad h(-2) = 20 - 6 + 6 = 20$

$\quad$ b) $\quad h(x) = 5x^2 - 3x - 6$

$\quad\quad h(0) = 5(0)^2 - 3(0) - 6$

$\quad\quad h(0) = -6$

31) $\left(6c^2 + 2c - 8\right) - \left(8c^2 + c - 13\right)$

$\quad = \left(6c^2 + 2c - 8\right) + \left(-8c^2 - c + 13\right)$

$\quad = -2c^2 + c + 5$

33) $\quad\quad 6.7j^3 - 1.4j^2 + \quad j - 5.3$

$\quad + \quad 3.1j^3 + 5.7j^2 + 2.4j + 4.8$

$\quad \overline{\quad\quad 9.8j^3 + 4.3j^2 + 3.4j - 0.5 \quad}$

35) $\left(\dfrac{3}{5}k^2 + \dfrac{1}{2}k + 4\right) - \left(\dfrac{1}{10}k^2 + \dfrac{3}{2}k - 2\right)$

$\quad = \left(\dfrac{3}{5}k^2 + \dfrac{1}{2}k + 4\right) + \left(-\dfrac{1}{10}k^2 - \dfrac{3}{2}k + 2\right)$

$\quad = \dfrac{5}{10}k^2 - k + 6 = \dfrac{1}{2}k^2 - k + 6$

37) $\left(x^2y^2 + 2x^2y - 4xy + 11\right)$

$\quad -\left(4x^2y^2 - 7x^2y + xy + 5\right)$

$\quad = \left(x^2y^2 + 2x^2y - 4xy + 11\right)$

$\quad + \left(-4x^2y^2 + 7x^2y - xy - 5\right)$

$\quad = -3x^2y^2 + 9x^2y - 5xy + 6$

39) $(6m + 2n - 17) + (-3m + 2n + 14) = 3m + 4n - 3$

41) $\left[(7x - 16) + \left(8x^2 - 15x + 6\right)\right] -$

$\quad \left(2x^2 + 3x + 18\right) = \left[8x^2 - 8x - 10\right]$

$\quad + \left(-2x^2 - 3x - 18\right)$

$\quad = 6x^2 - 11x - 28$

43) $\quad P = 2\left(d^2 + 6d + 2\right) + 2\left(d^2 - 3d + 1\right)$

$\quad\quad = 2d^2 + 12d + 4 + 2d^2 - 6d + 2$

$\quad\quad = 4d^2 + 6d + 6$ units

45) $\quad 3r(8r - 13) = 24r^2 - 39r$

47) $\quad (4w + 3)\left(-8w^3 - 2w + 1\right)$

$\quad\quad = -32w^4 - 8w^2 + 4w - 24w^3 - 6w + 3$

$\quad\quad = -32w^4 - 24w^3 - 8w^2 - 2w + 3$

49) $(y - 3)(y - 9) = y^2 - 9y - 3y + 27$

$\quad = y^2 - 12y + 27$

51) $(3n - 4)(2n - 7) = 6n^2 - 21n - 8n + 28$

$\quad = 6n^2 - 29n + 28$

53) $-(a - 13)(a + 10) = -\left(a^2 + 10a - 13a - 130\right)$

$\quad = -\left(a^2 - 3a - 130\right) = -a^2 + 3a + 130$

55) $6pq^2\left(7p^3q^2+11p^2q^2-pq+4\right)$

$\quad = 42p^4q^4+66p^3q^4-6p^2q^3+24pq^2$

57) $(2x-9y)(2x+y)$

$\quad = 4x^2+2xy-18xy-9y^2$

$\quad = 4x^2-16xy-9y^2$

59) $\left(x^2+5x-12\right)\left(10x^4-3x^2+6\right)$

$\quad = 10x^6-3x^4+6x^2+50x^5-15x^3$

$\quad\quad +30x-120x^4+36x^2-72$

$\quad = 10x^6+50x^5-123x^4-15x^3$

$\quad\quad +42x^2+30x-72$

61) $4f^2(2f-7)(f-6)=4f^2\left(2f^2-12f-7f+42\right)$

$\quad\quad\quad = 4f^2\left(2f^2-19f+42\right)$

$\quad\quad\quad = 8f^4-76f^3+168f^2$

63) $(z+3)(z+1)(z+4)$

$\quad = (z+3)\left(z^2+4z+z+4\right)$

$\quad = (z+3)\left(z^2+5z+4\right)$

$\quad = z^3+5z^2+4z+3z^2+15z+12$

$\quad = z^3+8z^2+19z+12$

65) $\left(\dfrac{2}{7}d+3\right)\left(\dfrac{1}{2}d-8\right)=\dfrac{1}{7}d^2-\dfrac{16}{7}d+\dfrac{3}{2}d-24$

$\quad\quad\quad\quad\quad = \dfrac{1}{7}d^2-\dfrac{32}{14}m+\dfrac{21}{14}d-24$

$\quad\quad\quad\quad\quad = \dfrac{1}{7}d^2-\dfrac{11}{14}d-24$

67) $(c+4)^2=c^2+8c+16$

69) $(4p-3)^2=16p^2-24p+9$

71) $(x-3)^3=(x-3)^2(x-3)$

$\quad\quad = \left(x^2-6x+9\right)(x-3)$

$\quad\quad = x^3-3x^2-6x^2+18x$

$\quad\quad\quad\quad +9x-27$

$\quad\quad = x^3-9x^2+27x-27$

73) $\left[(m-3)+n\right]^2$

$\quad = (m-3)^2+2(m-3)(n)+(n)^2$

$\quad = (m)^2-2(m)(3)+(3)^2+2mn-6n+n^2$

$\quad = m^2-6m+9+2mn-6n+n^2$

75) $(p-13)(p+13)=(p+13)(p-13)$

$\quad\quad\quad\quad\quad = (p)^2-(13)^2$

$\quad\quad\quad\quad\quad = p^2-169$

77) $\left(\dfrac{9}{2}+\dfrac{5}{6}x\right)\left(\dfrac{9}{2}-\dfrac{5}{6}x\right)=\dfrac{81}{4}-\dfrac{25}{36}x^2$

79) $\left(3a-\dfrac{1}{2}b\right)\left(3a+\dfrac{1}{2}b\right)$

$\quad = \left(3a+\dfrac{1}{2}b\right)\left(3a-\dfrac{1}{2}b\right)$

$\quad = 9a^2-\dfrac{1}{4}b^2$

81) $3u(u+4)^2=3u\left(u^2+8u+16\right)$

$\quad\quad\quad\quad = 3u^3+24u^2+48u$

83) a) $A = (2n+11)(n-2)$

$\qquad = 2n^2 - 4n + 11n - 22$

$\qquad = 2n^2 + 7n - 22$ sq. units

b) $P = 2(2n+11) + 2(n-2)$

$\qquad = 4n + 22 + 2n - 4$

$\qquad = 6n + 18$ units

85) $\dfrac{12t^6 - 30t^5 - 15t^4}{3t^4} = \dfrac{12t^6}{3t^4} - \dfrac{30t^5}{3t^4} - \dfrac{15t^4}{3t^4}$

$\qquad\qquad\qquad = 4t^2 - 10t - 5$

87)
$$
\begin{array}{r}
w+5 \\
w+4 \overline{\smash{\big)}\ w^2 + 9w + 20} \\
\underline{-(w^2 + 4w)} \\
5w + 20 \\
\underline{-(5w + 20)} \\
0
\end{array}
$$

89)
$$
\begin{array}{r}
4r^2 + r - 3 \\
2r+5 \overline{\smash{\big)}\ 8r^3 + 22r^2 - r - 15} \\
\underline{-(8r^3 + 20r^2)} \\
2r^2 - r \\
\underline{-(2r^2 + 5r)} \\
-6r - 15 \\
\underline{-(-6r - 15)} \\
0
\end{array}
$$

91) $\dfrac{14t^4 + 28t^3 - 21t^2 + 20t}{14t^3}$

$= \dfrac{14t^4}{14t^3} + \dfrac{28t^3}{14t^3} - \dfrac{21t^2}{14t^3} + \dfrac{20t}{14t^3}$

$= t + 2 - \dfrac{3}{2t} + \dfrac{10}{7t^2}$

93)
$$
\begin{array}{r}
2v\ -1 \\
4v+9 \overline{\smash{\big)}\ 8v^2 + 14v - 3} \\
\underline{-(8v^2 + 18v)} \\
-4v - 3 \\
\underline{-\ (-4v - 9)} \\
6
\end{array}
$$

$(14v + 8v^2 - 3) \div (4v + 9)$

$= 2v - 1 + \dfrac{6}{4v + 9}$

95)
$$
\begin{array}{r}
3v^2 - 7v + 8 \\
2v^2 + 3 \overline{\smash{\big)}\ 6v^4 - 14v^3 + 25v^2 - 21v + 24} \\
\underline{-\ (6v^4 + 9v^2)} \\
-14v^3 + 16v^2 - 21v \\
\underline{-\ (-14v^3 - 21v)} \\
16v^2 + 24 \\
\underline{-\ (16v^2 + 24)} \\
0
\end{array}
$$

97)
$$
\begin{array}{r}
c^2 + 2c + 4 \\
c-2 \overline{\smash{\big)}\ c^3 + 0c^2 + 0c - 8} \\
\underline{-(c^3 - 2c^2)} \\
2c^2 + 0c \\
\underline{-\ (2c^2 - 4c)} \\
4c - 8 \\
\underline{-\ (4c - 8)} \\
0
\end{array}
$$

99)
$$3k+2 \overline{)\,18k^3+0k^2+13k-4\,}$$

$$6k^2-4k+7$$

$$\underline{-(18k^3+12k^2)}$$
$$-12k^2+13k$$
$$\underline{-(-12k^2-8k)}$$
$$21k-\,4$$
$$\underline{-(-21k+14)}$$
$$-18$$

$$\frac{18k^3+13k-4}{3k+2}$$

$$=6k^2-4k+7-\frac{18}{3k+2}$$

101)
$$\left(20x^4y^4-48x^2y^4-12xy^2+15x\right)$$
$$\div\left(-12xy^2\right)$$

$$=\frac{20x^4y^4-48x^2y^4-12xy^2+15x}{-12xy^2}$$

$$=\frac{20x^4y^4}{-12xy^2}-\frac{48x^2y^4}{-12xy^2}-\frac{12xy^2}{-12xy^2}+\frac{15x}{-12xy^2}$$

$$=-\frac{5}{3}x^3y^2+4xy^2+1-\frac{5}{4y^2}$$

103) Let $b=$ the base.

$$12a^2+3a=\frac{1}{2}b(3a)$$
$$24a^2+6a=b(3a)$$
$$\frac{24a^2+6a}{3a}=b$$
$$\frac{24a^2}{3a}+\frac{6a}{3a}=b$$
$$8a+2=b$$

105)
$$\begin{array}{r}18c^3+7c^2-11c+2\\ +\ \ \underline{2c^3-19c^2-1}\\ 20c^3-12c^2-11c+1\end{array}$$

107) $(12-7w)(12+7w)$
$$=(12+7w)(12-7w)$$
$$=(12)^2-(7w)^2$$
$$=144-49w^2$$

109) $5\left(-2r^7t^9\right)^3$
$$=5(-2)^3\left(r^7\right)^3\left(t^9\right)^3$$
$$=5(-8)r^{7\cdot3}t^{9\cdot3}$$
$$=-40r^{21}t^{27}$$

111)
$$\left(39a^6b^6+21a^4b^5-5a^3b^4+a^2b\right)\div\left(3a^3b^3\right)$$

$$=\frac{39a^6b^6+21a^4b^5-5a^3b^4+a^2b}{3a^3b^3}$$

$$=\frac{39a^6b^6}{3a^3b^3}+\frac{21a^4b^5}{3a^3b^3}-\frac{5a^3b^4}{3a^3b^3}+\frac{a^2b}{3a^3b^3}$$

$$=13a^3b^3+7ab^2-\frac{5}{3}b+\frac{1}{3ab^2}$$

113) $(h-5)^3=(h-5)^2(h-5)$
$$=\left(h^2-10h+25\right)(h-5)$$
$$=h^3-5h^2-10h^2+50h$$
$$+25h-125$$
$$=h^3-15h^2+75h-125$$

115)
$$\begin{array}{r|rrrr}-4 & 2 & 0 & -23 & 41\\ & & -8 & 32 & -36\\ \hline & 2 & -8 & 9 & 5\end{array}$$

$$\left(2c^3-23c+41\right)\div(c+4)$$

$$=2c^2-8c+9+\frac{5}{c+4}$$

Chapter 6: Polynomials

117) $\left(\dfrac{5}{y^4}\right)^{-3} = \left(\dfrac{y^4}{5}\right)^3$

$= \dfrac{y^{12}}{125}$

119)

$$3p^2 + p - 4 \overline{\smash{\big)}\ 6p^4 + 11p^3 - 20p^2 - 17p + 20} \quad\quad 2p^2 + 3p - 5$$
$$\underline{-(6p^4 + 2p^3 - 8p^2)}$$
$$9p^3 - 12p^2 - 17p$$
$$\underline{-(9p^3 + 3p^2 - 12p)}$$
$$-15p^2 - 5p + 20$$
$$\underline{-(-15p^2 - 5p + 20)}$$
$$0$$

Chapter 6 Test

1) $\left(\dfrac{3}{4}\right)^{-3} = \left(\dfrac{4}{3}\right)^3 = \dfrac{64}{27}$

3) $\left(8p^3\right)\left(-4p^6\right) = -32p^9$

5) $\dfrac{g^{11}h^{-4}}{g^7 h^6} = g^{11-7}h^{-4-6} = \dfrac{g^4}{h^{10}}$

7) a) -1 b) 3

9) $-2r^2 + 7s, \ r = -4; \ s = 5$

$-2(-4)^2 + 7(5) = -32 + 35 = 3$

11) $\left(7a^3b^2 + 9a^2b^2 - 4ab + 8\right) +$

$\left(5a^3b^2 - 12a^2b^2 + ab + 1\right)$

$= 12a^3b^2 - 3a^2b^2 - 3ab + 9$

13) $3\left(-c^3 + 3c - 6\right) - 4\left(2c^3 + 3c^2 + 7c - 1\right)$

$= -3c^3 + 9c - 18 - 8c^3 - 12c^2 - 28c + 4$

$= -11c^3 - 12c^2 - 19c - 14$

15) $(4g+3)(2g+1) = 8g^2 + 4g + 6g + 3$

$= 8g^2 + 10g + 3$

17) $(3x - 7y)(2x + y) = 6x^2 + 3xy - 14xy - 7y^2$

$= 6x^2 - 11xy - 7y^2$

19) $2y(y+6)^2 = 2y\left(y^2 + 12y + 36\right)$

$= 2y^3 + 24y^2 + 72y$

21) $\left(\dfrac{4}{3}x + y\right)^2 = \dfrac{16}{9}x^2 + \dfrac{8}{3}xy + y^2$

23)

$$w+6 \overline{\smash{\big)}\ w^2 + 9w + 18} \quad\quad w+3$$
$$\underline{-(w^2 + 6w)}$$
$$3w + 18$$
$$\underline{-(3w + 18)}$$
$$0$$

25)

$$3p-7 \overline{\smash{\big)}\ 18p^3 - 45p^2 + 22p - 50} \quad\quad 6p^2 - p + 5$$
$$\underline{-(18p^3 - 42p^2)}$$
$$-3p^2 + 22p$$
$$\underline{-(-3p^2 + 7p)}$$
$$15p - 50$$
$$\underline{-(15p - 35)}$$
$$-15$$

$\left(18p^3 - 45p^2 + 22p - 50\right) \div (3p - 7)$

$= 6p^2 - p + 5 - \dfrac{15}{3p - 7}$

27) $r^2 + 5 \overline{\smash{\big)}\ 2r^4 + 3r^3 + 6r^2 + 15r - 20}$

Quotient: $2r^2 + 3r - 4$

$\quad -(2r^4 \qquad + 10r^2)$
$\quad \overline{\qquad 3r^3 - 4r^2 + 15r}$
$\quad -(3r^3 \qquad + 15r)$
$\quad \overline{\qquad -4r^2 \qquad -20}$
$\quad -(-4r^2 \qquad -20)$
$\quad \overline{\qquad\qquad\qquad 0}$

29) Let $b =$ the base.

$$20n^2 + 15n = \frac{1}{2}b(10n)$$

$$40n^2 + 30n = b(10n)$$

$$\frac{40n^2 + 30n}{10n} = b$$

$$\frac{40n^2}{10n} + \frac{30n}{10n} = b$$

$$4n + 3 = b$$

Cumulative Review: Chapters 1-6

1) a) $\{41, 0\}$ b) $\{-15, 41, 0\}$

   c) $\left\{\dfrac{3}{8}, -15, 2.1, 41, 0.\overline{52}, 0\right\}$

3) $3\dfrac{1}{8} \div 1\dfrac{7}{24} = \dfrac{25}{8} \div \dfrac{31}{24}$

$\qquad = \dfrac{25}{8} \cdot \dfrac{24}{31}$

$\qquad = \dfrac{25}{1} \cdot \dfrac{3}{31}$

$\qquad = \dfrac{75}{31} \text{ or } 2\dfrac{13}{31}$

5) $c^{10} \cdot c^7 = c^{10+7} = c^{17}$

7) $\quad -\dfrac{18}{7}m - 9 = 21$

$\qquad -\dfrac{18}{7}m = 30$

$\qquad m = -\dfrac{7}{18} \cdot 30$

$\qquad c = -\dfrac{7 \cdot 5}{3}$

$\qquad c = -\dfrac{35}{3}, \left\{-\dfrac{35}{3}\right\}$

9) $x =$ number of mL of 15% alcohol solution

   $70 - x =$ number of mL of 8% alcohol solution

Soln	Concn	No of mL of soln	No of mL of acid in the soln
15%	0.15	x	$0.15x$
8%	0.08	$70 - x$	$0.08(70 - x)$
12%	0.12	70	$0.12(70)$

$0.15x + 0.08(70 - x) = 0.12(70)$

$100\left[0.15x + 0.08(70 - x)\right]$
$= 100\left[0.12(70)\right]$

$15x + 8(70 - x) = 12(70)$

$15x + 560 - 8x = 840$

$\qquad 7x + 560 = 840$

$\qquad\qquad 7x = 280$

$\qquad\qquad\quad x = 40$

8% alcohol solution $= 70 - 40 = 30$

40 mL of 15% solution, 30 mL of 8% solution

11) $3x - 8y = 24$

x-int: Let $y = 0$, and solve for x.

$3x - 8(0) = 24$

$3x = 24$

$x = 8$ $\qquad$ (8, 0)

y-int: Let $x = 0$, and solve for y.

$3(0) - 8y = 24$

$-8y = 24$

$y = -3$ $\qquad$ (0, -3)

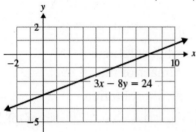

13) $m = \dfrac{-11 - 7}{2 - (-4)} = \dfrac{-18}{6} = -3,$

$(x_1,\ y_1) = (-4,\ 7)$

$(y - y_1) = m(x - x_1)$

$(y - 7) = -3(x - (-4))$

$y - 7 = -3(x + 4)$

$y - 7 = -3x - 12$

$3x + y = -5$

15) $\qquad x + 2y = -4$

$2(x + 2y) = 2(-4)$

$2x + 4y = -8$

Add the equations.

$3x - 4y = -17$

$\underline{+\quad 2x + 4y = -8}$

$\qquad\qquad 5x = -25$

$\qquad\qquad x = -5$

$x + 2y = -4$

$(-5) + 2y = -4$

$2y = 1$

$y = \dfrac{1}{2}$

$\left(-5,\ \dfrac{1}{2}\right)$

17) $\left(6q^2 + 7q - 1\right) - 4\left(2q^2 - 5q + 8\right)$

$+3(-9q - 4) = 6q^2 + 7q - 1 - 8q^2$

$+ 20q - 32 - 27q - 12$

$= -2q^2 - 45$

19) $(3a - 11)(3a + 11) = 9a^2 - 121$

21)

$$\begin{array}{r} 5p^2 + p - 7 \\ p - 3 \overline{)\ 5p^3 - 14p^2 - 10p + 5} \\ \underline{-(5p^3 - 15p^2)} \\ p^2 - 10p \\ \underline{-(p^2 - 3p)} \\ -7p + 5 \\ \underline{-(-7p + 21)} \\ -16 \end{array}$$

$\left(5p^3 - 14p^2 - 10p + 5\right) \div (p - 3)$

$= 5p^2 + p - 7 - \dfrac{16}{p - 3}$

23)
$$
\begin{array}{r}
4z^2 - 2z + 1 \\
2z+1 \overline{\smash{\big)}\ 8z^3 + 0z^2 + 0z + 1} \\
\underline{-(8z^3 + 4z^2)} \\
-4z^2 + 0z \\
\underline{-(-4z^2 - 2z)} \\
2z + 1 \\
\underline{-(2z+1)} \\
0
\end{array}
$$

25) $-3x^2 + 2x + 9,\ x = -2$

$-3(-2)^2 + 2(-2) + 9 = -12 - 4 + 9 = -7$

Section 7.1 Exercises

1) $28 = 7 \cdot 2 \cdot 2$
$21c = 7 \cdot 3 \cdot c$
GCF of 28 and $21c$ is 7.

3) $18p^3 = 2 \cdot 3 \cdot 3 \cdot p \cdot p \cdot p$,
$12p^2 = 2 \cdot 2 \cdot 3 \cdot p \cdot p$
GCF of $18p^3$ and $12p^2$
is $2 \cdot 3 \cdot p \cdot p = 6p^2$.

5) $4n^6$

7) $5a^2b$

9) $21r^3s^2$

11) ab

13) $(k-9)$

15) Answers may vary

17) Yes

19) *No*

21) Yes

23) $2(w+5)$

25) $9(2z^2 - 1)$

27) $10m(10m^2 - 3)$

29) $r^2(r^7 + 1)$

31) $\dfrac{1}{5}y(y+4)$

33) Does not factor

35) GCF is $5n^3$
$10n^5 - 5n^4 + 40n^3$
$= \left(5n^3\right)\left(2n^2\right) - \left(5n^3\right)(n) + (5n^3)(8)$
$= 5n^3\left(2n^2 - n + 8\right)$

37) GCF is $8p^3$
$40p^6 + 40p^5 - 8p^4 + 8p^3$
$= \left(8p^3\right)\left(5p^3\right) + \left(8p^3\right)\left(5p^2\right) - (8p^3)(1) + (8p^3)(1)$
$= 8p^3\left(5p^3 + 5p^2 - p + 1\right)$

39) GCF is $9a^2b$
$63a^3b^3 - 36a^3b^2 + 9a^2b$
$= (9a^2b)(7ab^2) - (9a^2b)(4ab) + (9a^2b)(1)$
$= 9a^2b(7ab^2 - 4ab + 1)$

41) GCF is 6
$-30m - 42 = -6(5m+7)$

43) GCF is $-4w^3$
$-12w^5 - 16w^3$
$= -4w^3(3w^2 + 4)$

45) GCF is -1
$-k + 3 = -1(k-3)$

47) $u(t-5) + 6(t-5)$
$= (u+6)(t-5)$

49) $y(6x+1) - z(6x+1)$
$= (y-z)(6x+1)$

51) $p(q+12) + (q+12)$
$= (p+1)(q+12)$

53) $5h^2(9k+8)-(9k+8)$

$= (5h^2-1)(9k+8)$

55) Factor out a from the first two terms.

Factor out 7 from the last two terms.

$ab+2a+7b+14$

$= a(b+2)+7(b+2)=(a+7)(b+2)$

57) Factor out r from the first two terms.

Factor out -9 from the last two terms.

$3rt+4r-27t-36$

$= r(3t+4)-9(3t+4)=(r-9)(3t+4)$

59) Factor out $4b$ from the first two terms.

Factor out c^2 from the last two terms.

$8b^2+20bc+2bc^2+5c^3$

$= 4b(2b+5c)+c^2(2b+5c)$

$= (4b+c^2)(2b+c)$

61) $fg-7f+4g-28$

$= f(g-7)+4(g-7)$

$= (f+4)(g-7)$

63) $st-10s-6t+60$

$= s(t-10)-6(t-10)$

$= (s-6)(t-10)$

65) $5tu+6t-5u-6$

$= t(5u+6)-1(5u+6)$

$= (t-1)(5u+6)$

67) $36g^4+3gh-96g^3h-8h^2$

$= 3g(12g^3+h)-8h(12g^3+h)$

$= (3g-8h)(12g^3+h)$

69) Answers may vary.

71) $4(xy+3x+5y+15)$

Group the terms and factor out the GCF from each group.

$4(x+5)(y+3)$

73) $3cd+6c+21d+42$

$= 3(cd+2c+7d+14)$

$= 3[c(d+2)+7(d+2)]$

$= 3(c+7)(d+2)$

75) $2p^2q-10p^2-8pq+40p$

$= 2p(pq-5p-4q+20)$

$= 2p[p(q-5)-4(q-5)]$

$= 2p(q-5)(p-4)$

77) $10st+5s-12t-6$

$= 5s(2t+1)-6(2t+1)$

$= (5s-6)(2t+1)$

79) $3a^3-21a^2b-2ab+14b^2$

$= 3a^2(a-7b)-2b(a-7b)$

$= (3a^2-2b)(a-7b)$

81) $8u^2v^2+16u^2v+10uv^2+20uv$

$= 2uv(4uv+8u+5v+10)$

$= 2uv[4u(v+2)+5(v+2)]$

$= 2uv(4u+5)(v+2)$

83) $3mn+21m+10n+70$

$= 3m(n+7)+10(n+7)$

$= (3m+10)(n+7)$

85) $16b-24=8(2b-3)$

213

87) $cd + 6c - 4d - 24$
$= c(d+6) - 4(d+6)$
$= (c-4)(d+6)$

89) $6a^4b + 12a^4 - 8a^3b - 16a^3$
$= 2a^3(3ab + 6a - 4b - 8)$
$= 2a^3[3a(b+2) - 4(b+2)]$
$= 2a^3(3a-4)(b+2)$

91) $7cd + 12 + 28c + 3d$
$= 7cd + 28c + 3d + 12$
$= 7c(d+4) + 3(d+4)$
$= (7c+3)(d+4)$

93) $dg - d + g - 1$
$= d(g-1) + 1(g-1)$
$= (d+1)(g-1)$

95) $x^4y^2 + 12x^3y^3$
$= x^3y^2(x + 12y)$

97) $4mn + 8m + 12n + 24$
$= 4(mn + 2m + 3n + 6)$
$= 4[m(n+2) + 3(n+2)]$
$= 4(m+3)(n+2)$

99) $-6p^2 - 20p + 2 = -2(3p^2 + 10p - 1)$

Section 7.2 Exercises

1) a) $5, 2$ b) $-8, 7$

    c) $-4, -1$ d) $-9, -4$

3) They are negative.

5) Can I factor out a GCF?

7) Can I factor again? If so, factor again.

9) $n^2 + 7n + 10 = (n+5)(n+2)$

11) $c^2 - 16c + 60 = (c-6)(c-10)$

13) $x^2 + x - 12 = (x-3)(x+4)$

15) $g^2 + 8g + 12 = (g+6)(g+2)$

17) $y^2 + 10y + 16 = (y+8)(y+2)$

19) $w^2 - 17w + 72 = (w-9)(w-8)$

21) $b^2 - 3b - 4 = (b-4)(b+1)$

23) $z^2 + 6z - 11$ prime

25) $c^2 - 13c + 36 = (c-9)(c-4)$

27) $m^2 + 4m - 60 = (m+10)(m-6)$

29) $r^2 - 4r - 96 = (r+8)(r-12)$

31) $q^2 + 12q + 42$ prime

33) $x^2 + 16x + 64 = (x+8)(x+8)$
$\quad\quad\quad\quad\quad$ or $(x+8)^2$

35) $n^2 - 2n + 1 = (n-1)(n-1)$ or $(n-1)^2$

37) $24 + 14d + d^2 = d^2 + 14d + 24$
$\quad\quad\quad\quad\quad = (d+12)(d+2)$

39) $-56 + 12a + a^2 = a^2 + 12a - 56$
$\quad\quad\quad\quad\quad$ prime

41) $2k^2 - 22k + 48 = 2(k^2 - 11k + 24)$
$$= 2(k - 8)(k - 3)$$

43) $50h + 35h^2 + 5h^3 = 5h(10 + 7h + h^2)$
$$= 5h(h^2 + 7h + 10)$$
$$= 5h(h + 2)(h + 5)$$

45) $r^4 + r^3 - 132r^2 = r^2(r^2 + r - 132)$
$$= r^2(r + 12)(r - 11)$$

47) $7q^3 - 49q^2 - 42q = 7q(q^2 - 7q - 6)$

49) $3z^4 + 24z^3 + 48z^2 = 3z^2(z^2 + 8z + 16)$
$$= 3z^2(z + 4)(z + 4)$$
$$\text{or } 3z^2(z + 4)^2$$

51) $xy^3 - 2xy^2 - 63xy = xy(y^2 - 2y - 63)$
$$= xy(y - 9)(y + 7)$$

53) $-m^2 - 12m - 35 = -(m^2 + 12m + 35)$
$$= -(m + 5)(m + 7)$$

55) $-c^2 - 3c + 28 = -(c^2 + 3c - 28)$
$$= -(c + 7)(c - 4)$$

57) $-z^2 + 13z - 30 = -(z^2 - 13 + 30)$
$$= -(z - 10)(z - 3)$$

59) $-p^2 + p + 56 = -(p^2 - p - 56)$
$$= -(p - 8)(p + 7)$$

61) $x^2 + 7xy + 12y^2 = (x + 4y)(x + 3y)$

63) $c^2 - 7cd - 8d^2 = (c - 8d)(c + d)$

65) $u^2 - 14uv + 45v^2 = (u - 9v)(u - 5v)$

67) $m^2 + 4mn - 21n^2 = (m + 7n)(m - 3n)$

69) $a^2 + 24ab + 144b^2 = (a + 12b)(a + 12b)$
$$= (a + 12b)^2$$

71) No;
$$3x^2 + 21x + 30 = 3(x^2 + 7x + 10)$$
$$= 3(x + 5)(x + 2)$$

73) yes

75) $2x^2 + 16x + 30 = 2(x^2 + 8x + 15)$
$$= 2(x + 3)(x + 5)$$

77) $n^2 - 6n + 8 = (n - 4)(n - 2)$

79) $m^2 + 7mn - 44n^2 = (m + 11n)(m - 4n)$

81) $h^2 - 10h + 32$ prime

83) $4q^3 - 28q^2 + 48q = 4q(q^2 - 7q + 12)$
$$= 4q(q - 4)(q - 3)$$

85) $-k^2 - 18k - 81 = -(k^2 + 18k + 81)$
$$= -(k + 9)(k + 9)$$
$$\text{or } -1(k + 9)^2$$

87) $4h^5 + 32h^4 + 28h^3 = 4h^3(h^2 + 8h + 7)$
$$= 4h^3(h + 7)(h + 1)$$

89) $k^2 + 21k + 108 = (k + 12)(k + 9)$

91) $p^3q - 17p^2q^2 + 70pq^3$
$$= pq(p^2 - 17pq + 70q^2)$$
$$= pq(p - 10q)(p - 7q)$$
$$a^2 + 9ab + 24b^2 \quad \text{prime}$$

93) $x^2 - 13xy + 12y^2 = (x - 12y)(x - y)$

95) $5v^5 + 55v^4 - 45v^3 = 5v^3(v^2 + 11v - 9)$

97) $6x^3y^2 - 48x^2y^2 - 54xy^2$
   $= 6xy^2(x^2 - 8x - 9)$
   $= 6xy^2(x - 9)(x + 1)$

99) $36 - 13z + z^2 = z^2 - 13z + 36$
   $= (z - 9)(z - 4)$

101) $a^2b^2 + 13ab + 42 = (ab + 7)(ab + 6)$

103) $(x + y)z^2 + 7(x + y)z - 30(x + y)$
   $= (x + y)(z^2 + 7z - 30)$
   $= (x + y)(z + 10)(z - 3)$

105) $(a - b)c^2 - 11(a - b)c + 28(a - b)$
   $= (a - b)(c^2 - 11c + 28)$
   $= (a - b)(c - 7)(c - 4)$

107) $(p + q)r^2 + 24(p + q)r + 144(p + q)$
   $= (p + q)(r^2 + 24r + 144)$
   $= (p + q)(r + 12)(r + 12)$ or
   $= (p + q)(r + 12)^2$

109) $n^2 + 2n - 24 = (n + 6)(n - 4)$

Section 7.3 Exercises

1) a) $10, -5$ b) $-27, -1$

   c) $6, 2$ d) $-12, 6$

3) $3c^2 + 12c + 8c + 32$
   $= 3c(c + 4) + 8(c + 4)$
   $= (3c + 8)(c + 4)$

5) $6k^2 - 6k - 7k + 7$
   $= 6k(k - 1) - 7(k - 1)$
   $= (6k - 7)(k - 1)$

7) $6x^2 - 27xy + 8xy - 36y^2$
   $= 3x(2x - 9y) + 4y(2x - 9y)$
   $= (3x + 4y)(2x - 9y)$

9) Can I factor out a GCF?

11) $(4k + 9)(k + 2)$
   $= 4k^2 + 8k + 9k + 18$
   $= 4k^2 + 17k + 18$

13) $5t^2 + 13t + 6$
   $= 5t^2 + 10t + 3t + 6$
   $= 5t(t + 2) + 3(t + 2)$
   $= (5t + 3)(t + 2)$

15) $6a^2 - 11a - 10 = (2a - 5)(3a + 2)$

17) $12x^2 - 25xy + 7y^2$
   $= (4x - 7y)(3x - y)$

19) $2h^2 + 13h + 15 = 2h^2 + 10h + 3h + 15$
   $= 2h(h + 5) + 3(h + 5)$
   $= (2h + 3)(h + 5)$

21) $7y^2 - 11y + 4 = 7y^2 - 7y - 4y + 4$
   $= 7y(y - 1) - 4(y - 1)$
   $= (7y - 4)(y - 1)$

23) $5b^2 + 9b - 18 = 5b^2 + 15b - 6b - 18$
$$= 5b(b+3) - 6(b+3)$$
$$= (5b-6)(b+3)$$

25) $6p^2 + p - 2 = 6p^2 - 3p + 4p - 2$
$$= 3p(2p-1) + 2(2p-1)$$
$$= (3p+2)(2p-1)$$

27) $4t^2 + 16t + 15 = 4t^2 + 10t + 6t + 15$
$$= 2t(2t+5) + 3(2t+5)$$
$$= (2t+5)(2t+3)$$

29) $9x^2 - 13xy + 4y^2 = 9x^2 - 9xy - 4xy + 4y^2$
$$= 9x(x-y) - 4y(x-y)$$
$$= (9x - 4y)(x-y)$$

31) because 2 can be factored out of $(2x-4)$, but 2cannot be factored out of $(2x^2 + 13x - 24)$

33) $2r^2 + 9r + 10 = (2r+5)(r+2)$

35) $3u^2 - 23u + 30 = (3u-5)(u-6)$

37) $7a^2 + 31a - 20 = (7a-4)(a+5)$

39) $6y^2 + 23y + 10 = (3y+10)(2y+1)$

41) $9w^2 + 20w - 21 = (9w-7)(w+3)$

43) $8c^2 - 42c + 27 = (4c-3)(2c-9)$

45) $4k^2 + 40k + 99 = (2k+11)(2k+9)$

47) $20b^2 - 32b - 45 = (10b+9)(2b-5)$

49) $2r^2 + 13rt - 24t^2 = (2r-3t)(r+8t)$

51) $6a^2 - 25ab + 4b^2 = (6a-b)(a-4b)$

53) $(4z-3)(z+2)$; the answer is the same.

55) $3p^2 - 16p - 12 = (3p+2)(p-6)$

57) $4k^2 + 15k + 9 = (4k+3)(k+3)$

59) $30w^3 + 76w^2 + 14w$
$$= 2w(15w^2 + 38w + 7)$$
$$= 2w(5w+1)(3w+7)$$

61) $21r^2 - 90r + 24 = 3(7r-2)(r-4)$

63) $6y^2 - 10y + 3$ prime

65) $42b^2 + 11b - 3 = (7b+3)(6b-1)$

67) $7x^2 - 17xy + 6y^2 = (7x - 3y)(x - 2y)$

69) $2d^2 + 2d - 40 = 2(d+5)(d-4)$

71) $30r^4t^2 + 23r^3t^2 + 3r^2t^2$
$$= r^2t^2(30r^2 + 23r + 3)$$
$$= r^2t^2(6r+1)(5r+3)$$

73) $9k^2 - 42k + 49 = (3k-7)^2$

75) $2m^2(n+9) - 5m(n+9) - 7(n+9)$
$$= (n+9)(2m^2 - 5m - 7)$$
$$= (n+9)(2m-7)(m+1)$$

77) $6v^2(u+4)^2 + 23v(u+4)^2 + 20(u+4)^2$
$$= (u+4)^2(6v^2 + 23v + 20)$$
$$= (u+4)^2(2v+5)(3v+4)$$

Chapter 7: Factoring Polynomials

79) $15b^2(2a-1)^4 - 28b(2a-1)^4 + 12(2a-1)^4$
$= (2a-1)^4(15b^2 - 28b + 12)$
$= (2a-1)^4(5b-6)(3b-2)$

81) $-n^2 - 8n + 48 = -(n^2 + 8n - 48)$
$= -(n+12)(n-4)$

83) $-7a^2 + 4a + 3 = -(7a^2 - 4a - 3)$
$= -(7a-3)(a+1)$

85) $-10z^2 + 19z - 6 = -(10z^2 - 19z + 6)$
$= -(5z-2)(2z-3)$

87) $-20m^3 - 120m^2 - 135m$
$= -5m(4m^2 + 24m + 27)$
$= -5m(2m+9)(2m+3)$

89) $-6a^3b + 11a^2b^2 + 2ab^3$
$= -ab(6a^2 - 11ab - 2b^2)$
$= -ab(6a+b)(a-2b)$

Section 7.4 Exercises

1) a) $7^2 = 49$ b) $9^2 = 81$
c) $6^2 = 36$ d) $10^2 = 100$
e) $5^2 = 25$ f) $4^2 = 16$
g) $11^2 = 121$ h) $\left(\frac{1}{3}\right)^2 = \frac{1}{9}$
i) $\left(\frac{3}{8}\right)^2 = \frac{9}{64}$

3) a) $\left(c^2\right)^2 = c^4$ b) $(3r)^2 = 9r^2$
c) $(9p)^2 = 81p^2$ d) $(6m^2)^2 = 36m^4$
e) $\left(\frac{1}{2}\right)^2 = \frac{1}{4}$ f) $\left(\frac{12}{5}\right)^2 = \frac{144}{5}$

5) $y^2 + 12y + 36$

7) The middle term does not equal $2(2a)(-3)$. It would have to equal $-12a$ to be a perfect square trinomial.

9) $h^2 + 10h + 25 = (h)^2 + (2 \cdot h \cdot 5) + (5)^2$
$= (h+5)^2$

11) $b^2 - 14b + 49 = (b)^2 - (2 \cdot b \cdot 7) + (7)^2$
$= (b-7)^2$

13) $4w^2 + 4w + 1$
$= (2w)^2 + (2 \cdot 2w \cdot 1) + (1)^2$
$= (2w+1)^2$

15) $9k^2 - 24k + 16$
$= (3k)^2 - (2 \cdot 3k \cdot 4) + (4)^2$
$= (3k-4)^2$

17) $c^2 + c + \frac{1}{4}$
$= (c)^2 + \left(2 \cdot c \cdot \frac{1}{2}\right) + \left(\frac{1}{2}\right)^2$
$= \left(c+\frac{1}{2}\right)^2$

19) $k^2 - \frac{14}{5}k + \frac{49}{25}$
$= (k)^2 - \left(2 \cdot k \cdot \frac{7}{5}\right) + \left(\frac{7}{5}\right)^2$
$= \left(k-\frac{7}{5}\right)^2$

218

21) $a^2 + 8ab + 16b^2$

    $= (a)^2 + (2 \cdot a \cdot 4b) + (4b)^2$

    $= (a + 4b)^2$

23) $25m^2 - 30mn + 9n^2$

    $= (5m)^2 - (2 \cdot 5m \cdot 3n) + (3n)^2$

    $= (5m - 3n)^2$

25) $4f^2 + 24f + 36$

    $= 4(f^2 + 6f + 9)$

    $= 4\left[(f)^2 + (2 \cdot f \cdot 3) + (3)^2\right]$

    $= 4(f + 3)^2$

27) $5a^4 - 30a^3 + 45a^2$

    $= 5a^2(a^2 - 6a + 9)$

    $= 5a^2\left[(a)^2 - (2 \cdot p \cdot 3) + (3)^2\right]$

    $= 5a^2(a - 3)^2$

29) $-16y^2 - 80y - 100$

    $= -4(4y^2 + 20y + 25)$

    $= -4\left[(2y)^2 + (2 \cdot 2y \cdot 5) + (5)^2\right]$

    $= -4(2y + 5)^2$

31) $75h^3 - 6h^2 + 12h = 3h(25^2 - 2h + 4)$

33) a) $x^2 - 81$ b) $81 - x^2$

35) $w^2 - 64 = (w)^2 - (8)^2$

    $= (w + 8)(w - 8)$

37) $121 - p^2 = (11)^2 - (p)^2$

    $= (11 + p)(11 - p)$

39) $64c^2 - 25b^2 = (8c)^2 - (5b)^2$

    $= (8c + 5b)(8c - 5b)$

41) $k^2 - 4 = (k + 2)(k - 2)$

43) $c^2 - 25 = (c + 5)(c - 5)$

45) $w^2 + 9$ prime

47) $x^2 - \dfrac{1}{9} = \left(x + \dfrac{1}{3}\right)\left(x - \dfrac{1}{3}\right)$

49) $a^2 - \dfrac{4}{49} = \left(a + \dfrac{2}{7}\right)\left(a - \dfrac{2}{7}\right)$

51) $144 - v^2 = (12 + v)(12 - v)$

53) $1 - h^2 = (1 + h)(1 - h)$

55) $\dfrac{36}{25} - b^2 = \left(\dfrac{6}{5} + b\right)\left(\dfrac{6}{5} - b\right)$

57) $100m^2 - 49 = (10m + 7)(10m - 7)$

59) $169k^2 - 1 = (13k + 1)(13k - 1)$

61) $4y^2 + 49$ prime

63) $\dfrac{1}{9}t^2 - \dfrac{25}{4} = \left(\dfrac{1}{3}t + \dfrac{5}{2}\right)\left(\dfrac{1}{3}t - \dfrac{5}{2}\right)$

65) $u^4 - 100 = (u^2 + 10)(u^2 - 10)$

67) $36c^2 - d^4 = (6c + d^2)(6c - d^2)$

69) $r^4 - 1 = (r^2 + 1)(r^2 - 1)$

    $= (r^2 + 1)(r + 1)(r - 1)$

71) $r^4 - 81t^4 = (r^2 + 9t^2)(r^2 - 9t^2)$

    $= (r^2 + 9t^2)(r + 3t)(r - 3t)$

73) $5u^2 - 45 = 5(u^2 - 9)$
$$= 5(u+3)(u-3)$$

75) $2n^2 - 288 = 2(n^2 - 144)$
$$= 2(n+12)(n-12)$$

77) $12z^4 - 75z^2 = 3z^2(4z^2 - 25)$
$$= 3z^2(2z+5)(2z-5)$$

79) a) $4^3 = 64$ b) $1^3 = 1$
   c) $10^3 = 1000$ d) $3^3 = 27$
   e) $5^3 = 125$ f) $2^3 = 8$

81) a) $(m)^3 = m^3$ b) $(3t)^3 = 27t^3$
   c) $(2b)^3 = 8b^3$ d) $(h^2)^3 = h^6$

83) $y^3 + 8 = (y)^3 + (2)^3$
$$= (y+2)(y^2 - 2y + 4)$$

85) $t^3 + 64 = (t)^3 + (4)^3$
$$= (t+4)(t^2 - 4t + 16)$$

87) $z^3 - 1 = (z)^3 - (1)^3$
$$= (z-1)(z^2 + z + 1)$$

89) $27m^3 - 125$
$$= (3m)^3 - (5)^3$$
$$= (3m-5)(9m^2 + 15m + 25)$$

91) $125y^3 - 8 = (5y)^3 - (2)^3$
$$= (5y-2)(25y^2 + 10y + 4)$$

93) $1000c^3 - d^3$
$$= (10c)^3 - (d)^3$$
$$= (10c - d)(100c^2 + 10cd + d^2)$$

95) $8j^3 + 27k^3$
$$= (2j)^3 + (3k)^3$$
$$= (2j+3k)(4j^2 - 6jk + 9k^2)$$

97) $64x^3 + 125y^3$
$$= (4x)^3 + (5y)^3$$
$$= (4x+5y)(16x^2 - 20xy + 25y^2)$$

99) $6c^3 + 48 = 6(c^3 + 8)$
$$= 6\left[(c)^3 + (2)^3\right]$$
$$= 6(c+2)(c^2 - 2c + 4)$$

101) $7v^3 - 7000w^3$
$$= 7(v^3 - 1000w^3)$$
$$= 7\left[(v)^3 - (10w)^3\right]$$
$$= 7(v-10w)(v^2 + 10vw + 100w^2)$$

103) $h^6 - 64$
$$= (h^3)^2 - (8)^2 = (h^3 + 8)(h^3 - 8)$$
$$= \left[(h)^3 + (2)^3\right]\left[(h)^3 - (2)^3\right]$$
$$= (h+2)(h^2 - 2h + 4)(h-2)(h^2 + 2h + 4)$$

105) $(d+4)^2 - (d-3)^2$
$$= \left[(d+4)+(d-3)\right]\left[(d+4)-(d-3)\right]$$
$$= (2d+1)(d+4-d+3)$$
$$= 7(2d+1)$$

107) $(3k+1)^2 - (k+5)^2$

$= \left[(3k+1) + (k+5) \right] \left[(3k+1) - (k+5) \right]$

$= (4k+6)(3k+1-k-5)$

$= (4k+6)(2k-4) = 4(2k+3)(k-2)$

109) $(r-2)^3 + 27$

$= (r-2)^3 + (3)^3$

$= (r-2+3)[(r-2)^2 - (r-2)(3) + 9]$

$= (r+1)(r^2 - 4r + 4 - 3r + 6 + 9)$

$= (r+1)(r^2 - 7r + 19)$

111) $(c+4)^3 - 125$

$= (c+4)^3 - (5)^3$

$= (c+4-5)[(c+4)^2 + 5(c+4) + 25]$

$= (c-1)(c^2 + 8c + 16 + 5c + 20 + 25)$

$= (c-1)(c^2 + 13c + 61)$

Chapter 7: Putting It All Together

1) $c^2 + 15c + 56 = (c+7)(c+8)$

3) $uv + 6u + 9v + 54$

$= u(v+6) + 9(v+6)$

$= (u+9)(v+6)$

5) $2p^2 - 13p + 21 = (2p-7)(p+3)$

7) $9v^5 + 90v^4 - 54v^3 = 9v^3(v^2 + 10v - 6)$

9) $24q^3 + 52q^2 - 32q = 4q(3q+8)(2q-1)$

11) $g^3 + 125 = (g+5)(g^2 - 5g + 25)$

13) $144 - w^2 = (12+w)(12-w)$

15) $9r^2 + 12rt + 4t^2 = (3r+2t)^2$

17) $7n^4 - 63n^3 - 70n^2$

$= 7n^2(n^2 - 9n - 10)$

$= 7n^2(n-10)(n+1)$

19) $9h^2 + 25 =$ prime

21) $40x^3 - 135 = 5(8x^3 - 27)$

$= 5(2x-3)(4x^2 + 6x + 9)$

23) $m^2 - \dfrac{1}{100} = \left(m + \dfrac{1}{10} \right)\left(m - \dfrac{1}{10} \right)$

25) $20x^2y + 6 - 24x^2 - 5y$

$= 20x^2y - 24x^2 - 5y + 6$

$= 4x^2(5y-6) - 1(5y-6)$

$= (4x^2 - 1)(5y-6)$

$= (2x+1)(2x-1)(5y-6)$

27) $p^2 + 17pq + 30q^2$

$= (p+15q)(p+2q)$

29) $t^2 - 2t - 16$ prime

31) $50n^2 - 40n + 8 = 2(5n-2)^2$

33) $36r^2 + 57rs + 21s^2$

$= 3(12r^2 + 19rs + 7s^2)$

$= 3(12r + 7s)(r+s)$

35) $81x^4 - y^4 = (9x^2 + y^2)(9x^2 - y^2)$

$= (9x^2 + y^2)(3x+y)(3x-y)$

37) $2a^2 - 10a - 72 = 2(a^2 - 5a - 36)$

$\qquad = 2(a-9)(a+4)$

39) $h^2 - \dfrac{2}{5}h + \dfrac{1}{25} = \left(h - \dfrac{1}{5}\right)^2$

41) $16uv + 24u - 10v - 15$

$\qquad = 8u(2v+3) - 5(2v+3)$

$\qquad = (8u-5)(2v+3)$

43) $8b^2 - 14b - 15 = (4b+3)(2b-5)$

45) $8y^4z^3 - 28y^3z^3 - 40y^3z^2 + 4y^2z^2$

$\qquad = 4y^2z^2(2y^2z - 7yz - 10y + 1)$

47) $2a^2 - 7a + 8 \quad$ prime

49) $16u^2 + 40uv + 25v^2 = (4u+5v)^2$

51) $24k^2 + 31k - 15 = (8k-3)(3k+5)$

53) $5s^3 - 320t^3 = 5(s^3 - 64t^3)$

$\qquad = 5(s-4t)(s^2 + 4st + 16t^2)$

55) $ab - a - b + 1 = a(b-1) - 1(b-1)$

$\qquad = (a-1)(b-1)$

57) $7h^2 - 7 = 7(h^2 - 1)$

$\qquad = 7\left[(h)^2 - (1)^2\right]$

$\qquad = 7(h+1)(h-1)$

59) $6m^2 - 60m + 150 = 6(m-5)^2$

61) $121z^2 - 169 = (11z+13)(11z-13)$

63) $-12r^2 - 75r - 18 = -3(4r+1)(r+6)$

65) $n^3 + 1 = (n+1)(n^2 - n + 1)$

67) $81u^4 - v^4$

$\qquad = (9u^2)^2 - (v^2)^2$

$\qquad = (9u^2 + v^2)(9u^2 - v^2)$

$\qquad = (9u^2 + v^2)\left[(3u)^2 + (v)^2\right]$

$\qquad = (9u^2 + v^2)(3u + v)(3u - v)$

69) $13h^2 + 15h + 2 = (13h+2)(h+1)$

71) $5t^7 - 8t^4 = t^4(5t^3 - 8)$

73) $d^2 - 7d - 30 = (d-10)(d+3)$

75) $z^2 + 144 \quad$ prime

77) $r^2 + 2r + 1 = (r+1)^2$

79) $49n^2 - 100 = (7n+10)(7n-10)$

81) $(2z+1)y^2 + 6(2z+1)y - 55(2z+1)$

$\qquad = (2z+1)(y^2 + 6y - 55)$

$\qquad = (2z+1)(y+11)(y-5)$

83) $(t-3)^2 + 3(t-3) - 4$

$\qquad = [(t-3)+4][(t-3)-1]$

$\qquad = (t+1)(t-4)$

85) $(z+7)^2 - 11(z+7) + 28$

$\qquad = [(z+7)-7][(z+7)-4]$

$\qquad = z(z+3)$

87) $(a+b)^2 - (a-b)^2$

$\qquad = [(a+b)+(a-b)][(a+b)-(a-b)]$

$\qquad = 4ab$

89) $(5p-2q)^2 - (2p+q)^2$

$= [(5p-2q)+(2p+q)][(5p-2q)-(2p+q)]$

$= (7p-q)(3p-3q)$

$= 3(7p-q)(p-q)$

91) $(r+2)^3 + 27$

$= (r+2+3)[(r+2)^2 - 3(r+2)+9]$

$= (r+5)(r^2 + 4r + 4 - 3r - 6 + 9)$

$= (r+5)(r^2 + r + 7)$

93) $(k-7)^3 - 1$

$= (k-7-1)[(k-7)^2 + (k-7)+1]$

$= (k-8)(k^2 - 14k + 49 + k - 7 + 1)$

$= (k-8)(k^2 - 13k + 43)$

95) $a^2 - 8a + 16 - b^2$

$= (a-4)^2 - b^2$

$= (a-4+b)(a-4-b)$

97) $s^2 + 18s + 81 - t^2$

$= (s+9)^2 - t^2$

$= (s+t+9)(s-t+9)$

Section 7.5 Exercises

1) $ax^2 + bx + c = 0$

3) a) $5x^2 + 3x - 7 = 0$ quadratic

    b) $6(p+1) = 0$ linear

    c) $(n+4)(n-9) = 8$ quadratic

    d) $2w + 3(w-5) = 4w + 9$ linear

5) It says that if the product of two quantities equals to 0, then one or both of the quantities must be zero.

7) $(z+11)(z-4) = 0$

$z + 11 = 0$ or $z - 4 = 0$

$z = -11$ or $z = 4$ $\{-11, 4\}$

9) $(2r-3)(r-10) = 0$

$2r - 3 = 0$ or $r - 10 = 0$

$r = \dfrac{3}{2}$ or $r = 10$ $\left\{\dfrac{3}{2}, 10\right\}$

11) $d(d-12) = 0$

$d = 0$ or $d - 12 = 0$

$d = 0$ or $d = 12$ $\{0, 12\}$

13) $(3x+5)^2 = 0$

$3x + 5 = 0$

$x = -\dfrac{5}{3}$ $\left\{-\dfrac{5}{3}\right\}$

15) $(9h+2)(2h+1) = 0$

$9h + 2 = 0$ or $2h + 1 = 0$

$h = -\dfrac{2}{9}$ or $h = -\dfrac{1}{2}$

$\left\{-\dfrac{2}{9}, -\dfrac{1}{2}\right\}$

17) $\left(m+\dfrac{1}{4}\right)\left(m-\dfrac{2}{5}\right)=0$

$m+\dfrac{1}{4}=0$ or $m-\dfrac{2}{5}=0$

$m=-\dfrac{1}{4}$

or $m=\dfrac{2}{5}$ $\left\{-\dfrac{1}{4},\dfrac{2}{5}\right\}$

19) $n(n-4.6)=0$

$n=0$ or $n-4.6=0$

$n=0$ or $n=4.6$ $\{0,4.6\}$

21) No, the product of the factors must equal zero.

23) $p^2+8p+12=0$

$(p+6)(p+2)=0$

$p=-6$ or

$p=-2$ $\{-6,-2\}$

25) $t^2-t-110=0$

$(t-11)(t+10)=0$

$t=11$ or $t=-10$ $\{11,-10\}$

27) $3a^2-10a+8=0\rightarrow$

$(3a-4)(a-2)=0$

$a=\dfrac{4}{3}$ or $a=2$

$\left\{\dfrac{4}{3},2\right\}$

29) $12z^2+z-6=0$

$(4z+3)(3z-2)=0$

$z=-\dfrac{3}{4}$ or $z=\dfrac{2}{3}$

$\left\{-\dfrac{3}{4},\dfrac{2}{3}\right\}$

31) $r^2=60-7r$

$r^2+7r-60=0$

$(r+12)(r-5)=0$

$r=-12$ or

$r=5$

$\{-12,5\}$

33) $d^2-15d=-54$

$d^2-15d+54=0$

$(d-9)(d-6)=0$

$d=9$

or $d=6$ $\{9,6\}$

35) $x^2-64=0$

$(x+8)(x-8)=0$

$x=-8$

or $x=8$ $\{-8,8\}$

37) $49=100u^2$

$100u^2-49=0$

$(10u+7)(10u-7)=0$

$u=\dfrac{7}{10}$ or $u=-\dfrac{7}{10}$ $\left\{\dfrac{7}{10},-\dfrac{7}{10}\right\}$

39)
$$22k = -10k^2 - 12$$
$$10k^2 + 22k + 12 = 0 \qquad \text{Divide by 2}$$
$$5k^2 + 11k + 6 = 0$$
$$(5k + 6)(k + 1) = 0$$
$$k = -\frac{6}{5} \text{ or } k = -1$$
$$\left\{ -\frac{5}{6}, -1 \right\}$$

41)
$$v^2 = 4v$$
$$v^2 - 4v = 0$$
$$v(v - 4) = 0$$
$$v = 0 \text{ or } v = 4 \qquad \{0, 4\}$$

43)
$$(z + 3)(z + 1) = 15$$
$$z^2 + z + 3z + 3 = 15$$
$$z^2 + 4z - 12 = 0$$
$$(z + 6)(z - 2) = 0$$
$$z = -6$$
$$\text{or } z = 2 \qquad \{-6, 2\}$$

45)
$$t(19 - t) = 84$$
$$19t - t^2 = 84$$
$$t^2 - 19t + 84 = 0$$
$$(t - 12)(t - 7) = 0$$
$$t = 12$$
$$\text{or } t = 7 \qquad \{12, 7\}$$

47)
$$6k(k + 4) + 3 = 5(k^2 - 12) + 8k$$
$$6k^2 + 24k + 3 = 5k^2 - 60 + 8k$$
$$k^2 + 16k + 63 = 0$$
$$(k + 9)(k + 7) = 0$$
$$k = -9 \text{ or } k = -7$$
$$\{-9, -7\}$$

49)
$$3(n^2 - 15) + 4n = 4n(n - 3) + 19$$
$$3n^2 - 45 + 4n = 4n^2 - 12n + 19$$
$$n^2 - 16n + 64 = 0$$
$$(n - 8)^2 = 0$$
$$n = 8 \qquad \{8\}$$

51)
$$\frac{1}{2}(m + 1)^2 = -\frac{3}{4}m(m + 5) - \frac{5}{2}$$
Multiply by 4
$$2(m + 1)^2 = -3m(m + 5) - 10$$
$$2m^2 + 4m + 2 = -3m^2 - 15m - 10$$
$$5m^2 + 19m + 12 = 0$$
$$(5m + 4)(m + 3) = 0$$
$$m = -\frac{4}{5} \text{ or } m = -3$$
$$\left\{ -\frac{4}{5}, -3 \right\}$$

53) No. You cannot divide an equation by a variable because you may eliminate a solution and may be dividing by zero.

55)
$$7w(8w - 9)(w + 6) = 0$$
$$w = 0 \text{ or } 8w - 9 = 0$$
$$\text{or } w + 6 = 0$$
$$w = 0 \text{ or } w = \frac{9}{8}$$
$$\text{or } w = -6 \qquad \left\{ 0, \frac{9}{8}, -6 \right\}$$

57)
$$(6m + 7)(m^2 - 5m + 6) = 0$$
$$(6m + 7)(m - 2)(m - 3) = 0$$
$$6m + 7 = 0 \text{ or } m - 2 = 0 \text{ or } m - 3 = 0$$
$$m = -\frac{7}{6} \text{ or } m = 2 \qquad \text{or } m = 3$$
$$\left\{ -\frac{7}{6}, 2, 3 \right\}$$

59)
$$49h = h^3$$
$$h^3 - 49h = 0$$
$$h(h^2 - 49) = 0$$
$$h(h+7)(h-7) = 0$$
$$h = 0 \text{ or } h = -7$$
$$\text{or } h = 7 \qquad \{0, -7, 7\}$$

61)
$$5w^2 + 36w = w^3$$
$$w^3 - 5w^2 - 36w = 0$$
$$w(w^2 - 5w - 36) = 0$$
$$w(w-9)(w+4) = 0$$
$$w = 0 \text{ or } w = 9$$
$$\text{or } w = -4 \qquad \{0, 9, -4\}$$

63)
$$60a = 44a^2 - 8a^3$$
$$8a^3 - 44a^2 + 60a = 0$$
$$4a(2a^2 - 11a + 15) = 0$$
$$4a(2a-5)(a-3) = 0$$
$$a = 0 \text{ or } a = \frac{5}{2}$$
$$\text{or } a = 3 \qquad \left\{0, \frac{5}{2}, 3\right\}$$

65)
$$162b^3 - 8b = 0$$
$$2b(81b^2 - 4) = 0$$
$$2b(9b+2)(9b-2) = 0$$
$$b = 0 \text{ or } b = -\frac{2}{9}$$
$$\text{or } b = \frac{2}{9} \qquad \left\{0, -\frac{2}{9}, \frac{2}{9}\right\}$$

67)
$$-63 = 4y(y-8)$$
$$4y^2 - 32y + 63 = 0$$
$$(2y-7)(2y-9) = 0$$
$$y = \frac{7}{2}$$
$$\text{or } y = \frac{9}{2} \qquad \left\{\frac{7}{2}, \frac{9}{2}\right\}$$

69) $\dfrac{1}{2}d(2-d) - \dfrac{3}{2} = \dfrac{2}{5}d(d+1) - \dfrac{7}{5}$

Multiply by 10
$$5d(2-d) - 15 = 4d(d+1) - 14$$
$$10d - 5d^2 - 15 = 4d^2 + 4d - 14$$
$$9d^2 - 6d + 1 = 0$$
$$(3d-1)^2 = 0$$
$$d = \frac{1}{3} \qquad \left\{\frac{1}{3}\right\}$$

71)
$$a^2 - a = 30$$
$$a^2 - a - 30 = 0$$
$$(a-6)(a+5) = 0$$
$$a = 6$$
$$\text{or } a = -5 \qquad \{6, -5\}$$

73)
$$48t = 3t^3$$
$$3t^3 - 48t = 0$$
$$3t(t^2 - 16) = 0$$
$$3t(t+4)(t-4) = 0$$
$$t = 0 \text{ or } t = -4$$
$$\text{or } t = 4 \qquad \{0, -4, 4\}$$

75) $\quad 104r + 36 = 12r^2$

$\quad 12r^2 - 104r - 36 = 0 \quad$ Divide by 4

$\quad 3r^2 - 26r - 9 = 0$

$\quad (3r+1)(r-9) = 0$

$$r = -\frac{1}{3}$$

$\quad$ or $r = 9 \qquad \left\{-\frac{1}{3}, 9\right\}$

77) $\quad w^2 - 121 = 0$

$\quad (w+11)(w-11) = 0$

$\qquad w = -11$ or $w = 11$

$\qquad \{-11, 11\}$

79) $(2n-5)(n^2 - 6n + 9) = 0$

$\quad (2n-5)(n-3)^2 = 0$

$$n = \frac{5}{2}$$

$\quad$ or $n = 3$

$$\left\{\frac{5}{2}, 3\right\}$$

81) $\qquad (2d-5)^2 - (d+6)^2 = 0$

$\quad (2d-5+d+6)(2d-5-d-6) = 0$

$\quad (3d+1)(d-11) = 0$

$$d = -\frac{1}{3}$$

$\quad$ or $d = 11$

$$\left\{-\frac{1}{3}, 11\right\}$$

83) $(11z-4)^2 - (2z+5)^2 = 0$

$\quad (11z-4+2z+5)(11z-4-2z-5) = 0$

$\quad (13z+1)(9z-9) = 0$

$\quad z = -\frac{1}{13}$ or $z = 1 \qquad \left\{-\frac{1}{13}, 1\right\}$

85) $2p^2(p-4) + 9p(p-4) + 9(p-4) = 0$

$\quad (p-4)(2p^2 + 9p + 9) = 0$

$\quad (p-4)(p+3)(2p+3) = 0$

$\quad p = 4$

$\quad$ or $p = -3$

$\quad$ or $p = -\frac{3}{2}$

$$\left\{4, -3, -\frac{3}{2}\right\}$$

87) $10c^2(2c-7) + 7(2c-7) = 37c(2c-7)$

$\quad 10c^2(2c-7) - 37c(2c-7) + 7(2c-7) = 0$

$\quad (2c-7)(10c^2 - 37c + 7) = 0$

$\quad (2c-7)(5c-1)(2c-7) = 0$

$$c = \frac{7}{2} \quad \text{or}$$

$$c = \frac{1}{5} \qquad \left\{\frac{1}{5}, \frac{7}{2}\right\}$$

89) $h^3 + 8h^2 - h - 8 = 0$

$\quad h^2(h+8) - 1(h+8) = 0$

$\quad (h+8)(h^2 - 1) = 0$

$\quad (h+8)(h+1)(h-1) = 0$

$\quad h = -8$ or $h = -1$ or $h = 1 \qquad \{-8, -1, 1\}$

91) $\qquad f(x) = x^2 + 10x + 16; \; f(x) = 0$

$\quad x^2 + 10x + 16 = 0$

$\quad (x+8)(x+2) = 0$

$\qquad x = -8$ or $x = -2 \qquad \{-8, -2\}$

93) $\quad g(a) = 2a^2 - 13a + 24; \quad g(a) = 4$

$2a^2 - 13a + 24 = 4$

$2a^2 - 13a + 20 = 0$

$(2a - 5)(a - 4) = 0$

$a = \dfrac{5}{2}$ or $a = 4 \qquad \left\{\dfrac{5}{2}, 4\right\}$

95) $\quad P(a) = a^2 - 12; \quad p(a) = 13$

$a^2 - 12 = 13$

$a^2 - 25 = 0$

$(a + 5)(a - 5) = 0$

$a = 5$ or $a = -5 \quad \{5, -5\}$

97) $h(t) = 3t^3 - 21t^2 + 18t; \quad h(t) = 0$

$3t^3 - 21t^2 + 18t = 0$

$3t(t^2 - 7t + 6) = 0$

$3t(t - 6)(t - 1) = 0$

$t = 0$ or $t = 6$ or $t = 1 \qquad \{0, 6, 1\}$

Section 7.6 Exercises

1) $x + 3 = $ length of rectangle

$\quad x = $ width of rectangle

Area $= ($length$)($width$)$

$28 = x(x + 3)$

$28 = x^2 + 3x$

$0 = x^2 + 3x - 28$

$0 = (x + 7)(x - 4)$

$x - 4 = 0$ or $x + 7 = 0$

$x = 4 \qquad x = -7$

length $= 4 + 3 = 7$ in.; width $= 4$ in.

3) $2x + 1 = $ base of triangle

$x + 3 = $ height of triangle

Area $= \dfrac{1}{2}($base$)($height$)$

$44 = \dfrac{1}{2}(2x + 1)(x + 3)$

$88 = 2x^2 + 7x + 3$

$0 = 2x^2 + 7x - 85$

$0 = (2x + 17)(x - 5)$

$2x + 17 = 0$ or $x - 5 = 0$

$2x = -17 \qquad x = 5$

$x = -\dfrac{17}{2}$

base $= 2(5) + 1 = 11$ cm;

height $= 5 + 3 = 8$ cm

5) $x - 1 = $ base of parallelogram

$\dfrac{1}{2}x - 1 = $ height of parallelogram

Area $= ($base$)($height$)$

$36 = (x - 1)\left(\dfrac{1}{2}x - 1\right)$

$72 = (x - 1)(x - 2)$

$72 = x^2 - 3x + 2$

$0 = x^2 - 3x - 70$

$0 = (x - 10)(x + 7)$

$x - 10 = 0$ or $x + 7 = 0$

$x = 10 \qquad x = -7$

base $= 10 - 1 = 9$ in.;

width $= \dfrac{1}{2}(10) - 1 = 4$ in.

7) $x+1 = $ width of box

$x-2 = $ height of box

Volume $= ($length$)($width$)($height$)$

$648 = 12(x+1)(x-2)$

$648 = 12(x^2 - x - 2)$

$54 = x^2 - x - 2$

$0 = x^2 - x - 56$

$0 = (x-8)(x+7)$

$x-8 = 0$ or $x+7 = 0$

$x = 8$ $x = -7$

width $= 8+1 = 9$ in.;

height $= 8-2 = 6$ in.

9) $w = $ the width of the sign

$2w = $ the length of the sign

Area $= ($length$)($width$)$

$8 = 2w \cdot w$

$8 = 2w^2$

$0 = 2w^2 - 8$

$0 = w^2 - 4$

$0 = (w+2)(w-2)$

$w+2 = 0$ or $w-2 = 0$

$w = -2$ $w = 2$

width $= 2$ ft; length $= 2 \cdot 2 = 4$ ft

11) $w+3.5 = $ the length of the granite

$w = $ the width of the granite

Area $= ($length$)($width$)$

$15 = w \cdot (w+3.5)$

$15 = w^2 + 3.5w$ Multiply by 2.

$30 = 2w^2 + 7w$

$0 = 2w^2 + 7w - 30$

$0 = (2w-5)(w+6)$

$2w-5 = 0$ or $w+6 = 0$

$w = 2.5$ $w = -6$

length $= 2.5 + 3.5 = 6.0$ ft; width $= 2.5$ ft

13) $l-1 = $ width of case

$l = $ length of case

Volume $= ($length$)($width$)($height$)$

$90 = (3l)(l-1)$

$90 = 3l^2 - 3l$

$30 = l^2 - l$

$0 = l^2 - l - 30$

$0 = (l+5)(l-6)$

$l+5 = 0$ or $l-6 = 0$

$l = -5$ $l = 6$

length $= 6$ in.; width $= 6-1 = 5$ in.

15) $b = $ base of triangle

$b+1 = $ height of triangle

Area $= \dfrac{1}{2}($base$)($height$)$

$21 = \dfrac{1}{2}(b)(b+1)$

$42 = b^2 + b$

$0 = b^2 + b - 42$

$0 = (b+7)(b-6)$

$b+7 = 0$ or $b-6 = 0$

$b = -7$ $b = 6$

base $= 6$ cm; height $= 6+1 = 7$ cm

$x = $ the first integer

17) $x+1 = $ the second integer

$x(x+1) = 5(x+x+1) - 13$

$x^2 + x = 5(2x+1) - 13$

$x^2 + x = 10x + 5 - 13$

$x^2 - 9x + 8 = 0$

$(x-8)(x-1) = 0$

$x-8=0$ or $x-1=0$

$x=8$ $x=1$

$x=8$, then $8+1=9$

$x=1$, then $1+1=2$

8 and 9;   1 and 2

19) $x =$ the first even integer

$x+2 =$ the second even integer

$x+4 =$ the third even integer

$$x(x+2)=2(x+x+2+x+4)$$

$$x^2+2x=2(3x+6)$$

$$x^2+2x=6x+12$$

$$x^2-4x-12=0$$

$$(x-6)(x+2)=0$$

$x=-2$ or $x=6$

$x=-2$, then $-2+2=0$ and $-2+4=2$

$x=6$, then $6+2=8$ and $6+4=10$

$\{-2,0,2\}$ or $\{6,8,10\}$

21) $x =$ the first odd integer

$x+2 =$ the second odd integer

$x+4 =$ the third odd integer

$$(x+2)(x+4)=3(x+x+2+x+4)+18$$

$$x^2+6x+8=3(3x+6)+18$$

$$x^2+6x+8=9x+36$$

$$x^2-3x-28=0$$

$$(x-7)(x+4)=0$$

$$x=7$$

$x=7$, then $7+2=9$ and $7+4=11$

7,9,11

23) Answers may vary.

25) $a^2+b^2=c^2$

$$a^2+(12)^2=(15)^2$$

$$a^2+144=225$$

$$a^2-81=0$$

$$(a+9)(a-9)=0$$

$a+9=0$ or $a-9=0$

$a=-9$ $a=9$

The length of the missing side is 9.

27) $a^2+b^2=c^2$

$$a^2+(12)^2=(13)^2$$

$$a^2+144=169$$

$$a^2-25=0$$

$$(a+5)(a-5)=0$$

$a+5=0$ or $a-5=0$

$a=-5$ $a=5$

The length of the missing side is 5.

29) $a^2+b^2=c^2$

$$(16)^2+(12)^2=c^2$$

$$256+144=c^2$$

$$400=c^2$$

$$0=c^2-400$$

$$0=(c+20)(c-20)$$

$c+20=0$ or $c-20=0$

$c=-20$ $c=20$

The length of the missing side is 20.

31)
$$a^2 + b^2 = c^2$$
$$x^2 + (x-1)^2 = (x+1)^2$$
$$x^2 + x^2 - 2x + 1 = x^2 + 2x + 1$$
$$x^2 - 4x = 0$$
$$x(x-4) = 0$$
$$x = 4$$
$$x = 4; x - 1 = 3; x + 1 = 5$$
$$3, 4, 5$$

33)
$$a^2 + b^2 = c^2$$
$$\left(\frac{1}{2}x\right)^2 + (x+2)^2 = (x+3)^2$$
$$\frac{1}{4}x^2 + x^2 + 4x + 4 = x^2 + 6x + 9$$

Multiply by 4.

$$x^2 + 4x^2 + 16x + 16 = 4x^2 + 24x + 36$$
$$x^2 - 8x - 20 = 0$$
$$(x-10)(x+2) = 0$$
$$x = 10$$
$$\frac{1}{2}x = 5; x + 2 = 12; x + 3 = 13$$
$$5, 12, 13$$

35) $x =$ Shorter leg

$x + 2 =$ Longer leg

$x + 4 =$ Hypotenuse

$$a^2 + b^2 = c^2$$
$$(x+2)^2 + x^2 = (x+4)^2$$
$$x^2 + 4x + 4 + x^2 = x^2 + 8x + 16$$
$$x^2 - 4x - 12 = 0$$
$$(x-6)(x+2) = 0$$
$$x = 6$$

$x + 4 = 10$

The hypotenuse is 10 in.

37) $x =$ distance from bottom of the ladder to the wall

$x + 7 =$ distance from top of the ladder to bottom

$13 =$ the length of the ladder, hypotenuse

$$a^2 + b^2 = c^2$$
$$x^2 + (x+7)^2 = 13^2$$
$$x^2 + x^2 + 14x + 49 = 169$$
$$2x^2 + 14x - 120 = 0$$
$$x^2 + 7x - 60 = 0$$
$$(x+12)(x-5) = 0$$
$$x = -12 \text{ or } x = 5$$

The distance from the bottom of the ladder to the wall is 5 ft.

39) $x =$ distance traveled by Lance

$x + 2 =$ distance between Lance and Alberto

$4 =$ distance traveled by Alberto

$a^2 + b^2 = c^2$

$x^2 + 4^2 = (x+2)^2$

$x^2 + 16 = x^2 + 4x + 4$

$12 = 4x$

$x = 3$

The distance between Alberto and Lance is 5 miles.

41) a) Let $t = 0$ and solve for h.

$h = -16(0)^2 + 144 = 144$

The initial height of the rock is 144 ft.

b) Let $h = 80$ and solve for t.

$80 = -16t^2 + 144$

$-64 = -16t^2$

$4 = t^2$

$0 = t^2 - 4$

$0 = (t+2)(t-2)$

$t + 2 = 0$ or $t - 2 = 0$

$t = -2$ $t = 2$

The rock will be 80 ft above the water after 2 seconds.

c) Let $h = 0$ and solve for t.

$0 = -16t^2 + 144$

$-144 = -16t^2$

$9 = t^2$

$0 = t^2 - 9$

$0 = (t+3)(t-3)$

$t + 3 = 0$ or $t - 3 = 0$

$t = -3$ $t = 3$

The rock will hit the water after 3 seconds.

43) a) Let $t = 3$.

$y = -16(3)^2 + 144(3)$

$y = -16(9) + 432$

$y = 288$ ft

b) Let $t = 3$.

$x = 39(3)$

$x = 117$ ft

c) Let $t = 4.5$.

$y = -16(4.5)^2 + 144(4.5)$

$y = -16(20.25) + 648$

$y = 324$ ft

d) Let $t = 4.5$.

$x = 39(4.5)$

$x = 175.5 \approx 176$ ft

45) a) Let $t = 0$

$h = -16t^2 + 96t$

$h = -16(0)^2 + 96(0)$

$h = 0$ ft

b) Let $h = 128$

$128 = -16t^2 + 96t$

Divide by 16.

$8 = -t^2 + 6t$

$t^2 - 6t + 8 = 0$

$(t-4)(t-2) = 0$

$t = 4$ sec or $t = 2$ sec

c) Let $t = 3$

$h = -16(3)^2 + 96(3)$

$h = -144 + 288$

$h = 144$ ft

d) Let $h = 0$

$0 = -16t^2 + 96t$

$16t^2 - 96t = 0$

$16t(t-6) = 0$

$t = 6$ sec

47) a) Let $p = 15$

$R(p) = -9p^2 + 324p$

$R(15) = -9(15)^2 + 324(15)$

$R(15) = -9(225) + 4860$

$R(15) = -2025 + 4860 = \$2835$

b) Let $p = 20$

$R(20) = -9(20)^2 + 324(20)$

$R(20) = -9(400) + 6480$

$R(20) = -3600 + 6480 = \$2880$

c) Let $R = 2916$

$2916 = -9p^2 + 324p$

Divde by 9.

$324 = -p^2 + 36p$

$p^2 - 36p + 324 = 0$

$(p-18)^2 = 0$

$p = \$18$

Chapter 7 Review

1) GCF of 50 and 56 is 8.

3) GCF of $15h^2$, $45h^5$ and $20h^3$ is $5h^3$

5) GCF $= 9$

7) GCF $= 2p^4$

$2p^6 - 20p^5 + 2p^4$

$= 2p^4(p^2) - 2p^4(10p) + 2p^4(1)$

$= 2p^4\left(p^2 - 10p + 1\right)$

9) GCF $= (n-5)$

$n(m+8) - 5(m+8)$

$(n-5)(m+8)$

11) $-15r^3 - 40r^2 + 5r$

$= -5r(3r^2 + 8r - 1)$

13) $ab + 2a + 9b + 18$

$= a(b+2) + 9(b+2)$

$= (a+9)(b+2)$

15) $4xy - 28y - 3x + 21$

$= 4y(x-7) - 3(x-7)$

$= (4y-3)(x-7)$

17) $q^2 + 10q + 24 = (q+4)(q+6)$

19) $z^2 - 6z - 72 = (z-12)(z+6)$

21) $m^2 - 13mn + 30n^2 = (m-10n)(m-3n)$

23) $4v^2 - 24v - 64 = 4(v-8)(v+2)$

25) $9w^4 + 9w^3 - 18w^2$
$= 9w^2(w+2)(w-1)$

27) $3r^2 - 23r + 14 = (3r-2)(r-7)$

29) $4p^2 - 8p - 5 = (2p-5)(2p+1)$

31) $12c^2 + 38c + 20 = 2(3c+2)(2c+5)$

33) $10x^2 + 39xy - 27y^2$
$= (5x-3y)(2x+9y)$

35) $w^2 - 49 = (w+7)(w-7)$

37) $64t^2 - 25u^2 = (8t+5u)(8t-5u)$

39) $4b^2 + 9$ prime

41) $64x^3 - 4x = 4x(4+x)(4-x)$

43) $r^2 + 12r + 36 = (r+6)^2$

45) $20k^2 - 60k + 45 = 5(2k-3)^2$

47) $v^3 - 27 = (v-3)(v^2+3v+9)$

49) $125x^3 + 64y^3$
$= (5x+4y)(25x^2 - 20xy + 16y^2)$

51) $10z^2 - 7z - 12 = (5z+4)(2z-3)$

53) $9k^4 - 16k^2 = k^2(3k+4)(3k-4)$

55) $d^2 - 17d + 60 = (d-12)(d-5)$

57) $3a^2b + a^2 - 12b - 4$
$= a^2(3b+1) - 4(3b+1)$
$= (a^2-4)(3b+1)$
$= (a+2)(a-2)(3b+1)$

59) $48p^3 - 6q^3$
$= 6(2p-q)(4p^2 + 2pq + q^2)$

61) $(x+4)^2 - (y-5)^2$
$= (x+4+y-5)(x+4-y+5)$
$= (x+y-1)(x-y+9)$

63) $25c^2 - 20c + 4 = (5c-2)^2$

65) $y(3y+7) = 0$
$y = 0$ or $3y + 7 = 0$
$y = 0$ or $y = -\dfrac{7}{3}$ $\quad \left\{0, -\dfrac{7}{3}\right\}$

67) $\quad 2k^2 + 18 = 13k$
$2k^2 - 13k + 18 = 0$
$(2k-9)(k-2) = 0$
$k = \dfrac{9}{2}$ or $k = 2$
$\left\{\dfrac{9}{2}, 2\right\}$

69) $h^2 + 17h + 72 = 0$
$(h+9)(h+8) = 0$
$h = -9$ or $h = -8$ $\{-9, -8\}$

71)
$$121 = 81r^2$$
$$0 = 81r^2 - 121$$
$$(9r+11)(9r-11) = 0$$
$$r = \frac{11}{9} \text{ or } r = -\frac{11}{9}$$
$$\left\{ \frac{11}{9}, -\frac{11}{9} \right\}$$

73) $3m^2 - 120 = 18m$
$$3m^2 - 18m - 120 = 0$$
$$3(m+4)(m-10) = 0$$
$$m = -4 \text{ or } m = 10 \qquad \{-4, 10\}$$

75) $(w+3)(w+8) = -6$
$$w^2 + 8w + 3w + 24 = -6$$
$$w^2 + 11w + 30 = 0$$
$$(w+6)(w+5) = 0$$
$$w = -6 \text{ or } w = -5$$
$$\{-6, -5\}$$

77) $(5z+4)(3z^2 - 7z + 4) = 0$
$$(5z+4)(z-1)(3z-4) = 0$$
$$z = -\frac{4}{5} \text{ or } z = 1$$
$$\text{or } z = \frac{4}{3}$$
$$\left\{ -\frac{4}{5}, 1, \frac{4}{3} \right\}$$

79) $3v + (v-3)^2 = 5(v^2 - 4v + 1) + 8$
$$3v + v^2 - 6v + 9 = 5v^2 - 20v + 5 + 8$$
$$0 = 4v^2 - 17v + 4$$
$$(4v-1)(v-4) = 0$$
$$v = \frac{1}{4} \text{ or } v = 4 \qquad \left\{ \frac{1}{4}, 4 \right\}$$

81) $45p^3 - 20p = 0$
$$5p(9p^2 - 4) = 0$$
$$5p(3p+2)(3p-2) = 0$$
$$p = 0 \text{ or } p = -\frac{2}{3}$$
$$\text{or } p = \frac{2}{3}$$
$$\left\{ 0, -\frac{2}{3}, \frac{2}{3} \right\}$$

83) $4x + 1 =$ Base of the triangle
$x + 2 =$ Height of the triangle
$18 =$ Area of the triangle
$$\frac{1}{2}(4x+1)(x+2) = 18$$
$$4x^2 + 8x + x + 2 = 36$$
$$4x^2 + 9x - 34 = 0$$
$$(x-2)(4x+17) = 0$$
$$x = 2 \text{ or } x = -17$$
Base of Triangle: 9 cm; Height: 4 cm

85)
$$2x = \text{Length}$$
$$x - 1 = \text{Width}$$
$$60 = \text{Area}$$
$$2x(x-1) = 60$$
$$2x^2 - 2x - 60 = 0$$
$$2(x-6)(x+5) = 0$$
$$x = 6 \text{ or } x = -5$$
Length: 12 in; Width: 5 in

87)
$$a^2 + b^2 = c^2$$
$$a^2 + (8)^2 = (17)^2$$
$$a^2 + 64 = 289$$
$$a^2 - 225 = 0$$
$$(a+15)(a-15) = 0$$

$a = -15$ or $a = 15$

The missing side is 15.

89) $x =$ width

$x + 1.5 =$ length

$10 =$ area

$$x(x+1.5) = 10$$
$$x^2 + 1.5x = 10$$
$$2x^2 + 3x - 20 = 0$$
$$(2x-5)(x+4) = 0$$
$$x = 2.5 \text{ or } x = -4$$

Length: 4 ft; Width: 2.5 ft

91) $x =$ First Integer

$x + 1 =$ Second Integer

$x + 2 =$ Third Integer

$$x + (x+1) + (x+2) = x^2 - 1$$
$$0 = x^2 - 3x - 4$$
$$(x-4)(x+1) = 0$$
$$x = 4 \text{ or } x = -1$$

The integers are; $-1, 0, 1$ or $4, 5, 6$

93) $x =$ Distance traveled by Desmond

$x + 1 =$ Distance traveled by Marcus

$x + 2 =$ Distance between Marcus and Desmond

$$x^2 + (x+1)^2 = (x+2)^2$$
$$x^2 + x^2 + 2x + 1 = x^2 + 4x + 4$$
$$x^2 - 2x - 3 = 0$$
$$(x-3)(x+1) = 0$$

$x = 3$ or $x = -1$

distance traveled by Desmond: 3 Miles

Chapter 7 Test

1) See whether you can factor out a GCF.

3) $36 - v^2 = (6+v)(6-v)$

5) $20a^3b^4 + 36a^2b^3 + 4ab^2$
$= 4ab^2(5a^2b^2 + 9ab + 1)$

7) $64t^3 - 27u^3$
$= (4t - 3u)(16t^2 + 12tu + 9u^2)$

9) $36r^2 - 60r + 25 = (6r-5)^2$

11) $x^2 - 3xy - 18y^2 = (x - 6y)(x + 3y)$

13) $p^2(q-4)^2 + 17p(q-4)^2 + 30(q-4)^2$
$= (q-4)^2(p^2 + 17p + 30)$
$= (q-4)^2(p+15)(p+2)$

15) $k^8 + k^5 = k^5(k^3 + 1)$
$= k^5(k+1)(k^2 - k + 1)$

17) $$144r = r^3$$
$$r(r^2 - 144) = 0$$
$$r(r+12)(r-12) = 0$$
$$r = 0 \text{ or } r = 12 \text{ or}$$
$$r = -12$$
$$\{0, 12, -12\}$$

19) $$(y-7)(y-5) = 3$$
$$y^2 - 5y - 7y + 35 = 3$$
$$y^2 - 12y + 32 = 0$$
$$(y-8)(y-4) = 0$$
$$y = 8 \text{ or } y = 4 \quad \{8, 4\}$$

21) $20k^2 - 52k = 24$
$$20k^2 - 52k - 24 = 0$$
$$4(5k+2)(k-3) = 0$$
$$k = -\frac{2}{5} \text{ or } k = 3 \quad \left\{-\frac{2}{5}, 3\right\}$$

23) $x =$ the first odd integer

$x + 2 =$ the second odd integer

$x + 4 =$ the third odd integer

$$x + x + 2 + x + 4 = (x+4)(x+2) - 110$$
$$3x + 6 = (x^2 + 6x + 8) - 110$$
$$3x + 6 = x^2 + 6x - 102$$
$$0 = x^2 + 3x - 108$$
$$0 = (x+12)(x-9)$$

$x + 12 = 0$ or $x - 9 = 0$

$x = -12$ or $x = 9$

The odd integers are $\{9, 11, 13\}$

25) $x =$ width

$7x =$ length

$252 =$ area

$$7x^2 = 252$$
$$7(x^2 - 36) = 0$$
$$7(x+6)(x-6) = 0$$
$$x = 6 \text{ or } x = -6$$

Width: 6 ft.; Length: 42 ft.

Cumulative Review: Chapters 1-7

1) $\dfrac{2}{9} - \dfrac{5}{6} + \dfrac{1}{3} = \dfrac{4}{18} - \dfrac{15}{18} + \dfrac{6}{18}$

$\qquad = \dfrac{4 - 15 + 6}{18}$

$\qquad = -\dfrac{5}{18}$

3) $2(-3p^4 q)^2 = 18 p^8 q^2$

5) $0.0000839 = 8.39 \times 10^{-5}$

7) $A = \dfrac{1}{2} h(b_1 + b_2)$

$\dfrac{2A}{h} = b_1 + b_2$

$b_2 = \dfrac{2A}{h} - b_1$

$b_2 = \dfrac{2A - b_1 h}{h}$

9)

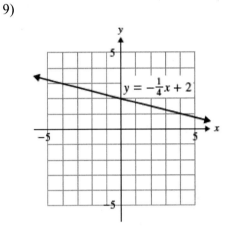

$y = -\frac{1}{4}x + 2$

11) $4 + 2(1 - 3x) + 7y = 3x - 5y$

$\qquad 4 + 2 - 6x + 7y = 3x - 5y$

I $9x - 12y = -6$

$\qquad 10y + 9 = 3(2x + 5) + 2y$

$\qquad 10y + 9 = 6x + 15 + 2y$

II $6x - 8y = 6$

Add Equations (I)$\cdot 2$ and (II)$\cdot(-3)$

$\qquad 18x - 24y = -12$

$\qquad \underline{-18x + 24y = -18}$

$\qquad\qquad\qquad 0 \neq -30$

No Solution $\varnothing$

13) $(4w - 7)(2w + 3)$

$\qquad = 8w^2 + 12w - 14w - 21$

$\qquad = 8w^2 - 2w - 21$

15) $(6z-5)(2z^2+7z-3)$

$12z^3+42z^2-18z$

$\underline{\quad\quad -10z^2-35z+15}$

$12z^3+32z^2-53z+15$

17)

$$\begin{array}{r} 4x^2 \quad\quad +7x-2 \\ 4x-7\overline{)\ 16x^3 \quad\quad -57x+14} \\ \underline{-(16x^3-28x^2)} \\ 0 \quad\quad 28x^2-57x \\ \underline{-\quad (\ 28x^2-49x)} \\ -8x+14 \\ \underline{-\quad (8x+14)} \\ 0 \end{array}$$

19) $6c^2+15c-54=3(2c+9)(c-2)$

21) xy^2+4y^2-x-4

$\quad =y^2(x+4)-1(x+4)$

$\quad =(y^2-1)(x+4)$

$\quad =(y+1)(y-1)(x+4)$

23) $h^3+125=(h+5)(h^2-5h+25)$

25) $24n^3=54n$

$24n^3-54n=0$

$6n(2n+3)(2n-3)=0$

$n=0$ or $n=\dfrac{3}{2}$ or $n=-\dfrac{3}{2}$

$\left\{0,\dfrac{3}{2},-\dfrac{3}{2}\right\}$

Section 8.1: Exercises

1) when its numerator equals zero

3) a) $\dfrac{2(3)-1}{5(3)+2} = \dfrac{6-1}{15+2} = \dfrac{5}{17}$

   b) $\dfrac{2(-2)-1}{5(-2)+2} = \dfrac{-4-1}{-10+2} = \dfrac{-5}{-8} = \dfrac{5}{8}$

5) a) $\dfrac{[4(1)]^2}{(1)^2-(1)-12} = \dfrac{(4)^2}{1-1-12}$

   $= -\dfrac{16}{12} = -\dfrac{4}{3}$

   b) $\dfrac{[4(-3)]^2}{(-3)^2-(-3)-12}$

   $= \dfrac{(-12)^2}{9+3-12}$

   $= \dfrac{144}{0} = $ Undefined

7) a) $\dfrac{15+5(1)}{16-(1)^2} = \dfrac{15+5}{16-1} = \dfrac{20}{15} = \dfrac{4}{3}$

   b) $\dfrac{15+5(-3)}{16-(-3)^2} = \dfrac{15-15}{16-9} = \dfrac{0}{7} = 0$

9) Set the denominator equal to zero and solve for the variable. That value cannot be substituted into the expression because it will make the denominator equal to zero.

11) a) $m+4=0$ b) $3m=0$

    $m=-4$ $m=0$

13) a) $2w-7=0$ b) $4w+1=0$

    $2w=7$ $4w=-1$

    $w=\dfrac{7}{2}$ $w=-\dfrac{1}{4}$

15) a) $11v-v^2=0$ b) $5v-9=0$

    $v(11-v)=0$ $5v=9$

    $v=0$ or $v=\dfrac{9}{5}$

    $11-v=0$

    $v=11$

17) a) It never equals zero.

    b) $p=0$

19) a) $7k=0$

    $k=0$

    b) $k^2+9k+20=0$

    $(k+4)(k+5)=0$

    $k+4=0$ or $k+5=0$

    $k=-4$ $k=-5$

21) a) $c+20=0$

    $c=-20$

    b) $2c^2+3c-9=0$

    $(2c-3)(c+3)=0$

    $2c-3=0$ or $c+3=0$

    $2c=3$ $c=-3$

    $c=\dfrac{3}{2}$

23) a) $g^2+9g+18=0$

    $(g+6)(g+3)=0$

    $g+6=0$ or $g+3=0$

    $g=-6$ $g=-3$

    b) $9g=0$

    $g=0$

25) a) $4y = 0$

$y = 0$

b) $y^2 + 9$ never equals zero.

Never undefined.

Any real number

can be substituted for y.

27) $\dfrac{7x(x-11)}{3(x-11)} = \dfrac{7x}{3}$

29) $\dfrac{24g^2}{56g^4} = \dfrac{3}{7g^2}$

31) $\dfrac{4d-20}{5d-25} = \dfrac{4(d-5)}{5(d-5)} = \dfrac{4}{5}$

33) $\dfrac{-14h-56}{6h+24} = \dfrac{-14(h+4)}{6(h+4)} = -\dfrac{14}{6} = -\dfrac{7}{3}$

35) $\dfrac{39u^2+26}{30u^2+20} = \dfrac{13(3u^2+2)}{10(3u^2+2)} = \dfrac{13}{10}$

37) $\dfrac{g^2-g-56}{g+7} = \dfrac{(g-8)(g+7)}{g+7} = g-8$

39) $\dfrac{t-5}{t^2-25} = \dfrac{t-5}{(t-5)(t+5)} = \dfrac{1}{t+5}$

41) $\dfrac{3c^2+28c+32}{c^2+10c+16} = \dfrac{(3c+4)(c+8)}{(c+2)(c+8)}$

$= \dfrac{3c+4}{c+2}$

43) $\dfrac{q^2-25}{2q^2-7q-15} = \dfrac{(q+5)(q-5)}{(2q+3)(q-5)}$

$= \dfrac{q+5}{2q+3}$

45) $\dfrac{w^3+125}{5w^2-25w+125}$

$= \dfrac{(w+5)(w^2-5w+25)}{5(w^2-5w+25)} = \dfrac{w+5}{5}$

47) $\dfrac{9c^2-27c+81}{c^3+27}$

$= \dfrac{9(c^2-3c+9)}{(c+3)(c^2-3c+9)} = \dfrac{9}{c+3}$

49) $\dfrac{4u^2-20u+4uv-20v}{13u+13v}$

$= \dfrac{4(u^2-5u+uv-5v)}{13(u+v)}$

$= \dfrac{4\left[u(u-5)+v(u-5)\right]}{13(u+v)}$

$= \dfrac{4(u-5)(u+v)}{13(u+v)} = \dfrac{4(u-5)}{13}$

51) $\dfrac{m^2-n^2}{m^3-n^3} = \dfrac{(m+n)(m-n)}{(m-n)(m^2+mn+n^2)}$

$= \dfrac{m+n}{m^2+mn+n^2}$

53) -1

55) $\dfrac{8-q}{q-8} = -1$

57) $\dfrac{m^2-121}{11-m} = \dfrac{(m+11)(m-11)}{11-m}$

$= -(m+11)$

$= -m-11$

59) $\dfrac{36-42x}{7x^2+8x-12} = \dfrac{6(6-7x)}{(7x-6)(x+2)}$

$= -\dfrac{6}{x+2}$

61) $\dfrac{16-4b^2}{b-2}=\dfrac{4\left(4-b^2\right)}{b-2}$

$\qquad=\dfrac{4\left(2+b\right)\left(2-b\right)}{b-2}$

$\qquad=-\dfrac{4\left(b+2\right)\left(b-2\right)}{b-2}$

$\qquad=-4\left(b+2\right)$

63) $\dfrac{y^3-3y^2+2y-6}{21-7y}$

$\quad=\dfrac{y^2\left(y-3\right)+2\left(y-3\right)}{7\left(3-y\right)}=-\dfrac{y^2+2}{7}$

65) $\dfrac{18c+45}{12c^2+18c-30}=\dfrac{9\left(2c+5\right)}{6\left(c-1\right)\left(2c+5\right)}$

$\qquad\qquad=\dfrac{3}{2\left(c-1\right)}$

67) $\dfrac{r^3-t^3}{t^2-r^2}=-\dfrac{\left(r-t\right)\left(r^2+rt+t^2\right)}{\left(r+t\right)\left(r-t\right)}$

$\qquad\quad=-\dfrac{r^2+rt+t^2}{r+t}$

69) $\dfrac{b^2+6b-72}{4b^2+52b+48}=\dfrac{\left(b+12\right)\left(b-6\right)}{4\left(b^2+13b+12\right)}$

$\qquad\qquad=\dfrac{\left(b+12\right)\left(b-6\right)}{4\left(b+12\right)\left(b+1\right)}$

$\qquad\qquad=\dfrac{b-6}{4\left(b+1\right)}$

71) $\dfrac{28h^4-56h^3+7h}{7h}=\dfrac{7h\left(4h^3-8h^2+1\right)}{7h}$

$\qquad\qquad=4h^3-8h^2+1$

73) $\dfrac{14-6w}{12w^3-28w^2}=\dfrac{2\left(7-3w\right)}{4w^2\left(3w-7\right)}$

$\qquad\qquad=-\dfrac{2\left(3w-7\right)}{4w^2\left(3w-7\right)}$

$\qquad\qquad=-\dfrac{1}{2w^2}$

75) $\dfrac{-5v-10}{v^3-v^2-4v+4}=\dfrac{-5\left(v+2\right)}{v^2\left(v-1\right)-4\left(v-1\right)}$

$\qquad\qquad=\dfrac{-5\left(v+2\right)}{\left(v^2-4\right)\left(v-1\right)}$

$\qquad\qquad=\dfrac{-5\left(v+2\right)}{\left(v+2\right)\left(v-2\right)\left(v-1\right)}$

$\qquad\qquad=-\dfrac{5}{\left(v-2\right)\left(v-1\right)}$

77) Possible answers:

$\dfrac{-u-7}{u-2},\dfrac{-\left(u+7\right)}{u-2},\dfrac{u+7}{2-u},$

$\dfrac{u+7}{-\left(u-2\right)},\dfrac{u+7}{-u+2}$

79) Possible answers:

$\dfrac{-9+5t}{2t-3},\dfrac{5t-9}{2t-3},\dfrac{-\left(9-5t\right)}{2t-3},$

$\dfrac{9-5t}{-2t+3},\dfrac{9-5t}{3-2t},\dfrac{9-5t}{-\left(2t-3\right)}$

81) Possible answers:

$-\dfrac{12m}{m^2-3},\dfrac{12m}{-\left(m^2-3\right)},\dfrac{12m}{-m^2+3},\dfrac{12m}{3-m^2}$

83) a)
$$y-2{\overline{\smash{\big)}\,4y^2-11y+6}} \\ 4y-3$$

$$- \underline{(4y^2-8y)}$$
$$-3y+6$$
$$- \underline{(-3y+6)}$$
$$0$$

b) $\dfrac{4y^2-11y+6}{y-2}=\dfrac{(4y-3)(y-2)}{y-2}$
$$=4y-3$$

85) a)
$$2a+5{\overline{\smash{\big)}\,8a^3+0a^2+0a+125}} \\ 4a^2-10a+25$$

$$- \underline{(8a^3+20a^2)}$$
$$-20a^2+0a$$
$$- \underline{(-20a^2-50a)}$$
$$50a+125$$
$$- \underline{(50a+125)}$$
$$0$$

b) $\dfrac{8a^3+125}{2a+5}$
$$=\dfrac{(2a+5)(4a^2-10a+125)}{2a+5}$$
$$=4a^2-10a+125$$

87) $l=\dfrac{3x^2+8x+4}{x+2}$
$$=\dfrac{(3x+2)(x+2)}{x+2}=3x+2$$

89) $w=\dfrac{2c^3+4c^2+8c+16}{c^2+4}$
$$=\dfrac{c^2(2c+4)+4(2c+4)}{c^2+4}$$
$$=\dfrac{(2c+4)(c^2+4)}{c^2+4}=2c+4$$

91) $h=\dfrac{2(3k^2+13k+4)}{2k+8}$
$$=\dfrac{2(k+4)(3k+1)}{2(k+4)}=3k+1$$

93) $p-7=0$
$$p=7$$
The domain contains all real numbers except 7. Domain: $(-\infty,7)\cup(7,\infty)$

95) $5r+2=0$
$$5r=-2$$
$$r=-\dfrac{2}{5}$$
The domain contains all real numbers except $-\dfrac{2}{5}$.
Domain: $\left(-\infty,-\dfrac{2}{5}\right)\cup\left(-\dfrac{2}{5},\infty\right)$

97) $t^2-9t+8=0$
$$(t-8)(t-1)=0$$
$$t-8=0 \quad\text{or}\quad t-1=0$$
$$t=8 \qquad\qquad t=1$$
The domain contains all real numbers except 1 and 8.
Domain: $(-\infty,1)\cup(1,8)\cup(8,\infty)$

99) $w^2-81=0$
$$(w+9)(w-9)=0$$
$$w+9=0 \quad\text{or}\quad w-9=0$$
$$w=-9 \qquad\qquad w=9$$
The domain contains all real numbers except -9 and 9.
Domain: $(-\infty,-9)\cup(-9,9)\cup(9,\infty)$

101) $\dfrac{8}{c^2+6}$ is defined for all real

numberssince c^2+6 never

equals zero.

Domain: $(-\infty, \infty)$

103) Answers may vary.

Section 8.2: Exercises

1) $\dfrac{5}{6}\cdot\dfrac{7}{9}=\dfrac{35}{54}$

3) $\dfrac{6}{15}\cdot\dfrac{25}{42}=\dfrac{\overset{1}{\cancel{6}}}{\underset{3}{\cancel{15}}}\cdot\dfrac{\overset{5}{\cancel{25}}}{\underset{7}{\cancel{42}}}=\dfrac{5}{21}$

5) $\dfrac{16b^5}{3}\cdot\dfrac{4}{36b}=\dfrac{\overset{4b^4}{\cancel{16b^5}}}{3}\cdot\dfrac{4}{\underset{9}{\cancel{36b}}}=\dfrac{16b^4}{27}$

7) $\dfrac{21s^4}{15t^2}\cdot\dfrac{5t^4}{42s^{10}}=\dfrac{\overset{1}{\cancel{21s^4}}}{\underset{3}{\cancel{15t^2}}}\cdot\dfrac{\overset{t^2}{\cancel{5t^4}}}{\underset{2s^6}{\cancel{42s^{10}}}}=\dfrac{t^2}{6s^6}$

9) $\dfrac{9c^4}{42c}\cdot\dfrac{35}{3c^3}=\dfrac{\overset{3c}{\cancel{9c^4}}}{\underset{6c}{\cancel{42c}}}\cdot\dfrac{\overset{5}{\cancel{35}}}{\underset{1}{\cancel{3c^3}}}=\dfrac{\overset{5}{\cancel{15c}}}{\underset{2}{\cancel{6c}}}=\dfrac{5}{2}$

11) $\dfrac{5t^2}{(3t-2)^2}\cdot\dfrac{3t-2}{10t^3}=\dfrac{\cancel{5t^2}}{(3t-2)^{\cancel{2}}}\cdot\dfrac{\cancel{3t-2}}{\underset{2t}{\cancel{10t^3}}}$

$=\dfrac{1}{2t(3t-2)}$

13) $\dfrac{4u-5}{9u^3}\cdot\dfrac{6u^5}{(4u-5)^3}=\dfrac{\cancel{4u-5}}{\underset{3}{\cancel{9u^3}}}\cdot\dfrac{\overset{2u^2}{\cancel{6u^5}}}{(4u-5)^{\cancel{3}\,2}}$

$=\dfrac{2u^2}{3(4u-5)^2}$

15) $\dfrac{6}{n+5}\cdot\dfrac{n^2+8n+15}{n+3}$

$=\dfrac{6}{\cancel{n+5}}\cdot\dfrac{\cancel{(n+5)}\,\cancel{(n+3)}}{\cancel{n+3}}=6$

17) $\dfrac{18y-12}{4y^2}\cdot\dfrac{y^2-4y-5}{3y^2+y-2}$

$=\dfrac{\overset{3}{\cancel{6}}\,(\cancel{3y-2})}{\underset{2}{\cancel{4}}\,y^2}\cdot\dfrac{(y-5)\,\cancel{(y+1)}}{\cancel{(3y-2)}\,\cancel{(y+1)}}$

$=\dfrac{3(y-5)}{2y^2}$

19) $(c-6)\cdot\dfrac{5}{c^2-6c}$

$=\cancel{(c-6)}\cdot\dfrac{5}{c\cancel{(c-6)}}$

$=\dfrac{5}{c}$

21) $\dfrac{7x}{11-x}\cdot(x^2-121)$

$=-\dfrac{7x}{\cancel{(x-11)}}\cdot(x+11)\cancel{(x-11)}$

$=-7x(x+11)$

23) $\dfrac{20}{9}\div\dfrac{10}{27}=\dfrac{\overset{2}{\cancel{20}}}{\underset{1}{\cancel{9}}}\cdot\dfrac{\overset{3}{\cancel{27}}}{\underset{1}{\cancel{10}}}=6$

25) $42\div\dfrac{9}{2}=\overset{14}{\cancel{42}}\cdot\dfrac{2}{\underset{3}{\cancel{9}}}=\dfrac{28}{3}$

27) $\dfrac{12}{5m^5}\div\dfrac{21}{8m^{12}}=\dfrac{\overset{4}{\cancel{12}}}{5\cancel{m^5}}\cdot\dfrac{8\,\overset{m^7}{\cancel{m^{12}}}}{\underset{7}{\cancel{21}}}=\dfrac{32m^7}{35}$

29) $-\dfrac{50g}{7h^3} \div \dfrac{15g^4}{14h} = -\dfrac{\overset{10}{\cancel{50g}}}{\underset{h^2}{\cancel{7h^3}}} \cdot \dfrac{\overset{2}{\cancel{14h}}}{\underset{3g^3}{\cancel{15g^4}}}$

$\qquad = -\dfrac{20}{3g^3 h^2}$

31) $\dfrac{2(k-2)}{21k^6} \div \dfrac{(k-2)^2}{28}$

$\quad = \dfrac{2\cancel{(k-2)}}{\underset{3}{\cancel{21}}k^6} \cdot \dfrac{\overset{4}{\cancel{28}}}{(k-2)^{\cancel{2}}} = \dfrac{8}{3k^6(k-2)}$

33) $\dfrac{16q^5}{p+7} \div \dfrac{2q^4}{(p+7)^2} = \dfrac{\overset{8}{\cancel{16}}q^{\cancel{5}}}{\cancel{p+7}} \cdot \dfrac{(p+7)^{\cancel{2}}}{\cancel{2}\,\cancel{q^4}}$

$\qquad = 8q(p+7)$

35) $\dfrac{q+8}{q} \div \dfrac{q^2+q-56}{5}$

$\quad = \dfrac{q+8}{q} \cdot \dfrac{5}{q^2+q-56}$

$\quad = \dfrac{\cancel{q+8}}{q} \cdot \dfrac{5}{(q-7)\cancel{(q+8)}}$

$\quad = \dfrac{5}{q(q-7)}$

37) $\dfrac{z^2+18z+80}{2z+1} \div (z+8)^2$

$\quad = \dfrac{z^2+18z+80}{2z+1} \cdot \dfrac{1}{(z+8)^2}$

$\quad = \dfrac{\cancel{(z+8)}(z+10)}{2z+1} \cdot \dfrac{1}{(z+8)\cancel{(z+8)}}$

$\quad = \dfrac{z+10}{(2z+1)(z+8)}$

39) $\dfrac{36a-12}{16} \div (9a^2-1)$

$\quad = \dfrac{36a-12}{16} \cdot \dfrac{1}{9a^2-1}$

$\quad = \dfrac{12\cancel{(3a-1)}}{16} \cdot \dfrac{1}{(3a+1)\cancel{(3a-1)}}$

$\quad = \dfrac{\overset{3}{\cancel{12}}}{\underset{4}{\cancel{16}}(3a+1)} = \dfrac{3}{4(3a+1)}$

41) $\dfrac{7n^2-14n}{8n} \div \dfrac{n^2+4n-12}{4n+24}$

$\quad = \dfrac{7n^2-14n}{8n} \cdot \dfrac{4n+24}{n^2+4n-12}$

$\quad = \dfrac{7\cancel{n}\cancel{(n-2)}}{\underset{2}{\cancel{8}}\cancel{n}} \cdot \dfrac{\cancel{4}\cancel{(n+6)}}{\cancel{(n+6)}\cancel{(n-2)}} = \dfrac{7}{2}$

43) $\dfrac{4c-9}{2c^2-8c} \div \dfrac{12c-27}{c^2-3c-4}$

$\quad = \dfrac{4c-9}{2c^2-8c} \cdot \dfrac{c^2-3c-4}{12c-27}$

$\quad = \dfrac{\cancel{4c-9}}{2c\cancel{(c-4)}} \cdot \dfrac{\cancel{(c-4)}(c+1)}{3\cancel{(4c-9)}} = \dfrac{c+1}{6c}$

45) Answers may vary.

47) Let ? be the missing polynomial. Then

$\quad \dfrac{9h+45}{\cancel{h}} \cdot \dfrac{\cancel{h}}{?} = \dfrac{9}{h(h-2)}$

$\quad \dfrac{9h+45}{h(?)} = \dfrac{9}{h(h-2)}$

$$\frac{9(h+5)}{h(?)} = \frac{9}{h(h-2)}$$

$$\frac{\cancel{9}(h+5)\,\cancel{h}(h-2)}{\cancel{9}\,\cancel{h}} = ?$$

$$? = (h+5)(h-2)$$

$$? = h^2 - 2h + 5h - 10$$

$$? = h^2 + 3h - 10$$

49) Let ? be the missing binomial. Then

$$\frac{4z^2 - 49}{z^2 - 3z - 40} \div \frac{?}{z+5} = \frac{2z+7}{8-z}$$

$$\frac{4z^2 - 49}{z^2 - 3z - 40} \cdot \frac{z+5}{?} = \frac{2z+7}{8-z}$$

$$\frac{\cancel{(z+5)}(2z+7)(2z-7)}{(z-8)\cancel{(z+5)}\,?} = \frac{2z+7}{8-z}$$

$$\frac{(2z+7)(2z-7)}{(z-8)?} = \frac{2z+7}{8-z}$$

$$(2z+7)(2z-7)(8-z) = (2z+7)(z-8)?$$

$$\frac{\cancel{(2z+7)}(2z-7)(8-z)}{\cancel{(2z+7)}(z-8)} = ?$$

$$-\frac{(2z-7)\cancel{(z-8)}}{\cancel{(z-8)}} = ?$$

$$? = 7 - 2z$$

51) $\dfrac{\frac{25}{42}}{\frac{8}{21}} = \dfrac{25}{42} \div \dfrac{8}{21} = \dfrac{25}{\underset{2}{\cancel{42}}} \cdot \dfrac{\overset{1}{\cancel{21}}}{8} = \dfrac{25}{16}$

53) $\dfrac{\frac{5}{24}}{\frac{15}{4}} = \dfrac{5}{24} \div \dfrac{15}{4} = \dfrac{\overset{1}{\cancel{5}}}{\underset{6}{\cancel{24}}} \cdot \dfrac{\overset{1}{\cancel{4}}}{\underset{3}{\cancel{15}}} = \dfrac{1}{18}$

55) $\dfrac{\frac{3d+7}{24}}{\frac{3d+7}{6}} = \dfrac{3d+7}{24} \div \dfrac{3d+7}{6}$

$$= \frac{\cancel{3d+7}}{\underset{4}{\cancel{24}}} \cdot \frac{\cancel{6}}{\cancel{3d+7}} = \frac{1}{4}$$

57) $\dfrac{\frac{16r+24}{r^3}}{\frac{12r+18}{r}} = \dfrac{16r+24}{r^3} \div \dfrac{12r+18}{r}$

$$= \frac{16r+24}{r^3} \cdot \frac{r}{12r+18}$$

$$= \frac{\overset{4}{\cancel{8}}(2r+3)}{\underset{r^2}{\cancel{r^3}}} \cdot \frac{\cancel{r}}{\underset{3}{\cancel{6}}(2r+3)}$$

$$= \frac{4}{3r^2}$$

59) $\dfrac{\frac{a^2-25}{3a^{11}}}{\frac{4a+20}{a^3}}$

$$= \frac{a^2-25}{3a^{11}} \div \frac{4a+20}{a^3}$$

$$= \frac{a^2-25}{3a^{11}} \cdot \frac{a^3}{4a+20}$$

$$= \frac{(a-5)\cancel{(a+5)}}{3\underset{a^8}{\cancel{a^{11}}}} \cdot \frac{\cancel{a^3}}{4\cancel{(a+5)}} = \frac{a-5}{12a^8}$$

61) $\dfrac{\frac{16x^2-25}{x^7}}{\frac{36x-45}{6x^3}}$

$$= \frac{16x^2-25}{x^7} \div \frac{36x-45}{6x^3}$$

$$= \frac{16x^2-25}{x^7} \cdot \frac{6x^3}{36x-45}$$

$$= \frac{(4x+5)\,\cancel{(4x-5)}}{\cancel{x}\!\!\!\!_{x^4}} \cdot \frac{\cancel{6}^{\,2}\,\cancel{x}}{\cancel{9}_{\,3}\,\cancel{(4x-5)}} = \frac{2(4x+5)}{3x^4}$$

63) $\dfrac{c^2+c-30}{9c+9} \cdot \dfrac{c^2+2c+1}{c^2-25}$

$$= \frac{\cancel{(c-5)}\,(c+6)}{9\,\cancel{(c+1)}} \cdot \frac{(c+1)^{\cancel{2}}}{(c+5)\,\cancel{(c-5)}}$$

$$= \frac{(c+6)(c+1)}{9(c+5)}$$

65) $\dfrac{3x+2}{9x^2-4} \div \dfrac{4x}{15x^2-7x-2}$

$$= \frac{3x+2}{9x^2-4} \cdot \frac{15x^2-7x-2}{4x}$$

$$= \frac{\cancel{3x+2}}{\cancel{(3x+2)}\,(3x-2)} \cdot \frac{(5x+1)\,\cancel{(3x-2)}}{4x}$$

$$= \frac{5x+1}{4x}$$

67) $\dfrac{3k^2-12k}{12k^2-30k-72} \cdot (2k+3)^2$

$$= \frac{3k(k-4)}{6(2k^2-5k-12)} \cdot (2k+3)^2$$

$$= \frac{\cancel{3}k\,\cancel{(k-4)}}{\cancel{6}_{\,2}\,(2k+3)\,\cancel{(k-4)}} \cdot (2k+3)^{\cancel{2}}$$

$$= \frac{k(2k+3)}{2}$$

69) $\dfrac{7t^6}{t^2-4} \div \dfrac{14t^2}{3t^2-7t+2}$

$$= \frac{7t^6}{t^2-4} \cdot \frac{3t^2-7t+2}{14t^2}$$

$$= \frac{\cancel{7t^6}^{\,t^4}}{(t+2)\,\cancel{(t-2)}} \cdot \frac{(3t-1)\,\cancel{(t-2)}}{\cancel{14t^2}_{\,2}}$$

$$= \frac{t^4(3t-1)}{2(t+2)}$$

71) $\dfrac{4h^3}{h^2-64} \cdot \dfrac{8h-h^2}{12}$

$$= \frac{\cancel{4}^{\,1}\,h^3}{(h+8)\,\cancel{(h-8)}} \cdot \frac{h\,\cancel{(8-h)}^{\,-1}}{\cancel{12}_{\,3}}$$

$$= -\frac{h^4}{3(h+8)}$$

73) $\dfrac{54x^8}{22x^3y^2} \div \dfrac{36xy^5}{11x^2y} = \dfrac{54x^8}{22x^3y^2} \cdot \dfrac{11x^2y}{36xy^5}$

with cancellations $\cancel{54x^8}^{\,3x^7}$, $\cancel{22x^3y^2}_{\,2xy^2}$, $\cancel{11x^2y}^{\,1}$, $\cancel{36xy^5}_{\,2y^5}$

$$= \frac{3x^7y}{4xy^7}$$

$$= \frac{3x^6}{4y^6}$$

75) $\dfrac{r^3+8}{r+2} \cdot \dfrac{7}{3r^2-6r+12}$

$$= \frac{\cancel{(r+2)}\,\cancel{(r^2-2r+4)}}{\cancel{r+2}} \cdot \frac{7}{3\,\cancel{(r^2-2r+4)}}$$

$$= \frac{7}{3}$$

77) $\dfrac{a^2-4a}{6a+54}\cdot\dfrac{a^2+13a+36}{16-a^2}$

$=\dfrac{a\,(a-4)}{6(a+9)}\cdot\dfrac{(a+9)\,(a+4)}{(4+a)\,(4-a)^{-1}}=-\dfrac{a}{6}$

79) $\dfrac{2a^2}{a^2+a-20}\cdot\dfrac{a^3+5a^2+4a+20}{2a^2+8}$

$=\dfrac{2a^2}{(a-4)\,(a+5)}\cdot\dfrac{(a+5)\,(a^2+4)}{2\,(a^2+4)}$

$=\dfrac{a^2}{a-4}$

81) $\dfrac{30}{4y^2-4x^2}\div\dfrac{10x^2+10xy+10y^2}{x^3-y^3}$

$=\dfrac{30}{4\left(y^2-x^2\right)}\cdot\dfrac{x^3-y^3}{10x^2+10xy+10y^2}$

$=\dfrac{\overset{3}{30}}{4(y+x)\,(y-x)}\cdot\dfrac{(x-y)\,\left(x^2+xy+y^2\right)^{-1}}{10\left(x^2+xy+y^2\right)}$

$=-\dfrac{3}{4(x+y)}$

83) $\dfrac{3m^2+8m+4}{4}\div(12m+8)=\dfrac{3m^2+8m+4}{4}\cdot\dfrac{1}{12m+8}=\dfrac{(3m+2)\,(m+2)}{4}\cdot\dfrac{1}{4(3m+2)}=\dfrac{m+2}{16}$

85) $\dfrac{4j^2-21j+5}{j^3}\div\left(\dfrac{3j+2}{j^3-j^2}\cdot\dfrac{j^2-6j+5}{j}\right)=\dfrac{4j^2-21j+5}{j^3}\div\left(\dfrac{3j+2}{j^2\,(j-1)}\cdot\dfrac{(j-5)\,(j-1)}{j}\right)$

$=\dfrac{(4j-1)(j-5)}{j^3}\div\dfrac{(3j+2)(j-5)}{j^3}$

$=\dfrac{(4j-1)\,(j-5)}{j^3}\cdot\dfrac{j^3}{(3j+2)\,(j-5)}=\dfrac{4j-1}{3j+2}$

87) $\dfrac{x}{3x^2-15x+75}\div\left(\dfrac{4x+20}{x+9}\cdot\dfrac{x^2-81}{x^3+125}\right)$

$=\dfrac{x}{3\left(x^2-5x+25\right)}\div\left(\dfrac{4(x+5)}{x+9}\cdot\dfrac{(x+9)\,(x-9)}{(x+5)\left(x^2-5x+25\right)}\right)$

$=\dfrac{x}{3\left(x^2-5x+25\right)}\div\dfrac{4(x-9)}{x^2-5x+25}=\dfrac{x}{3\left(x^2-5x+25\right)}\cdot\dfrac{x^2-5x+25}{4(x-9)}=\dfrac{x}{12(x-9)}$

89) $l = \dfrac{\dfrac{3x}{2y^2}}{\dfrac{y}{8x^4}} = \dfrac{3x}{\cancel{2}y^2} \cdot \dfrac{\cancel{8}^4 x^4}{y} = \dfrac{12x^5}{y^3}$

Section 8.3: Exercises

1) $\quad 12 = 2 \cdot 2 \cdot 3$

$\quad 15 = 3 \cdot 5$

$\quad \text{LCD} = 2^2 \cdot 3 \cdot 5 = 4 \cdot 3 \cdot 5 = 60$

3) $\quad 40 = 2 \cdot 2 \cdot 2 \cdot 5$

$\quad 10 = 2 \cdot 5$

$\quad 12 = 2 \cdot 2 \cdot 3$

$\quad \text{LCD} = 2^3 \cdot 5 \cdot 3 = 8 \cdot 15 = 120$

5) $\quad \text{LCD} = n^{11}$

7) $\quad \text{LCD} = 28r^7$

9) $\quad \text{LCD} = 36z^5$

11) $\text{LCD} = 110m^4$

13) $\text{LCD} = 24x^3y^2$

15) $\text{LCD} = 11(z-3)$

17) $\text{LCD} = w(2w+1)$

19) Factor the denominators.

21) $5c - 5 = 5(c-1)$

$\quad 2c - 2 = 2(c-1)$

$\quad \text{LCD} = 10(c-1)$

23) $9p^4 - 6p^3 = 3p^3(3p-2)$

$\quad 3p^6 - 2p^5 = p^5(3p-2)$

$\quad \text{LCD} = 3p^5(3p-2)$

25) $m-7$ and $m-3$ are different factors. The LCD will be the product of these factors. $\text{LCD} = (m-7)(m-3)$

27) $z^2 + 11z + 24 = (z+3)(z+8)$

$\quad z^2 + 5z - 24 = (z-3)(z+8)$

$\quad \text{LCD} = (z+3)(z+8)(z-3)$

29) $t^2 - 3t - 18 = (t-6)(t+3)$

$\quad t^2 - 36 = (t+6)(t-6)$

$\quad t^2 + 9t + 18 = (t+6)(t+3)$

$\quad \text{LCD} = (t-6)(t+3)(t+6)$

31) $\text{LCD} = a-8$ or $8-a$

33) $\text{LCD} = x-y$ or $y-x$

35) Answers may vary.

37) $\dfrac{7}{12} \cdot \dfrac{4}{4} = \dfrac{28}{48}$

39) $\dfrac{8}{z} \cdot \dfrac{9}{9} = \dfrac{72}{9z}$

41) $\dfrac{3}{8k} \cdot \dfrac{7k^3}{7k^3} = \dfrac{21k^3}{56k^4}$

43) $\dfrac{6}{5t^5u^2} \cdot \dfrac{2t^2u^3}{2t^2u^3} = \dfrac{12t^2u^3}{10t^7u^5}$

45) $\dfrac{7}{3r+4} \cdot \dfrac{r}{r} = \dfrac{7r}{r(3r+4)}$

47) $\dfrac{v}{4(v-3)} \cdot \dfrac{4v^5}{4v^5} = \dfrac{4v^6}{16v^5(v-3)}$

49) $\dfrac{9x}{x+6} \cdot \dfrac{x-5}{x-5} = \dfrac{9x(x-5)}{(x+6)(x-5)}$

$\qquad = \dfrac{9x^2-45x}{(x+6)(x-5)}$

51) $\dfrac{z-3}{2z-5} \cdot \dfrac{z+8}{z+8} = \dfrac{(z-3)(z+8)}{(2z-5)(z+8)}$

$\qquad = \dfrac{z^2+5z-24}{(2z-5)(z+8)}$

53) $\dfrac{5}{3-p} \cdot \dfrac{-1}{-1} = \dfrac{-5}{-(3-p)} = -\dfrac{5}{p-3}$

55) $-\dfrac{8c}{6c-7} \cdot \dfrac{-1}{-1} = -\dfrac{-8c}{-(6c-7)} = \dfrac{8c}{7-6c}$

57) LCD $= 30$

$\dfrac{8}{15} \cdot \dfrac{2}{2} = \dfrac{16}{30}$

$\dfrac{1}{6} \cdot \dfrac{5}{5} = \dfrac{5}{30}$

59) LCD $= u^3$

$\dfrac{4}{u} \cdot \dfrac{u^2}{u^2} = \dfrac{4u^2}{u^3}$

$\dfrac{8}{u^3}$ is already written with the LCD.

61) LCD $= 24n^6$

$\dfrac{9}{8n^6} \cdot \dfrac{3}{3} = \dfrac{27}{24n^6}$

$\dfrac{2}{3n^2} \cdot \dfrac{8n^4}{8n^4} = \dfrac{16n^4}{24n^6}$

63) LCD $= 4a^4b^5$

$\dfrac{6}{4a^3b^5} \cdot \dfrac{a}{a} = \dfrac{6a}{4a^4b^5}$

$\dfrac{6}{a^4b} \cdot \dfrac{4b^4}{4b^4} = \dfrac{24b^4}{4a^4b^5}$

65) LCD $= 5(r-4)$

$\dfrac{r}{5} \cdot \dfrac{r-4}{r-4} = \dfrac{r^2-4r}{5(r-4)}$

$\dfrac{2}{r-4} \cdot \dfrac{5}{5} = \dfrac{10}{5(r-4)}$

67) LCD $= d(d-9)$

$\dfrac{3}{d} \cdot \dfrac{d-9}{d-9} = \dfrac{3d-27}{d(d-9)}$

$\dfrac{7}{d-9} \cdot \dfrac{d}{d} = \dfrac{7d}{d(d-9)}$

69) LCD $= m(m+7)$

$\dfrac{m}{m+7} \cdot \dfrac{m}{m} = \dfrac{m^2}{m(m+7)}$

$\dfrac{3}{m} \cdot \dfrac{m+7}{m+7} = \dfrac{3m+21}{m(m+7)}$

71) $\dfrac{a}{30a-15} = \dfrac{a}{15(2a-1)}$

$\dfrac{1}{12a-6} = \dfrac{1}{6(2a-1)}$

LCD $= 30(2a-1)$

$\dfrac{a}{15(2a-1)} \cdot \dfrac{2}{2} = \dfrac{2a}{30(2a-1)}$

$\dfrac{1}{6(2a-1)} \cdot \dfrac{5}{5} = \dfrac{5}{30(2a-1)}$

73) $\text{LCD} = (k-9)(k+3)$

$$\frac{8}{k-9} \cdot \frac{k+3}{k+3} = \frac{8k+24}{(k-9)(k+3)}$$

$$\frac{5k}{k+3} \cdot \frac{k-9}{k-9} = \frac{5k^2-45k}{(k-9)(k+3)}$$

75) $\text{LCD} = (a+2)(3a+4)$

$$\frac{3}{a+2} \cdot \frac{3a+4}{3a+4} = \frac{9a+12}{(a+2)(3a+4)}$$

$$\frac{2a}{3a+4} \cdot \frac{a+2}{a+2} = \frac{2a^2+4a}{(a+2)(3a+4)}$$

77) $\dfrac{9y}{y^2-y-42} = \dfrac{9y}{(y-7)(y+6)}$

$$\frac{3}{2y^2+12y} = \frac{3}{2y(y+6)}$$

$$\text{LCD} = 2y(y+6)(y-7)$$

$$\frac{9y}{(y-7)(y+6)} \cdot \frac{2y}{2y} = \frac{18y^2}{2y(y-7)(y+6)}$$

$$\frac{3}{2y(y+6)} \cdot \frac{(y-7)}{(y-7)} = \frac{3y-21}{2y(y-7)(y+6)}$$

79) $\dfrac{c}{c^2+9c+18} = \dfrac{c}{(c+6)(c+3)}$

$$\frac{11}{c^2+12c+36} = \frac{11}{(c+6)^2}$$

$$\text{LCD} = (c+6)^2(c+3)$$

$$\frac{c}{(c+6)(c+3)} \cdot \frac{c+6}{c+6} = \frac{c^2+6c}{(c+6)^2(c+3)}$$

$$\frac{11}{(c+6)^2} \cdot \frac{c+3}{c+3} = \frac{11c+33}{(c+6)^2(c+3)}$$

81) $\dfrac{11}{g-3}$

$$\frac{4}{9-g^2} = \frac{4}{(3+g)(3-g)}$$

$$= -\frac{4}{(g+3)(g-3)}$$

$$\text{LCD} = (g+3)(g-3)$$

$$\frac{11}{g-3} \cdot \frac{g+3}{g+3} = \frac{11g+33}{(g-3)(g+3)}$$

$$-\frac{4}{(g+3)(g-3)} \text{ is written with the LCD.}$$

83) $\dfrac{4}{3x-4}$

$$\frac{7x}{16-9x^2} = \frac{7x}{(4+3x)(4-3x)}$$

$$= -\frac{7x}{(3x+4)(3x-4)}$$

$$\text{LCD} = (3x+4)(3x-4)$$

$$\frac{4}{3x-4} \cdot \frac{3x+4}{3x+4} = \frac{12x+16}{(3x+4)(3x-4)}$$

$$-\frac{7x}{(3x+4)(3x-4)} \text{ is written with the LCD.}$$

85) $\dfrac{2}{z^2+3z}=\dfrac{2}{z(z+3)}$

$\dfrac{6}{3z^2+9z}=\dfrac{6}{3z(z+3)}$

$\dfrac{8}{z^2+6z+9}=\dfrac{8}{(z+3)^2}$

$\text{LCD}=3z(z+3)^2$

$\dfrac{2}{z(z+3)}\cdot\dfrac{3(z+3)}{3(z+3)}=\dfrac{6(z+3)}{3z(z+3)^2}$

$=\dfrac{6z+18}{3z(z+3)^2}$

$\dfrac{6}{3z(z+3)}\cdot\dfrac{z+3}{z+3}=\dfrac{6z+18}{3z(z+3)^2}$

$\dfrac{8}{(z+3)^2}\cdot\dfrac{3z}{3z}=\dfrac{24z}{3z(z+3)^2}$

87) $\dfrac{t}{t^2-13t+30}=\dfrac{t}{(t-10)(t-3)}$

$\dfrac{6}{t-10}$

$\dfrac{7}{t^2-9}=\dfrac{7}{(t+3)(t-3)}$

$\text{LCD}=(t+3)(t-3)(t-10)$

$\dfrac{t}{(t-10)(t-3)}\cdot\dfrac{t+3}{t+3}=\dfrac{t^2+3t}{(t+3)(t-3)(t-10)}$

$\dfrac{6}{t-10}\cdot\dfrac{(t+3)(t-3)}{(t+3)(t-3)}=\dfrac{6(t^2-9)}{(t+3)(t-3)(t-10)}$

$=\dfrac{6t^2-54}{(t+3)(t-3)(t-10)}$

$\dfrac{7}{(t+3)(t-3)}\cdot\dfrac{t-10}{t-10}=\dfrac{7t-70}{(t+3)(t-3)(t-10)}$

89) $-\dfrac{9}{h^3+8},\dfrac{2h}{5h^2-10h+20}$

$h^3+8=(h+2)(h^2-2h+4)$

$5h^2-10h+20=5(h^2-2h+4)$

$\text{LCD}=5(h+2)(h^2-2h+4)$

$-\dfrac{9}{h^3+8}=-\dfrac{9}{(h+2)(h^2-2h+4)}\cdot\dfrac{5}{5}$

$=-\dfrac{45}{5(h+2)(h^2-2h+4)}$

$\dfrac{2h}{5h^2-10h+20}=\dfrac{2h}{5(h^2-2h+4)}\cdot\dfrac{h+2}{h+2}$

$\dfrac{2h^2+4h}{5(h+2)(h^2-2h+4)}$

Section 8.4: Exercises

1) $\dfrac{5}{16}+\dfrac{9}{16}=\dfrac{5+9}{16}=\dfrac{14}{16}=\dfrac{7}{8}$

3) $\dfrac{11}{14}-\dfrac{3}{14}=\dfrac{11-3}{14}=\dfrac{8}{14}=\dfrac{4}{7}$

5) $\dfrac{5}{p}-\dfrac{23}{p}=\dfrac{5-23}{p}=-\dfrac{18}{p}$

7) $\dfrac{7}{3c}+\dfrac{8}{3c}=\dfrac{7+8}{3c}=\dfrac{15}{3c}=\dfrac{5}{c}$

9) $\dfrac{6}{z-1}+\dfrac{z}{z-1}=\dfrac{z+6}{z-1}$

11) $\dfrac{8}{x+4}+\dfrac{2x}{x+4}=\dfrac{8+2x}{x+4}$

$=\dfrac{2(\cancel{4+x})}{\cancel{x+4}}$

$=2$

13) $\dfrac{25t+17}{t(4t+3)}-\dfrac{5t+2}{t(4t+3)}=\dfrac{25t+17-5t-2}{t(4t+3)}$

$=\dfrac{20t+15}{t(4t+3)}$

$=\dfrac{5\cancel{(4t+3)}}{t\cancel{(4t+3)}}=\dfrac{5}{t}$

21) $\dfrac{4t}{3}+\dfrac{3}{2}=\dfrac{8t}{6}+\dfrac{9}{6}$

$=\dfrac{8t+9}{6}$

23) $\dfrac{10}{3h^3}+\dfrac{2}{5h}=\dfrac{50}{15h^3}+\dfrac{6h^2}{15h^3}=\dfrac{6h^2+50}{15h^3}$

15) $\dfrac{d^2+15}{(d+5)(d+2)}+\dfrac{8d-3}{(d+5)(d+2)}$

$=\dfrac{d^2+15+8d-3}{(d+5)(d+2)}$

$=\dfrac{d^2+8d+12}{(d+5)(d+1)}$

$=\dfrac{(d+6)\cancel{(d+2)}}{(d+5)\cancel{(d+2)}}=\dfrac{d+6}{d+5}$

25) $\dfrac{3}{2f^2}-\dfrac{7}{f}=\dfrac{3}{2f^2}-\dfrac{14f}{2f^2}=\dfrac{3-14f}{2f^2}$

27) $\dfrac{13}{y+3}+\dfrac{3}{y}=\dfrac{13y}{y(y+3)}+\dfrac{3(y+3)}{y(y+3)}$

$=\dfrac{13y+3y+9}{y(y+3)}$

$=\dfrac{16y+9}{y(y+3)}$

17) a) $18b^4$

b) Multiply the numerator and denominator of $\dfrac{4}{9b^2}$ by $2b^2$, and multiply the numerator and denominator of $\dfrac{5}{6b^4}$ by 3.

29)

$\dfrac{15}{d-8}-\dfrac{4}{d}=\dfrac{15d}{d(d-8)}-\dfrac{4(d-8)}{d(d-8)}$

$=\dfrac{15d-4d+32}{d(d-8)}$

$=\dfrac{11d+32}{d(d-8)}$

c) $\dfrac{4}{9b^2}\cdot\dfrac{2b^2}{2b^2}=\dfrac{8b^2}{18b^4}$

$\dfrac{5}{6b^4}\cdot\dfrac{3}{3}=\dfrac{15}{18b^4}$

19) $\dfrac{3}{8}+\dfrac{2}{5}=\dfrac{15}{40}+\dfrac{16}{40}=\dfrac{31}{40}$

31) $\dfrac{9}{c-4}+\dfrac{6}{c+8}$

$=\dfrac{9(c+8)}{(c-4)(c+8)}+\dfrac{6(c-4)}{(c-4)(c+8)}$

$=\dfrac{9c+72+6c-24}{(c-4)(c+8)}=\dfrac{15c+48}{(c-4)(c+8)}$

$=\dfrac{3(5c+16)}{(c-4)(c+8)}$

33) $\dfrac{m}{3m+5}-\dfrac{2}{m-10}$

$=\dfrac{m(m-10)}{(3m+5)(m-10)}-\dfrac{2(3m+5)}{(3m+5)(m-10)}$

$=\dfrac{m^2-10m-6m-10}{(3m+5)(m-10)}=\dfrac{m^2-16m-10}{(3m+5)(m-10)}$

35) $\dfrac{8u+2}{u^2-1}+\dfrac{3u}{u+1}=\dfrac{8u+2}{(u+1)(u-1)}+\dfrac{3u}{u+1}$

$=\dfrac{8u+2}{(u+1)(u-1)}+\dfrac{3u(u-1)}{(u+1)(u-1)}=\dfrac{8u+2+3u^2-3u}{(u+1)(u-1)}=\dfrac{3u^2+5u+2}{(u+1)(u-1)}=\dfrac{(u+1)(3u+2)}{(u+1)(u-1)}=\dfrac{3u+2}{u-1}$

37) $\dfrac{7g}{g^2+10g+16}+\dfrac{3}{g^2-64}=\dfrac{7g}{(g+8)(g+2)}+\dfrac{3}{(g+8)(g-8)}$

$=\dfrac{7g(g-8)}{(g+2)(g+8)(g-8)}+\dfrac{3(g+2)}{(g+2)(g+8)(g-8)}=\dfrac{7g^2-56g+3g+6}{(g+2)(g+8)(g-8)}=\dfrac{7g^2-53g+6}{(g+2)(g+8)(g-8)}$

39) $\dfrac{5a}{a^2-6a-27}-\dfrac{2a+1}{a^2+2a-3}=\dfrac{5a}{(a-9)(a+3)}-\dfrac{2a+1}{(a-1)(a+3)}$

$=\dfrac{5a(a-1)}{(a-9)(a+3)(a-1)}-\dfrac{(2a+1)(a-9)}{(a-9)(a+3)(a-1)}=\dfrac{5a^2-5a-(2a^2-17a-9)}{(a-9)(a+3)(a-1)}=\dfrac{3a^2+12a+9}{(a-9)(a+3)(a-1)}$

$=\dfrac{3(a^2+4a+3)}{(a-9)(a+3)(a-1)}=\dfrac{3(a+3)(a+1)}{(a-9)(a+3)(a-1)}=\dfrac{3(a+1)}{(a-9)(a-1)}$

41) $\dfrac{2x}{x^2+x-20}-\dfrac{4}{x^2+2x-15}=\dfrac{2x}{(x-4)(x+5)}-\dfrac{4}{(x-3)(x+5)}$

$=\dfrac{2x(x-3)}{(x-4)(x+5)(x-3)}-\dfrac{4(x-4)}{(x-4)(x+5)(x-3)}$

$=\dfrac{2x^2-6x-4x+16}{(x-4)(x+5)(x-3)}=\dfrac{2x^2-10x+16}{(x-4)(x+5)(x-3)}=\dfrac{2(x^2-5x+8)}{(x-4)(x+5)(x-3)}$

43) $\dfrac{4b+1}{3b-12}+\dfrac{5b}{b^2-b-12}=\dfrac{4b+1}{3(b-4)}+\dfrac{5b}{(b-4)(b+3)}$

$=\dfrac{(4b+1)(b+3)}{3(b-4)(b+3)}+\dfrac{15b}{3(b-4)(b+3)}=\dfrac{4b^2+13b+3+15b}{3(b-4)(b+3)}=\dfrac{4b^2+28b+3}{3(b-4)(b+3)}$

45) No. If the sum is rewritten as $\dfrac{9}{x-6}-\dfrac{4}{x-6}$, then the LCD $=x-6$.

If the sum is rewritten as $\dfrac{-9}{6-x}+\dfrac{4}{6-x}$, then the LCD $=6-x$.

47) $\dfrac{16}{q-4}+\dfrac{10}{4-q}=\dfrac{16}{q-4}-\dfrac{10}{q-4}=\dfrac{6}{q-4}$ or $-\dfrac{6}{4-q}$

49) $\dfrac{11}{f-7}-\dfrac{15}{7-f}=\dfrac{11}{f-7}+\dfrac{15}{f-7}=\dfrac{26}{f-7}$ or $-\dfrac{26}{7-f}$

51) $\dfrac{7}{x-4}+\dfrac{x-1}{4-x}=\dfrac{7}{x-4}-\dfrac{x-1}{x-4}=\dfrac{7-x+1}{x-4}=\dfrac{8-x}{x-4}$ or $\dfrac{x-8}{4-x}$

53) $\dfrac{8}{3-a}+\dfrac{a+5}{a-3}=\dfrac{8}{3-a}-\dfrac{a+5}{3-a}=\dfrac{8-a-5}{3-a}=\dfrac{3-a}{3-a}=1$

55) $\dfrac{3}{2u-3v}-\dfrac{6u}{3v-2u}=\dfrac{3}{2u-3v}+\dfrac{6u}{2u-3v}=\dfrac{3+6u}{2u-3v}=\dfrac{3(1+2u)}{2u-3v}$ or $-\dfrac{3(1+2u)}{3v-2u}$

57) $\dfrac{8}{x^2-9}+\dfrac{2}{3-x}=\dfrac{8}{(x+3)(x-3)}-\dfrac{2}{x-3}=\dfrac{8}{(x+3)(x-3)}-\dfrac{2(x+3)}{(x+3)(x-3)}$

$=\dfrac{8-2x-6}{(x+3)(x-3)}=\dfrac{2-2x}{(x+3)(x-3)}=-\dfrac{2(x-1)}{(x+3)(x-3)}$

59) $\dfrac{a}{4a^2-9}-\dfrac{4}{3-2a}=\dfrac{a}{(2a+3)(2a-3)}+\dfrac{4}{(2a-3)}=\dfrac{a}{(2a+3)(2a-3)}+\dfrac{4(2a+3)}{(2a+3)(2a-3)}$

$=\dfrac{a+8a+12}{(2a+3)(2a-3)}=\dfrac{9a+12}{(2a+3)(2a-3)}=\dfrac{3(3a+4)}{(2a+3)(2a-3)}$

61) $\dfrac{5}{a^2-2a}+\dfrac{8}{a}-\dfrac{10a}{a-2}=\dfrac{5}{a(a-2)}+\dfrac{8}{a}-\dfrac{10a}{a-2}=\dfrac{5}{a(a-2)}+\dfrac{8(a-2)}{a(a-2)}-\dfrac{a\cdot10a}{a(a-2)}$

$=\dfrac{5+8a-16-10a^2}{a(a-2)}=\dfrac{-10a^2+8a-11}{a(a-2)}$

63) $\dfrac{3b-1}{b^2+8b}+\dfrac{b}{3b^2+25b+8}+\dfrac{2}{3b^2+b}=\dfrac{3b-1}{b(b+8)}+\dfrac{b}{(3b+1)(b+8)}+\dfrac{2}{b(3b+1)}$

$=\dfrac{(3b-1)(3b+1)}{b(b+8)(3b+1)}+\dfrac{b\cdot b}{b(3b+1)(b+8)}+\dfrac{2(b+8)}{b(3b+1)(b+8)}$

$=\dfrac{(9b^2-1)+b^2+(2b+16)}{b(b+8)(3b+1)}=\dfrac{10b^2+2b+15}{b(b+8)(3b+1)}$

65) $\dfrac{c}{c^2-8c+16}-\dfrac{5}{c^2-c-12}=\dfrac{c}{(c-4)^2}-\dfrac{5}{(c-4)(c+3)}$

$=\dfrac{c(c+3)}{(c-4)^2(c+3)}-\dfrac{5(c-4)}{(c-4)^2(c+3)}=\dfrac{c^2+3c-5c+20}{(c-4)^2(c+3)}=\dfrac{c^2-2c+20}{(c-4)^2(c+3)}$

67) $\dfrac{9}{4a+4b}+\dfrac{8}{a-b}-\dfrac{6a}{a^2-b^2}=\dfrac{9}{4(a+b)}+\dfrac{8}{a-b}-\dfrac{6a}{(a+b)(a-b)}$

$=\dfrac{9(a-b)}{4(a+b)(a-b)}+\dfrac{8\cdot4(a+b)}{4(a+b)(a-b)}-\dfrac{4\cdot6a}{4(a+b)(a-b)}$

$=\dfrac{9a-9b+32a+32b-24a}{4(a+b)(a-b)}=\dfrac{17a+23b}{4(a+b)(a-b)}$

69) $\dfrac{2v+1}{6v^2-29v-5}-\dfrac{v-2}{3v^2-13v-10}=\dfrac{2v+1}{(6v+1)(v-5)}-\dfrac{v-2}{(3v+2)(v-5)}$

$=\dfrac{(2v+1)(3v+2)}{(6v+1)(3v+2)(v-5)}-\dfrac{(6v+1)(v-2)}{(6v+1)(3v+2)(v-5)}$

$=\dfrac{6v^2+4v+3v+2-(6v^2-12v+v-2)}{(6v+1)(3v+2)(v-5)}=\dfrac{\cancel{6v^2}+7v+2-\cancel{6v^2}+11v+2}{(6v+1)(3v+2)(v-5)}$

$=\dfrac{18v+4}{(6v+1)(3v+2)(v-5)}=\dfrac{2(9v+2)}{(6v+1)(3v+2)(v-5)}$

71) $\dfrac{g-5}{5g^2-30g}+\dfrac{g}{2g^2-17g+30}-\dfrac{6}{2g^2-5g}=\dfrac{g-5}{5g(g-6)}+\dfrac{g}{(2g-5)(g-6)}-\dfrac{6}{g(2g-5)}$

$$=\dfrac{(g-5)(2g-5)}{5g(g-6)(2g-5)}+\dfrac{5g\cdot g}{5g(g-6)(2g-5)}-\dfrac{6\cdot5(g-6)}{5g(g-6)(2g-5)}$$

$$=\dfrac{2g^2-15g+25+5g^2-30g+180}{5g(g-6)(2g-5)}=\dfrac{7g^2-45g+205}{5g(g-6)(2g-5)}$$

73) a) $A=\left(\dfrac{k-4}{\cancel{4}}\right)\left(\dfrac{\cancel{8}^{2}}{k+1}\right)=\dfrac{2(k-4)}{(k+1)}$ sq. units

b) $P=2\left(\dfrac{k-4}{4}\right)+2\left(\dfrac{8}{x+1}\right)=\dfrac{k-4}{2}+\dfrac{16}{k+1}$

$$=\dfrac{(k-4)(k+1)}{2(k+1)}+\dfrac{2\cdot16}{2(k+1)}=\dfrac{(k-4)(k+1)+32}{2(k+1)}=\dfrac{k^2-3k+28}{2(k+1)}\text{ units}$$

75) a) $A=\left(\dfrac{6}{h^2+9h+20}\right)\left(\dfrac{h}{h+5}\right)=\left(\dfrac{6}{(h+5)(h+4)}\right)\left(\dfrac{h}{h+5}\right)=\dfrac{6h}{(h+5)^2(h+4)}$ sq. units

b) $P=2\left(\dfrac{6}{h^2+9h+20}\right)+2\left(\dfrac{h}{h+5}\right)=\dfrac{12}{(h+5)(h+4)}+\dfrac{2h}{h+5}$

$$=\dfrac{12}{(h+5)(h+4)}+\dfrac{2h(h+4)}{(h+5)(h+4)}=\dfrac{12+2h^2+8h}{(h+5)(h+4)}$$

$$=\dfrac{2h^2+8h+12}{(h+5)(h+4)}=\dfrac{2(h^2+4h+6)}{(h+5)(h+4)}\text{ units}$$

77) $P=\dfrac{1}{4x}+\dfrac{3}{2x^2}+\dfrac{12}{x}=\dfrac{1\cdot x}{4x^2}+\dfrac{3\cdot2}{4x^2}+\dfrac{12\cdot4x}{4x^2}=\dfrac{x+6+48x}{4x^2}=\dfrac{49x+6}{4x^2}\text{ units}$

Putting It All Together

1) a) $\dfrac{-3+3}{3(-3)+4}=\dfrac{0}{-9+4}=\dfrac{0}{-5}=0$

b) $\dfrac{2+3}{3(2)+4}=\dfrac{5}{6+4}=\dfrac{5}{10}=\dfrac{1}{2}$

3) a)

$\dfrac{5(-3)-3}{(-3)^2+10(-3)+21}=\dfrac{-15-3}{9-30+21}$

$=\dfrac{-18}{-21+21}=\dfrac{-18}{0}$

undefined

b)

$\dfrac{5(2)-3}{(2)^2+10(2)+21}=\dfrac{10-3}{4+20+21}=\dfrac{7}{45}$

5) a) $\qquad w^2-36=0$

$\qquad (w+6)(w-6)=0$

$\qquad w+6=0 \ \text{ or } \ w-6=0$

$\qquad\quad w=-6 \qquad\quad w=6$

b) $5w=0$

$\quad w=0$

7) a) $\qquad b^2+2b-8=0$

$\qquad (b+4)(b-2)=0$

$\qquad b+4=0 \ \text{ or } \ b-2=0$

$\qquad\quad b=-4 \qquad\quad b=2$

b) $3-5b=0$

$\quad -5b=-3$

$\qquad b=\dfrac{3}{5}$

9) a) $5r=0$

$\quad r=0$

b) It never equals zero.

11) $\dfrac{12w^{16}}{3w^5}=4w^{11}$

13) $\dfrac{m^2+6m-27}{2m^2+2m-24}=\dfrac{(m+9)\,\cancel{(m-3)}}{2(m+4)\,\cancel{(m-3)}}$

$\qquad\qquad =\dfrac{m+9}{2(m+4)}$

15) $\dfrac{12-15n}{5n^2+6n-8}=\dfrac{3(4-5n)}{(5n-4)(n+2)}$

$\qquad\qquad =\dfrac{3\,\overset{-1}{\cancel{(4-5n)}}}{\cancel{(5n-4)}\,(n+2)}$

$\qquad\qquad =-\dfrac{3}{n+2}$

17) $\dfrac{4c^2+4c-24}{c+3}\div\dfrac{3c-6}{8}$

$\qquad =\dfrac{4(c^2+c-6)}{c+3}\cdot\dfrac{8}{3c-6}$

$\qquad =\dfrac{4\,\cancel{(c+3)}\,\cancel{(c-2)}}{\cancel{c+3}}\cdot\dfrac{8}{3\cancel{(c-2)}}=\dfrac{32}{3}$

19) $\dfrac{4j}{j^2-81}+\dfrac{3}{j^2-3j-54}$

$\qquad =\dfrac{4j}{(j+9)(j-9)}+\dfrac{3}{(j-9)(j+6)}$

$\qquad =\dfrac{4j(j+6)}{(j+9)(j-9)(j+6)}$

$\qquad\qquad +\dfrac{3(j+9)}{(j+9)(j-9)(j+6)}$

$\qquad =\dfrac{4j^2+24j+3j+27}{(j+9)(j-9)(j+6)}$

$\qquad =\dfrac{4j^2+27j+27}{(j+9)(j-9)(j+6)}$

21) $\dfrac{12\overset{y}{\cancel{y^7}}}{\underset{z^2}{\cancel{4z^6}}}\cdot\dfrac{\overset{2}{\cancel{8z^4}}}{\underset{6}{\cancel{72y^6}}}=\dfrac{2y}{6z^2}=\dfrac{y}{3z^2}$

23) $\dfrac{x}{2x^2-7x-4}-\dfrac{x+3}{4x^2+4x+1}$

$=\dfrac{x}{(2x+1)(x-4)}-\dfrac{x+3}{(2x+1)^2}$

$=\dfrac{x(2x+1)}{(2x+1)^2(x-4)}-\dfrac{(x+3)(x-4)}{(2x+1)^2(x-4)}$

$=\dfrac{2x^2+x-(x^2-x-12)}{(2x+1)^2(x-4)}=\dfrac{x^2+2x+12}{(2x+1)^2(x-4)}$

25) $\dfrac{16-m^2}{m+4}\div\dfrac{8m-32}{m+7}$

$=\dfrac{16-m^2}{m+4}\cdot\dfrac{m+7}{8m-32}$

$=\dfrac{(4+m)(4-m)}{m+4}\cdot\dfrac{m+7}{8(m-4)}$

$=\dfrac{\cancel{(4+m)}\overset{-1}{\cancel{(4-m)}}}{\cancel{m+4}}\cdot\dfrac{m+7}{8\cancel{(m-4)}}$

$=-\dfrac{m+7}{8}$

27) $\dfrac{3xy-24x-5y+40}{y^2-64}\div\dfrac{27x^3-125}{9x}$

$=\dfrac{3xy-24x-5y+40}{y^2-64}\cdot\dfrac{9x}{27x^3-125}$

$=\dfrac{\cancel{(y-8)}\cancel{(3x-5)}}{(y+8)\cancel{(y-8)}}\cdot\dfrac{9x}{\cancel{(3x-5)}(9x^2+15x+25)}$

$=\dfrac{9x}{(y+8)(9x^2+15x+25)}$

29) $\dfrac{9}{d+3}+\dfrac{8}{d^2}=\dfrac{9d^2}{d^2(d+3)}+\dfrac{8(d+3)}{d^2(d+3)}$

$=\dfrac{9d^2+8d+24}{d^2(d+3)}$

31) $\dfrac{\dfrac{9k^2-1}{14k}}{\dfrac{3k-1}{21k^4}}=\dfrac{9k^2-1}{14k}\div\dfrac{3k-1}{21k^4}$

$=\dfrac{9k^2-1}{14k}\cdot\dfrac{21k^4}{3k-1}$

$=\dfrac{\cancel{(3k-1)}(3k+1)}{\underset{2}{\cancel{14k}}}\cdot\dfrac{\overset{3k^3}{\cancel{21k^4}}}{\cancel{3k-1}}$

$=\dfrac{3k^3(3k+1)}{2}$

33) $\dfrac{2w}{25-w^2}+\dfrac{w-3}{w^2-12w+35}$

$=\dfrac{2w}{(5+w)(5-w)}+\dfrac{w-3}{(w-5)(w-7)}$

$=\dfrac{-2w}{(w+5)(w-5)}+\dfrac{w-3}{(w-5)(w-7)}$

$=\dfrac{-2w(w-7)}{(w+5)(w-5)(w-7)}$

$\qquad +\dfrac{(w+5)(w-3)}{(w+5)(w-5)(w-7)}$

$=\dfrac{-2w^2+14w+w^2+2w-15}{(w+5)(w-5)(w-7)}$

$=\dfrac{-w^2+16w-15}{(w+5)(w-5)(w-7)}$

$=-\dfrac{w^2-16w+15}{(w+5)(w-5)(w-7)}$

$=-\dfrac{(w-15)(w-1)}{(w+5)(w-5)(w-7)}$

35) $\dfrac{10}{x-8}+\dfrac{4}{x+3}$

$=\dfrac{10(x+3)}{(x-8)(x+3)}+\dfrac{4(x-8)}{(x-8)(x+3)}$

$=\dfrac{10x+30+4x-32}{(x-8)(x+3)}$

$=\dfrac{14x-2}{(x-8)(x+3)}=\dfrac{2(7x-1)}{(x-8)(x+3)}$

37) $\dfrac{2h^2+11h+5}{8}\div(2h+1)^2$

$=\dfrac{2h^2+11h+5}{8}\cdot\dfrac{1}{(2h+1)^2}$

$=\dfrac{\cancel{(2h+1)}(h+5)}{8}\cdot\dfrac{1}{(2h+1)\cancel{(2h+1)}}$

$=\dfrac{(h+5)}{8(2h+1)}$

39) $\dfrac{3m}{7m-4n}-\dfrac{20n}{4n-7m}$

$=\dfrac{3m}{7m-4n}+\dfrac{20n}{7m-4n}=\dfrac{3m+20n}{7m-4n}$

41) $\dfrac{2p+3}{p^2+7p}-\dfrac{4p}{p^2-p-56}+\dfrac{5}{p^2-8p}$

$=\dfrac{2p+3}{p(p+7)}-\dfrac{4p}{(p+7)(p-8)}+\dfrac{5}{p(p-8)}$

$=\dfrac{(2p+3)(p-8)}{p(p+7)(p-8)}-\dfrac{4p^2}{p(p+7)(p-8)}$

$+\dfrac{5(p+7)}{p(p+7)(p-8)}$

$=\dfrac{2p^2-13p-24-4p^2+5p+35}{p(p+7)(p-8)}$

$=\dfrac{-2p^2-8p+11}{p(p+7)(p-8)}$

43) $\dfrac{6t+6}{3t^2-24t}\cdot(t^2-7t-8)$

$=\dfrac{6t+6}{3t^2-24t}\cdot(t^2-7t-8)$

$=\dfrac{\overset{2}{\cancel{6}}(t+1)}{\cancel{3}t\cancel{(t-8)}}\cdot\cancel{(t-8)}(t+1)$

$=\dfrac{2(t+1)^2}{t}$

45) $\dfrac{\dfrac{3c^3}{8c+40}}{\dfrac{9c}{c+5}}=\dfrac{3c^3}{8c+40}\div\dfrac{9c}{c+5}=\dfrac{3c^3}{8c+40}\cdot\dfrac{c+5}{9c}$

$=\dfrac{\overset{c^2}{\cancel{3c^3}}}{8\cancel{(c+5)}}\cdot\dfrac{\cancel{c+5}}{\underset{3}{\cancel{9c}}}=\dfrac{c^2}{24}$

47) $\dfrac{f-8}{f-4}-\dfrac{4}{4-f}=\dfrac{f-8}{f-4}+\dfrac{4}{f-4}=\dfrac{f-4}{f-4}=1$

49) $\left(\dfrac{3m}{3m-1}-\dfrac{4}{m+4}\right)\cdot\dfrac{9m^2-1}{21m^2+28}$

$=\left(\dfrac{3m(m+4)}{(3m-1)(m+4)}-\dfrac{4(3m-1)}{(3m-1)(m+4)}\right)\bullet$

$\left[\dfrac{(3m+1)(3m-1)}{7(3m^2+4)}\right]$

$=\left(\dfrac{3m^2+12m-12m+4}{(3m-1)(m+4)}\right)\cdot\dfrac{(3m+1)(3m-1)}{7(3m^2+4)}$

$=\dfrac{\cancel{3m^2+4}}{\cancel{(3m-1)}(m+4)}\cdot\dfrac{(3m+1)\cancel{(3m-1)}}{7\cancel{(3m^2+4)}}$

$=\dfrac{3m+1}{7(m+4)}$

51) $\dfrac{3}{k^2+3k}-\dfrac{4}{3k}+\dfrac{1}{k+3}$

$=\dfrac{3}{k(k+3)}-\dfrac{4}{3k}+\dfrac{1}{k+3}$

$=\dfrac{9}{3k(k+3)}-\dfrac{4(k+3)}{3k(k+3)}+\dfrac{3k}{3k(k+3)}$

$=\dfrac{9-4k-12+3k}{3k(k+3)}$

$=\dfrac{-k-3}{3k(k+3)}=\dfrac{-1(k+3)}{3k(k+3)}=-\dfrac{1}{3k}$

53) a) $A=\left(\dfrac{z}{z+5}\right)\left(\dfrac{6}{z+2}\right)$

$=\dfrac{6z}{(z+5)(z+2)}$ sq. units

b) $P=2\left(\dfrac{z}{z+5}\right)+2\left(\dfrac{6}{z+2}\right)$

$=\dfrac{2z}{z+5}+\dfrac{12}{z+2}$

$=\dfrac{2z(z+2)}{(z+5)(z+2)}+\dfrac{12(z+5)}{(z+5)(z+2)}$

$=\dfrac{2z^2+4z+12z+60}{(z+5)(z+2)}=\dfrac{2z^2+16z+60}{(z+5)(z+2)}$

$=\dfrac{2(z^2+8z+30)}{(z+5)(z+2)}$ units

55) $\qquad 4t^2-1=0$

$(2t+1)(2t-1)=0$

$2t+1=0 \ \text{ or } \ 2t-1=0$

$w=-\dfrac{1}{2} \qquad w=\dfrac{1}{2}$

The domain contains all

real numbers except $-\dfrac{1}{2}$ and $\dfrac{1}{2}$.

57) Never undefined. Any real number
can be substituted for x.
The domain contains all
real numbers.
Domain: $(-\infty,\infty)$

59) $4n=0$

$n=0$

The domain contains all
real numbers except 0.
Domain: $(-\infty,0)\cup(0,\infty)$

Section 8.5: Exercises

1) i) Rewrite it as a division problem,
then simplify.

$\dfrac{2}{9}\div\dfrac{5}{18}=\dfrac{2}{\cancel{9}}\cdot\dfrac{\cancel{18}^{2}}{5}=\dfrac{4}{5}$

ii) Multiply the numerator and
denominator by 18, the LCD

of $\dfrac{2}{9}$ and $\dfrac{5}{18}$. Then, simplify

$\dfrac{\cancel{18}^{2}\left(\dfrac{2}{\cancel{9}}\right)}{\cancel{18}\left(\dfrac{5}{\cancel{18}}\right)}=\dfrac{4}{5}$

3) $\dfrac{\dfrac{5}{9}}{\dfrac{7}{4}}=\dfrac{5}{9}\cdot\dfrac{4}{7}=\dfrac{20}{63}$

5) $\dfrac{\dfrac{u^4}{v^2}}{\dfrac{u^3}{v}}=\dfrac{u^{4}}{v^{2}}\cdot\dfrac{v}{u^{3}}=\dfrac{u}{v}$

7) $\dfrac{\dfrac{x^4}{y}}{\dfrac{x^2}{y^2}} = \dfrac{x^4}{y} \div \dfrac{x^2}{y^2} = \dfrac{x^4}{y} \cdot \dfrac{y^2}{x^2} = x^2 y$

9) $\dfrac{\dfrac{14m^5n^4}{9}}{\dfrac{35mn^6}{3}} = \dfrac{14m^5n^4}{9} \div \dfrac{35mn^6}{3}$

$= \dfrac{\overset{2m^4}{\cancel{14m^5n^4}}}{\cancel{9}_{3}} \cdot \dfrac{\cancel{3}}{\underset{5n^2}{\cancel{35mn^6}}} = \dfrac{2m^4}{15n^2}$

11) $\dfrac{\dfrac{m-7}{m}}{\dfrac{m-7}{18}} = \dfrac{m-7}{m} \div \dfrac{m-7}{18}$

$= \dfrac{m-7}{m} \cdot \dfrac{18}{m-7} = \dfrac{18}{m}$

13) $\dfrac{\dfrac{g^2-36}{20}}{\dfrac{g+6}{60}} = \dfrac{g^2-36}{20} \div \dfrac{g+6}{60}$

$= \dfrac{(g-6)(\cancel{g+6})}{\cancel{20}} \cdot \dfrac{\overset{3}{\cancel{60}}}{\cancel{g+6}}$

$= 3(g-6)$

15) $\dfrac{\dfrac{d^3}{16d-24}}{\dfrac{d}{40d-60}} = \dfrac{d^3}{16d-24} \div \dfrac{d}{40d-60}$

$= \dfrac{\overset{d^2}{\cancel{d^3}}}{\underset{2}{\cancel{8(2d-3)}}} \cdot \dfrac{\overset{5}{\cancel{20(2d-3)}}}{\cancel{d}}$

$= \dfrac{5d^2}{2}$

17) $\dfrac{\dfrac{c^2-7c-8}{11c}}{\dfrac{c+1}{c}} = \dfrac{c^2-7c-8}{11c} \div \dfrac{c+1}{c}$

$= \dfrac{(\cancel{c+1})(c-8)}{11\cancel{c}} \cdot \dfrac{\cancel{c}}{\cancel{c+1}}$

$= \dfrac{c-8}{11}$

19) $\dfrac{\dfrac{7}{9}-\dfrac{2}{3}}{3+\dfrac{1}{9}} = \dfrac{\dfrac{7}{9}-\dfrac{6}{9}}{\dfrac{27}{9}+\dfrac{1}{9}} = \dfrac{\dfrac{1}{9}}{\dfrac{28}{9}}$

$= \dfrac{1}{9} \div \dfrac{28}{9} = \dfrac{1}{9} \cdot \dfrac{9}{28} = \dfrac{1}{28}$

21) $\dfrac{\dfrac{r}{s}-4}{\dfrac{3}{s}+\dfrac{1}{r}} = \dfrac{\dfrac{r}{s}-\dfrac{4s}{s}}{\dfrac{3r}{rs}+\dfrac{s}{rs}} = \dfrac{\dfrac{r-4s}{s}}{\dfrac{3r+s}{rs}}$

$= \dfrac{r-4s}{s} \div \dfrac{3r+s}{rs}$

$= \dfrac{r-4s}{\cancel{s}} \cdot \dfrac{r\cancel{s}}{3r+s}$

$= \dfrac{r(r-4s)}{3r+s}$

23) $\dfrac{\dfrac{8}{r^2t}}{\dfrac{3}{r}-\dfrac{r}{t}} = \dfrac{\dfrac{8}{r^2t}}{\dfrac{3t}{rt}-\dfrac{r^2}{rt}} = \dfrac{\dfrac{8}{r^2t}}{\dfrac{3t-r^2}{rt}}$

$= \dfrac{8}{r^2t} \div \dfrac{3t-r^2}{rt}$

$= \dfrac{8}{\underset{r}{\cancel{r^2}}\cancel{t}} \cdot \dfrac{\cancel{r}\cancel{t}}{3t-r^2}$

$= \dfrac{8}{r(3t-r^2)}$

25) $\dfrac{\dfrac{5}{w-1}+\dfrac{3}{w+4}}{\dfrac{6}{w+4}+\dfrac{4}{w-1}}$

$\dfrac{\dfrac{5(w+4)}{(w+4)(w-1)}+\dfrac{3(w-1)}{(w+4)(w-1)}}{\dfrac{6(w-1)}{(w+4)(w-1)}+\dfrac{4(w+4)}{(w+4)(w-1)}}$

$=\dfrac{\dfrac{5w+20+3w-3}{(w+4)(w-1)}}{\dfrac{6w-6+4w+16}{(w+4)(w-1)}}$

$=\dfrac{\dfrac{8w+17}{(w+4)(w-1)}}{\dfrac{10w+10}{(w+4)(w-1)}}$

$=\dfrac{8w+17}{(w+4)(w-1)}\div\dfrac{10w+10}{(w+4)(w-1)}$

$=\dfrac{8w+17}{(w+4)(w-1)}\cdot\dfrac{(w+4)(w-1)}{10(w+1)}$

$=\dfrac{8w+17}{10(w+1)}$

27) $\dfrac{\dfrac{7}{9}-\dfrac{2}{3}}{3+\dfrac{1}{9}}=\dfrac{9\left(\dfrac{7}{9}-\dfrac{2}{3}\right)}{9\left(3+\dfrac{1}{9}\right)}=\dfrac{7-6}{27+1}=\dfrac{1}{28}$

29) $\dfrac{\dfrac{r}{s}-4}{\dfrac{3}{s}+\dfrac{1}{r}}=\dfrac{rs\left(\dfrac{r}{s}-4\right)}{rs\left(\dfrac{3}{s}+\dfrac{1}{r}\right)}=\dfrac{r^2-4rs}{3r+s}$

$=\dfrac{r(r-4s)}{3r+s}$

31) $\dfrac{\dfrac{8}{r^2t}}{\dfrac{3}{r}-\dfrac{r}{t}}=\dfrac{r^2t\left(\dfrac{8}{r^2t}\right)}{r^2t\left(\dfrac{3}{r}-\dfrac{r}{t}\right)}=\dfrac{8}{3rt-r^3}=\dfrac{8}{r(3t-r^2)}$

33) $\dfrac{\dfrac{5}{w-1}+\dfrac{3}{w+4}}{\dfrac{6}{w+4}+\dfrac{4}{w-1}}$

$=\dfrac{(w+4)(w-1)\left(\dfrac{5}{w-1}+\dfrac{3}{w+4}\right)}{(w+4)(w-1)\left(\dfrac{6}{w+4}+\dfrac{4}{w-1}\right)}$

$=\dfrac{5(w+4)+3(w-1)}{6(w-1)+4(w+4)}$

$=\dfrac{5w+20+3w-3}{6w-6+4w+16}=\dfrac{8w+17}{10w+10}$

$=\dfrac{8w+17}{10(w+1)}$

35) Answers may vary.

37) $\dfrac{\dfrac{a-4}{12}}{\dfrac{a-4}{a}}=\dfrac{a-4}{12}\div\dfrac{a-4}{a}$

$=\dfrac{a-4}{12}\cdot\dfrac{a}{a-4}=\dfrac{a}{12}$

39) $\dfrac{\dfrac{3}{n}-\dfrac{4}{n-2}}{\dfrac{1}{n-2}+\dfrac{5}{n}}$

$=\dfrac{n(n-2)\left(\dfrac{3}{n}-\dfrac{4}{n-2}\right)}{n(n-2)\left(\dfrac{1}{n-2}+\dfrac{5}{n}\right)}=\dfrac{3(n-2)-4n}{n+5(n-2)}$

$=\dfrac{3n-6-4n}{n+5n-10}=\dfrac{-n-6}{6n-10}=-\dfrac{n+6}{2(n-5)}$

41) $\dfrac{\dfrac{6}{w}-w}{1+\dfrac{6}{w}}=\dfrac{\dfrac{6-w^2}{w}}{\dfrac{w+6}{w}}=\dfrac{6-w^2}{w}\div\dfrac{w+6}{w}$

$=\dfrac{6-w^2}{w}\cdot\dfrac{w}{w+6}=\dfrac{6-w^2}{w+6}$

43) $\dfrac{\dfrac{6}{5}}{\dfrac{9}{15}}=\dfrac{6}{5}\div\dfrac{9}{15}=\dfrac{6}{\cancel{5}}\cdot\dfrac{\cancel{15}^{3}}{9}=\dfrac{18}{9}=2$

45) $\dfrac{1-\dfrac{4}{t+5}}{\dfrac{4}{t^2-25}+\dfrac{t}{t-5}}$

$=\dfrac{(t+5)(t-5)\left(1-\dfrac{4}{t+5}\right)}{(t+5)(t-5)\left(\dfrac{4}{(t+5)(t-5)}+\dfrac{t}{t-5}\right)}$

$=\dfrac{(t+5)(t-5)-4(t-5)}{4+t(t+5)}$

$=\dfrac{t^2-25-4t+20}{4+t^2+5t}=\dfrac{t^2-4t-5}{t^2+5t+4}$

$=\dfrac{(t-5)(t+1)}{(t+4)(t+1)}=\dfrac{t-5}{t+4}$

47) $\dfrac{\dfrac{9}{x}-\dfrac{9}{y}}{\dfrac{2}{x^2}-\dfrac{2}{y^2}}=\dfrac{x^2y^2\left(\dfrac{9}{x}-\dfrac{9}{y}\right)}{x^2y^2\left(\dfrac{2}{x^2}-\dfrac{2}{y^2}\right)}$

$=\dfrac{9xy^2-9x^2y}{2y^2-2x^2}$

$=\dfrac{9xy(y-x)}{2(y+x)(y-x)}$

$=\dfrac{9xy}{2(x+y)}$

49) $\dfrac{\dfrac{24c-60}{5}}{\dfrac{8c-20}{c^2}}=\dfrac{24c-60}{5}\div\dfrac{8c-20}{c^2}$

$=\dfrac{\cancel{12}^{3}(2\cancel{c-5})}{5}\cdot\dfrac{c^2}{\cancel{4}(2\cancel{c-5})}=\dfrac{3c^2}{5}$

51) $\dfrac{\dfrac{4}{9}+\dfrac{2}{5}}{\dfrac{1}{5}-\dfrac{2}{3}}=\dfrac{45\left(\dfrac{4}{9}+\dfrac{2}{5}\right)}{45\left(\dfrac{1}{5}-\dfrac{2}{3}\right)}=\dfrac{20+18}{9-30}=-\dfrac{38}{21}$

53) $\dfrac{\dfrac{1}{10}}{\dfrac{7}{8}}=\dfrac{1}{10}\div\dfrac{7}{8}=\dfrac{1}{\cancel{10}_{5}}\cdot\dfrac{\cancel{8}^{4}}{7}=\dfrac{4}{35}$

55) $\dfrac{\dfrac{2}{uv^2}}{\dfrac{6}{v}-\dfrac{4v}{u}}=\dfrac{uv^2\left(\dfrac{2}{uv^2}\right)}{uv^2\left(\dfrac{6}{v}-\dfrac{4v}{u}\right)}=\dfrac{2}{\left(6uv-4v^3\right)}$

$=\dfrac{2}{2v\left(3u-2v^2\right)}=\dfrac{1}{v\left(3u-2v^2\right)}$

57) $\dfrac{1+\dfrac{b}{a-b}}{\dfrac{b}{a^2-b^2}+\dfrac{1}{a+b}}$

$=\dfrac{(a+b)(a-b)\left(1+\dfrac{b}{a-b}\right)}{(a+b)(a-b)\left(\dfrac{b}{(a+b)(a-b)}+\dfrac{1}{a+b}\right)}$

$=\dfrac{(a+b)(a-b)+b(a+b)}{b+a-b}$

$=\dfrac{a^2-b^2+ab+b^2}{a}=\dfrac{a^2+ab}{a}$

$=\dfrac{a(a+b)}{a}=a+b$

59) $\dfrac{\dfrac{x^2-x-42}{2x-14}}{\dfrac{x^2-36}{8x+16}} = \dfrac{x^2-x-42}{2x-14} \div \dfrac{x^2-36}{8x+16}$

$= \dfrac{(x-7)(x+6)}{2(x-7)} \cdot \dfrac{\overset{4}{8}(x+2)}{(x+6)(x-6)}$

$= \dfrac{4(x+2)}{x-6}$

63) $\dfrac{\dfrac{7-\dfrac{8}{m}}{7m-8}}{11} = \dfrac{\dfrac{7m-8}{m}}{\dfrac{7m-8}{11}}$

$= \dfrac{7m-8}{m} \div \dfrac{7m-8}{11}$

$= \dfrac{7m-8}{m} \cdot \dfrac{11}{7m-8} = \dfrac{11}{m}$

61) $\dfrac{\dfrac{y^4}{z^3}}{\dfrac{y^6}{z^4}} = \dfrac{y^4}{z^3} \div \dfrac{y^6}{z^4} = \dfrac{\overset{}{y^4}}{z^3} \cdot \dfrac{z^4}{\underset{y^2}{y^6}} = \dfrac{z}{y^2}$

65) $\dfrac{\dfrac{1}{h^2-4}+\dfrac{2}{h+2}}{h-\dfrac{3}{2}} = \dfrac{\dfrac{1}{(h+2)(h-2)}+\dfrac{2}{h+2}}{h-\dfrac{3}{2}}$

$= \dfrac{2(h+2)(h-2)\left(\dfrac{1}{(h+2)(h-2)}+\dfrac{2}{h+2}\right)}{2(h+2)(h-2)\left(h-\dfrac{3}{2}\right)} = \dfrac{2+4(h-2)}{2h(h+2)(h-2)-3(h+2)(h-2)}$

$= \dfrac{2+4h-8}{(2h-3)(h+2)(h-2)} = \dfrac{4h-6}{(2h-3)(h+2)(h-2)} = \dfrac{2(2h-3)}{(2h-3)(h+2)(h-2)} = \dfrac{2}{(h+2)(h-2)}$

67) $\dfrac{\dfrac{6}{v+3}-\dfrac{4}{v-1}}{\dfrac{2}{v-1}+\dfrac{1}{v+2}}$

$= \dfrac{(v+3)(v+2)(v-1)\left(\dfrac{6}{v+3}-\dfrac{4}{v-1}\right)}{(v+3)(v+2)(v-1)\left(\dfrac{2}{v-1}+\dfrac{1}{v+2}\right)} = \dfrac{6(v+2)(v-1)-4(v+3)(v+2)}{2(v+3)(v+2)+1(v+3)(v-1)}$

$= \dfrac{(v+2)(6v-6-4v-12)}{(v+3)(2v+4+v-1)} = \dfrac{(v+2)(2v-18)}{(v+3)(3v+3)} = \dfrac{2(v+2)(v-9)}{3(v+3)(v+1)}$

Section 8.6: Exercises

1) Eliminate the denominators.

3) difference; $\dfrac{3r+5}{2} - \dfrac{r}{6}$

$$= \dfrac{3(3r+5)}{6} - \dfrac{r}{6}$$

$$= \dfrac{9r+15-r}{6} = \dfrac{8r+15}{6}$$

5) equation; $\dfrac{3h}{2} + \dfrac{4}{3} = \dfrac{2h+3}{3}$

$$6\left(\dfrac{3h}{2} + \dfrac{4}{3}\right) = 6\left(\dfrac{2h+3}{3}\right)$$

$$9h+8 = 2(2h+3)$$

$$9h+8 = 4h+6$$

$$5h = -2$$

$$h = -\dfrac{2}{5} \qquad \left\{-\dfrac{2}{5}\right\}$$

7) sum; $\dfrac{3}{a^2} + \dfrac{1}{a+11}$

$$= \dfrac{3(a+11)}{a^2(a+11)} + \dfrac{a^2}{a^2(a+11)}$$

$$= \dfrac{3a+33+a^2}{a^2(a+11)} = \dfrac{a^2+3a+33}{a^2(a+11)}$$

9) equation; $\dfrac{8}{b-11} - 5 = \dfrac{3}{b-11}$

$$(b-11)\left(\dfrac{8}{b-11} - 5\right) = (b-11)\cdot\dfrac{3}{b-11}$$

$$8 - 5(b-11) = 3$$

$$8 - 5b + 55 = 3$$

$$-5b + 63 = 3$$

$$-5b = -60$$

$$b = 12 \qquad \{12\}$$

11) $k-2=0 \qquad\qquad k=0$

$$k=2$$

13) $p+3=0 \quad p=0$

$$p=-3$$

$$p^2-9=0$$

$$(p+3)(p-3)=0$$

$$p+3=0 \text{ or } p-3=0$$

$$p=-3 \qquad\qquad p=3$$

15) $h^2-5h-36=0$

$$(h-9)(h+4)=0$$

$$h-9=0 \text{ or } h+4=0$$

$$h=9 \qquad\qquad h=-4$$

$$h+4=0 \qquad\qquad 3h-27=0$$

$$h=-4 \qquad\qquad 3h=27$$

$$h=9$$

17) $\qquad \dfrac{a}{3} + \dfrac{7}{12} = \dfrac{1}{4}$

$$12\left(\dfrac{a}{3} + \dfrac{7}{12}\right) = 12\cdot\dfrac{1}{4}$$

$$4a+7 = 3$$

$$4a = -4$$

$$a = -1 \qquad \{-1\}$$

19) $\qquad \dfrac{1}{4}j - j = -4$

$$4\left(\dfrac{1}{4}j - j\right) = 4(-4)$$

$$j - 4j = -16$$

$$-3j = -16$$

$$j = \dfrac{16}{3} \qquad \left\{\dfrac{16}{3}\right\}$$

21) $\dfrac{8m-5}{24} = \dfrac{m}{6} - \dfrac{7}{8}$

$24\left(\dfrac{8m-5}{24}\right) = 24\left(\dfrac{m}{6} - \dfrac{7}{8}\right)$

$8m-5 = 4m-21$

$4m = -16$

$m = -4 \qquad \{-4\}$

23) $\dfrac{8}{3x+1} = \dfrac{2}{x+3}$

$8(x+3) = 2(3x+1)$

$8x+24 = 6x+2$

$2x = -22$

$x = -11 \qquad \{-11\}$

25) $\dfrac{r+1}{2} = \dfrac{4r+1}{5}$

$5(r+1) = 2(4r+1)$

$5r+5 = 8r+2$

$-3r = -3$

$r = 1 \qquad \{1\}$

27) $\dfrac{23}{z} + 8 = -\dfrac{25}{z}$

$z\left(\dfrac{23}{z} + 8\right) = z\left(-\dfrac{25}{z}\right)$

$23 + 8z = -25$

$8z = -48$

$z = -6 \qquad \{-6\}$

29) $\dfrac{5q}{q+1} - 2 = \dfrac{5}{q+1}$

$(q+1)\left(\dfrac{5q}{q+1} - 2\right) = (q+1)\left(\dfrac{5}{q+1}\right)$

$5q - 2(q+1) = 5$

$5q - 2q - 2 = 5$

$3q = 7$

$q = \dfrac{7}{3} \qquad \left\{\dfrac{7}{3}\right\}$

31) $\dfrac{2}{s+6} + 4 = \dfrac{2}{s+6}$

$(s+6)\left(\dfrac{2}{s+6} + 4\right) = (s+6)\left(\dfrac{2}{s+6}\right)$

$2 + (s+6)4 = 2$

$2 + 4s + 24 = 2$

$4s + 26 = 2$

$4s = -24$

$s = -6$

If $s = -6$, the denominators $= 0$. $\varnothing$

33) $\dfrac{3b}{b+7} - 6 = \dfrac{3}{b+7}$

$(b+7)\left(\dfrac{3b}{b+7} - 6\right) = (b+7)\left(\dfrac{3}{b+7}\right)$

$3b - 6(b+7) = 3$

$3b - 6b - 42 = 3$

$-3b = -45$

$b = -15 \quad \{-15\}$

35) $\dfrac{8}{r} - 1 = \dfrac{6}{r}$

$r\left(\dfrac{8}{r} - 1\right) = r\left(\dfrac{6}{r}\right)$

$8 - r = 6$

$-r = -2$

$r = 2 \qquad \{2\}$

37) $z + \dfrac{12}{z} = -8$

$z\left(z + \dfrac{12}{z}\right) = z(-8)$

$z^2 + 12 = -8z$

$z^2 + 8z + 12 = 0$

$(z+6)(z+2) = 0$

$z = -6$

$z = -2$ $\quad \{-6, -2\}$

39) $\dfrac{15}{b} = 8 - b$

$b\left(\dfrac{15}{b}\right) = b(8-b)$

$15 = 8b - b^2$

$b^2 - 8b + 15 = 0$

$(b-5)(b-3) = 0$

$b = 5$

$b = 3$ $\quad \{3, 5\}$

41) $\dfrac{8}{c+2} - \dfrac{12}{c-4} = \dfrac{2}{c+2}$

$(c+2)(c-4)\left(\dfrac{8}{c+2} - \dfrac{12}{c-4}\right)$

$\qquad = (c+2)(c-4)\left(\dfrac{2}{c+2}\right)$

$8(c-4) - 12(c+2) = 2(c-4)$

$8c - 32 - 12c - 24 = 2c - 8$

$-4c - 56 = 2c - 8$

$-6c = 48$

$c = -8$ $\quad \{-8\}$

43) $\dfrac{9}{c-8} - \dfrac{15}{c} = 1$

$(c-8)c\left(\dfrac{9}{c-8} - \dfrac{15}{c}\right) = (c-8)c(1)$

$9c - 15(c-8) = c^2 - 8c$

$9c - 15c + 120 = c^2 - 8c$

$c^2 - 2c + 120 = 0$

$(c-12)(c+10) = 0$

$c = -10$

$c = 12$ $\quad \{-10, 12\}$

45) $\dfrac{3}{p-4} + \dfrac{8}{p+4} = \dfrac{13}{p^2 - 16}$

$\dfrac{3}{p-4} + \dfrac{8}{p+4} = \dfrac{13}{(p+4)(p-4)}$

$(p+4)(p-4)\left(\dfrac{3}{p-4} + \dfrac{8}{p+4}\right)$

$= (p+4)(p-4)\left(\dfrac{13}{(p+4)(p-4)}\right)$

$3(p+4) + 8(p-4) = 13$

$3p + 12 + 8p - 32 = 13$

$11p - 20 = 13$

$11p = 33$

$p = 3$ $\quad \{3\}$

47) $\dfrac{9}{k+5} - \dfrac{4}{k+1} = \dfrac{10}{k^2+6k+5}$

$\dfrac{9}{k+5} - \dfrac{4}{k+1} = \dfrac{10}{(k+5)(k+1)}$

$(k+5)(k+1)\left(\dfrac{9}{k+5} - \dfrac{4}{k+1}\right)$

$= (k+5)(k+1)\left(\dfrac{10}{(k+5)(k+1)}\right)$

$9(k+1) - 4(k+5) = 10$

$9k+9-4k-20 = 10$

$5k-11 = 10$

$5k = 21$

$k = \dfrac{21}{5} \qquad \left\{\dfrac{21}{5}\right\}$

49) $\dfrac{12}{g^2-9} + \dfrac{2}{g+3} = \dfrac{7}{g-3}$

$\dfrac{12}{(g+3)(g-3)} + \dfrac{2}{g+3} = \dfrac{7}{g-3}$

$(g+3)(g-3)\left(\dfrac{12}{(g+3)(g-3)} + \dfrac{2}{g+3}\right)$

$= (g+3)(g-3)\left(\dfrac{7}{g-3}\right)$

$12 + 2(g-3) = 7(g+3)$

$12 + 2g - 6 = 7g + 21$

$2g + 6 = 7g + 21$

$-15 = 5g$

$-3 = g$

If $g = -3$, two denominators = 0. $\varnothing$

51) $\dfrac{5}{p-3} - \dfrac{7}{p^2-7p+12} = \dfrac{8}{p-4}$

$\dfrac{5}{p-3} - \dfrac{7}{(p-4)(p-3)} = \dfrac{8}{p-4}$

$(p-4)(p-3)\left(\dfrac{5}{p-3} - \dfrac{7}{(p-4)(p-3)}\right)$

$= (p-4)(p-3)\left(\dfrac{8}{p-4}\right)$

$5(p-4) - 7 = 8(p-3)$

$5p - 20 - 7 = 8p - 24$

$5p - 27 = 8p - 24$

$-3 = 3p$

$-1 = p \qquad \{-1\}$

53) $\dfrac{x^2}{2} = \dfrac{x^2-6x}{3}$

$3x^2 = 2(x^2 - 6x)$

$3x^2 = 2x^2 - 12x$

$x^2 + 12x = 0$

$x(x+12) = 0$

$x = 0 \ \text{ or } \ x + 12 = 0$

$x = -12 \quad \{-12, 0\}$

55) $\dfrac{3}{t^2} = \dfrac{6}{t^2+8t}$

$t^2(t+8)\left(\dfrac{3}{t^2}\right) = t^2(t+8)\left(\dfrac{6}{t(t+8)}\right)$

$3(t+8) = 6t$

$3t + 24 = 6t$

$-3t + 24 = 0$

$-3t = -24$

$t = 8 \qquad \{8\}$

57)
$$\frac{b+3}{3b-18}-\frac{b+2}{b-6}=\frac{b}{3}$$

$$3(b-6)\left(\frac{b+3}{3(b-6)}-\frac{b+2}{b-6}\right)=3(b-6)\left(\frac{b}{3}\right)$$

$$(b+3)-3(b+2)=b(b-6)$$

$$b+3-3b-6=b^2-6b$$

$$-2b-3=b^2-6b$$

$$0=b^2-6b+2b+3$$

$$0=b^2-4b+3$$

$$0=(b-3)(b-1)$$

$$b-3=0 \text{ or } b-1=0$$

$$b=3 \qquad b=1 \qquad \{3,1\}$$

59)
$$\frac{4}{n+1}=\frac{10}{n^2-1}-\frac{5}{n-1}$$

$$\frac{4}{n+1}=\frac{10}{(n+1)(n-1)}-\frac{5}{n-1}$$

$$(n+1)(n-1)\left(\frac{4}{n+1}\right)$$

$$=(n+1)(n-1)\left(\frac{10}{(n+1)(n-1)}-\frac{5}{n-1}\right)$$

$$4(n-1)=10-5(n+1)$$

$$4n-4=10-5n-5$$

$$4n-4=5-5n$$

$$-9=-9n$$

$$1=n$$

If $n=1$, two denominators $=0$. $\quad\varnothing$

61)
$$-\frac{a}{5}=\frac{3}{a+8}$$

$$-a(a+8)=3(5)$$

$$-a^2-8a=15$$

$$0=a^2+8a+15$$

$$0=(a+5)(a+3)$$

$$a+5=0 \text{ or } a+3=0$$

$$a=-5 \qquad a=-3 \qquad \{-5,-3\}$$

63)
$$\frac{8}{p+2}+\frac{p}{p+1}=\frac{5p+2}{p^2+3p+2}$$

$$(p+2)(p+1)\left(\frac{8}{p+2}+\frac{p}{p+1}\right)$$

$$=(p+2)(p+1)\left(\frac{5p+2}{(p+2)(p+1)}\right)$$

$$(p+1)8+(p+2)p=5p+2$$

$$8p+8+p^2+2p=5p+2$$

$$p^2+5p+6=0$$

$$(p+3)(p+2)=0$$

$$p+3=0 \text{ or } p+2=0$$

$$p=-3 \qquad p=-2$$

If $p=-2$, two denominators $=0$.

The solution set is $\{-3\}$.

65) $\dfrac{-14}{3a^2+15a-18}=\dfrac{a}{a-1}+\dfrac{2}{3a+18}$

$3(a+6)(a-1)\left(\dfrac{-14}{3(a+6)(a-1)}\right)$

$=3(a+6)(a-1)\left(\dfrac{a}{a-1}+\dfrac{2}{3(a+6)}\right)$

$-14=3a(a+6)+2(a-1)$

$-14=3a^2+18a+2a-2$

$0=3a^2+20a+12$

$0=(3a+2)(a+6)$

$3a+2=0 \quad a=-\dfrac{2}{3}$

$a+6=0 \quad a=-6$

If $a=-6$, two denominators $=0$.

The solution set is $\left\{-\dfrac{2}{3}\right\}$.

67) $\dfrac{3}{f+4}=\dfrac{f}{f+6}-\dfrac{2}{f^2+10f+24}$

$(f+4)(f+6)\left(\dfrac{3}{f+4}\right)$

$=(f+4)(f+6)\left(\dfrac{f}{f+6}-\dfrac{2}{(f+4)(f+6)}\right)$

$3(f+6)=f(f+4)-2$

$3f+18=f^2+4f-2$

$0=f^2+f-20$

$0=(f+5)(f-4)$

$f+5=0 \text{ or } f-4=0$

$\qquad f=-5 \qquad f=4 \qquad \{-5,4\}$

69)

$$\dfrac{b}{b^2+b-6}+\dfrac{3}{b^2+9b+18}=\dfrac{8}{b^2+4b-12}$$

$$\dfrac{b}{(b+3)(b-2)}+\dfrac{3}{(b+6)(b+3)}=\dfrac{8}{(b+6)(b-2)}$$

$$(b+6)(b+3)(b-2)\cdot\left(\dfrac{b}{(b+3)(b-2)}+\dfrac{3}{(b+6)(b+3)}\right)=(b+6)(b+3)(b-2)\left(\dfrac{8}{(b+6)(b-2)}\right)$$

$$b(b+6)+3(b-2)=8(b+3)$$

$$b^2+6b+3b-6=8b+24$$

$$b^2+9b-6=8b+24$$

$$b^2+b-30=0$$

$$(b+6)(b-5)=0$$

$$b+6=0 \text{ or } b-5=0$$

$$b=-6 \qquad b=5$$

If $b=-6$, two denominators $=0$. The solution set is $\{5\}$.

71)

$$\frac{r}{r^2+8r+15}-\frac{2}{r^2+r-6}=\frac{2}{r^2+3r-10}$$

$$\frac{r}{(r+5)(r+3)}-\frac{2}{(r+3)(r-2)}=\frac{2}{(r+5)(r-2)}$$

$$(r+5)(r+3)(r-2)\cdot\left(\frac{r}{(r+5)(r+3)}-\frac{2}{(r+3)(r-2)}\right)=(r+5)(r+3)(r-2)\left(\frac{2}{(r+5)(r-2)}\right)$$

$$r(r-2)-2(r+5)=2(r+3)$$

$$r^2-2r-2r-10=2r+6$$

$$r^2-4r-10=2r+6$$

$$r^2-6r-16=0$$

$$(r+2)(r-8)=0$$

$$r+2=0 \text{ or } r-8=0$$

$$r=-2 \qquad r=8 \qquad \{-2,8\}$$

73)

$$\frac{k}{k^2-6k-16}-\frac{12}{5k^2-65k+200}=\frac{28}{5k^2-15k-50}$$

$$\frac{k}{(k+2)(k-8)}-\frac{12}{5(k-5)(k-8)}=\frac{28}{5(k+2)(k-5)}$$

$$5(k+2)(k-8)(k-5)\cdot\left(\frac{k}{(k+2)(k-8)}-\frac{12}{5(k-5)(k-8)}\right)=5(k+2)(k-8)(k-5)\left(\frac{28}{5(k+2)(k-5)}\right)$$

$$5k(k-5)-12(k+2)=28(k-8)$$

$$5k^2-25k-12k-24=28k-224$$

$$5k^2-37k-24=28k-224$$

$$5k^2-65k+200=0$$

$$5(k^2-13k+40)=0$$

$$5(k-5)(k-8)=0$$

$$k-5=0 \quad k=5$$

$$k-8=0 \quad k=8$$

If $k=5$ or $k=8$, two denominators $=0$. $\varnothing$

Chapter 8: Rational Expressions

75) $W = \dfrac{CA}{\boxed{m}}$

$\boxed{m}W = CA$

$\boxed{m} = \dfrac{CA}{W}$

77) $a = \dfrac{rt}{2\boxed{b}}$

$2\boxed{b}a = rt$

$\boxed{b} = \dfrac{rt}{2a}$

79) $B = \dfrac{t+u}{3\boxed{x}}$

$3\boxed{x}B = t+u$

$\boxed{x} = \dfrac{t+u}{3B}$

81) $d = \dfrac{t}{z-\boxed{n}}$

$\left(z-\boxed{n}\right)d = t$

$dz - d\boxed{n} = t$

$dz - t = d\boxed{n}$

$\dfrac{dz-t}{d} = \boxed{n}$

83) $h = \dfrac{3A}{r+\boxed{s}}$

$\left(r+\boxed{s}\right)h = 3A$

$hr + h\boxed{s} = 3A$

$h\boxed{s} = 3A - hr$

$\boxed{s} = \dfrac{3A-hr}{h}$

85) $r = \dfrac{kx}{\boxed{y}-az}$

$\left(\boxed{y}-az\right)r = kx$

$r\boxed{y} - raz = kx$

$r\boxed{y} = kx + raz$

$\boxed{y} = \dfrac{kx+raz}{r}$

87) $\dfrac{1}{t} = \dfrac{1}{\boxed{r}} - \dfrac{1}{s}$

$\boxed{r}st\left(\dfrac{1}{t}\right) = \boxed{r}st\left(\dfrac{1}{\boxed{r}} - \dfrac{1}{s}\right)$

$\boxed{r}s = st - \boxed{r}t$

$\boxed{r}s + \boxed{r}t = st$

$\boxed{r}(s+t) = st$

$\boxed{r} = \dfrac{st}{s+t}$

89) $\dfrac{5}{x} = \dfrac{1}{y} - \dfrac{4}{\boxed{z}}$

$\left(xy\boxed{z}\right)\left(\dfrac{5}{x}\right) = \left(xy\boxed{z}\right)\left(\dfrac{1}{y} - \dfrac{4}{\boxed{z}}\right)$

$5y\boxed{z} = x\boxed{z} - 4xy$

$4xy = -5y\boxed{z} + x\boxed{z}$

$4xy = (x-5y)\boxed{z}$

$\dfrac{4xy}{x-5y} = \boxed{z}$

Section 8.7: Exercises

1) $\dfrac{8}{15} = \dfrac{32}{x}$

$8x = 15 \cdot 32$

$8x = 480$

$x = \dfrac{480}{8} = 60 \qquad \{60\}$

3) $\dfrac{4}{7} = \dfrac{n}{n+9}$

$4(n+9) = 7n$

$4n + 36 = 7n$

$36 = 3n$

$\dfrac{36}{3} = n$

$12 = n \qquad \{12\}$

5) $l =$ length of the room

$\dfrac{2.5}{10} = \dfrac{3}{l}$

$2.5l = 3(10)$

$2.5l = 30$

$l = \dfrac{30}{2.5} = 12$

The length of the room is 12 feet.

7) $n =$ employees that do not have direct deposit

$n + 14 =$ employees that have direct deposit

$\dfrac{9}{2} = \dfrac{n+14}{n}$

$9n = 2(n+14)$

$9n = 2n + 28$

$7n = 28$

$n = \dfrac{28}{7} = 4$

$n + 14 = 4 + 14 = 18$

The number of employees that do not have direct deposit is 4.

9) $n =$ number of students who graduated in more than four years.

$n - 1200 =$ number of students who graduated in four years.

$\dfrac{2}{5} = \dfrac{n-1200}{n}$

$2n = 5(n-1200)$

$2n = 5n - 6000$

$-3n = -6000$

$n = \dfrac{-6000}{-3} = 2000$

$n - 1200 = 2000 - 1200 = 800$

The number of students who graduated in four years is 800.

11) $l =$ length of the floor

$l - 18 =$ width of the floor

$\dfrac{8}{5} = \dfrac{l}{l-18}$

$8(l-18) = 5l$

$8l - 144 = 5l$

$3l = 144$

$l = 48$

$l - 18 = 48 - 18 = 30$

The length is 48 feet, and the width is 30 feet.

13) $n =$ number of students who use the tutoring service

$n + 15 =$ number of students who did not use the tutoring service

$$\frac{3}{8} = \frac{n}{n+15}$$
$$3(n+15) = 8 \cdot n$$
$$3n + 45 = 8n$$
$$45 = 5n$$
$$n = \frac{45}{5} = 9$$
$$n + 15 = 9 + 15 = 24$$

The number of students who used the tutoring service is 9.

The number of students who did not use is 24.

15) a) Speed against current $= 8 - 2$
$$= 6 \text{ mph}$$

   b) Speed with current $= 8 + 2$
$$= 10 \text{ mph}$$

17) a) Speed with wind $= x + 40$ mph

   b) Speed into wind $= x - 30$ mph

19) s = speed of boat in still water

	d	$= r \cdot$	t
Downstream	6	$s+2$	$\dfrac{6}{s+2}$
Upstream	4	$s-2$	$\dfrac{4}{s-2}$

Solve $d = rt$ for t to get $t = \dfrac{d}{r}$.

$$\frac{\text{Time with}}{\text{the current}} = \frac{\text{Time against}}{\text{the current}}$$
$$\frac{6}{s+2} = \frac{4}{s-2}$$
$$6(s-2) = 4(s+2)$$
$$6s - 12 = 4s + 8$$
$$2s = 20$$
$$s = 10$$

The speed of the boat in still water is 10 mph.

21) s = speed of the plane

	d	$= r \cdot$	t
with wind	500	$s+25$	$\dfrac{500}{s+25}$
against wind	400	$s-25$	$\dfrac{400}{s-25}$

Solve $d = rt$ for t to get $t = \dfrac{d}{r}$.

$$\frac{\text{Time with}}{\text{wind}} = \frac{\text{Time against}}{\text{wind}}$$
$$\frac{500}{s+25} = \frac{400}{s-25}$$
$$500(s-25) = 400(s+25)$$
$$500s - 12,500 = 400s + 10,000$$
$$100s = 22,500$$
$$s = 225$$

The speed of the plane is 225 mph.

23) s = speed of the current

	d	$= r \cdot$	t
downstream	32	$28+s$	$\dfrac{32}{28+s}$
upstream	24	$28-s$	$\dfrac{24}{28-s}$

Solve $d = rt$ for t to get $t = \dfrac{d}{r}$.

$$\frac{\text{Time with}}{\text{the current}} = \frac{\text{Time against}}{\text{the current}}$$

$$\frac{32}{28+s} = \frac{24}{28-s}$$

$$32(28-s) = 24(28+s)$$

$$896 - 32s = 672 + 24s$$

$$224 = 56s$$

$$4 = s$$

The speed of the current is 4 mph.

25) s = speed of the wind

with wind

against wind

$$d = r \cdot t$$

800	280 + s	$\frac{800}{280+s}$
600	280 − s	$\frac{600}{280-s}$

Solve $d = rt$ for t to get $t = \dfrac{d}{r}$.

$$\frac{\text{Time with}}{\text{wind}} = \frac{\text{Time against}}{\text{wind}}$$

$$\frac{800}{280+s} = \frac{600}{280-s}$$

$$800(280-s) = 600(280+s)$$

$$224{,}000s - 800s = 168{,}000s + 600s$$

$$-1400s = -56{,}000$$

$$s = \frac{-56{,}000}{-1400} = 40$$

The speed of the wind is 40 mph.

27) s = average speed of Bill

$d =$	r	$\cdot$	t

	d	r	t
San Diego to LA	120	s	$\dfrac{120}{s}$
LA to Las Vegas	240	s	$\dfrac{240}{s}$

Solve $d = rt$ for t to get $t = \dfrac{d}{r}$.

$$\begin{pmatrix}\text{Time} \\ \text{for San Diego} \\ \text{to LA}\end{pmatrix} = \begin{pmatrix}\text{Time for} \\ \text{LA to} \\ \text{Las Vegas}\end{pmatrix} - 2$$

$$\frac{120}{s} = \frac{240}{s} - 2$$

$$120 = 240 - 2s$$

$$-120 = -2s$$

$$s = \frac{-120}{-2}$$

$$s = 60$$

Average speed for Bill is 60 mph.

29) $\text{rate} = \dfrac{1 \text{ hw}}{3 \text{ hours}} = \dfrac{1}{3} \text{ hw/hour}$

31) $\text{rate} = \dfrac{1 \text{ job}}{t \text{ hours}} = \dfrac{1}{t} \text{ job/hour}$

33) t = hours to paint together

$$\begin{pmatrix}\text{fractional} \\ \text{part Rupinderjeet}\end{pmatrix} + \begin{pmatrix}\text{fractional} \\ \text{part Sana}\end{pmatrix} = 1 \text{ job}$$

$$\frac{1}{4}t + \frac{1}{5}t = 1$$

$$\frac{1}{4}t + \frac{1}{5}t = 1$$

$$20\left(\frac{1}{4}t + \frac{1}{5}t\right) = 20(1)$$

$$5t + 4t = 20$$

$$9t = 20$$

$$t = \frac{20}{9} = 2\frac{2}{9}$$

They could paint the room

together in $2\dfrac{2}{9}$ hours.

35) t = the number of hours to clean
   the carpets together

$$\underset{\text{part Wayne}}{\text{fractional}} + \underset{\text{part Garth}}{\text{fractional}} = 1\text{ job}$$

$$\dfrac{1}{4}t \quad + \quad \dfrac{1}{6}t \quad = 1$$

$$\dfrac{1}{4}t + \dfrac{1}{6}t = 1$$

$$24\left(\dfrac{1}{4}t + \dfrac{1}{6}t\right) = 24(1)$$

$$6t + 4t = 24$$

$$10t = 24$$

$$t = \dfrac{24}{10} = \dfrac{12}{5} = 2\dfrac{2}{5}$$

It would take $2\dfrac{2}{5}$ hours to

clean the carpets together.

37) t = number of minutes it would take
   to fill a tub that has a leaky drain.

$$\underset{\text{part faucet}}{\text{fractional}} + \underset{\text{part drain}}{\text{fractional}} = 1\text{ job}$$

$$\dfrac{1}{12}t \quad + \left(-\dfrac{1}{30}t\right) \quad = 1$$

$$\dfrac{1}{12}t - \dfrac{1}{30}t = 1$$

$$60\left(\dfrac{1}{12}t - \dfrac{1}{30}t\right) = 60(1)$$

$$5t - 2t = 60$$

$$3t = 60$$

$$t = 20$$

It would take 20 minutes to fill the tub.

39) t = hours for old machine to do a job

$$\underset{\text{part new machine}}{\text{fractional}} + \underset{\text{part old machine}}{\text{fractional}} = 1\text{ job}$$

$$\dfrac{1}{5}(3) \quad + \quad \dfrac{1}{t}(3) \quad = 1$$

$$\dfrac{3}{5} + \dfrac{3}{t} = 1$$

$$5t\left(\dfrac{3}{5} + \dfrac{3}{t}\right) = 1 \cdot 5t$$

$$3t + 15 = 5t$$

$$15 = 2t$$

$$t = \dfrac{15}{2} = 7.5$$

It would take old machine 7.5 hours
to do the job by itself.

41) t = number of hours for Ting to
   make decorations

$2t$ = number of hours for Mei.

$$\underset{\text{Ting}}{\text{fractional part}} + \underset{\text{Mei}}{\text{fractional part}} = 1\text{ job}$$

$$\dfrac{1}{t}(40) \quad + \quad \dfrac{1}{2t}(40) \quad = 1$$

$$\dfrac{40}{t} + \dfrac{40}{2t} = 1$$

$$2t\left(\dfrac{40}{t} + \dfrac{40}{2t}\right) = 2t \cdot 1$$

$$80 + 40 = 2t$$

$$120 = 2t$$

$$60 = t$$

$$t = 1\text{ hour}$$

It would take Ting 1 hour
to make decorations by herself.
It would take Mei 2 hours
to make decorations by herself.

Chapter 8 Review

1) a) $\dfrac{(5)^2 - 3(5) - 10}{3(5) + 2} = \dfrac{25 - 15 - 10}{15 + 2}$

$= \dfrac{0}{17} = 0$

b) $\dfrac{(-2)^2 - 3(-2) - 10}{4(-2) + 2}$

$= \dfrac{4 + 6 - 10}{-8 + 2} = \dfrac{0}{-6} = 0$

3) a) $2s = 0$ b) $4s + 11 = 0$

$s = 0$ $4s = -11$

$s = -\dfrac{11}{4}$

5) a) never equals zero

b) $4t^2 - 9 = 0$

$(2t + 3)(2t - 3) = 0$

$2t + 3 = 0$ or $2t - 3 = 0$

$2t = -3$ $2t = 3$

$t = -\dfrac{3}{2}$ $t = \dfrac{3}{2}$

7) a) $3m^2 - m - 10 = 0$

$(3m + 5)(m - 2) = 0$

$3m + 5 = 0$ or $m - 2 = 0$

$3m = -5$ $m = 2$

$m = -\dfrac{5}{3}$

b)   never undefined- any real
number can be substituted
for m

9) $\dfrac{77k^9}{7k^3} = 11k^6$

11) $\dfrac{r^2 - 14r + 48}{4r^2 - 24r} = \dfrac{(\cancel{r - 6})(r - 8)}{4r(\cancel{r - 6})}$

$= \dfrac{r - 8}{4r}$

13) $\dfrac{3z - 5}{6z^2 - 7z - 5} = \dfrac{\cancel{3z - 5}}{(2z + 1)(\cancel{3z - 5})}$

$= \dfrac{1}{2z + 1}$

15) $\dfrac{11 - x}{x^2 - 121} = \dfrac{\cancel{11 - x}^{-1}}{(x + 11)(\cancel{x - 11})}$

$= -\dfrac{1}{x + 11}$

17) $\dfrac{-4n - 1}{5 - 3n}, \dfrac{-(4n + 1)}{5 - 3n}, \dfrac{4n + 1}{3n - 5},$

$\dfrac{4n + 1}{-5 + 3n}, \dfrac{4n + 1}{-(5 - 3n)}$

19) $w = \dfrac{2b^2 + 13b + 21}{2b + 7}$

$= \dfrac{(2b + 7)(b + 3)}{2b + 7} = b + 3$

21) $x - 2 = 0$

$x = 2$

The domain contains all real
numbers except 2.

Domain: $(-\infty, 2) \cup (2, \infty)$

23) $t^2 - 64 = 0$

$(t + 8)(t - 8) = 0$

$t + 8 = 0$ or $t - 8 = 0$

$t = -8$ $t = 8$

The domain contains all real
numbers except -8 and 8.

Domain: $(-\infty, -8) \cup (-8, 8) \cup (8, \infty)$

Chapter 8: Rational Expressions

25) $\dfrac{64}{45}\cdot\dfrac{27}{56}=\dfrac{\cancel{64}^{8}}{\cancel{45}_{5}}\cdot\dfrac{\cancel{27}^{3}}{\cancel{56}_{7}}=\dfrac{24}{35}$

27) $\dfrac{t+6}{4}\cdot\dfrac{2(t+2)}{(t+6)^2}=\dfrac{\cancel{t+6}}{\cancel{4}_{2}}\cdot\dfrac{\cancel{2}(t+2)}{(t+6)^{\cancel{2}}}$

$=\dfrac{t+2}{2(t+6)}$

29) $\dfrac{3x^2+11x+8}{15x+40}\div\dfrac{9x+9}{x-3}$

$=\dfrac{3x^2+11x+8}{15x+40}\cdot\dfrac{x-3}{9x+9}$

$=\dfrac{\cancel{(3x+8)}\,\cancel{(x+1)}}{5\cancel{(3x+8)}}\cdot\dfrac{x-3}{9\cancel{(x+1)}}$

$=\dfrac{x-3}{45}$

31) $\dfrac{r^2-16r+63}{2r^3-18r^2}\div(r-7)^2$

$=\dfrac{r^2-16r+63}{2r^3-18r^2}\cdot\dfrac{1}{(r-7)^2}$

$=\dfrac{\cancel{(r-7)}\,\cancel{(r-9)}}{2r^2\cancel{(r-9)}}\cdot\dfrac{1}{(r-7)\cancel{(r-7)}}$

$=\dfrac{1}{2r^2(r-7)}$

33) $\dfrac{3p^5}{20q^2}\cdot\dfrac{4q^3}{21p^7}=\dfrac{3\cancel{p^5}}{\cancel{20}_5 \cancel{q^2}}\cdot\dfrac{\cancel{4}\cancel{q^3}^{q}}{\cancel{21}\cancel{p^7}_{7p^2}}=\dfrac{q}{35p^2}$

35) $\dfrac{\frac{3s+8}{12}}{\frac{3s+8}{4}}=\dfrac{3s+8}{12}\div\dfrac{3s+8}{4}$

$=\dfrac{\cancel{3s+8}}{\cancel{12}_3}\cdot\dfrac{\cancel{4}}{\cancel{3s+8}}=\dfrac{1}{3}$

37) $\dfrac{\frac{9}{8}}{\frac{15}{4}}=\dfrac{9}{8}\div\dfrac{15}{4}=\dfrac{\cancel{9}^{3}}{\cancel{8}_{2}}\cdot\dfrac{\cancel{4}}{\cancel{15}_{5}}=\dfrac{3}{10}$

39) $LCD=30$

41) $LCD=k^5$

43) $LCD=(4x+9)(x-7)$

45) $LCD=w-5$ or $5-w$

47) $c^2+9c+20=(c+4)(c+5)$
$c^2-2c-35=(c+5)(c-7)$
$LCD=(c+4)(c+5)(c-7)$

49) $\dfrac{3}{5y}\cdot\dfrac{4y^2}{4y^2}=\dfrac{12y^2}{20y^3}$

51) $\dfrac{6}{2z+5}\cdot\dfrac{z}{z}=\dfrac{6z}{z(2z+5)}$

53) $\dfrac{t-3}{3t+1}\cdot\dfrac{t+4}{t+4}=\dfrac{t^2+t-12}{(3t+1)(t+4)}$

55) $\dfrac{8c}{c^2+5c-24}=\dfrac{8c}{(c+8)(c-3)}$

$\dfrac{5}{c^2-6c+9}=\dfrac{5}{(c-3)^2}$

$LCD=(c-3)^2(c+8)$

$\dfrac{8c}{(c+8)(c-3)}\cdot\dfrac{(c-3)}{(c-3)}=\dfrac{8c^2-24c}{(c-3)^2(c+8)}$

$\dfrac{5}{(c-3)^2}\cdot\dfrac{(c+8)}{(c+8)}=\dfrac{5c+40}{(c-3)^2(c+8)}$

278

57) $\dfrac{7}{2q^2-12q}=\dfrac{7}{2q(q-6)}$

$\dfrac{3q}{36-q^2}=\dfrac{3q}{(6+q)(6-q)}$

$\dfrac{q-5}{2q^2+12q}=\dfrac{q-5}{2q(q+6)}$

$\text{LCD}=2q(q+6)(q-6)$

$\dfrac{7}{2q(q-6)}\cdot\dfrac{(q+6)}{(q+6)}=\dfrac{7q+42}{2q(q+6)(q-6)}$

$\dfrac{3q}{(6+q)(6-q)}\cdot\dfrac{-2q}{-2q}=-\dfrac{6q^2}{2q(q+6)(q-6)}$

$\dfrac{q-5}{2q(q+6)}\cdot\dfrac{(q-6)}{(q-6)}=\dfrac{q^2-11q+30}{2q(q+6)(q-6)}$

59) $\dfrac{5}{9c}+\dfrac{7}{9c}=\dfrac{12}{9c}=\dfrac{4}{3c}$

61) $\dfrac{9}{10u^2v^2}-\dfrac{1}{8u^3v}=\dfrac{36u}{40u^3v^2}-\dfrac{5v}{40u^3v^2}$

$=\dfrac{36u-5v}{40u^3v^2}$

67) $\dfrac{k-3}{k^2+14k+49}-\dfrac{2}{k^2+7k}$

$=\dfrac{k-3}{(k+7)^2}-\dfrac{2}{k(k+7)}=\dfrac{k(k-3)}{k(k+7)^2}-\dfrac{2(k+7)}{k(k+7)^2}$

$=\dfrac{k^2-3k-2k-14}{k(k+7)^2}=\dfrac{k^2-5k-14}{k(k+7)^2}=\dfrac{(k-7)(k+2)}{k(k+7)^2}$

69) $\dfrac{t+9}{t-18}-\dfrac{11}{18-t}=\dfrac{t+9}{t-18}+\dfrac{11}{t-18}=\dfrac{t+20}{t-18}$

63) $\dfrac{n}{3n-5}-\dfrac{4}{n}=\dfrac{n^2}{n(3n-5)}-\dfrac{4(3n-5)}{n(3n-5)}$

$=\dfrac{n^2-12n+20}{n(3n-5)}$

65) $\dfrac{9}{y+2}-\dfrac{5}{y-3}$

$=\dfrac{9(y-3)}{(y+2)(y-3)}-\dfrac{5(y+2)}{(y+2)(y-3)}$

$=\dfrac{9y-27-5y-10}{(y+2)(y-3)}$

$=\dfrac{4y-37}{(y+2)(y-3)}$

71) $\dfrac{4w}{w^2+11w+24}-\dfrac{3w-1}{2w^2-w-21}=\dfrac{4w}{(w+3)(w+8)}-\dfrac{3w-1}{(w+3)(2w-7)}$

$$=\dfrac{4w(2w-7)}{(w+3)(w+8)(2w-7)}-\dfrac{(3w-1)(w+8)}{(w+3)(w+8)(2w-7)}$$

$$=\dfrac{8w^2-28w-3w^2-24w+w+8}{(w+3)(w+8)(2w-7)}=\dfrac{5w^2-51w+8}{(w+3)(w+8)(2w-7)}$$

73) $\dfrac{b}{9b^2-4}+\dfrac{b+1}{6b^2-4b}+\dfrac{1}{6b+4}=\dfrac{b}{(3b-2)(3b+2)}+\dfrac{b+1}{2b(3b-2)}-\dfrac{1}{2(3b+2)}$

$$=\dfrac{2b\cdot b}{2b(3b-2)(3b+2)}+\dfrac{(3b+2)(b+1)}{2b(3b-2)(3b+2)}-\dfrac{b(3b-2)}{2b(3b-2)(3b+2)}$$

$$=\dfrac{2b^2+3b^2+3b+2b+2-3b^2+2b}{2b(3b-2)(3b+2)}=\dfrac{2b^2+7b+2}{2b(3b-2)(3b+2)}$$

75) a) $A=\left(\dfrac{x}{x+2}\right)\left(\dfrac{2}{x^2}\right)=\dfrac{2x}{x^2(x+2)}=\dfrac{2}{x(x+2)}$ sq. units

b) $P=2\left(\dfrac{x}{x+2}\right)+2\left(\dfrac{2}{x^2}\right)=\dfrac{2x}{x+2}+\dfrac{4}{x^2}=\dfrac{x^2\cdot 2x}{x^2(x+2)}+\dfrac{4(x+2)}{x^2(x+2)}=\dfrac{2x^3+4x+8}{x^2(x+2)}$ units

77) $\dfrac{\dfrac{x}{y}}{\dfrac{x^3}{y^2}}=\dfrac{y^2\left(\dfrac{x}{y}\right)}{y^2\left(\dfrac{x^3}{y^2}\right)}=\dfrac{xy}{x^3}=\dfrac{y}{x^2}$

79) $\dfrac{p+\dfrac{4}{p}}{\dfrac{9}{p}+p}=\dfrac{p\left(p+\dfrac{4}{p}\right)}{p\left(\dfrac{9}{p}+p\right)}=\dfrac{p^2+4}{9+p^2}$

$$=\dfrac{p^2+4}{p^2+9}$$

81) $\dfrac{\dfrac{4}{5}-\dfrac{2}{3}}{\dfrac{1}{2}+\dfrac{1}{6}}=\dfrac{30\left(\dfrac{4}{5}-\dfrac{2}{3}\right)}{30\left(\dfrac{1}{2}+\dfrac{1}{6}\right)}=\dfrac{24-20}{15+5}$

$$=\dfrac{4}{20}=\dfrac{1}{5}$$

83)
$$\frac{1-\dfrac{1}{y-8}}{\dfrac{2}{y+4}+1} = \frac{(y+4)(y-8)\left(1-\dfrac{1}{y-8}\right)}{(y+4)(y-8)\left(\dfrac{2}{y+4}+1\right)}$$

$$= \frac{(y+4)(y-8)-(y+4)}{2(y-8)+(y+4)(y-8)}$$

$$= \frac{y^2-4y-32-y-4}{2y-16+y^2-4y-32}$$

$$= \frac{y^2-5y-36}{y^2-2y-48}$$

$$= \frac{(y+4)(y-9)}{(y-8)(y+6)}$$

85)
$$\frac{1+\dfrac{1}{r-t}}{\dfrac{1}{r^2-t^2}+\dfrac{1}{r+t}}$$

$$= \frac{(r+t)(r-t)\left(1+\dfrac{1}{r-t}\right)}{(r+t)(r-t)\left(\dfrac{1}{(r+t)(r-t)}+\dfrac{1}{r+t}\right)}$$

$$= \frac{(r+t)(r-t)+(r+t)}{1+r-t}$$

$$= \frac{(r+t)(r-t+1)}{r-t+1} = r+t$$

87)
$$\frac{5a+4}{15} = \frac{a}{5}+\frac{4}{5}$$

$$15\left(\frac{5a+4}{15}\right) = 15\left(\frac{a}{5}+\frac{4}{5}\right)$$

$$5a+4 = 3a+12$$

$$2a = 8$$

$$a = 4 \qquad \{4\}$$

89)
$$\frac{m}{7} = \frac{5}{m+2}$$

$$m(m+2) = 5\cdot 7$$

$$m^2+2m = 35$$

$$m^2+2m-35 = 0$$

$$(m+7)(m-5) = 0$$

$$m+7 = 0 \ \text{ or } \ m-5 = 0$$

$$m = -7 \qquad m = 5 \qquad \{-7,5\}$$

91)
$$\frac{r}{r+5}+4 = \frac{5}{r+5}$$

$$(r+5)\left(\frac{r}{r+5}+4\right) = (r+5)\left(\frac{5}{r+5}\right)$$

$$r+4(r+5) = 5$$

$$r+4r+20 = 5$$

$$5r = -15$$

$$r = -3 \qquad \{-3\}$$

93)
$$\frac{5}{t^2+10t+24}+\frac{5}{t^2+3t-18}=\frac{t}{t^2+t-12}$$

$$\frac{5}{(t+6)(t+4)}+\frac{5}{(t+6)(t-3)}=\frac{t}{(t+4)(t-3)}$$

$$(t+6)(t+4)(t-3)\cdot\left(\frac{5}{(t+6)(t+4)}+\frac{5}{(t+6)(t-3)}\right)=(t+6)(t+4)(t-3)\left(\frac{t}{(t+4)(t-3)}\right)$$

$$5(t-3)+5(t+4)=t(t+6)$$

$$5t-15+5t+20=t^2+6t$$

$$10t+5=t^2+6t$$

$$t^2-4t-5=0$$

$$(t+1)(t-5)=0$$

$$t=-1 \quad\text{or}\quad t=5 \qquad \{-1,5\}$$

95) $\quad\dfrac{3}{x+1}=\dfrac{6x}{x^2-1}-\dfrac{4}{x-1}$

$$\frac{3}{x+1}=\frac{6x}{(x+1)(x-1)}-\frac{4}{x-1}$$

$$(x+1)(x-1)\left(\frac{3}{x+1}\right)=(x+1)(x-1)\left(\frac{6x}{(x+1)(x-1)}-\frac{4}{x-1}\right)$$

$$3(x-1)=6x-4(x+1)$$

$$3x-3=6x-4x-4$$

$$3x-3=2x-4$$

$$x=-1$$

If $x=-1$, two denominators = 0. $\quad\varnothing$

97) $\quad R=\dfrac{s+T}{\boxed{D}}$

$$\boxed{D}R=s+T$$

$$\boxed{D}=\frac{s+T}{R}$$

99) $\quad w=\dfrac{N}{c-a\boxed{k}}$

$$w\left(c-a\boxed{k}\right)=N$$

$$wc-wa\boxed{k}=N$$

$$-wa\boxed{k}=N-wc$$

$$\boxed{k}=\frac{N-wc}{-wa}$$

$$\boxed{k}=\frac{wc-N}{wa}$$

101)

$$\frac{1}{\boxed{R_1}}+\frac{1}{R_2}=\frac{1}{R_3}$$

$$\left(\boxed{R_1}R_2R_3\right)\left(\frac{1}{\boxed{R_1}}+\frac{1}{R_2}\right)=\left(\boxed{R_1}R_2R_3\right)\left(\frac{1}{R_3}\right)$$

$$R_2R_3+\boxed{R_1}R_3=\boxed{R_1}R_2$$

$$R_2R_3=\boxed{R_1}R_2-\boxed{R_1}R_3$$

$$R_2R_3=\boxed{R_1}\left(R_2-R_3\right)$$

$$\frac{R_2R_3}{R_2-R_3}=\boxed{R_1}$$

103) $s=$ speed of the current

$$d = r \cdot t$$

downstream	8	$14+s$	$\dfrac{8}{14+s}$
upstream	6	$14-s$	$\dfrac{6}{14-s}$

Solve $d=rt$ for t to get $t=\dfrac{d}{r}$.

$$\frac{\text{Time}}{\text{dowstream}}=\frac{\text{Time}}{\text{upstream}}$$

$$\frac{8}{14+s}=\frac{6}{14-s}$$

$$8(14-s)=6(14+s)$$

$$112-8s=84+6s$$

$$28=14s$$

$$2=s$$

The speed of the current is 2 mph.

105) $t=$ the number of hours to
assemble the notebooks together

$$\frac{\text{fractional}}{\text{part Crayton}}+\frac{\text{fractional}}{\text{part Flow}}=1\text{ job}$$

$$\frac{1}{5}t \quad + \quad \frac{1}{8}t \quad =1$$

$$40\left(\frac{1}{5}t+\frac{1}{8}t\right)=40\cdot1$$

$$8t+5t=40$$

$$13t=40$$

$$t=\frac{40}{13}=3\frac{1}{13}$$

It would take $3\dfrac{1}{13}$ hours to
assemble the notebooks together.

107) $$\frac{5n}{2n-1}-\frac{2n+3}{n+2}$$

$$=\frac{5n(n+2)}{(2n-1)(n+2)}-\frac{(2n+3)(2n-1)}{(2n-1)(n+2)}$$

$$=\frac{5n^2+10n-\left(4n^2-2n+6n-3\right)}{(2n-1)(n+2)}$$

$$=\frac{5n^2+10n-4n^2-4n+3}{(2n-1)(n+2)}$$

$$=\frac{n^2+6n+3}{(2n-1)(n+2)}$$

109) $$\frac{2a^2+9a+10}{4a-7}\div(2a+5)^2$$

$$=\frac{2a^2+9a+10}{4a-7}\cdot\frac{1}{(2a+5)^2}$$

$$=\frac{\cancel{(2a+5)}(a+2)}{(4a-7)}\cdot\frac{1}{(2a+5)\cancel{(2a+5)}}$$

$$=\frac{(a+2)}{(4a-7)(2a+5)}$$

Chapter 8: Rational Expressions

111) $\dfrac{c^2}{c^2-d^2}+\dfrac{c}{d-c}$

$=\dfrac{c^2}{(c+d)(c-d)}-\dfrac{c}{c-d}$

$=\dfrac{c^2}{(c+d)(c-d)}-\dfrac{c(c+d)}{(c+d)(c-d)}$

$=\dfrac{c^2-c^2-cd}{(c+d)(c-d)}=-\dfrac{cd}{(c+d)(c-d)}$

113) $\dfrac{h}{5}=\dfrac{h-3}{h+1}+\dfrac{12}{5h+5}$

$\dfrac{h}{5}=\dfrac{h-3}{h+1}+\dfrac{12}{5(h+1)}$

$5(h+1)\dfrac{h}{5}=5(h+1)\left(\dfrac{h-3}{h+1}+\dfrac{12}{5(h+1)}\right)$

$(h+1)h=5(h-3)+12$

$h^2+h=5h-15+12$

$h^2+h=5h-3$

$h^2-4h+3=0$

$(h-3)(h-1)=0$

$h-3=0$ or $h-1=0$

$h=3$ $\qquad h=1$ $\qquad \{1,3\}$

115) $\dfrac{8}{3g^2-7g-6}-\dfrac{8}{3g+2}=-\dfrac{4}{g-3}$

$\dfrac{8}{(3g+2)(g-3)}-\dfrac{8}{3g+2}=-\dfrac{4}{g-3}$

$(3g+2)(g-3)\cdot\left(\dfrac{8}{(3g+2)(g-3)}-\dfrac{8}{3g+2}\right)$

$=(3g+2)(g-3)\left(-\dfrac{4}{g-3}\right)$

$8-8(g-3)=-4(3g+2)$

$8-8g+24=-12g-8$

$32-8g=-12g-8$

$32+4g=-8$

$4g=-40$

$g=-10 \quad \{-10\}$

Chapter 8 Test

1) $\dfrac{5(-4)+8}{(-4)^2+16}=\dfrac{-20+8}{16+16}=\dfrac{-12}{32}=-\dfrac{3}{8}$

3) a) $n^2-5n-36=0$

$(k-9)(k+4)=0$

$k-9=0$ or $k+4=0$

$k=9 \qquad k=-4$

b)   It never equals zero.

5) $\dfrac{3h^2-25h+8}{27h^3-1}=\dfrac{(h-8)\,(3h-1)}{(3h-1)\,(9h^2+3h+1)}$

$=\dfrac{h-8}{9h^2+3h+1}$

7)   z and z+6 are different factors. The LCD will be the product of these factors. LCD $=z(z+6)$

284

9) $\dfrac{28a^9}{b^2} \div \dfrac{20a^{15}}{b^3} = \dfrac{\overset{7}{\cancel{28a^9}}}{\cancel{b^2}} \cdot \dfrac{\overset{b}{\cancel{b^3}}}{\underset{5a^6}{\cancel{20a^{15}}}}$

$= \dfrac{7b}{5a^6}$

11) $\dfrac{6}{c+2} + \dfrac{c}{3c+5} = \dfrac{6(3c+5)+c(c+2)}{(c+2)(3c+5)}$

$= \dfrac{18c+30+c^2+2c}{(c+2)(3c+5)}$

$= \dfrac{c^2+20c+30}{(c+2)(3c+5)}$

13) $\dfrac{8d^2+24d}{20} \div (d+3)^2$

$= \dfrac{8d(d+3)}{20} \div (d+3)^2$

$= \dfrac{\overset{2}{\cancel{8}}\,d\,\cancel{(d+3)}}{\underset{5}{\cancel{20}}} \cdot \dfrac{1}{(d+3)\,\cancel{(d+3)}}$

$= \dfrac{2d}{5(d+3)}$

15) $\dfrac{3}{2v^2-7v+6} - \dfrac{v+4}{v^2+7v-18}$

$= \dfrac{3}{(2v-3)(v-2)} - \dfrac{v+4}{(v+9)(v-2)}$

$= \dfrac{3(v+9)-(v+4)(2v-3)}{(2v-3)(v-2)(v+9)}$

$= \dfrac{3v+27-\left(2v^2-3v+8v-12\right)}{(2v-3)(v-2)(v+9)}$

$= \dfrac{3v+27-2v^2-5v+12}{(2v-3)(v-2)(v+9)}$

$= \dfrac{-2v^2-2v+39}{(2v-3)(v-2)(v+9)}$

17) $\dfrac{\dfrac{5x+5y}{x^2y^2}}{\dfrac{20}{xy}} = \dfrac{x^2y^2\left(\dfrac{5x+5y}{x^2y^2}\right)}{x^2y^2\left(\dfrac{20}{xy}\right)} = \dfrac{5(x+y)}{20xy} = \dfrac{x+y}{4xy}$

19) $\dfrac{28}{w^2-4} = \dfrac{7}{w-2} - \dfrac{5}{w+2}$

$\dfrac{28}{(w+2)(w-2)} = \dfrac{7}{w-2} - \dfrac{5}{w+2}$

$28 = 7(w+2)-5(w-2)$

$28 = 7w+14-5w+10$

$28 = 2w+24$

$4 = 2w$

$2 = w$

If $w=2$, two denominators equal 0.

$\varnothing$

21) $\dfrac{1}{a}+\dfrac{1}{\boxed{b}}=\dfrac{1}{c}$

$a\boxed{b}c\left(\dfrac{1}{a}+\dfrac{1}{\boxed{b}}\right)=a\boxed{b}c\left(\dfrac{1}{c}\right)$

$\boxed{b}c+ac=a\boxed{b}$

$ac=\boxed{b}a-\boxed{b}c$

$ac=\boxed{b}(a-c)$

$\dfrac{ac}{a-c}=\boxed{b}$

23) $P=2\left(\dfrac{10}{x+1}\right)+2\left(\dfrac{x-2}{5}\right)$

$=\dfrac{20}{x+1}+\dfrac{2(x-2)}{5}$

$=\dfrac{5\cdot 20}{5(x+1)}+\dfrac{2(x-2)(x+1)}{5(x+1)}$

$=\dfrac{100+2(x^2+x-2x-2)}{5(x+1)}$

$=\dfrac{2x^2-2x+96}{5(x+1)}$

$=\dfrac{2(x^2-x+48)}{5(x+1)}$ units

25) $s=$ speed of boat in still water

	$d=$	r $\cdot$	t
Downstream	12	$s+4$	$\dfrac{12}{s+4}$
Upstream	6	$s-4$	$\dfrac{6}{s-4}$

Solve $d=rt$ for t to get $t=\dfrac{d}{r}$.

$\dfrac{\text{Time with}}{\text{the current}}=\dfrac{\text{Time against}}{\text{the current}}$

$\dfrac{12}{s+4}=\dfrac{6}{s-4}$

$12(s-4)=6(s+4)$

$12s-48=6s+24$

$6s=72$

$s=12$

The speed of the boat in still water is 12 mph.

Cumulative Review: Chapters 1-8

1) $A=\dfrac{1}{2}bh$

$=\dfrac{1}{2}(18)(5)$

$=45$ cm^2

3) $(2p^3)^5=2^5 p^{15}=32p^{15}$

5) Let $l=$ the length of the rectangle.
Let $w=$ the width of the rectangle.
$l=w+4$
$2l+2w=28$
Use substitution.
$2(w+4)+2w=28$
$2w+8+2w=28$
$4w+8=28$
$4w=20$
$w=5$
$l=w+4; \quad w=5$
$l=5+4$
$l=9$
width: 5 ft, length : 9 ft

7) $\quad 4 \le \dfrac{3}{5}t + 4 \le 13$

$\quad 20 \le 3t + 20 \le 65$

$\quad 0 \le 3t \le 45$

$\quad 0 \le t \le 15$

$\quad [0, 15]$

9) $\quad (x_1, y_1) = (4,1)$

$\quad (x_2, y_2) = (-2,9)$

$\quad m = \dfrac{y_2 - y_1}{x_2 - x_1} = \dfrac{9-1}{-2-4} = \dfrac{8}{-6} = -\dfrac{4}{3}$

11) $(2n-3)^2 = 4n^2 - 12n + 9$

13) $\dfrac{45h^4 - 25h^3 + 15h^2 - 10}{15h^2}$

$= \dfrac{\overset{3h^2}{\cancel{45h^4}}}{\cancel{15h^2}} - \dfrac{\overset{5h}{\cancel{25h^3}}}{\underset{3}{\cancel{15h^2}}} + \dfrac{\cancel{15h^2}}{\cancel{15h^2}} - \dfrac{\overset{2}{\cancel{10}}}{\underset{3h^2}{\cancel{15h^2}}}$

$= 3h^2 - \dfrac{5}{3}h + 1 - \dfrac{2}{3h^2}$

15) $4d^2 + 4d - 15 = (2d+5)(2d-3)$

17) $3m^4 - 24m = 3m(m^3 - 8)$

$\quad = 3m(m-2)(m^2 + 2m + 4)$

19) $\qquad x(x+16) = x - 36$

$\qquad x^2 + 16x = x - 36$

$\qquad x^2 + 15x + 36 = 0$

$\qquad (x+3)(x+12) = 0$

$\qquad x+3 = 0 \ \text{ or } \ x+12 = 0$

$\qquad x = -3 \qquad x = -12 \qquad \{-12, -3\}$

21) $\dfrac{3c^2 + 21c - 54}{c^2 + 3c - 54} = \dfrac{3(c^2 + 7c - 18)}{(c-6)(c+9)}$

$\qquad = \dfrac{3\cancel{(c+9)}(c-2)}{(c-6)\cancel{(c+9)}}$

$\qquad = \dfrac{3(c-2)}{c-6}$

23) $\dfrac{6}{y+5} - \dfrac{3}{y} = \dfrac{6 \cdot y - 3(y+5)}{y(y+5)}$

$\qquad = \dfrac{6y - 3y - 15}{y(y+5)}$

$\qquad = \dfrac{3(y-5)}{y(y+5)}$

25) $\dfrac{1}{v-1} + \dfrac{2}{5v-3} = \dfrac{37}{5v^2 - 8v + 3}$

$(v-1)(5v-3)\left(\dfrac{1}{v-1} + \dfrac{2}{5v-3}\right)$

$\qquad = (v-1)(5v-3)\left(\dfrac{37}{(v-1)(5v-3)}\right)$

$(5v-3) + 2(v-1) = 37$

$\qquad 5v - 3 + 2v - 2 = 37$

$\qquad 7v - 5 = 37$

$\qquad 7v = 42$

$\qquad v = 6$

$\qquad \{6\}$

Section 9.1: Exercises

1) Answers may vary.

3) Answers may vary.

5) $|q| = 6$

   $q = 6$ or $q = -6$ $\{-6, 6\}$

7) $|q - 5| = 3$

   $q - 5 = 3$ or $q - 5 = -3$

   $q = 8$ or $q = 2$ $\{2, 8\}$

9) $|4t - 5| = 7$

   $4t - 5 = 7$ or $4t - 5 = -7$

   $4t = 12$ $4t = -2$

   $t = 3$ or $t = -\dfrac{1}{2}$ $\left\{-\dfrac{1}{2}, 3\right\}$

11) $|12c + 5| = 1$

   $12c + 5 = 1$ or $12c + 5 = -1$

   $12c = -4$ $12c = -6$

   $c = \dfrac{-4}{12}$ $c = \dfrac{-6}{12}$

   $c = -\dfrac{1}{3}$ or $c = -\dfrac{1}{2}$

   $\left\{-\dfrac{1}{2}, -\dfrac{1}{3}\right\}$

13) $|1 - 8m| = 9$

   $1 - 8m = 9$ or $1 - 8m = -9$

   $-8m = 8$ $-8m = -10$

   $m = \dfrac{8}{-8}$ $m = \dfrac{-10}{-8}$

   $m = -1$ or $m = \dfrac{5}{4}$

   $\left\{-1, \dfrac{5}{4}\right\}$

15) $\left|\dfrac{2}{3}b + 3\right| = 13$

   $\dfrac{2}{3}b + 3 = 13$ or $\dfrac{2}{3}b + 3 = -13$

   $2b + 9 = 39$ $2b + 9 = -39$

   $2b = 30$ $2b = -48$

   $b = 15$ or $b = -24$

   $\{-24, 15\}$

17) $9 = |9 - 1.5d|$

   $9 = 9 - 1.5d$ or $-9 = 9 - 1.5d$

   $9 - 9 = -1.5d$ or $-9 - 9 = -1.5d$

   $0 = -1.5d$ or $-18 = -1.5d$

   $d = 0$ or $d = 12$

   $\{0, 12\}$

19) $\left|\dfrac{3}{4}y - 2\right| = \dfrac{3}{5}$

   $\dfrac{3}{4}y - 2 = \dfrac{3}{5}$ or $\dfrac{3}{4}y - 2 = -\dfrac{3}{5}$

   $15y - 40 = 12$ $15y - 40 = -12$

   $15y = 52$ $15y = 28$

   $y = \dfrac{52}{15}$ or $y = \dfrac{28}{15}$

   $\left\{\dfrac{28}{15}, \dfrac{52}{15}\right\}$

21) $|x| = \dfrac{1}{2}$, answers may vary

23) $|z - 6| + 4 = 20$

   $|z - 6| = 16$

   $z - 6 = 16$ or $z - 6 = -16$

   $z = 22$ or $z = -10$ $\{-10, 22\}$

25) $|5b+3|+6=19$

$|5b+3|=13$

$5b+3=13$ or $5b+3=-13$

$5b=10$ or $5b=-16$

$b=2$ or $b=-\dfrac{16}{5}$ $\left\{-\dfrac{16}{5},2\right\}$

27) $|6t-11|+5=10$

$|6t-11|=5$

$6t-11=5$ or $6t-11=-5$

$6t=16$ or $6t=6$

$t=\dfrac{8}{3}$ or $t=1$ $\left\{1,\dfrac{8}{3}\right\}$

29) $1=|7-8x|-4$

$5=|7-8x|$

$7-8x=5$ or $7-8x=-5$

$8x=2$ or $8x=12$

$x=\dfrac{1}{4}$ or $x=\dfrac{3}{2}$ $\left\{\dfrac{1}{4},\dfrac{3}{2}\right\}$

31) $-3=\left|\dfrac{2}{3}c+4\right|-7$

$4=\left|\dfrac{2}{3}c+4\right|$

$\dfrac{2}{3}c+4=4$ or $\dfrac{2}{3}c+4=-4$

$\dfrac{2}{3}c=0$ or $\dfrac{2}{3}c=-8$

$c=0$ or $c=-12$ $\{-12,0\}$

33) $|s+9|=|2s+5|$

$s+9=2s+5$ or $s+9=-2s-5$

$3s=-14$

$4=s$ or $s=-\dfrac{14}{3}$

$\left\{-\dfrac{14}{3},4\right\}$

35) $|3z+2|=|6-5z|$

$3z+2=6-5z$ or $3z+2=-6+5z$

$8z=4$ $8=2z$

$z=\dfrac{1}{2}$ or $4=z$

$\left\{\dfrac{1}{2},4\right\}$

37) $\left|\dfrac{3}{2}x-1\right|=|x|$

$\dfrac{3}{2}x-1=x$ or $\dfrac{3}{2}x-1=-x$

$3x-2=2x$ $3x-2=-2x$

$5x=2$

$x=2$ or $x=\dfrac{2}{5}$ $\left\{\dfrac{2}{5},2\right\}$

39) $|7c+10|=|5c+2|$

$7c+10=5c+2$ or $7c+10=-5c-2$

$2c=-8$ $12c=-12$

$c=-4$ or $c=-1$

$\{-4,-1\}$

41) $\left|\dfrac{1}{4}t-\dfrac{5}{2}\right|=\left|5-\dfrac{1}{2}t\right|$

$\dfrac{1}{4}t-\dfrac{5}{2}=5-\dfrac{1}{2}t$ or $\dfrac{1}{4}t-\dfrac{5}{2}=-5+\dfrac{1}{2}t$

$\quad t-10=20-2t \qquad t-10=-20+2t$

$\qquad 3t=30 \qquad\qquad -t=-10$

$\qquad\quad t=10 \quad$ or $\quad\; t=10$

$\{10\}$

43) $|1.6-0.3p|=|0.7p+0.4|$

$\quad 1.6-0.3p=0.7p+0.4$ or

$\quad 1.6-0.3p=-0.7p-0.4$

$\quad p=1.2$ or $0.4p=-2.0$

$\quad p=1.2$ or $p=-5.0 \qquad \{-5.0,1.2\}$

45) $|m-5|=-3 \qquad\qquad \varnothing$

47) $|10p+2|=0$

$\quad 10p=-2$

$\quad p=-\dfrac{1}{5}$

49) $|w+14|=0$

$\quad w=-14$

51) $|8n+11|=-1 \qquad\qquad \varnothing$

53) $|3m-1|+5=2$

$\quad |3m-1|=-3 \qquad\qquad \varnothing$

55) $11=|7-v|+4$

$\quad 7=|7-v|$

$\quad 7-v=7$ or $7-v=-7$

$\quad v=0$ or $v=14 \qquad\qquad \{0,14\}$

57) $\left|\dfrac{3}{5}p+3\right|-7=-5$

$\quad \left|\dfrac{3}{5}p+3\right|=2$

$\quad \dfrac{3}{5}p+3=2$ or $\dfrac{3}{5}p+3=-2$

$\quad p=-\dfrac{5}{3}$ or $p=-\dfrac{25}{3} \quad \left\{-\dfrac{5}{3},-\dfrac{25}{3}\right\}$

59) $|10h-3|=0$

$\quad h=\dfrac{3}{10}$

61) $|1.8a-3|=|4.2-1.2a|$

$\quad 1.8a-3=4.2-1.2a$ or

$\quad 1.8a-3=-4.2+1.2a$

$\quad 3a=7.2$ or $0.6a=-1.2$

$\quad a=2.4$ or $a=-2 \qquad \{2.4,-2\}$

63) $15+|2k+1|=6$

$\quad |2k+1|=-9 \qquad\qquad \varnothing$

65) $|9-1.5n|=1$

$\quad 9-1.5n=1$ or $9-1.5n=-1$

$\quad 1.5n=8$ or $1.5n=10$

$\quad n=\dfrac{16}{3}$ or $n=\dfrac{20}{3} \qquad \left\{\dfrac{16}{3},\dfrac{20}{3}\right\}$

67) $\left|\dfrac{1}{3}g-2\right|=\left|\dfrac{7}{9}g+\dfrac{1}{6}\right|$

$\quad |6g-36|=|14g+3|$

$\quad 6g-36=14g+3$ or

$\quad 6g-36=-14g-3$

$\quad -8g=39$ or $20g=33$

$\quad g=-\dfrac{39}{8}$ or $g=\dfrac{33}{20} \qquad \left\{-\dfrac{39}{8},\dfrac{33}{20}\right\}$

69) $7.6 = |2.8d + 3.5| + 7.6$

$|2.8d + 3.5| = 0$

$2.8d = -3.5$

$d = -1.25$

Section 9.2: Exercises

1) $[-1, 5]$

3) $(-\infty, 2) \cup (9, \infty)$

5) $\left(-\infty, -\dfrac{9}{2}\right] \cup \left[\dfrac{3}{5}, \infty\right)$

7) $\left(4, \dfrac{17}{3}\right)$

9) $|m| \leq 7$

$-7 \leq m \leq 7 \quad [-7, 7]$

11) $|3k| < 12$

$-12 < 3k < 12$

$-4 < k < 4 \quad (-4, 4)$

13) $|w - 2| < 4$

$-4 < w - 2 < 4$

$-2 < w < 6 \quad (-2, 6)$

15) $|3r + 10| \leq 4$

$-4 \leq 3r + 10 \leq 4$

$-14 \leq 3r \leq -6$

$-\dfrac{14}{3} \leq r \leq -2 \quad \left[-\dfrac{14}{3}, -2\right]$

17) $|7 - 6p| \leq 3$

$-3 \leq 7 - 6p \leq 3$

$-10 \leq -6p \leq -4$

$\dfrac{-10}{-6} \geq p \geq \dfrac{-4}{-6}$

$\dfrac{5}{3} \geq p \geq \dfrac{2}{3} \quad \left[\dfrac{2}{3}, \dfrac{5}{3}\right]$

19) $|8c - 3| + 15 < 20$

$|8c - 3| < 5$

$-5 < 8c - 3 < 5$

$-2 < 8c < 8$

$-\dfrac{1}{4} < c < 1$

21) $|\dfrac{3}{2}h + 6| - 2 \leq 10$

$|\dfrac{3}{2}h + 6| \leq 12$

$-12 \leq \dfrac{3}{2}h + 6 \leq 12$

$-18 \leq \dfrac{3}{2}h \leq 6$

$-12 \leq h \leq 4$

23) $|t| \geq 7$

$t \geq 7$ or $t \leq -7$

$(-\infty, -7] \cup [7, \infty)$

25) $|5a| > 2$

$5a > 2$ or $5a < -2$

$a > \dfrac{2}{5}$ $\qquad a < -\dfrac{2}{5}$

$\left(-\infty, -\dfrac{2}{5}\right) \cup \left(\dfrac{2}{5}, \infty\right)$

27) $|d + 10| \geq 4$

$d + 10 \geq 4$ or $d + 10 \leq -4$

$d \geq -6$ $\qquad d \leq -14$

$(-\infty, -14] \cup [-6, \infty)$

29) $|4v - 3| \geq 9$

$4v - 3 \geq 9$ or $4v - 3 \leq -9$

$4v \geq 12$ $\qquad 4v \leq -6$

$v \geq 3$ $\qquad v \leq -\dfrac{3}{2}$

$\left(-\infty, -\dfrac{3}{2}\right] \cup [3, \infty)$

31) $|17 - 6x| > 5$

$17 - 6x > 5$ or $17 - 6x < -5$

$-6x > -12$ $\qquad -6x < -22$

$x < \dfrac{-12}{-6}$ $\qquad x > \dfrac{-22}{-6}$

$x < 2$ $\qquad x > \dfrac{11}{3}$

$(-\infty, 2) \cup \left(\dfrac{11}{3}, \infty\right)$

33) $|2m - 1| + 4 > 5$

$|2m - 1| > 1$

$2m - 1 > 1$ or $2m - 1 < -1$

$2m > 2$ $\quad$ or $2m < 0$

$m > 1$ or $m < 0$

$(-\infty, 0) \cup (1, \infty)$

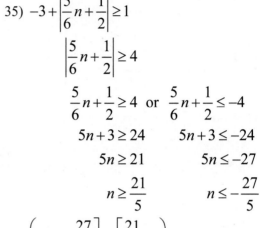

35) $-3 + \left|\dfrac{5}{6}n + \dfrac{1}{2}\right| \geq 1$

$\left|\dfrac{5}{6}n + \dfrac{1}{2}\right| \geq 4$

$\dfrac{5}{6}n + \dfrac{1}{2} \geq 4$ or $\dfrac{5}{6}n + \dfrac{1}{2} \leq -4$

$5n + 3 \geq 24$ $\qquad 5n + 3 \leq -24$

$5n \geq 21$ $\qquad 5n \leq -27$

$n \geq \dfrac{21}{5}$ $\qquad n \leq -\dfrac{27}{5}$

$\left(-\infty, -\dfrac{27}{5}\right] \cup \left[\dfrac{21}{5}, \infty\right)$

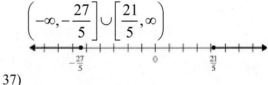

37)

$|5q + 11| < 0$

The absolute value of a quantity is always 0 or positive; it cannot be less than 0. $\varnothing$

39) $|2x + 7| \leq -12$

The absolute value of a quantity is always 0 or positive; it cannot be less than 0. $\varnothing$

41) $|8k + 5| \geq 0$ $(-\infty, \infty)$

43) $|z - 3| \geq -5$ $(-\infty, \infty)$

292

45) The absolute value of a quantity is always 0 or positive; it cannot be less than 0.

47) The absolute value of a quantity is always 0 or positive; so for any real number, x, the quantity $|2x+1|$ will be greater than -3.

49) $|2v+9|>3$

$2v+9>3$ or $2v+9<-3$

$2v>-6 \qquad 2v<-12$

$v>-3 \qquad v<-6$

$(-\infty,-6)\cup(-3,\infty)$

51) $3=|4t+5|$

$3=4t+5$ or $-3=4t+5$

$-2=4t \qquad -8=4t$

$-\dfrac{1}{2}=t \qquad -2=t$

$\left\{-2,-\dfrac{1}{2}\right\}$

53) $9\le|7-8q|$

$9\le7-8q$ or $-9\ge7-8q$

$2\le-8q \qquad -16\ge-8q$

$\dfrac{2}{-8}\ge q \qquad \dfrac{-16}{-8}\le q$

$-\dfrac{1}{4}\ge q \qquad 2\le q$

$\left(-\infty,-\dfrac{1}{4}\right]\cup[2,\infty)$

55) $2(x-8)+10<4x$

$2x-16+10<4x$

$-6<2x$

$-3<x \quad (-3,\infty)$

57) $|8-r|\le5$

$-5\le8-r\le5$

$-13\le-r\le-3$

$13\ge r\ge3 \quad [3,13]$

59) $\varnothing$, the absolute value of a quantity cannot be less than or equal to a negative number.

61) $\left|\dfrac{4}{3}x+1\right|=\left|\dfrac{5}{3}x+8\right|$

$\dfrac{4}{3}x+1=\dfrac{5}{3}x+8$ or $\dfrac{4}{3}x+1=-\dfrac{5}{3}x-8$

$4x+3=5x+24 \qquad 4x+3=-5x-24$

$-21=x \qquad\qquad 9x=-27$

$\qquad\qquad\qquad x=-3$

$\{-21,-3\}$

63) $|3m-8|-11>-3$

$|3m-8|>8$

$3m-8>8$ or $3m-8<-8$

$3m>16 \qquad 3m<0$

$m>\dfrac{16}{3} \qquad m<0$

$(-\infty,0)\cup\left(\dfrac{16}{3},\infty\right)$

65) $|4-9t|+2=1$

$|4-9t|=-1$

$\varnothing$, the absolute value of a quantity cannot be negative.

67) $-\dfrac{3}{5}\ge\dfrac{5}{2}a-\dfrac{1}{2}$

$-6\ge25a-5$

$-1\ge25a$

$-\dfrac{1}{25}\ge a \qquad \left(-\infty,-\dfrac{1}{25}\right]$

69) $|6k+17| > -4$ $(-\infty, \infty)$

71) $5 \ge |c+8| - 2$

$7 \ge |c+8|$

$7 \ge c+8 \ge -7$

$-1 \ge c \ge -15$ $[-15, -1]$

73) $|5h-8| > 7$

$5h-8 > 7$ or $5h-8 < -7$

$5h > 15$ $5h < 1$

$h > 3$ $h < \dfrac{1}{5}$

$\left(-\infty, \dfrac{1}{5}\right) \cup (3, \infty)$

75) $\left|\dfrac{1}{2}d - 4\right| + 7 = 13$

$\left|\dfrac{1}{2}d - 4\right| = 6$

$\dfrac{1}{2}d - 4 = 6$ or $\dfrac{1}{2}d - 4 = -6$

$\dfrac{1}{2}d = 10$ $\dfrac{1}{2}d = -2$

$d = 20$ $d = -4$

$\{-4, 20\}$

77) $|5j+3| + 1 \le 9$

$|5j+3| \le 8$

$-8 \le 5j+3 \le 8$

$-11 \le 5j+3 \le 5$

$-\dfrac{11}{5} \le j \le 1$ $\left[-\dfrac{11}{5}, 1\right]$

79) $|a-128| \le 0.75$

$127.25 \le a \le 128.75$

There is between 127.25 oz and 128.75 oz of milk in the container.

81) $|b-38| \le 5$

$33 \le b \le 43$

He will spend between \$33 and \$43 on his daughter's gift.

Section 9.3: Exercises

1)   Answers may vary.

3)   Answers may vary.

5)   Answers may vary.

7)   dotted

9)

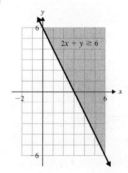

11)

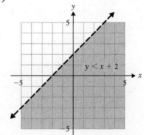

13)

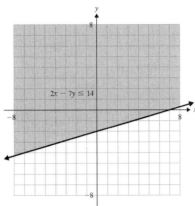

$2x - 7y \leq 14$

23)

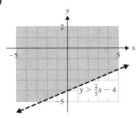

$y > \frac{2}{5}x - 4$

25)

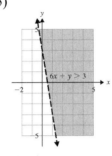

$6x + y > 3$

15)

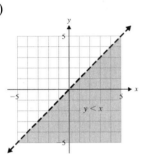

$y < x$

17)

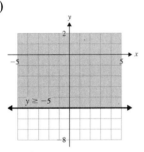

$y \geq -5$

27)

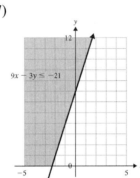

$9x - 3y \leq -21$

19) below

21)

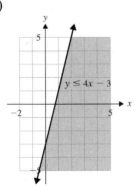

$y \leq 4x - 3$

29)

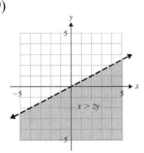

$x > 2y$

31) Answers may vary

33)

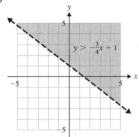

$y > -\frac{3}{4}x + 1$

35)

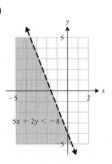

$5x + 2y < -8$

37)

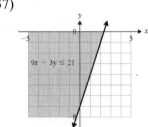

$9x - 3y \leq 21$

39)

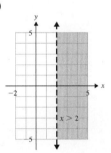

$x > 2$

41)

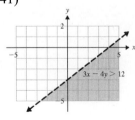

$3x - 4y > 12$

43) No; $(3,5)$ satisfies $x - y \geq -6$ but not $2x + y < 7$. Since the inequality contains *and*, it must satisfy *both* inequalities.

45)

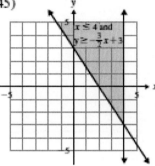

$x \leq 4$ and $y \geq -\frac{3}{2}x + 3$

47)

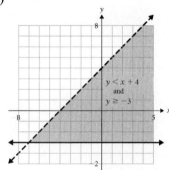

$y < x + 4$ and $y \geq -3$

49)

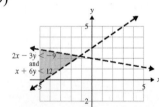

$2x - 3y < -9$ and $x + 6y < 12$

51)

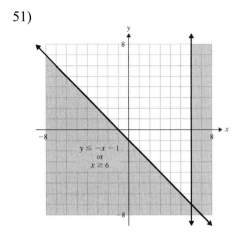

$y \leq -x - 1$
or
$x \geq 6$

53)

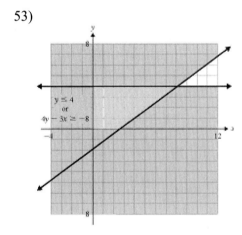

$y \leq 4$
or
$4y - 3x \geq -8$

55)

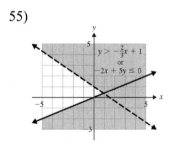

$y > -\frac{2}{3}x + 1$
or
$-2x + 5y \leq 0$

57)

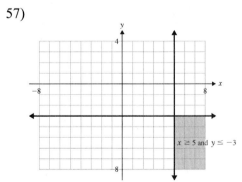

$x \geq 5$ and $y \leq -3$

59)

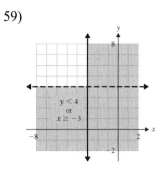

$y < 4$
or
$x \geq -3$

61)

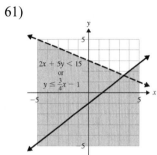

$2x + 5y < 15$
or
$y \leq \frac{3}{4}x - 1$

63)

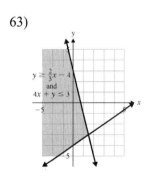

$y \geq \frac{2}{3}x - 4$
and
$4x + y \leq 3$

65) a) $x \geq 0$

$y \geq 0$

$x + y \leq 15$

b)

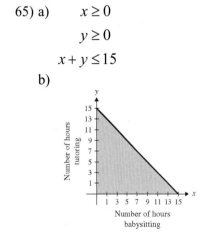

c-d)   Answers may vary.

67) a) $150 \le p \le 250$ and $100 \le r \le 200$ and
$p + r \ge 300$

b)

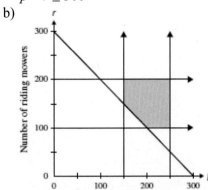

c) It represents the production of 175 push
mowers and 110 riding mowers per day.
This does not meet the level of production
needed because it is not a total of at least
300 mowers per day and is not in the
feasible region.
d) Answers may vary. e) Answers may vary.

Section 9.4: Exercises

1) $\begin{bmatrix} 1 & -7 & | & 15 \\ 4 & 3 & | & -1 \end{bmatrix}$

3) $\begin{bmatrix} 1 & 6 & -1 & | & -2 \\ 3 & 1 & 4 & | & 7 \\ -1 & -2 & 3 & | & 8 \end{bmatrix}$

5) $3x + 10y = -4$
   $x - 2y = 5$

7) $x - 6y = 8$
   $y = -2$

9) $x - 3y + 2z = 7$
   $4x - y + 3z = 0$
   $-2x + 2y - 3z = -9$

11) $x + 5y + 2z = 14$
    $y - 8z = 2$
    $z = -3$

298

13) $\begin{bmatrix} 1 & 4 & | & -1 \\ 3 & 5 & | & 4 \end{bmatrix} \xrightarrow{-3R_1+R_2 \to R_2}$
$\quad y=-1 \qquad x+4y=-1$
$\qquad\qquad\qquad x+4(-1)=-1$
$\begin{bmatrix} 1 & 4 & | & -1 \\ 0 & -7 & | & 7 \end{bmatrix} \xrightarrow{-\frac{1}{7}R_2 \to R_2} \begin{bmatrix} 1 & 4 & | & -1 \\ 0 & 1 & | & -1 \end{bmatrix}$
$\qquad\qquad\qquad x-4=-1$
$\qquad\qquad\qquad x=3 \quad (3,-1)$

15) $\begin{bmatrix} 1 & -3 & | & 9 \\ -6 & 5 & | & 11 \end{bmatrix} \xrightarrow{6R_1+R_2 \to R_2}$
$\quad y=-5 \qquad x-3y=9$
$\qquad\qquad\qquad x-3(-5)=9$
$\begin{bmatrix} 1 & -3 & | & 9 \\ 0 & -13 & | & 65 \end{bmatrix} \xrightarrow{-\frac{1}{13}R_2 \to R_2} \begin{bmatrix} 1 & -3 & | & 9 \\ 0 & 1 & | & -5 \end{bmatrix}$
$\qquad\qquad\qquad x+15=9$
$\qquad\qquad\qquad x=-6 \quad (-6,-5)$

17) $\begin{bmatrix} 4 & -3 & | & 6 \\ 1 & 1 & | & -2 \end{bmatrix} \xrightarrow{R_1 \leftrightarrow R_2} \begin{bmatrix} 1 & 1 & | & -2 \\ 4 & -3 & | & 6 \end{bmatrix} \xrightarrow{-4R_1+R_2 \to R_2} \begin{bmatrix} 1 & 1 & | & -2 \\ 0 & -7 & | & 14 \end{bmatrix} \xrightarrow{-\frac{1}{7}R_2 \to R_2} \begin{bmatrix} 1 & 1 & | & -2 \\ 0 & 1 & | & -2 \end{bmatrix}$

$y=-2 \qquad x+y=-2$
$\qquad\qquad x+(-2)=-2$
$\qquad\qquad x=0 \quad (0,-2)$

19) $\begin{bmatrix} 1 & 1 & -1 & | & -5 \\ 4 & 5 & -2 & | & 0 \\ 8 & -3 & 2 & | & -4 \end{bmatrix} \xrightarrow[-8R_1+R_3 \to R_3]{-4R_1+R_2 \to R_2} \begin{bmatrix} 1 & 1 & -1 & | & -5 \\ 0 & 1 & 2 & | & 20 \\ 0 & -11 & 10 & | & 36 \end{bmatrix}$

$\xrightarrow{11R_2+R_3 \to R_3} \begin{bmatrix} 1 & 1 & -1 & | & -5 \\ 0 & 1 & 2 & | & 20 \\ 0 & 0 & 32 & | & 256 \end{bmatrix} \xrightarrow{\frac{1}{32}R_3 \to R_3} \begin{bmatrix} 1 & 1 & -1 & | & -5 \\ 0 & 1 & 2 & | & 20 \\ 0 & 0 & 1 & | & 8 \end{bmatrix}$

$x+y-z=-5 \qquad y+2z=20 \qquad x+y-z=-5$
$y+2z=20 \qquad y+2(8)=20 \qquad x+4-8=-5$
$z=8 \qquad y+16=20 \qquad x-4=-5$
$\qquad\qquad y=4 \qquad x=-1 \quad (-1,4,8)$

21) $\begin{bmatrix} 1 & -3 & 2 & | & -1 \\ 3 & -8 & 4 & | & 6 \\ -2 & -3 & -6 & | & 1 \end{bmatrix} \xrightarrow[2R_1+R_3 \to R_3]{-3R_1+R_2 \to R_2} \begin{bmatrix} 1 & -3 & 2 & | & -1 \\ 0 & 1 & -2 & | & 9 \\ 0 & -9 & -2 & | & -1 \end{bmatrix} \xrightarrow{9R_2+R_3 \to R_3}$

$\begin{bmatrix} 1 & -3 & 2 & | & -1 \\ 0 & 1 & -2 & | & 9 \\ 0 & 0 & -20 & | & 80 \end{bmatrix} \xrightarrow{-\frac{1}{20}R_3 \to R_3} \begin{bmatrix} 1 & -3 & 2 & | & -1 \\ 0 & 1 & -2 & | & 9 \\ 0 & 0 & 1 & | & -4 \end{bmatrix}$

$$x - 3y + 2z = -1 \qquad y - 2z = 9 \qquad x - 3y + 2z = -1$$
$$y - 2z = 9 \qquad y - 2(-4) = 9 \quad x - 3(1) + 2(-4) = -1$$
$$z = -4 \qquad\quad y + 8 = 9 \qquad\quad x - 3 - 8 = -1$$
$$y = 1 \qquad\qquad x - 11 = -1$$
$$x = 10 \quad (10, 1, -4)$$

23) $\begin{bmatrix} -4 & -3 & 1 & | & 5 \\ 1 & 1 & -1 & | & -7 \\ 6 & 4 & 1 & | & 12 \end{bmatrix} \xrightarrow{R_1 \leftrightarrow R_2} \begin{bmatrix} 1 & 1 & -1 & | & -7 \\ -4 & -3 & 1 & | & 5 \\ 6 & 4 & 1 & | & 12 \end{bmatrix}$

$\xrightarrow[-6R_1 + R_3 \to R_3]{4R_1 + R_2 \to R_2} \begin{bmatrix} 1 & 1 & -1 & | & -7 \\ 0 & 1 & -3 & | & -23 \\ 0 & -2 & 7 & | & 54 \end{bmatrix} \xrightarrow{2R_2 + R_3 \to R_3} \begin{bmatrix} 1 & 1 & -1 & | & -7 \\ 0 & 1 & -3 & | & -23 \\ 0 & 0 & 1 & | & 8 \end{bmatrix}$

$$x + y - z = -7 \qquad y - 3z = -23 \qquad x + y - z = -7$$
$$y - 3z = -23 \qquad y - 3(8) = -23 \qquad x + 1 - 8 = -7$$
$$z = 8 \qquad\quad y - 24 = -23 \qquad\quad x - 7 = -7$$
$$y = 1 \qquad\qquad x = 0 \quad (0, 1, 8)$$

25) $\begin{bmatrix} 1 & -3 & 1 & | & -4 \\ 4 & 5 & -1 & | & 0 \\ 2 & -6 & 2 & | & 1 \end{bmatrix} \xrightarrow[-2R_1 + R_3 \to R_3]{-4R_1 + R_2 \to R_2} \begin{bmatrix} 1 & -3 & 1 & | & -4 \\ 0 & 17 & -5 & | & 16 \\ 0 & 0 & 0 & | & 9 \end{bmatrix}$

$\varnothing$

27) $\begin{bmatrix} 1 & 1 & 3 & 1 & | & -1 \\ -1 & 0 & 1 & -1 & | & 7 \\ 2 & 3 & 9 & -2 & | & 7 \\ 1 & -2 & 1 & 3 & | & -11 \end{bmatrix} \xrightarrow[\substack{-2R_1 + R_3 \to R_3 \\ -R_1 + R_4 \to R_4}]{R_1 + R_2 \to R_2} \begin{bmatrix} 1 & 1 & 3 & 1 & | & -1 \\ 0 & 1 & 4 & 0 & | & 6 \\ 0 & 1 & 3 & -4 & | & 9 \\ 1 & -3 & -2 & 2 & | & -10 \end{bmatrix}$

$\xrightarrow[3R_2 + R_4 \to R_4]{-R_2 + R_3 \to R_3} \begin{bmatrix} 1 & 1 & 3 & 1 & | & -1 \\ 0 & 1 & 4 & 0 & | & 6 \\ 0 & 0 & -1 & -4 & | & 3 \\ 0 & 0 & 10 & 2 & | & 8 \end{bmatrix} \xrightarrow{-R_3 \to R_3} \begin{bmatrix} 1 & 1 & 3 & 1 & | & -1 \\ 0 & 1 & 4 & 0 & | & 6 \\ 0 & 0 & 1 & 4 & | & -3 \\ 0 & 0 & 10 & 2 & | & 8 \end{bmatrix}$

$\xrightarrow{-10R_3 + R_4 \to R_4} \begin{bmatrix} 1 & 1 & 3 & 1 & | & -1 \\ 0 & 1 & 4 & 0 & | & 6 \\ 0 & 0 & -1 & -4 & | & -3 \\ 0 & 0 & 0 & 38 & | & 8 \end{bmatrix} \xrightarrow{-\frac{1}{38}R_4 \to R_4} \begin{bmatrix} 1 & 1 & 3 & -1 & | & -1 \\ 0 & 1 & 4 & 0 & | & 6 \\ 0 & 0 & 1 & 4 & | & -3 \\ 0 & 0 & 0 & 1 & | & -1 \end{bmatrix}$

$$a+b+3c+d=-1 \qquad c+4d=-3 \qquad b+4c=6 \qquad a+b+3c+d=-1$$

$$b+4c=6 \qquad c+4(-1)=-3 \qquad b+4(-1)=6 \qquad a+2+3(1)+(-1)=-1$$

$$c+4d=-3 \qquad c=1 \qquad b=2 \qquad a+4=-1$$

$$d=-1 \qquad\qquad\qquad\qquad\qquad a=-5$$

$$(-5,2,1,-1)$$

29) $\begin{bmatrix} 1 & -3 & 2 & -1 & | & -2 \\ -3 & 8 & -5 & 1 & | & 2 \\ 2 & -1 & 1 & 3 & | & 7 \\ 1 & -2 & 1 & 2 & | & 3 \end{bmatrix} \xrightarrow[\substack{3R_1+R_2\to R_2 \\ -2R_1+R_3\to R_3 \\ -R_1+R_4\to R_4}]{} \begin{bmatrix} 1 & -3 & 2 & -1 & | & -2 \\ 0 & -1 & 1 & -2 & | & -4 \\ 0 & 5 & -3 & 5 & | & 11 \\ 0 & 1 & -1 & 3 & | & 5 \end{bmatrix}$

$\xrightarrow[-R_2\to R_2]{} \begin{bmatrix} 1 & -3 & 2 & -1 & | & -2 \\ 0 & 1 & -1 & 2 & | & 4 \\ 0 & 5 & -3 & 5 & | & 11 \\ 0 & 1 & -1 & 3 & | & 5 \end{bmatrix} \xrightarrow[\substack{-5R_2+R_3\to R_3 \\ -R_2+R_4\to R_4}]{} \begin{bmatrix} 1 & -3 & 2 & -1 & | & -2 \\ 0 & 1 & -1 & 2 & | & -8 \\ 0 & 0 & 2 & -5 & | & -28 \\ 0 & 0 & 0 & 1 & | & 12 \end{bmatrix} \xrightarrow[\frac{1}{2}R_3+R_4\to R_4]{} \begin{bmatrix} 1 & -3 & 2 & -1 & | & -2 \\ 0 & 1 & -1 & 2 & | & 4 \\ 0 & 0 & 1 & -\frac{5}{2} & | & -\frac{9}{2} \\ 0 & 0 & 0 & 1 & | & 1 \end{bmatrix}$

$$a-3b+2c-d=-2 \qquad c-\frac{5}{2}d=-\frac{9}{2} \qquad b-c+2d=4 \qquad a-3b+2c-d=-2$$

$$b-4c+2d=4 \qquad c-\frac{5}{2}(1)=-\frac{9}{2} \qquad b-(-2)+2(1)=4 \qquad a-3(0)-2(-2)-(1)=-2$$

$$c-\frac{5}{2}d=-\frac{9}{2} \qquad c=-2 \qquad b=0 \qquad a-5=-2$$

$$d=1 \qquad\qquad\qquad\qquad\qquad\qquad a=3$$

$$(3,0,-2,1)$$

Chapter 9 Review

1) $|m|=9$

$\quad m=9$ or $m=-9 \quad \{-9,9\}$

5) $|8p+11|-7=-3$

$\quad |8p+11|=4$

$\quad 8p+11=4$ or $8p+11=-4$

$\qquad 8p=-7 \qquad\qquad 8p=-15$

$\qquad p=-\dfrac{7}{8}$ or $\qquad p=-\dfrac{15}{8}$

3) $|7t+3|=4$

$\quad 7t+3=4$ or $7t+3=-4$

$\quad 7t=1$ or $7t=-7$

$\quad t=\dfrac{1}{7}$ or $t=-1 \qquad \left\{-1,\dfrac{1}{7}\right\}$

$\left\{-\dfrac{15}{8},-\dfrac{7}{8}\right\}$

7) $\left|4-\dfrac{5}{3}x\right|=\dfrac{1}{3}$

$4-\dfrac{5}{3}x=\dfrac{1}{3}$ or $4-\dfrac{5}{3}x=-\dfrac{1}{3}$

$12-5x=1 \qquad 12-5x=-1$

$-5x=-11 \qquad -5x=-13$

$x=\dfrac{11}{5}$ or $x=\dfrac{13}{5}$

$\left\{\dfrac{11}{5},\dfrac{13}{5}\right\}$

9) $|7r-6|=|8r+2|$

$7r-6=8r+2$ or $7r-6=-8r-2$

$15r=4$

$-8=r$ or $r=\dfrac{4}{15}$

$\left\{-8,\dfrac{4}{15}\right\}$

11) $\varnothing$, the absolute value of a quantity cannot be negative.

13) $|9d+4|=0$

$9d+4=0$

$9d=-4$

$d=-\dfrac{4}{9} \qquad \left\{-\dfrac{4}{9}\right\}$

15) $|a|=4$, Answers may vary

17) $|c|\le3$

$-3\le c\le3 \quad [-3,3]$

19) $|4t|>8$

$4t>8$ or $4t<-8$

$t>2 \qquad t<-2$

$(-\infty,-2)\cup(2,\infty)$

21) $|12r+5|\ge7$

$12r+5\ge7$ or $12r+5\le-7$

$12r\ge2 \qquad\qquad 12r\le-12$

$r\ge\dfrac{1}{6} \qquad\qquad r\le-1$

$(-\infty,-1]\cup\left[\dfrac{1}{6},\infty\right)$

23) $|4-a|<9$

$-9<4-a<9$

$-13<-a<5$

$13>a>-5 \quad (-5,13)$

25) $|4c+9|-8\le-2$

$|4c+9|\le6$

$-6\le4c+9\le6$

$-15\le4c\le-3$

$-\dfrac{15}{4}\le c\le-\dfrac{3}{4} \quad \left[-\dfrac{15}{4},-\dfrac{3}{4}\right]$

27) $|5y+12|-15 \geq -8$

$\quad\quad |5y+12| \geq 7$

$\quad\quad 5y+12 \geq 7$ or $5y+12 \leq -7$

$\quad\quad\quad 5y \geq -5 \quad\quad 5y \leq -19$

$\quad\quad\quad\quad y \geq -1 \quad\quad y \leq -\dfrac{19}{5}$

$\left(-\infty, -\dfrac{19}{5}\right] \cup [-1, \infty)$

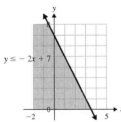

29) $(-\infty, \infty)$

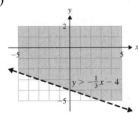

31) $|12s+1| \leq 0$

$\quad 12s+1 = 0$

$\quad 12s = -1$

$\quad s = -\dfrac{1}{12} \quad \left\{-\dfrac{1}{12}\right\}$

33)

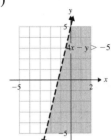

$y \leq -2x+7$

35)

$y > -\dfrac{1}{3}x - 4$

37)

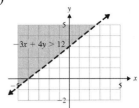

$-3x + 4y > 12$

39)

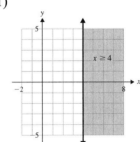

$4x - y > -5$

41)

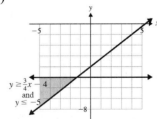

$x \geq 4$

43)

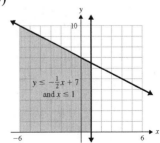

$y \geq \dfrac{3}{4}x - 4$
and
$y \leq -5$

45)

$y \leq -\dfrac{1}{2}x + 7$
and $x \leq 1$

47)

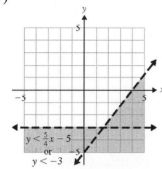

51)

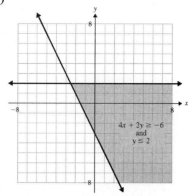

49)

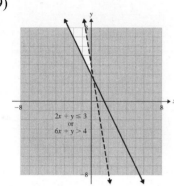

53) $\begin{bmatrix} 1 & -1 & | & -11 \\ 2 & 9 & | & 0 \end{bmatrix} \xrightarrow{-2R_1+R_2 \to R_2} \begin{bmatrix} 1 & -1 & | & -11 \\ 0 & 11 & | & 22 \end{bmatrix} \xrightarrow{\frac{1}{11}R_2 \to R_2} \begin{bmatrix} 1 & -1 & | & -11 \\ 0 & 1 & | & 2 \end{bmatrix}$

$y = 2 \quad x - y = -11$

$\qquad\qquad x - 2 = -11$

$\qquad\qquad\quad x = -9 \quad (-9, 2)$

55) $\begin{bmatrix} 5 & 3 & | & 5 \\ -1 & 8 & | & -1 \end{bmatrix} \xrightarrow{R_1 \leftrightarrow R_2} \begin{bmatrix} -1 & 8 & | & -1 \\ 5 & 3 & | & 5 \end{bmatrix} \xrightarrow{-R_1 \to R_1} \begin{bmatrix} 1 & -8 & | & 1 \\ 5 & 3 & | & 5 \end{bmatrix}$

$\xrightarrow{-5R_1+R_2 \to R_2} \begin{bmatrix} 1 & -8 & | & 1 \\ 0 & 43 & | & 0 \end{bmatrix} \xrightarrow{\frac{1}{43}R_2 \to R_2} \begin{bmatrix} 1 & -8 & | & 1 \\ 0 & 1 & | & 0 \end{bmatrix}$

$y = 0 \quad x - 8y = 1$

$\qquad\quad x - 8(0) = 1$

$\qquad\qquad\quad x = 1 \quad (1, 0)$

57) $\begin{bmatrix} 1 & -3 & -3 & | & -7 \\ 2 & -5 & -3 & | & 2 \\ -3 & 5 & 4 & | & -1 \end{bmatrix} \xrightarrow[3R_1+R_3 \to R_3]{-2R_1+R_2 \to R_2} \begin{bmatrix} 1 & -3 & -3 & | & -7 \\ 0 & 1 & 3 & | & 16 \\ 0 & -4 & -5 & | & -22 \end{bmatrix} \xrightarrow{4R_2+R_3 \to R_3}$

$\begin{bmatrix} 1 & -3 & -3 & | & -7 \\ 0 & 1 & 3 & | & 16 \\ 0 & 0 & 7 & | & 42 \end{bmatrix} \xrightarrow{\frac{1}{7}R_3 \to R_3} \begin{bmatrix} 1 & -3 & -3 & | & -7 \\ 0 & 1 & 3 & | & 16 \\ 0 & 0 & 1 & | & 6 \end{bmatrix}$

$$x-3y-3z=-7 \qquad y+3z=16 \qquad x-3y-3z=-7$$
$$y+3z=16 \qquad y+3(6)=16 \qquad x-3(-2)-3(6)=-7$$
$$z=6 \qquad y+18=16 \qquad x+6-18=-7$$
$$y=-2 \qquad x-12=-7$$
$$x=5 \qquad (5,-2,6)$$

Chapter 9 Test

1) $|4y-9|=11$

$4y-9=11$ or $4y-9=-11$

$4y=20 \qquad 4y=-2$

$y=5$ or $y=-\dfrac{1}{2}$

$\left\{-\dfrac{1}{2}, 5\right\}$

3) $|3k+5|=|k-11|$

$3k+5=k-11$ or $3k+5=-k+11$

$2k=-16 \qquad 4k=6$

$k=-8$ or $k=\dfrac{3}{2}$

$\left\{-8, \dfrac{3}{2}\right\}$

5) $|x|=8$

7) $|c|>4$

$c>4$ or $c<-4$ $(-\infty,-4) \cup (4,\infty)$

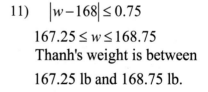

9) $|4m+9|-8 \ge 5$

$|4m+9| \ge 13$

$4m+9 \ge 13$ or $4m+9 \le -13$

$4m \ge 4 \qquad 4m \le -22$

$m \ge 1 \qquad m \le -\dfrac{11}{2}$

$\left(-\infty, -\dfrac{11}{2}\right] \cup [1,\infty)$

11) $|w-168| \le 0.75$

$167.25 \le w \le 168.75$

Thanh's weight is between
167.25 lb and 168.75 lb.

13)

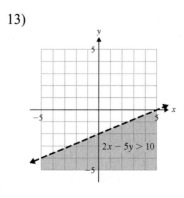

15)

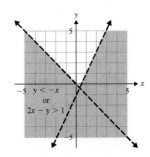

$$y < -x$$
$$\text{or}$$
$$2x - y > 1$$

17) $\begin{bmatrix} -3 & 5 & 8 & | & 0 \\ 1 & -3 & 4 & | & 8 \\ 2 & -4 & -3 & | & 3 \end{bmatrix} \xrightarrow{R_1 \leftrightarrow R_2} \begin{bmatrix} 1 & -3 & 4 & | & 8 \\ -3 & 5 & 8 & | & 0 \\ 2 & -4 & -3 & | & 3 \end{bmatrix} \xrightarrow[-2R_1+R_3 \to R_3]{3R_1+R_2 \to R_2} \begin{bmatrix} 1 & -3 & 4 & | & 8 \\ 0 & -4 & 20 & | & 24 \\ 0 & 2 & -11 & | & -13 \end{bmatrix}$

$\xrightarrow{-\frac{1}{4}R_2 \to R_2} \begin{bmatrix} 1 & -3 & 4 & | & 8 \\ 0 & 1 & -5 & | & -6 \\ 0 & 2 & -11 & | & -13 \end{bmatrix} \xrightarrow{-2R_2+R_3 \to R_3} \begin{bmatrix} 1 & -3 & 4 & | & 8 \\ 0 & 1 & -5 & | & -6 \\ 0 & 0 & -1 & | & -1 \end{bmatrix} \xrightarrow{-R_3 \to R_3} \begin{bmatrix} 1 & -3 & 4 & | & 8 \\ 0 & 1 & -5 & | & -6 \\ 0 & 0 & 1 & | & 1 \end{bmatrix}$

$$x - 3y + 4z = 8 \qquad y - 5z = -6 \qquad x - 3y + 4z = 8$$
$$y - 5z = -6 \qquad y - 5(1) = -6 \qquad x - 3(-1) + 4(1) = 8$$
$$z = 1 \qquad y - 5 = -6 \qquad x + 3 + 4 = 8$$
$$y = -1 \qquad x + 7 = 8$$
$$x = 1 \quad (1, -1, 1)$$

Cumulative Review Chapters 1-9

1) $5 \times 6 - 36 \div 3^2 = 5 \times 6 - 36 \div 9$
$$= 30 - 4 = 26$$

3) $3^4 = 3 \cdot 3 \cdot 3 \cdot 3 = 81$

5) $\left(\dfrac{1}{8}\right)^2 = \dfrac{1^2}{8^2} = \dfrac{1 \cdot 1}{8 \cdot 8} = \dfrac{1}{64}$

7) $0.00000914 = 9.14 \times 10^{-6}$

9) $3 - \dfrac{2}{7}n \geq 9$
$$-\dfrac{2}{7}n \geq 6$$
$$n \leq -\dfrac{7}{2} \cdot 6$$
$$n \leq -21 \quad (-\infty, -21]$$

11) $y = mx + b$
$$2 = \left(\dfrac{1}{3}\right)7 + b$$
$$2 = \dfrac{7}{3} + b$$
$$-\dfrac{1}{3} = b \qquad y = \dfrac{1}{3}x - \dfrac{1}{3}$$

13) $-4p^2\left(3p^2-7p-1\right)$

$\quad = -12p^4+28p^3+4p^2$

15) $(t+8)^2 = t^2+16t+64$

17) $9m^2-121 = (3m+11)(3m-11)$

19) $a^2+6a+9=0$

$\quad\quad (a+3)^2 = 0$

$\quad\quad\quad\quad a = -3 \quad\quad \{-3\}$

21) $\dfrac{1}{r^2-25} - \dfrac{r+3}{2r+10}$

$\quad = \dfrac{1}{(r+5)(r-5)} - \dfrac{r+3}{2(r+5)}$

$\quad = \dfrac{2-(r-5)(r+3)}{2(r+5)(r-5)}$

$\quad = \dfrac{2-\left(r^2-2r-15\right)}{2(r+5)(r-5)}$

$\quad = \dfrac{-r^2+2r+17}{2(r+5)(r-5)}$

23) $\left|\dfrac{1}{4}q-7\right|-8 = -5$

$\quad\quad \left|\dfrac{1}{4}q-7\right| = 3$

$\quad \dfrac{1}{4}q-7 = 3 \ \text{ or } \ \dfrac{1}{4}q-7 = -3$

$\quad\quad\quad \dfrac{1}{4}q = 10 \quad\quad\quad \dfrac{1}{4}q = 4$

$\quad\quad q = 40 \ \text{ or } \ q = 16 \quad \{16,40\}$

25)

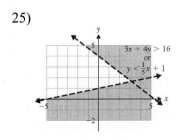

Section 10.1: Exercises

1) False; the $\sqrt{}$ symbol means to find only the positive square root of 121.
$$\sqrt{121} = 11$$

3) False; the square root of a negative number is not a real number.

5) 7 and -7

7) 1 and -1

9) 30 and -30

11) $\dfrac{2}{3}$ and $-\dfrac{2}{3}$

13) $\dfrac{1}{9}$ and $-\dfrac{1}{9}$

15) 0.5 and -0.5

17) $\sqrt{49} = 7$

19) $\sqrt{1} = 1$

21) $\sqrt{169} = 13$

23) $\sqrt{-4}$ is not real

25) $\sqrt{\dfrac{81}{25}} = \dfrac{9}{5}$

27) $-\sqrt{36} = -6$

29) $-\sqrt{\dfrac{1}{121}} = -\dfrac{1}{11}$

31) $\sqrt{0.04} = 0.2$

33) Since 11 is between 9 and 16,
$$\sqrt{9} < \sqrt{11} < \sqrt{16}$$
$$3 < \sqrt{11} < 4$$
$$\sqrt{11} \approx 3.3$$

35) Since 46 is between 36 and 49,
$$\sqrt{36} < \sqrt{46} < \sqrt{49}$$
$$6 < \sqrt{46} < 7$$
$$\sqrt{46} \approx 6.8$$

37) Since 17 is between 16 and 25,
$$\sqrt{16} < \sqrt{17} < \sqrt{25}$$
$$4 < \sqrt{17} < 5$$
$$\sqrt{17} \approx 4.1$$

39) Since 5 is between 4 and 9,
$$\sqrt{4} < \sqrt{5} < \sqrt{9}$$
$$2 < \sqrt{5} < 3$$
$$\sqrt{5} \approx 2.2$$

41) Since 61 is between 49 and 64,
$$\sqrt{49} < \sqrt{61} < \sqrt{64}$$
$$7 < \sqrt{61} < 8$$
$$\sqrt{61} \approx 7.8$$

43) True

45) False; the odd root of a negative number is a negative number.

47) $\sqrt[3]{64}$ is the number you cube to get 64. $\sqrt[3]{64} = 4$

49) No; the even root of a negative number is not a real number.

51) $\sqrt[3]{8} = 2$

53) $\sqrt[3]{125} = 5$

55) $\sqrt[3]{-1} = -1$

57) $\sqrt[4]{81} = 3$

59) $\sqrt[4]{-1}$ is not real

61) $-\sqrt[4]{16} = -2$

63) $\sqrt[5]{-32} = -2$

65) $-\sqrt[3]{-27} = -(-3) = 3$

67) $\sqrt[6]{-64}$ is not real

69) $\sqrt[3]{\dfrac{8}{125}} = \dfrac{2}{5}$

71) $\sqrt{60-11} = \sqrt{49} = 7$

73) $\sqrt[3]{100+25} = \sqrt[3]{125} = 5$

75) $\sqrt{1-9} = \sqrt{-8}$; not real

77) $\sqrt{5^2+12^2} = \sqrt{25+144} = \sqrt{169} = 13$

79) If a is negative and we didn't use the absolute value, the result would be negative. This is incorrect because if a is negative and n is even, then $a^n > 0$. Using absolute values ensures a positive result.

81) $\sqrt{8^2} = \sqrt{64} = 8$

83) $\sqrt{(-6)^2} = \sqrt{36} = 6$

85) $\sqrt{y^2} = |y|$

87) $\sqrt{m^2} = |m|$

89) $\sqrt[3]{5^3} = \sqrt[3]{125} = 5$

91) $\sqrt[3]{(-4)^3} = \sqrt[3]{-64} = -4$

93) $\sqrt[3]{z^3} = z$

95) $\sqrt[4]{h^4} = |h|$

97) $\sqrt[9]{p^9} = p$

99) $\sqrt{(x+7)^2} = |x+7|$

101) $\sqrt[3]{(2t-1)^3} = 2t-1$

103) $\sqrt[4]{(3n+2)^4} = |3n+2|$

105) $\sqrt[7]{(d-8)^7} = d-8$

Section 10.2: Exercises

1) The denominator of 2 becomes the index of the radical. $25^{1/2} = \sqrt[2]{25} = \sqrt{25}$

3) $9^{1/2} = \sqrt{9} = 3$

5) $1000^{1/3} = \sqrt[3]{1000} = 10$

7) $32^{1/5} = \sqrt[5]{32} = 2$

9) $-125^{1/3} = -\sqrt[3]{125} = -5$

11) $\left(\dfrac{4}{121}\right)^{1/2} = \sqrt{\dfrac{4}{121}} = \dfrac{2}{11}$

13) $\left(\dfrac{125}{64}\right)^{1/3} = \sqrt[3]{\dfrac{125}{64}} = \dfrac{5}{4}$

15) $-\left(\dfrac{36}{169}\right)^{1/2} = -\sqrt{\dfrac{36}{169}} = -\dfrac{6}{13}$

17) $(-81)^{1/4} = \sqrt[4]{-81} = $ not a real number.

$(-1)^{1/7} = \sqrt[7]{-1} = -1$

21) The denominator of 4 becomes the index of the radical. The numerator of 3 is the power to which we raise the radical expression.

$16^{3/4} = \left(\sqrt[4]{16}\right)^3$

23) $8^{4/3} = \left(8^{1/3}\right)^4 = \left(\sqrt[3]{8}\right)^4 = 2^4 = 16$

25) $64^{5/6} = \left(64^{1/6}\right)^5 = \left(\sqrt[6]{64}\right)^5 = 2^5 = 32$

27) $(-125)^{2/3} = \left((-125)^{1/3}\right)^2 = \left(\sqrt[3]{-125}\right)^2$

$= (-5)^2 = 25$

29) $-36^{3/2} = -\left(\sqrt{36}\right)^3 = -(6)^3 = -216$

31) $(-81)^{3/4} = \left((-81)^{1/4}\right)^3 = \left(\sqrt[4]{-81}\right)^3$

$= $ not a real number

33) $\left(\dfrac{16}{81}\right)^{3/4} = \left(\sqrt[4]{\dfrac{16}{81}}\right)^3 = \left(\dfrac{2}{3}\right)^3 = \dfrac{8}{27}$

35) $-\left(\dfrac{1000}{27}\right)^{2/3} = -\left(\sqrt[3]{\dfrac{1000}{27}}\right)^2$

$= -\left(\dfrac{10}{3}\right)^2 = -\dfrac{100}{9}$

37) False; the negative exponent does not make the result negative.

$81^{-1/2} = \dfrac{1}{9}$

39) $64^{-1/2} = \left(\boxed{\dfrac{1}{64}}\right)^{1/2}$ The reciprocal

of 64 is $\boxed{\dfrac{1}{64}}$

$= \sqrt{\dfrac{1}{64}}$

The denominator of the fractional
exponent is the index of the radical

$= \boxed{\dfrac{1}{8}}$ Simplify

41) $49^{-1/2} = \left(\dfrac{1}{49}\right)^{1/2} = \sqrt{\dfrac{1}{49}} = \dfrac{1}{7}$

43) $1000^{-1/3} = \left(\dfrac{1}{1000}\right)^{1/3} = \sqrt[3]{\dfrac{1}{1000}} = \dfrac{1}{10}$

45) $\left(\dfrac{1}{81}\right)^{-1/4} = (81)^{1/4} = \sqrt[4]{81} = 3$

47) $-\left(\dfrac{1}{64}\right)^{-1/3} = -(64)^{1/3} = -\sqrt[3]{64} = -4$

49) $64^{-5/6} = \left(\dfrac{1}{64}\right)^{5/6} = \left(\sqrt[6]{\dfrac{1}{64}}\right)^5$

$= \left(\dfrac{1}{2}\right)^5 = \dfrac{1}{32}$

51) $125^{-2/3} = \left(\dfrac{1}{125}\right)^{2/3} = \left(\sqrt[3]{\dfrac{1}{125}}\right)^2$

$= \left(\dfrac{1}{5}\right)^2 = \dfrac{1}{25}$

53) $\left(\dfrac{25}{4}\right)^{-3/2} = \left(\dfrac{4}{25}\right)^{3/2} = \left(\sqrt{\dfrac{4}{25}}\right)^3$

$= \left(\dfrac{2}{5}\right)^3 = \dfrac{8}{125}$

55) $\left(\dfrac{64}{125}\right)^{-2/3} = \left(\dfrac{125}{64}\right)^{2/3} = \left(\sqrt[3]{\dfrac{125}{64}}\right)^2$

$= \left(\dfrac{5}{4}\right)^2 = \dfrac{25}{16}$

57) $2^{2/3} \cdot 2^{7/3} = 2^{2/3+7/3} = 2^{9/3} = 2^3 = 8$

59) $\left(9^{1/4}\right)^2 = 9^{\frac{1}{4}\cdot 2} = 9^{1/2} = \sqrt{9} = 3$

61) $8^{7/5} \cdot 8^{-3/5} = 8^{7/5+(-3/5)} = 8^{4/5}$

63) $\dfrac{2^{23/4}}{2^{3/4}} = 2^{23/4-3/4} = 2^{20/4} = 2^5 = 32$

65) $\dfrac{4^{2/5}}{4^{6/5}\cdot 4^{3/5}} = \dfrac{4^{2/5}}{4^{6/5+3/5}} = \dfrac{4^{2/5}}{4^{9/5}}$

$= 4^{2/5-9/5} = 4^{-7/5} = \dfrac{1}{4^{7/5}}$

67) $z^{1/6}\cdot z^{5/6} = z^{1/6+5/6} = z^{6/6} = z^1 = z$

69) $\left(-9v^{5/8}\right)\left(8v^{3/4}\right) = -72v^{5/8+3/4}$

$= -72v^{5/8+6/8}$

$= -72v^{11/8}$

71) $\dfrac{a^{5/9}}{a^{4/9}} = a^{5/9-4/9} = a^{1/9}$

73) $\dfrac{20c^{-2/3}}{72c^{5/6}} = \dfrac{5}{18}c^{-2/3-5/6} = \dfrac{5}{18}c^{-4/6-5/6}$

$= \dfrac{5}{18}c^{-9/6} = \dfrac{5}{18}c^{-3/2} = \dfrac{5}{18c^{3/2}}$

75) $\left(x^{-2/9}\right)^3 = x^{-\frac{2}{9}\cdot 3} = x^{-2/3} = \dfrac{1}{x^{2/3}}$

77) $\left(z^{1/5}\right)^{2/3} = z^{\frac{1}{5}\cdot\frac{2}{3}} = z^{2/15}$

79) $\left(81u^{8/3}v^4\right)^{3/4}$

$= 81^{3/4}\cdot\left(u^{8/3}\right)^{3/4}\cdot\left(v^4\right)^{3/4}$

$= \left(\sqrt[4]{81}\right)^3\cdot u^2\cdot v^3$

$= (3)^3\cdot u^2 v^3 = 27u^2 v^3$

81) $\left(32r^{1/3}s^{4/9}\right)^{3/5}$

$= 32^{3/5}\cdot\left(r^{1/3}\right)^{3/5}\cdot\left(s^{4/9}\right)^{3/5}$

$= \left(\sqrt[5]{32}\right)^3\cdot r^{1/5}\cdot s^{4/15}$

$= (2)^3\cdot r^{1/5}s^{4/15} = 8r^{1/5}s^{4/15}$

83) $\left(\dfrac{f^{6/7}}{27g^{-5/3}}\right)^{1/3} = \dfrac{\left(f^{6/7}\right)^{1/3}}{(27)^{1/3}\left(g^{-5/3}\right)^{1/3}}$

$= \dfrac{f^{2/7}}{3g^{-5/9}} = \dfrac{f^{2/7}g^{5/9}}{3}$

85) $\left(\dfrac{x^{-5/3}}{w^{3/2}}\right)^{-6} = \left(\dfrac{w^{3/2}}{x^{-5/3}}\right)^{6} = \dfrac{\left(w^{3/2}\right)^{6}}{\left(x^{-5/3}\right)^{6}}$

$= \dfrac{w^{9}}{x^{-10}} = w^{9}x^{10}$

87) $\dfrac{y^{1/2} \cdot y^{-1/3}}{y^{5/6}} = \dfrac{y^{3/6} \cdot y^{-2/6}}{y^{5/6}} = \dfrac{y^{1/6}}{y^{5/6}}$

$= y^{1/6-5/6} = y^{-4/6}$

$= y^{-2/3} = \dfrac{1}{y^{2/3}}$

89) $\left(\dfrac{a^{4}b^{3}}{32a^{-2}b^{4}}\right)^{2/5}$

$= \left(\dfrac{1}{32}a^{4-(-2)}b^{3-4}\right)^{2/5}$

$= \left(\dfrac{1}{32}a^{6}b^{-1}\right)^{2/5}$

$= \left(\dfrac{1}{32}\right)^{2/5} \cdot \left(a^{6}\right)^{2/5} \cdot \left(b^{-1}\right)^{2/5}$

$= \left(\sqrt[5]{\dfrac{1}{32}}\right)^{2} \cdot a^{12/5} \cdot b^{-2/5}$

$= \left(\dfrac{1}{2}\right)^{2} \cdot a^{12/5}b^{-2/5} = \dfrac{a^{12/5}}{4b^{2/5}}$

91) $\left(\dfrac{r^{4/5}t^{-2}}{r^{2/3}t^{5}}\right)^{-3/2}$

$= \left(\dfrac{r^{2/3}t^{5}}{r^{4/5}t^{-2}}\right)^{3/2}$

$= \left(r^{\frac{2}{3}-\frac{4}{5}}t^{5+2}\right)^{3/2}$

$= \left(r^{\frac{10-12}{15}}\right)^{3/2} \cdot \left(t^{7}\right)^{3/2}$

$= r^{-\frac{6}{30}}t^{\frac{21}{2}}$

$= \dfrac{t^{21/2}}{r^{1/5}}$

93) $\left(\dfrac{h^{-2}k^{5/2}}{h^{-8}k^{5/6}}\right)^{-5/6}$

$= \left(\dfrac{h^{-8}k^{5/6}}{h^{-2}k^{5/2}}\right)^{5/6}$

$= \left(h^{-6}k^{\frac{5}{6}-\frac{5}{2}}\right)^{5/6}$

$= \left(h^{-6}\right)^{5/6}\left(k^{-10/6}\right)^{5/6}$

$= h^{-5}k^{-25/18}$

$= \dfrac{1}{h^{5}k^{25/18}}$

95) $p^{1/2}\left(p^{2/3} + p^{1/2}\right)$

$= p^{1/2}p^{2/3} + p^{1/2}p^{1/2}$

$= p^{\frac{1}{2}+\frac{2}{3}} + p^{\frac{1}{2}+\frac{1}{2}}$

$= p^{7/6} + p$

97) $\sqrt[12]{25^{6}} = \boxed{25^{6/12}}$ Write with a rational exponent

$= \boxed{25^{1/2}}$ Reduce the exponent

$= 5$ $\boxed{\text{Evaluate}}$

99) $\sqrt[6]{49^{3}} = \left(49^{3}\right)^{1/6} = 49^{1/2} = \sqrt{49} = 7$

101) $\sqrt[4]{81^{2}} = \left(81^{2}\right)^{1/4} = 81^{1/2} = \sqrt{81} = 9$

103) $\left(\sqrt{5}\right)^{2} = 5$

105) $\left(\sqrt[3]{12}\right)^{3} = 12$

107) $\left(\sqrt[3]{x^{12}}\right) = x^{12/3} = x^{4}$

109) $\left(\sqrt[6]{k^{2}}\right) = k^{2/6} = k^{1/3} = \sqrt[3]{k}$

111) $\left(\sqrt[4]{z^{2}}\right) = z^{2/4} = z^{1/2} = \sqrt{z}$

113) $\sqrt{d^{4}} = d^{4/2} = d^{2}$

115) a) $WC = 35.74 + 0.6215(20)$

$\qquad -35.7(5)^{4/25}$

$\qquad +0.4275(20)(5)^{4/25}$

$\qquad = 13 \text{ degrees}$

b) $WC = 35.74 + 0.6215(20)$

$\qquad -35.75(15)^{4/25}$

$\qquad +0.4275(20)(15)^{4/25}$

$\qquad = 6 \text{ degrees}$

Section 10.3: Exercises

1) $\sqrt{3} \cdot \sqrt{7} = \sqrt{3 \cdot 7} = \sqrt{21}$

3) $\sqrt{10} \cdot \sqrt{3} = \sqrt{10 \cdot 3} = \sqrt{30}$

5) $\sqrt{6} \cdot \sqrt{y} = \sqrt{6 \cdot y} = \sqrt{6y}$

7) False; 20 contains the factor 4 which is a perfect square.

9) True; 42 does not have any factors (other than 1) that are perfect squares.

11) $\sqrt{60} = \sqrt{4 \cdot 15}$ $\boxed{\text{Factor}}$

$\qquad = \boxed{\sqrt{4} \cdot \sqrt{15}}$ Product Rule

$\qquad = \boxed{2\sqrt{15}}$ Simplify

13) $\sqrt{20} = \sqrt{4 \cdot 5} = \sqrt{4} \cdot \sqrt{5} = 2\sqrt{5}$

15) $\sqrt{54} = \sqrt{9 \cdot 6} = \sqrt{9} \cdot \sqrt{6} = 3\sqrt{6}$

17) $\sqrt{33}$; simplified

19) $\sqrt{80} = \sqrt{16 \cdot 5} = \sqrt{16} \cdot \sqrt{5} = 4\sqrt{5}$

21) $\sqrt{98} = \sqrt{49 \cdot 2} = \sqrt{49} \cdot \sqrt{2} = 7\sqrt{2}$

23) $\sqrt{38}$; simplified

25) $\sqrt{400} = 20$

27) $\sqrt{750} = \sqrt{25}\sqrt{30} = 5\sqrt{30}$

29) $\sqrt{\dfrac{144}{25}} = \dfrac{\sqrt{144}}{\sqrt{25}} = \dfrac{12}{5}$

31) $\dfrac{\sqrt{4}}{\sqrt{49}} = \dfrac{2}{7}$

33) $\dfrac{\sqrt{54}}{\sqrt{6}} = \sqrt{\dfrac{54}{6}} = \sqrt{9} = 3$

35) $\sqrt{\dfrac{60}{5}} = \sqrt{12} = \sqrt{4 \cdot 3} = \sqrt{4} \cdot \sqrt{3} = 2\sqrt{3}$

37) $\dfrac{\sqrt{120}}{\sqrt{6}} = \sqrt{\dfrac{120}{6}} = \sqrt{20} = \sqrt{4 \cdot 5}$

$\qquad = \sqrt{4} \cdot \sqrt{5} = 2\sqrt{5}$

39) $\dfrac{\sqrt{30}}{\sqrt{2}} = \sqrt{\dfrac{30}{2}} = \sqrt{15}$

41) $\sqrt{\dfrac{6}{49}} = \dfrac{\sqrt{6}}{\sqrt{49}} = \dfrac{\sqrt{6}}{7}$

43) $\sqrt{\dfrac{45}{16}} = \dfrac{\sqrt{45}}{\sqrt{16}} = \dfrac{\sqrt{9 \cdot 5}}{4} = \dfrac{\sqrt{9} \cdot \sqrt{5}}{4}$

$\qquad = \dfrac{3\sqrt{5}}{4}$

45) $\sqrt{x^8} = x^{8/2} = x^4$

47) $\sqrt{w^{14}} = w^{14/2} = w^7$

49) $\sqrt{100c^2} = \sqrt{100} \cdot \sqrt{c^2} = 10c^{2/2} = 10c$

51) $\sqrt{64k^6 m^{10}} = \sqrt{64} \cdot \sqrt{k^6}\sqrt{m^{10}}$

$\qquad = 8k^{6/2}m^{10/2} = 8k^3 m^5$

53) $\sqrt{28r^4} = \sqrt{28} \cdot \sqrt{r^4} = \sqrt{4} \cdot \sqrt{7} \cdot r^{4/2}$

$\qquad = 2\sqrt{7} \cdot r^2 = 2r^2\sqrt{7}$

55) $\sqrt{300q^{22}t^{16}} = \sqrt{300} \cdot \sqrt{q^{22}}\sqrt{t^{16}}$

$\qquad = \sqrt{100} \cdot \sqrt{3} \cdot q^{22/2} \cdot t^{16/2}$

$\qquad = 10\sqrt{3} \cdot q^{11}t^8 = 10q^{11}t^8\sqrt{3}$

57) $\sqrt{\dfrac{81}{c^6}} = \dfrac{\sqrt{81}}{\sqrt{c^6}} = \dfrac{9}{c^{6/2}} = \dfrac{9}{c^3}$

59) $\dfrac{\sqrt{40}}{\sqrt{t^8}} = \dfrac{\sqrt{4} \cdot \sqrt{10}}{t^{8/2}} = \dfrac{2\sqrt{10}}{t^4}$

61) $\sqrt{\dfrac{75x^2}{y^{12}}} = \dfrac{\sqrt{75}\sqrt{x^2}}{\sqrt{y^{12}}} = \dfrac{\sqrt{25} \cdot x \cdot \sqrt{3}}{y^{12/2}} = \dfrac{5x\sqrt{3}}{y^6}$

63) $\sqrt{w^9} = \sqrt{w^8 \cdot w^1}$ $\boxed{\text{Factor}}$

    $= \boxed{\sqrt{w^8} \cdot \sqrt{w^1}}$ Product Rule

    $= w^4 \sqrt{w}$ $\boxed{\text{Simplify}}$

65) $\sqrt{a^5} = \sqrt{a^4} \cdot \sqrt{a} = a^{4/2} \cdot \sqrt{a} = a^2 \sqrt{a}$

67) $\sqrt{g^{13}} = \sqrt{g^{12}} \cdot \sqrt{g} = g^6 \sqrt{g}$

69) $\sqrt{b^{25}} = \sqrt{b^{24}} \cdot \sqrt{b} = b^{12} \sqrt{b}$

71) $\sqrt{72x^3} = \sqrt{72} \cdot \sqrt{x^3}$

    $= \sqrt{36} \cdot \sqrt{2} \cdot \sqrt{x^2} \cdot \sqrt{x}$

    $= 6\sqrt{2} \cdot x\sqrt{x} = 6x\sqrt{2x}$

73) $\sqrt{13q^7} = \sqrt{13} \cdot \sqrt{q^7}$

    $= \sqrt{13} \cdot \sqrt{q^6} \cdot \sqrt{q}$

    $= \sqrt{13} \cdot q^3 \sqrt{q} = q^3 \sqrt{13q}$

75) $\sqrt{75t^{11}} = \sqrt{75} \cdot \sqrt{t^{11}}$

    $= \sqrt{25} \cdot \sqrt{3} \cdot \sqrt{t^{10}} \cdot \sqrt{t}$

    $= 5\sqrt{3} \cdot t^5 \sqrt{t} = 5t^5 \sqrt{3t}$

77) $\sqrt{c^8 d^2} = \sqrt{c^8} \cdot \sqrt{d^2} = c^4 d$

79) $\sqrt{a^4 b^3} = \sqrt{a^4} \cdot \sqrt{b^3} = a^2 \cdot \sqrt{b^2} \cdot \sqrt{b}$

    $= a^2 b \sqrt{b}$

81) $\sqrt{u^5 v^7} = \sqrt{u^5} \cdot \sqrt{v^7}$

    $= \sqrt{u^4} \cdot \sqrt{u} \cdot \sqrt{v^6} \cdot \sqrt{v}$

    $= u^2 \sqrt{u} \cdot v^3 \sqrt{v} = u^2 v^3 \sqrt{uv}$

83) $\sqrt{36m^9 n^4} = \sqrt{36} \cdot \sqrt{m^9} \cdot \sqrt{n^4}$

    $= 6 \cdot \sqrt{m^8} \cdot \sqrt{m} \cdot n^2$

    $= 6m^4 n^2 \sqrt{m}$

85) $\sqrt{44x^{12} y^5} = \sqrt{44} \cdot \sqrt{x^{12}} \cdot \sqrt{y^5}$

    $= \sqrt{4} \cdot \sqrt{11} \cdot x^6 \cdot \sqrt{y^4} \cdot \sqrt{y}$

    $= 2\sqrt{11} \cdot x^6 \cdot y^2 \sqrt{y}$

    $= 2x^6 y^2 \sqrt{11y}$

87) $\sqrt{32t^5 u^7}$

    $= \sqrt{32} \cdot \sqrt{t^5} \cdot \sqrt{u^7}$

    $= \sqrt{16} \cdot \sqrt{2} \cdot \sqrt{t^4} \cdot \sqrt{t} \cdot \sqrt{u^6} \cdot \sqrt{u}$

    $= 4\sqrt{2} \cdot t^2 \sqrt{t} \cdot u^3 \sqrt{u} = 4t^2 u^3 \sqrt{2tu}$

89) $\sqrt{\dfrac{a^7}{81b^6}} = \dfrac{\sqrt{a^6} \cdot \sqrt{a}}{\sqrt{81b^6}} = \dfrac{a^3 \sqrt{a}}{9b^3}$

91) $\sqrt{\dfrac{3r^9}{s^2}} = \dfrac{\sqrt{3r^9}}{\sqrt{s^2}} = \dfrac{\sqrt{3} \cdot \sqrt{r^9}}{s}$

    $= \dfrac{\sqrt{3} \cdot \sqrt{r^8} \cdot \sqrt{r}}{s} = \dfrac{r^4 \sqrt{3r}}{s}$

93) $\sqrt{5} \cdot \sqrt{10} = \sqrt{50} = \sqrt{25} \cdot \sqrt{2} = 5\sqrt{2}$

95) $\sqrt{21} \cdot \sqrt{3} = \sqrt{63} = \sqrt{9} \cdot \sqrt{7} = 3\sqrt{7}$

97) $\sqrt{w} \cdot \sqrt{w^5} = \sqrt{w^6} = w^3$

99) $\sqrt{n^3} \cdot \sqrt{n^4} = \sqrt{n^2} \cdot \sqrt{n} \cdot n^2$

    $= n \cdot \sqrt{n} \cdot n^2 = n^3 \sqrt{n}$

101) $\sqrt{2k} \cdot \sqrt{8k^5} = \sqrt{16k^6} = \sqrt{16} \cdot \sqrt{k^6}$

    $= 4k^3$

103) $\sqrt{6x^4 y^3} \cdot \sqrt{2x^5 y^2}$

    $= \sqrt{12x^9 y^5} = \sqrt{12} \cdot \sqrt{x^9} \cdot \sqrt{y^5}$

    $= \sqrt{4} \cdot \sqrt{3} \cdot \sqrt{x^8} \cdot \sqrt{x} \cdot \sqrt{y^4} \cdot \sqrt{y}$

    $= 2\sqrt{3} \cdot x^4 \sqrt{x} \cdot y^2 \sqrt{y} = 2x^4 y^2 \sqrt{3xy}$

105) $\sqrt{8c^9 d^2} \cdot \sqrt{5cd^7}$

    $= \sqrt{40c^{10} d^9} = \sqrt{40} \cdot \sqrt{c^{10}} \cdot \sqrt{d^9}$

    $= \sqrt{4} \cdot \sqrt{10} \cdot c^5 \cdot \sqrt{d^8} \cdot \sqrt{d}$

    $= 2\sqrt{10} \cdot c^5 \cdot d^4 \sqrt{d} = 2c^5 d^4 \sqrt{10d}$

107) $\dfrac{\sqrt{18k^{11}}}{\sqrt{2k^3}} = \sqrt{\dfrac{18k^{11}}{2k^3}} = \sqrt{9k^8}$

    $= \sqrt{9} \cdot \sqrt{k^8} = 3k^4$

109) $\dfrac{\sqrt{120h^8}}{\sqrt{3h^2}}=\sqrt{\dfrac{120h^8}{3h^2}}=\sqrt{40h^6}$

$=\sqrt{4}\cdot\sqrt{10}\cdot\sqrt{h^6}$

$=2\sqrt{10}\cdot h^3=2h^3\sqrt{10}$

111) $\dfrac{\sqrt{50a^{16}b^9}}{\sqrt{5a^7b^4}}=\sqrt{\dfrac{50a^{16}b^9}{5a^7b^4}}=\sqrt{10a^9b^5}$

$=\sqrt{10}\cdot\sqrt{a^8}\cdot\sqrt{a}\cdot\sqrt{b^4}\cdot\sqrt{b}$

$=\sqrt{10}\cdot a^4\sqrt{a}\cdot b^2\sqrt{b}=a^4b^2\sqrt{10ab}$

113) $v=\sqrt{\dfrac{2KE}{m}}=\sqrt{\dfrac{2\cdot120,000}{600}}$

$=\sqrt{400}$

$=20$ m/s

Section 10.4: Exercises

1) To multiply radicals with the same indices, multiply the radicands and put the product under a radical with the same index.

3) i) Its radicand will not contain any factors that are perfect cubes.

ii) There will be no radical in the denominator of a fraction.

iii) The radicand will not contain fractions.

5) $\sqrt[3]{5}\cdot\sqrt[3]{4}=\sqrt[3]{20}$

7) $\sqrt[5]{9}\cdot\sqrt[5]{m^2}=\sqrt[5]{9m^2}$

9) $\sqrt[3]{a^2}\cdot\sqrt[3]{b}=\sqrt[3]{a^2b}$

11) $\sqrt[3]{56}=\sqrt[3]{8\cdot7}$ ⬚Factor⬚

$=\boxed{\sqrt[3]{8}\cdot\sqrt[3]{7}}$ Product Rule

$=\boxed{2\sqrt[3]{7}}$ Simplify

13) $\sqrt[3]{24}=\sqrt[3]{8}\cdot\sqrt[3]{3}=2\sqrt[3]{3}$

15) $\sqrt[4]{64}=\sqrt[4]{16}\cdot\sqrt[4]{4}=2\sqrt[4]{4}$

17) $\sqrt[3]{54}=\sqrt[3]{27}\cdot\sqrt[3]{2}=3\sqrt[3]{2}$

19) $\sqrt[3]{2000}=\sqrt[3]{1000}\cdot\sqrt[3]{2}=10\sqrt[3]{2}$

21) $\sqrt[5]{64}=\sqrt[5]{32}\cdot\sqrt[5]{2}=2\sqrt[5]{2}$

23) $\sqrt[4]{\dfrac{1}{16}}=\dfrac{\sqrt[4]{1}}{\sqrt[4]{16}}=\dfrac{1}{2}$

25) $\sqrt[3]{-\dfrac{54}{2}}=\sqrt[3]{-27}=-3$

27) $\dfrac{\sqrt[3]{48}}{\sqrt[3]{2}}=\sqrt[3]{\dfrac{48}{2}}=\sqrt[3]{24}=\sqrt[3]{8}\cdot\sqrt[3]{3}=2\sqrt[3]{3}$

29) $\dfrac{\sqrt[4]{240}}{\sqrt[4]{3}}=\sqrt[4]{\dfrac{240}{3}}=\sqrt[4]{80}$

$=\sqrt[4]{16}\cdot\sqrt[4]{5}=2\sqrt[4]{5}$

31) $\sqrt[3]{d^6}=d^{6/3}=d^2$

33) $\sqrt[4]{n^{20}}=n^{20/4}=n^5$

35) $\sqrt[5]{x^5y^{15}}=x^{5/5}y^{15/5}=xy^3$

37) $\sqrt[3]{w^{14}}=\sqrt[3]{w^{12}}\cdot\sqrt[3]{w^2}=w^{12/3}\cdot\sqrt[3]{w^2}$

$=w^4\sqrt[3]{w^2}$

39) $\sqrt[4]{y^9}=\sqrt[4]{y^8}\cdot\sqrt[4]{y}=y^{8/4}\cdot\sqrt[4]{y}=y^2\sqrt[4]{y}$

41) $\sqrt[3]{d^5}=\sqrt[3]{d^3}\cdot\sqrt[3]{d^2}=d\sqrt[3]{d^2}$

43) $\sqrt[3]{u^{10}v^{15}}=\sqrt[3]{u^{10}}\cdot\sqrt[3]{v^{15}}$

$=\sqrt[3]{u^9}\cdot\sqrt[3]{u}\cdot v^{15/3}$

$=u^{9/3}\sqrt[3]{u}\cdot v^5=u^3v^5\sqrt[3]{u}$

45) $\sqrt[3]{b^{16}c^5}=\sqrt[3]{b^{16}}\cdot\sqrt[3]{c^5}$

$=\sqrt[3]{b^{15}}\cdot\sqrt[3]{b}\cdot\sqrt[3]{c^3}\cdot\sqrt[3]{c^2}$

$=b^{15/3}\cdot\sqrt[3]{b}\cdot c^{3/3}\cdot\sqrt[3]{c^2}$

$=b^5c\sqrt[3]{bc^2}$

47) $\sqrt[4]{m^3n^{18}}=\sqrt[4]{m^3}\cdot\sqrt[4]{n^{18}}$

$=\sqrt[4]{m^3}\cdot\sqrt[4]{n^{16}}\cdot\sqrt[4]{n^2}$

$=\sqrt[4]{m^3}\cdot n^{16/4}\cdot\sqrt[4]{n^2}$

$=n^4\sqrt[4]{m^3n^2}$

49) $\sqrt[3]{24x^{10}y^{12}}=\sqrt[3]{24}\cdot\sqrt[3]{x^{10}}\cdot\sqrt[3]{y^{12}}$

$=\sqrt[3]{8}\cdot\sqrt[3]{3}\cdot\sqrt[3]{x^9}\cdot\sqrt[3]{x}\cdot y^{12/3}$

$=2\sqrt[3]{3}\cdot x^{9/3}\sqrt[3]{x}\cdot y^4$

$=2x^3y^4\sqrt[3]{3x}$

51) $\sqrt[3]{250w^4x^{16}}$

$= \sqrt[3]{250} \cdot \sqrt[3]{w^4} \cdot \sqrt[3]{x^{16}}$

$= \sqrt[3]{125} \cdot \sqrt[3]{2} \cdot \sqrt[3]{w^3} \cdot \sqrt[3]{w} \cdot \sqrt[3]{x^{15}} \cdot \sqrt[3]{x}$

$= 5\sqrt[3]{2} \cdot w^{3/3}\sqrt[3]{w} \cdot x^{15/3}\sqrt[3]{x}$

$= 5wx^5\sqrt[3]{2wx}$

53) $\sqrt[4]{\dfrac{m^8}{81}} = \dfrac{\sqrt[4]{m^8}}{\sqrt[4]{81}} = \dfrac{m^{8/4}}{3} = \dfrac{m^2}{3}$

55) $\sqrt[5]{\dfrac{32a^{23}}{b^{15}}} = \dfrac{\sqrt[5]{32a^{23}}}{\sqrt[5]{b^{15}}}$

$= \dfrac{\sqrt[5]{32} \cdot \sqrt[5]{a^{20}} \cdot \sqrt[5]{a^3}}{b^3}$

$= \dfrac{2a^4\sqrt[5]{a^3}}{b^3}$

57) $\sqrt[4]{\dfrac{t^9}{81s^{24}}} = \dfrac{\sqrt[4]{t^9}}{\sqrt[4]{81s^{24}}} = \dfrac{\sqrt[4]{t^8} \cdot \sqrt[4]{t}}{\sqrt[4]{81} \cdot \sqrt[4]{s^{24}}}$

$= \dfrac{t^2\sqrt[4]{t}}{3s^6}$

59) $\sqrt[3]{\dfrac{u^{28}}{v^3}} = \dfrac{\sqrt[3]{u^{28}}}{\sqrt[3]{v^3}} = \dfrac{\sqrt[3]{u^{27}} \cdot \sqrt[3]{u}}{v} = \dfrac{u^9\sqrt[3]{u}}{v}$

61) $\sqrt[3]{6} \cdot \sqrt[3]{4} = \sqrt[3]{24} = \sqrt[3]{8} \cdot \sqrt[3]{3} = 2\sqrt[3]{3}$

63) $\sqrt[3]{9} \cdot \sqrt[3]{12} = \sqrt[3]{9} \cdot \sqrt[3]{3} \cdot \sqrt[3]{4} = \sqrt[3]{27} \cdot \sqrt[3]{4}$

$= 3\sqrt[3]{4}$

65) $\sqrt[3]{20} \cdot \sqrt[3]{4} = \sqrt[3]{80} = \sqrt[3]{8} \cdot \sqrt[3]{10} = 2\sqrt[3]{10}$

67) $\sqrt[3]{m^4} \cdot \sqrt[3]{m^5} = \sqrt[3]{m^9} = m^3$

69) $\sqrt[4]{k^7} \cdot \sqrt[4]{k^9} = \sqrt[4]{k^{16}} = k^4$

71) $\sqrt[3]{r^7} \cdot \sqrt[3]{r^4} = \sqrt[3]{r^{11}} = \sqrt[3]{r^9} \cdot \sqrt[3]{r^2}$

$= r^3\sqrt[3]{r^2}$

73) $\sqrt[5]{p^{14}} \cdot \sqrt[5]{p^9} = \sqrt[5]{p^{23}} = \sqrt[5]{p^{20}} \cdot \sqrt[5]{p^3}$

$= p^4\sqrt[5]{p^3}$

75) $\sqrt[3]{9z^{11}} \cdot \sqrt[3]{3z^8} = \sqrt[3]{27z^{19}} = \sqrt[3]{27} \cdot \sqrt[3]{z^{19}}$

$= 3 \cdot \sqrt[3]{z^{18}} \cdot \sqrt[3]{z} = 3z^6\sqrt[3]{z}$

77) $\sqrt[3]{\dfrac{h^{14}}{h^2}} = \sqrt[3]{h^{12}} = h^4$

79) $\sqrt[3]{\dfrac{c^{11}}{c^4}} = \sqrt[3]{c^7} = \sqrt[3]{c^6} \cdot \sqrt[3]{c} = c^2\sqrt[3]{c}$

81) $\sqrt[4]{\dfrac{162d^{21}}{2d^2}} = \sqrt[4]{81d^{19}} = \sqrt[4]{81} \cdot \sqrt[4]{d^{19}}$

$= 3 \cdot \sqrt[4]{d^{16}} \cdot \sqrt[4]{d^3} = 3d^4\sqrt[4]{d^3}$

83) $\sqrt{a}\sqrt[4]{a^3} = a^{1/2} \cdot a^{3/4}$

Change radicals to fractional exponents.

$= a^{2/4} \cdot a^{3/4}$

Rewrite exponents with a common denominator.

$= a^{5/4}$ Add exponents

$= \sqrt[4]{a^5}$ Rewrite in radical form

$= a\sqrt[4]{a}$ Simplify

85) $\sqrt{p} \cdot \sqrt[3]{p} = p^{1/2} \cdot p^{1/3} = p^{3/6} \cdot p^{2/6} = p^{5/6} = \sqrt[6]{p^5}$

87) $\sqrt[4]{n^3} \cdot \sqrt{n} = n^{3/4} \cdot n^{1/2} = n^{3/4} \cdot n^{2/4}$

$= n^{5/4} = n^{4/4} \cdot n^{1/4} = n\sqrt[4]{n}$

89) $\dfrac{\sqrt{w}}{\sqrt[4]{w}} = \dfrac{w^{1/2}}{w^{1/4}} = \dfrac{w^{2/4}}{w^{1/4}} = w^{1/4} = \sqrt[4]{w}$

91) $\dfrac{\sqrt[5]{t^4}}{\sqrt[3]{t^2}} = \dfrac{t^{4/5}}{t^{2/3}} = \dfrac{t^{12/15}}{t^{10/15}} = t^{2/15} = \sqrt[15]{t^2}$

93) $s = \sqrt[3]{V} = \sqrt[3]{64} = 4$ inches

Section 10.5: Exercises

1) They have the same index and
the same radicand.

3) $5\sqrt{2} + 9\sqrt{2} = 14\sqrt{2}$

5) $7\sqrt[3]{4} + 8\sqrt[3]{4} = 15\sqrt[3]{4}$

7) $6 - \sqrt{13} + 5 - 2\sqrt{13} = 11 - 3\sqrt{13}$

9) $15\sqrt[3]{z^2} - 20\sqrt[3]{z^2} = -5\sqrt[3]{z^2}$

11) $2\sqrt[3]{n^2} + 9\sqrt[5]{n^2} - 11\sqrt[3]{n^2} + \sqrt[5]{n^2}$

$= -9\sqrt[3]{n^2} + 10\sqrt[5]{n^2}$

13) $\sqrt{5c} - 8\sqrt{6c} + \sqrt{5c} + 6\sqrt{6c}$

$= 2\sqrt{5c} - 2\sqrt{6c}$

15) i) Write each radical expression
   in simplest form.
   ii) Combine like radicals.

17) $\sqrt{48} + \sqrt{3} = \sqrt{16 \cdot 3} + \sqrt{3}$ $\boxed{\text{Factor}}$

$= \boxed{\sqrt{16} \cdot \sqrt{3} + \sqrt{3}}$ Product Rule

$= 4\sqrt{3} + \sqrt{3}$ $\boxed{\text{Simplify}}$

$= \boxed{5\sqrt{3}}$ Add like radicals

19) $6\sqrt{3} - \sqrt{12} = 6\sqrt{3} - 2\sqrt{3} = 4\sqrt{3}$

21) $\sqrt{32} - 3\sqrt{8} = 4\sqrt{2} - 3\left(2\sqrt{2}\right)$

$= 4\sqrt{2} - 6\sqrt{2} = -2\sqrt{2}$

23) $\sqrt{12} + \sqrt{75} - \sqrt{3}$

$= 2\sqrt{3} + 5\sqrt{3} - \sqrt{3} = 6\sqrt{3}$

25) $8\sqrt[3]{9} + \sqrt[3]{72} = 8\sqrt[3]{9} + \sqrt[3]{8} \cdot \sqrt[3]{9}$

$= 8\sqrt[3]{9} + 2\sqrt[3]{9} = 10\sqrt[3]{9}$

27) $\sqrt[3]{6} - \sqrt[3]{48} = \sqrt[3]{6} - \left(2\sqrt[3]{6}\right) = -\sqrt[3]{6}$

29) $6q\sqrt{q} + 7\sqrt{q^3} = 6q\sqrt{q} + 7q\sqrt{q}$

$= 13q\sqrt{q}$

31) $4d^2\sqrt{d} - 24\sqrt{d^5}$

$= 4d^2\sqrt{d} - 24d^2\sqrt{d} = -20d^2\sqrt{d}$

33) $9t^3\sqrt[3]{t} - 5\sqrt[3]{t^{10}} = 9t^3\sqrt[3]{t} - 5t^3\sqrt[3]{t}$

$= 4t^3\sqrt[3]{t}$

35) $5a\sqrt[4]{a^7} + \sqrt[4]{a^{11}}$

$= 5a\left(a\sqrt[4]{a^3}\right) + \left(a^2\sqrt[4]{a^3}\right)$

$= 5a^2\sqrt[4]{a^3} + a^2\sqrt[4]{a^3} = 6a^2\sqrt[4]{a^3}$

37) $2\sqrt{8p} - 6\sqrt{2p} = 2\left(2\sqrt{2p}\right) - 6\sqrt{2p}$

$= 4\sqrt{2p} - 6\sqrt{2p} = -2\sqrt{2p}$

39) $7\sqrt[3]{81a^5} + 4a\sqrt[3]{3a^2}$

$= 7\left(3a\sqrt[3]{3a^2}\right) + 4a\sqrt[3]{3a^2}$

$= 21a\sqrt[3]{3a^2} + 4a\sqrt[3]{3a^2} = 25a\sqrt[3]{3a^2}$

41) $\sqrt{xy^3} + 3y\sqrt{xy} = y\sqrt{xy} + 3y\sqrt{xy}$

$= 4y\sqrt{xy}$

43) $6c^2\sqrt{8d^3} - 9d\sqrt{2c^4d}$

$= 6c^2\left(2d\sqrt{2d}\right) - 9d\left(c^2\sqrt{2d}\right)$

$= 12c^2d\sqrt{2d} - 9c^2d\sqrt{2d} = 3c^2d\sqrt{2d}$

45) $18a^5\sqrt[3]{7a^2b} + 2a^3\sqrt[3]{7a^8b}$

$= 18a^5\sqrt[3]{7a^2b} + 2a^3\left(a^2\sqrt[3]{7a^2b}\right)$

$= 18a^5\sqrt[3]{7a^2b} + 2a^5\sqrt[3]{7a^2b}$

$= 20a^5\sqrt[3]{7a^2b}$

47) $15cd\sqrt[4]{9cd} - \sqrt[4]{9c^5d^5}$

$= 15cd\sqrt[4]{9cd} - \left(\sqrt[4]{c^4d^4} \cdot \sqrt[4]{9cd}\right)$

$= 15cd\sqrt[4]{9cd} - cd\sqrt[4]{9cd}$

$= 14cd\sqrt[4]{9cd}$

49) $\sqrt[3]{a^9b} - \sqrt[3]{b^7} = a^3\sqrt[3]{b} - b^2\sqrt[3]{b}$

$= \sqrt[3]{b}\left(a^3 - b^2\right)$

51) $3(x+5) = 3x + 15$

53) $7\left(\sqrt{6} + 2\right) = 7\sqrt{6} + 14$

55) $\sqrt{10}\left(\sqrt{3} - 1\right) = \sqrt{30} - \sqrt{10}$

57) $-6\left(\sqrt{32} + \sqrt{2}\right) = -6\left(4\sqrt{2} + \sqrt{2}\right)$

$= -6\left(5\sqrt{2}\right) = -30\sqrt{2}$

59) $4\left(\sqrt{45} - \sqrt{20}\right) = 4\left(3\sqrt{5} - 2\sqrt{5}\right)$

$= 4\left(\sqrt{5}\right) = 4\sqrt{5}$

60) $\sqrt{5}\left(\sqrt{24}-\sqrt{54}\right)=\sqrt{5}\left(2\sqrt{6}-3\sqrt{6}\right)$

$=\sqrt{5}\left(-\sqrt{6}\right)$

$=-\sqrt{30}$

63) $\sqrt[4]{3}\left(5-\sqrt[4]{27}\right)=\sqrt[4]{3}\cdot5-\sqrt[4]{3}\cdot\sqrt[4]{27}$

$=5\sqrt[4]{3}-3$

65) $\sqrt{t}\left(\sqrt{t}-\sqrt{81u}\right)=\sqrt{t}\left(\sqrt{t}-9\sqrt{u}\right)$

$=t-9\sqrt{tu}$

67) $\sqrt{ab}\left(\sqrt{5a}+\sqrt{27b}\right)$

$=\sqrt{ab}\left(\sqrt{5a}+3\sqrt{3b}\right)$

$=\sqrt{5a^2b}+3\sqrt{3ab^2}=a\sqrt{5b}+3b\sqrt{3a}$

69) $\sqrt[3]{c^2}\left(\sqrt[3]{c^2}+\sqrt[3]{125cd}\right)$

$=\sqrt[3]{c^2}\cdot\sqrt[3]{c^2}+\sqrt[3]{c^2}\cdot\sqrt[3]{125cd}$

$=\sqrt[3]{c^4}+\sqrt[3]{125c^3d}$

$=c\sqrt[3]{c}+5c\sqrt[3]{d}$

71) Both are examples of multiplication of two binomials. They can be multiplied using FOIL.

73) $(a+b)(a-b)=a^2-b^2$

75) $(p+7)(p+6)=p^2+6p+7p+42$

$=p^2+13p+42$

77) $\left(6+\sqrt{7}\right)\left(2+\sqrt{7}\right)$

$=6\cdot2+6\sqrt{7}+2\sqrt{7}+\sqrt{7}\cdot\sqrt{7}$ $\boxed{\text{Use FOIL}}$

$=12+6\sqrt{7}+2\sqrt{7}+7$ $\boxed{\text{Multiply}}$

$=\boxed{19+8\sqrt{7}}$ $\boxed{\text{Combine like terms}}$

79) $\left(\sqrt{2}+8\right)\left(\sqrt{2}-3\right)$

$=2-3\sqrt{2}+8\sqrt{2}-24=-22+5\sqrt{2}$

81) $\left(\sqrt{5}-4\sqrt{3}\right)\left(2\sqrt{5}-\sqrt{3}\right)$

$=2(5)-\sqrt{15}-8\sqrt{15}+4(3)$

$=10-9\sqrt{15}+12=22-9\sqrt{15}$

83) $\left(5+2\sqrt{3}\right)\left(\sqrt{7}+\sqrt{2}\right)$

$=5\sqrt{7}+5\sqrt{2}+2\sqrt{21}+2\sqrt{6}$

85) $\left(\sqrt[3]{25}-3\right)\left(\sqrt[3]{5}-\sqrt[3]{6}\right)$

$=\sqrt[3]{25}\sqrt[3]{5}-\sqrt[3]{25}\sqrt[3]{6}-3\sqrt[3]{5}+3\sqrt[3]{6}$

$=\sqrt[3]{125}-\sqrt[3]{150}-3\sqrt[3]{5}+3\sqrt[3]{6}$

$=5-\sqrt[3]{150}-3\sqrt[3]{5}+3\sqrt[3]{6}$

87) $\left(\sqrt{6p}-2\sqrt{q}\right)\left(8\sqrt{q}+5\sqrt{6p}\right)$

$=8\sqrt{6pq}+5(6p)-16q-10\sqrt{6pq}$

$=-2\sqrt{6pq}+30p-16q$

89) $\left(\sqrt{3}+1\right)^2=\left(\sqrt{3}\right)^2+2\left(\sqrt{3}\right)(1)+(1)^2$

$=3+2\sqrt{3}+1=4+2\sqrt{3}$

91) $\left(\sqrt{11}-\sqrt{5}\right)^2$

$=\left(\sqrt{11}\right)^2-2\left(\sqrt{11}\right)\left(\sqrt{5}\right)+\left(\sqrt{5}\right)^2$

$=11-2\sqrt{55}+5=16-2\sqrt{55}$

93) $\left(\sqrt{h}+\sqrt{7}\right)^2$

$=\left(\sqrt{h}\right)^2+2\left(\sqrt{h}\right)\left(\sqrt{7}\right)+\left(\sqrt{7}\right)^2$

$=h+2\sqrt{7h}+7$

95) $\left(\sqrt{x}-\sqrt{y}\right)^2$

$=\left(\sqrt{x}\right)^2-2\left(\sqrt{x}\right)\left(\sqrt{y}\right)+\left(\sqrt{y}\right)^2$

$=x-2\sqrt{xy}+y$

97) $(c+9)(c-9)=c^2-(9)^2=c^2-81$

99) $\left(6-\sqrt{5}\right)\left(6+\sqrt{5}\right)=6^2-\left(\sqrt{5}\right)^2$

$=36-5=31$

101) $\left(4\sqrt{3}+\sqrt{2}\right)\left(4\sqrt{3}-\sqrt{2}\right)$

$\qquad =\left(4\sqrt{3}\right)^2-\left(\sqrt{2}\right)^2$

$\qquad =16(3)-2=48-2=46$

103) $\left(\sqrt[3]{2}-3\right)\left(\sqrt[3]{2}+3\right)=\left(\sqrt[3]{2}\right)^2-3^2$

$\qquad\qquad =\sqrt[3]{4}-9$

105) $\left(\sqrt{c}+\sqrt{d}\right)\left(\sqrt{c}-\sqrt{d}\right)$

$\qquad =\left(\sqrt{c}\right)^2-\left(\sqrt{d}\right)^2=c-d$

107) $\left(8\sqrt{f}-\sqrt{g}\right)\left(8\sqrt{f}+\sqrt{g}\right)$

$\qquad =\left(8\sqrt{f}\right)^2-\left(\sqrt{g}\right)^2=64f-g$

109) $\left(1+2\sqrt[3]{5}\right)\left(1-2\sqrt[3]{5}+4\sqrt[3]{25}\right)$

$\qquad =1-2\sqrt[3]{5}+4\sqrt[3]{25}+2\sqrt[3]{5}$

$\qquad\qquad -4\sqrt[3]{25}+8\sqrt[3]{125}$

$\qquad =1+8\sqrt[3]{125}=1+8(5)=1+40=41$

111) $f\left(\sqrt{7}+2\right)=\left(\sqrt{7}+2\right)^2$

$\qquad =\left(\sqrt{7}\right)^2+2\left(\sqrt{7}\right)(2)+(2)^2$

$\qquad =7+4\sqrt{7}+4=11+4\sqrt{7}$

113) $f\left(1-2\sqrt{3}\right)=\left(1-2\sqrt{3}\right)^2$

$\qquad =(1)^2-2(1)\left(2\sqrt{3}\right)+\left(2\sqrt{3}\right)^2$

$\qquad =1-4\sqrt{3}+12=13-4\sqrt{3}$

Section 10.6: Exercises

1) Eliminate the radical of the denominator.

3) $\dfrac{1}{\sqrt{5}}=\dfrac{1}{\sqrt{5}}\cdot\dfrac{\sqrt{5}}{\sqrt{5}}=\dfrac{\sqrt{5}}{\sqrt{25}}=\dfrac{\sqrt{5}}{5}$

5) $\dfrac{9}{\sqrt{6}}=\dfrac{9}{\sqrt{6}}\cdot\dfrac{\sqrt{6}}{\sqrt{6}}=\dfrac{9\sqrt{6}}{6}=\dfrac{3\sqrt{6}}{2}$

7) $-\dfrac{20}{\sqrt{8}}=-\dfrac{20}{2\sqrt{2}}=-\dfrac{10}{\sqrt{2}}=-\dfrac{10}{\sqrt{2}}\cdot\dfrac{\sqrt{2}}{\sqrt{2}}$

$\qquad =-\dfrac{10\sqrt{2}}{2}=-5\sqrt{2}$

9) $\dfrac{\sqrt{3}}{\sqrt{28}}=\dfrac{\sqrt{3}}{2\sqrt{7}}=\dfrac{\sqrt{3}}{2\sqrt{7}}\cdot\dfrac{\sqrt{7}}{\sqrt{7}}=\dfrac{\sqrt{21}}{2\sqrt{49}}$

$\qquad =\dfrac{\sqrt{21}}{2(7)}=\dfrac{\sqrt{21}}{14}$

11) $\sqrt{\dfrac{20}{60}}=\sqrt{\dfrac{1}{3}}=\dfrac{1}{\sqrt{3}}=\dfrac{1}{\sqrt{3}}\cdot\dfrac{\sqrt{3}}{\sqrt{3}}=\dfrac{\sqrt{3}}{3}$

13) $\sqrt{\dfrac{56}{48}}=\dfrac{2\sqrt{14}}{4\sqrt{3}}=\dfrac{\sqrt{14}}{2\sqrt{3}}=\dfrac{\sqrt{14}}{2\sqrt{3}}\cdot\dfrac{\sqrt{3}}{\sqrt{3}}$

$\qquad =\dfrac{\sqrt{42}}{2(3)}=\dfrac{\sqrt{42}}{6}$

15) $\sqrt{\dfrac{10}{7}}\cdot\sqrt{\dfrac{7}{3}}=\sqrt{\dfrac{10}{7}\cdot\dfrac{7}{3}}=\sqrt{\dfrac{10}{3}}$

$\qquad =\dfrac{\sqrt{10}}{\sqrt{3}}\cdot\dfrac{\sqrt{3}}{\sqrt{3}}=\dfrac{\sqrt{30}}{3}$

17) $\sqrt{\dfrac{6}{5}}\cdot\sqrt{\dfrac{1}{8}}=\sqrt{\dfrac{6}{5}\cdot\dfrac{1}{8}}=\sqrt{\dfrac{3}{20}}=\dfrac{\sqrt{3}}{\sqrt{20}}$

$\qquad =\dfrac{\sqrt{3}}{2\sqrt{5}}\cdot\dfrac{\sqrt{5}}{\sqrt{5}}=\dfrac{\sqrt{15}}{2(5)}=\dfrac{\sqrt{15}}{10}$

19) $\dfrac{8}{\sqrt{y}}=\dfrac{8}{\sqrt{y}}\cdot\dfrac{\sqrt{y}}{\sqrt{y}}=\dfrac{8\sqrt{y}}{\sqrt{y^2}}=\dfrac{8\sqrt{y}}{y}$

21) $\dfrac{\sqrt{5}}{\sqrt{t}}=\dfrac{\sqrt{5}}{\sqrt{t}}\cdot\dfrac{\sqrt{t}}{\sqrt{t}}=\dfrac{\sqrt{5t}}{\sqrt{t^2}}=\dfrac{\sqrt{5t}}{t}$

23) $\sqrt{\dfrac{64v^7}{5w}}=\dfrac{8v^3\sqrt{v}}{\sqrt{5w}}=\dfrac{8v^3\sqrt{v}}{\sqrt{5w}}\cdot\dfrac{\sqrt{5w}}{\sqrt{5w}}$

$\qquad =\dfrac{8v^3\sqrt{5vw}}{5w}$

25) $\sqrt{\dfrac{a^3b^3}{3ab^4}}=\sqrt{\dfrac{a^2}{3b}}=\dfrac{a}{\sqrt{3b}}$

$\qquad =\dfrac{a}{\sqrt{3b}}\cdot\dfrac{\sqrt{3b}}{\sqrt{3b}}=\dfrac{a\sqrt{3b}}{3b}$

27) $-\dfrac{\sqrt{75}}{\sqrt{b^3}} = -\dfrac{5\sqrt{3}}{b\sqrt{b}} = -\dfrac{5\sqrt{3}}{b\sqrt{b}} \cdot \dfrac{\sqrt{b}}{\sqrt{b}}$

$\qquad = -\dfrac{5\sqrt{3b}}{b(b)} = -\dfrac{5\sqrt{3b}}{b^2}$

29) $\dfrac{\sqrt{13}}{\sqrt{j^5}} = \dfrac{\sqrt{13}}{j^2\sqrt{j}} = \dfrac{\sqrt{13}}{j^2\sqrt{j}} \cdot \dfrac{\sqrt{j}}{\sqrt{j}}$

$\qquad = \dfrac{\sqrt{13j}}{j^2(j)} = \dfrac{\sqrt{13j}}{j^3}$

31) 2^2 or 4

33) 3

35) c^2

37) 2^3 or 8

39) m

41) $\dfrac{4}{\sqrt[3]{3}} = \dfrac{4}{\sqrt[3]{3}} \cdot \dfrac{\sqrt[3]{3^2}}{\sqrt[3]{3^2}} = \dfrac{4\sqrt[3]{9}}{\sqrt[3]{3^3}} = \dfrac{4\sqrt[3]{9}}{3}$

43) $\dfrac{12}{\sqrt[3]{2}} = \dfrac{12}{\sqrt[3]{2}} \cdot \dfrac{\sqrt[3]{2^2}}{\sqrt[3]{2^2}} = \dfrac{12\sqrt[3]{4}}{\sqrt[3]{2^3}}$

$\qquad = \dfrac{12\sqrt[3]{4}}{2} = 6\sqrt[3]{4}$

45) $\dfrac{9}{\sqrt[3]{25}} = \dfrac{9}{\sqrt[3]{5^2}} = \dfrac{9}{\sqrt[3]{5^2}} \cdot \dfrac{\sqrt[3]{5}}{\sqrt[3]{5}}$

$\qquad = \dfrac{9\sqrt[3]{5}}{\sqrt[3]{5^3}} = \dfrac{9\sqrt[3]{5}}{5}$

47) $\sqrt[4]{\dfrac{5}{9}} = \dfrac{\sqrt[4]{5}}{\sqrt[4]{3^2}} = \dfrac{\sqrt[4]{5}}{\sqrt[4]{3^2}} \cdot \dfrac{\sqrt[4]{3^2}}{\sqrt[4]{3^2}}$

$\qquad = \dfrac{\sqrt[4]{5} \cdot \sqrt[4]{9}}{\sqrt[4]{3^4}} = \dfrac{\sqrt[4]{45}}{3}$

49) $\sqrt[5]{\dfrac{3}{8}} = \dfrac{\sqrt[5]{3}}{\sqrt[5]{2^3}} = \dfrac{\sqrt[5]{3}}{\sqrt[5]{2^3}} \cdot \dfrac{\sqrt[5]{2^2}}{\sqrt[5]{2^2}}$

$\qquad = \dfrac{\sqrt[5]{3} \cdot \sqrt[5]{4}}{\sqrt[5]{2^5}} = \dfrac{\sqrt[5]{12}}{2}$

51) $\dfrac{10}{\sqrt[3]{z}} = \dfrac{10}{\sqrt[3]{z}} \cdot \dfrac{\sqrt[3]{z^2}}{\sqrt[3]{z^2}} = \dfrac{10\sqrt[3]{z^2}}{\sqrt[3]{z^3}} = \dfrac{10\sqrt[3]{z^2}}{z}$

53) $\sqrt[3]{\dfrac{3}{n^2}} = \dfrac{\sqrt[3]{3}}{\sqrt[3]{n^2}} \cdot \dfrac{\sqrt[3]{n}}{\sqrt[3]{n}} = \dfrac{\sqrt[3]{3n}}{\sqrt[3]{n^3}} = \dfrac{\sqrt[3]{3n}}{n}$

55) $\dfrac{\sqrt[3]{7}}{\sqrt[3]{2k^2}} = \dfrac{\sqrt[3]{7}}{\sqrt[3]{2k^2}} \cdot \dfrac{\sqrt[3]{2^2 k}}{\sqrt[3]{2^2 k}} = \dfrac{\sqrt[3]{7} \cdot \sqrt[3]{4k}}{\sqrt[3]{2^3 k^3}}$

$\qquad = \dfrac{\sqrt[3]{28k}}{2k}$

57) $\dfrac{9}{\sqrt[5]{a^3}} = \dfrac{9}{\sqrt[5]{a^3}} \cdot \dfrac{\sqrt[5]{a^2}}{\sqrt[5]{a^2}} = \dfrac{9\sqrt[5]{a^2}}{\sqrt[5]{a^5}} = \dfrac{9\sqrt[5]{a^2}}{a}$

59) $\sqrt[4]{\dfrac{5}{2m}} = \dfrac{\sqrt[4]{5}}{\sqrt[4]{2m}} \cdot \dfrac{\sqrt[4]{2^3 m^3}}{\sqrt[4]{2^3 m^3}} = \dfrac{\sqrt[4]{5} \cdot \sqrt[4]{8m^3}}{\sqrt[4]{2^4 m^4}}$

$\qquad = \dfrac{\sqrt[4]{40m^3}}{2m}$

61) Change the sign between the two terms.

63) $\left(5+\sqrt{2}\right)\left(5-\sqrt{2}\right) = \left(5\right)^2 - \left(\sqrt{2}\right)^2$

$\qquad\qquad\qquad = 25 - 2 = 23$

65) $\left(\sqrt{2}+\sqrt{6}\right)\left(\sqrt{2}-\sqrt{6}\right)$

$\qquad = \left(\sqrt{2}\right)^2 - \left(\sqrt{6}\right)^2 = 2 - 6 = -4$

67) $\left(\sqrt{t}-8\right)\left(\sqrt{t}+8\right) = \left(\sqrt{t}\right)^2 - \left(8\right)^2$

$\qquad\qquad\qquad = t - 64$

69) $\dfrac{6}{4-\sqrt{5}} = \dfrac{6}{4-\sqrt{5}} \cdot \dfrac{4+\sqrt{5}}{4+\sqrt{5}}$

$\boxed{\text{Multiply by the conjugate.}}$

$= \dfrac{6\left(4+\sqrt{5}\right)}{\left(4\right)^2 - \left(\sqrt{5}\right)^2}$ $\boxed{(a+b)(a-b)=a^2-b^2}$

$= \boxed{\dfrac{24+6\sqrt{5}}{16-5}}$ Multiply terms in numerator,

square terms in denominator

$= \boxed{\dfrac{24+6\sqrt{5}}{11}}$ Simplify

71) $\dfrac{3}{2+\sqrt{3}} = \dfrac{3}{2+\sqrt{3}} \cdot \dfrac{2-\sqrt{3}}{2-\sqrt{3}}$

$ = \dfrac{3\left(2-\sqrt{3}\right)}{\left(2\right)^2 - \left(\sqrt{3}\right)^2} = \dfrac{3\left(2-\sqrt{3}\right)}{4-3}$

$ = 6 - 3\sqrt{3}$

73) $\dfrac{10}{9-\sqrt{2}} = \dfrac{10}{9-\sqrt{2}} \cdot \dfrac{9+\sqrt{2}}{9+\sqrt{2}}$

$ = \dfrac{10\left(9+\sqrt{2}\right)}{\left(9\right)^2 - \left(\sqrt{2}\right)^2} = \dfrac{10\left(9+\sqrt{2}\right)}{81-2}$

$ = \dfrac{90 + 10\sqrt{2}}{79}$

75) $\dfrac{\sqrt{8}}{\sqrt{3}+\sqrt{2}} = \dfrac{2\sqrt{2}}{\sqrt{3}+\sqrt{2}} \cdot \dfrac{\sqrt{3}-\sqrt{2}}{\sqrt{3}-\sqrt{2}}$

$ = \dfrac{2\sqrt{2}\left(\sqrt{3}-\sqrt{2}\right)}{\left(\sqrt{3}\right)^2 - \left(\sqrt{2}\right)^2}$

$ = \dfrac{2\sqrt{6} - 2(2)}{3-2} = 2\sqrt{6} - 4$

77) $\dfrac{\sqrt{3}-\sqrt{5}}{\sqrt{10}-\sqrt{3}} = \dfrac{\sqrt{3}-\sqrt{5}}{\sqrt{10}-\sqrt{3}} \cdot \dfrac{\sqrt{10}+\sqrt{3}}{\sqrt{10}+\sqrt{3}}$

$ = \dfrac{\left(\sqrt{3}-\sqrt{5}\right)\left(\sqrt{10}+\sqrt{3}\right)}{\left(\sqrt{10}\right)^2 - \left(\sqrt{3}\right)^2}$

$ = \dfrac{\sqrt{30}+3-\sqrt{50}-\sqrt{15}}{10-3}$

$ = \dfrac{\sqrt{30}+3-5\sqrt{2}-\sqrt{15}}{7}$

79) $\dfrac{\sqrt{m}}{\sqrt{m}+\sqrt{n}} = \dfrac{\sqrt{m}}{\sqrt{m}+\sqrt{n}} \cdot \dfrac{\sqrt{m}-\sqrt{n}}{\sqrt{m}-\sqrt{n}}$

$ = \dfrac{\sqrt{m}\left(\sqrt{m}-\sqrt{n}\right)}{\left(\sqrt{m}\right)^2 - \left(\sqrt{m}\right)^2} = \dfrac{m-\sqrt{mn}}{m-n}$

81) $\dfrac{b-25}{\sqrt{b}-5} = \dfrac{b-25}{\sqrt{b}-5} \cdot \dfrac{\sqrt{b}+5}{\sqrt{b}+5}$

$ = \dfrac{\left(b-25\right)\left(\sqrt{b}+5\right)}{\left(\sqrt{b}\right)^2 - \left(5\right)^2}$

$ = \dfrac{\left(b-25\right)\left(\sqrt{b}+5\right)}{b-25} = \sqrt{b}+5$

83) $\dfrac{\sqrt{x}+\sqrt{y}}{\sqrt{x}-\sqrt{y}} = \dfrac{\sqrt{x}+\sqrt{y}}{\sqrt{x}-\sqrt{y}} \cdot \dfrac{\sqrt{x}+\sqrt{y}}{\sqrt{x}+\sqrt{y}}$

$ = \dfrac{\left(\sqrt{x}+\sqrt{y}\right)^2}{\left(\sqrt{x}\right)^2 - \left(\sqrt{y}\right)^2}$

$ = \dfrac{x + 2\sqrt{xy} + y}{x-y}$

85) $\dfrac{\sqrt{5}}{3} = \dfrac{\sqrt{5}}{3} \cdot \dfrac{\sqrt{5}}{\sqrt{5}} = \dfrac{\sqrt{25}}{3\sqrt{5}} = \dfrac{5}{3\sqrt{5}}$

87) $\dfrac{\sqrt{x}}{\sqrt{7}} = \dfrac{\sqrt{x}}{\sqrt{7}} \cdot \dfrac{\sqrt{x}}{\sqrt{x}} = \dfrac{\sqrt{x^2}}{\sqrt{7x}} = \dfrac{x}{\sqrt{7x}}$

89) $\dfrac{2+\sqrt{3}}{6} = \dfrac{2+\sqrt{3}}{6} \cdot \dfrac{2-\sqrt{3}}{2-\sqrt{3}} = \dfrac{4-3}{6\left(2-\sqrt{3}\right)} = \dfrac{1}{12-6\sqrt{3}}$

91) $\dfrac{\sqrt{x}-2}{x-4} = \dfrac{\sqrt{x}-2}{x-4} \cdot \dfrac{\sqrt{x}+2}{\sqrt{x}+2}$

$ = \dfrac{x-4}{\left(x-4\right)\left(\sqrt{x}+2\right)} = \dfrac{1}{\sqrt{x}+2}$

93) $\dfrac{4-\sqrt{c+11}}{c-5} = \dfrac{4-\sqrt{c+11}}{c-5} \cdot \dfrac{4+\sqrt{c+11}}{4+\sqrt{c+11}}$

$ = \dfrac{16 - \left(c+11\right)}{\left(c-5\right)\left(4+\sqrt{c+11}\right)}$

$ = \dfrac{16-c-11}{\left(c-5\right)\left(4+\sqrt{c+11}\right)} = \dfrac{5-c}{\left(c-5\right)\left(4+\sqrt{c+11}\right)}$

$ = -\dfrac{1}{4+\sqrt{c+11}}$

Chapter 10 Putting It All Together

95) No, because when we multiply the numerator and denominator by the conjugate of the denominator, we are multiplying the original expression by 1.

97) $\dfrac{5+10\sqrt{3}}{5} = \dfrac{5\left(1+2\sqrt{3}\right)}{5} = 1+2\sqrt{3}$

99) $\dfrac{30-18\sqrt{5}}{4} = \dfrac{6\left(5-3\sqrt{5}\right)}{4}$

$= \dfrac{3\left(5-3\sqrt{5}\right)}{2} = \dfrac{15-9\sqrt{5}}{2}$

101) $\dfrac{\sqrt{45}+6}{9} = \dfrac{3\sqrt{5}+6}{9} = \dfrac{3\left(\sqrt{5}+2\right)}{9}$

$= \dfrac{\sqrt{5}+2}{3}$

103) $\dfrac{-10-\sqrt{50}}{5} = \dfrac{-10-5\sqrt{2}}{5}$

$= \dfrac{5\left(-2-\sqrt{2}\right)}{5} = -2-\sqrt{2}$

105)

a) $r(A) = \sqrt{\dfrac{A}{\pi}}$

$r(8\pi) = \sqrt{\dfrac{8\pi}{\pi}} = \sqrt{8} = 2\sqrt{2}$

When the area of the circle is 8π in^2, the radius is $2\sqrt{2}$ in.

b) $r(A) = \sqrt{\dfrac{A}{\pi}}$

$r(7) = \sqrt{\dfrac{7}{\pi}} = \dfrac{\sqrt{7}}{\sqrt{\pi}}\cdot\dfrac{\sqrt{\pi}}{\sqrt{\pi}} = \dfrac{\sqrt{7\pi}}{\pi}$

When the area of the circle is 7 in^2,

the radius is $\dfrac{\sqrt{7\pi}}{\pi}$ in.

(This is approximately 1.5 in.)

c) $r(A) = \sqrt{\dfrac{A}{\pi}} = \dfrac{\sqrt{A}}{\sqrt{\pi}}\cdot\dfrac{\sqrt{\pi}}{\sqrt{\pi}} = \dfrac{\sqrt{A\pi}}{\pi}$

Putting It All Together:

1) $\sqrt[4]{81} = 3$

3) $-\sqrt[6]{64} = -2$

5) $\sqrt{-169}$ is not a real number.

7) $(144)^{1/2} = \sqrt{144} = 12$

9) $-1000^{2/3} = -\left(1000^{1/3}\right)^2 = -\left(\sqrt[3]{1000}\right)^2$

$= -(10)^2 = -100$

11) $125^{-1/3} = \left(\dfrac{1}{125}\right)^{1/3} = \sqrt[3]{\dfrac{1}{125}} = \dfrac{1}{5}$

13) $k^{-3/5}\cdot k^{3/10} = k^{-6/10+3/10} = k^{-3/10} = \dfrac{1}{k^{3/10}}$

15) $\left(\dfrac{27a^{-8}}{b^9}\right)^{2/3} = \dfrac{(27)^{2/3}\left(a^{-8}\right)^{2/3}}{\left(b^9\right)^{2/3}}$

$= \dfrac{\left(27^{1/3}\right)^2 a^{-16/3}}{b^{18/3}} = \dfrac{9a^{-16/3}}{b^6}$

$= \dfrac{9}{a^{16/3}b^6}$

17) $\sqrt{24} = \sqrt{4\cdot6} = \sqrt{4}\cdot\sqrt{6} = 2\sqrt{6}$

19) $\sqrt[3]{72} = \sqrt[3]{8}\cdot\sqrt[3]{9} = 2\sqrt[3]{9}$

21) $\sqrt[4]{243} = \sqrt[4]{81\cdot3} = \sqrt[4]{81}\cdot\sqrt[4]{3} = 3\sqrt[4]{3}$

23) $\sqrt[3]{96m^7n^{15}} = \sqrt[3]{96}\cdot\sqrt[3]{m^7}\cdot\sqrt[3]{n^{15}}$

$= \sqrt[3]{8\cdot12}\cdot\sqrt[3]{m^6\cdot m}\cdot\sqrt[3]{n^{15}}$

$= 2\sqrt[3]{12}\cdot m^2\sqrt[3]{m}\cdot n^5$

$= 2m^2n^5\sqrt[3]{12m}$

25) $\sqrt[3]{12}\cdot\sqrt[3]{2} = \sqrt[3]{24} = \sqrt[3]{8\cdot3} = 2\sqrt[3]{3}$

321

Chapter 10: Radicals and Rational Exponents

27) $\left(6+\sqrt{7}\right)\left(2+\sqrt{7}\right)$

$=12+6\sqrt{7}+2\sqrt{7}+7=19+8\sqrt{7}$

29) $\dfrac{18}{\sqrt{6}}=\dfrac{18}{\sqrt{6}}\cdot\dfrac{\sqrt{6}}{\sqrt{6}}=\dfrac{18\sqrt{6}}{\sqrt{36}}=\dfrac{18\sqrt{6}}{6}=3\sqrt{6}$

31) $3\sqrt{75m^3n}+m\sqrt{12mn}$

$=3\left(5m\sqrt{3mn}\right)+m\left(2\sqrt{3mn}\right)$

$=15m\sqrt{3mn}+2m\sqrt{3mn}=17m\sqrt{3mn}$

33) $\dfrac{\sqrt{60t^8u^3}}{\sqrt{5t^2u}}=\sqrt{\dfrac{60t^8u^3}{5t^2u}}=\sqrt{12t^6u^2}$

$=\sqrt{12}\cdot\sqrt{t^6}\cdot\sqrt{u^2}$

$=2\sqrt{3}\cdot t^3\cdot u=2t^3u\sqrt{3}$

35) $\left(2\sqrt{3}+10\right)^2$

$=\left(2\sqrt{3}\right)^2+2\left(2\sqrt{3}\right)(10)+(10)^2$

$=4(3)+40\sqrt{3}+100=12+40\sqrt{3}+100$

$=112+40\sqrt{3}$

37) $\dfrac{\sqrt{2}}{4+\sqrt{10}}=\dfrac{\sqrt{2}}{4+\sqrt{10}}\cdot\dfrac{4-\sqrt{10}}{4-\sqrt{10}}$

$=\dfrac{\sqrt{2}\left(4-\sqrt{10}\right)}{(4)^2-\left(\sqrt{10}\right)^2}=\dfrac{\left(4\sqrt{2}-\sqrt{20}\right)}{16-10}$

$=\dfrac{\left(4\sqrt{2}-\sqrt{4\cdot5}\right)}{6}=\dfrac{\left(4\sqrt{2}-2\sqrt{5}\right)}{6}$

$=\dfrac{2\left(2\sqrt{2}-\sqrt{5}\right)}{6}=\dfrac{2\sqrt{2}-\sqrt{5}}{3}$

39) $\sqrt[3]{\dfrac{b^2}{9c}}=\dfrac{\sqrt[3]{b^2}}{\sqrt[3]{9c}}=\dfrac{\sqrt[3]{b^2}}{\sqrt[3]{9c}}\cdot\dfrac{\sqrt[3]{3c^2}}{\sqrt[3]{3c^2}}=\dfrac{\sqrt[3]{3b^2c^2}}{\sqrt[3]{27c^3}}=\dfrac{\sqrt[3]{3b^2c^2}}{3c}$

Section 10.7: Exercises

1) Sometimes these are extraneous solutions.

3) $\sqrt{q}=7$

$\left(\sqrt{q}\right)^2=7^2$

$q=49$

Check $\sqrt{49}=7$ $\{7\}$

5) $\sqrt{w}-\dfrac{2}{3}=0$

$\sqrt{w}=\dfrac{2}{3}$

$\left(\sqrt{w}\right)^2=\left(\dfrac{2}{3}\right)^2$

$w=\dfrac{4}{9}$ $\left\{\dfrac{4}{9}\right\}$

Check is left to the student.

7) $\sqrt{a}+5=3$

$\sqrt{a}=-2$

$\left(\sqrt{a}\right)^2=(-2)^2$

$a=4$

Check $\sqrt{4}+5=3$

$2+5\neq3$ $\varnothing$

9) $\sqrt{b-11}-3=0$

$\sqrt{b-11}=3$

$\left(\sqrt{b-11}\right)^2=(3)^2$

$b-11=9$

$b=20$

Check is left to the student. $\{20\}$

11) $\sqrt{4g-1}+7=1$

$$\sqrt{4g-1}=-6$$

$$\left(\sqrt{4g-1}\right)^2=(-6)^2$$

$$4g-1=36$$

$$4g=37$$

$$g=\frac{37}{4}$$

$\frac{37}{4}$ is an extraneous solution. $\varnothing$

13) $\sqrt{3f+2}+9=11$

$$\sqrt{3f+2}=2$$

$$\left(\sqrt{3f+2}\right)^2=(2)^2$$

$$3f+2=4$$

$$3f=2$$

$$f=\frac{2}{3}$$

Check is left to the student. $\left\{\frac{2}{3}\right\}$

15) $m=\sqrt{m^2-3m+6}$

$$m^2=\left(\sqrt{m^2-3m+6}\right)^2$$

$$m^2=m^2-3m+6$$

$$0=-3m+6$$

$$3m=6$$

$$m=2$$

Check is left to the student. $\{2\}$

17) $\sqrt{9r^2-2r+10}=3r$

$$\left(\sqrt{9r^2-2r+10}\right)^2=(3r)^2$$

$$9r^2-2r+10=9r^2$$

$$-2r+10=0$$

$$-2r=-10$$

$$r=5$$

Check is left to the student. $\{5\}$

19) $(n+5)^2=n^2+2(n)(5)+5^2$

$$=n^2+10n+25$$

21) $(c-6)^2=c^2-2(c)(6)+6^2$

$$=c^2-12c+36$$

23) $p+6=\sqrt{12+p}$

$$(p+6)^2=\left(\sqrt{12+p}\right)^2$$

$$p^2+12p+36=12+p$$

$$p^2+11p+24=0$$

$$(p+8)(p+3)=0$$

$$p+8=0 \text{ or } p+3=0$$

$$p=-8 \qquad p=-3$$

-8 is an extraneous solution. $\{-3\}$

25) $6+\sqrt{c^2+3c-9}=c$

$$\sqrt{c^2+3c-9}=c-6$$

$$\left(\sqrt{c^2+3c-9}\right)^2=(c-6)^2$$

$$c^2+3c-9=c^2-12c+36$$

$$3c-9=-12c+36$$

$$15c=45$$

$$c=3$$

3 is an extraneous solution. $\varnothing$

27) $w - \sqrt{10w + 6} = -3$

$$w + 3 = \sqrt{10w + 6}$$

$$(w + 3)^2 = \left(\sqrt{10w + 6}\right)^2$$

$$w^2 + 6w + 9 = 10w + 6$$

$$w^2 - 4w + 3 = 0$$

$$(w - 3)(w - 1) = 0$$

$$w - 3 = 0 \quad \text{or} \quad w - 1 = 0$$

$$w = 3 \qquad \qquad w = 1$$

Check is left to the student. $\{1, 3\}$

29) $\qquad\qquad 3v = 8 + \sqrt{3v + 4}$

$$3v - 8 = \sqrt{3v + 4}$$

$$(3v - 8)^2 = \left(\sqrt{3v + 4}\right)^2$$

$$9v^2 - 48v + 64 = 3v + 4$$

$$9v^2 - 51v + 60 = 0 \qquad \text{Divide by 3.}$$

$$3v^2 - 17v + 20 = 0$$

$$(3v - 5)(v - 4) = 0$$

$$3v - 5 = 0 \quad \text{or} \quad v - 4 = 0$$

$$3v = 5 \qquad \qquad v = 4$$

$$v = \frac{5}{3}$$

$\frac{5}{3}$ is an extraneous solution. $\{4\}$

31) $\qquad\qquad m + 4 = 5\sqrt{m}$

$$(m + 4)^2 = \left(5\sqrt{m}\right)^2$$

$$m^2 + 8m + 16 = 25m$$

$$m^2 + 8m - 25m + 16 = 0$$

$$m^2 - 17m + 16 = 0$$

$$(m - 16)(m - 1) = 0$$

$$m - 16 = 0 \quad \text{or} \quad m - 1 = 0$$

$$m = 16 \qquad \qquad m = 1$$

Check is left to the student. $\{1, 16\}$

33) $\qquad y + 2\sqrt{6 - y} = 3$

$$2\sqrt{6 - y} = 3 - y$$

$$\left(2\sqrt{6 - y}\right)^2 = (3 - y)^2$$

$$4(6 - y) = 9 - 6y + y^2$$

$$24 - 4y = y^2 - 6y + 9$$

$$0 = y^2 - 6y + 4y - 24 + 9$$

$$y^2 - 2y - 15 = 0$$

$$(y - 5)(y + 3) = 0$$

$$y - 5 = 0 \qquad\qquad y + 3 = 0$$

$$y = 5 \qquad\qquad y = -3$$

5 is an extraneous solution. $\{-3\}$

35) $\qquad \sqrt{r^2 - 8r - 19} = r - 9$

$$\left(\sqrt{r^2 - 8r - 19}\right)^2 = (r - 9)^2$$

$$r^2 - 8r - 19 = r^2 - 18r + 81$$

$$-8r - 19 = -18r + 81$$

$$10r = 100$$

$$r = 10$$

Check is left to the student. $\{10\}$

37) $\qquad 5\sqrt{1 - 5h} = 4\sqrt{1 - 8h}$

$$\left(5\sqrt{1 - 5h}\right)^2 = \left(4\sqrt{1 - 8h}\right)^2$$

$$25(1 - 5h) = 16(1 - 8h)$$

$$25 - 125h = 16 - 128h$$

$$3h = -9$$

$$h = -3$$

Check is left to the student. $\{-3\}$

39) $3\sqrt{3x+6} - 2\sqrt{9x-9} = 0$

$$3\sqrt{3x+6} = 2\sqrt{9x-9}$$

$$\left(3\sqrt{3x+6}\right)^2 = \left(2\sqrt{9x-9}\right)^2$$

$$9(3x+6) = 4(9x-9)$$

$$27x+54 = 36x-36$$

$$90 = 9x$$

$$10 = x$$

Check is left to the student. $\{10\}$

41) $\sqrt{m} = 3\sqrt{7}$

$$\left(\sqrt{m}\right)^2 = \left(3\sqrt{7}\right)^2$$

$$m = 9(7)$$

$$m = 63$$

Check is left to the student. $\{63\}$

43) $\sqrt{2w-1} + 2\sqrt{w+4} = 0$

$$\left(\sqrt{2w-1}\right)^2 = \left(-2\sqrt{w+4}\right)^2$$

$$2w-1 = 4w+16$$

$$2w-1-4w = 16$$

$$-2w-1 = 16$$

$$-2w = 17$$

$\dfrac{-17}{2}$ is an extraneous solution. $\varnothing$

45) $\left(\sqrt{x}+5\right)^2 = \left(\sqrt{x}\right)^2 + 2\left(\sqrt{x}\right)(5) + (5)^2$

$$= x + 10\sqrt{x} + 25$$

47) $\left(9 - \sqrt{a+4}\right)^2$

$$= (9)^2 - 2(9)\left(\sqrt{a+4}\right) + \left(\sqrt{a+4}\right)^2$$

$$= 81 - 18\sqrt{a+4} + a + 4$$

$$= 85 - 18\sqrt{a+4} + a$$

49) $\left(2\sqrt{3n-1}+7\right)^2$

$$= \left(2\sqrt{3n-1}\right)^2 + 2\left(2\sqrt{3n-1}\right)(7) + 7^2$$

$$= 4(3n-1) + 28\sqrt{3n-1} + 49$$

$$= 12n - 4 + 28\sqrt{3n-1} + 49$$

$$= 12n + 28\sqrt{3n-1} + 45$$

51) $\sqrt{2y-1} = 2 + \sqrt{y-4}$

$$\left(\sqrt{2y-1}\right)^2 = \left(2+\sqrt{y-4}\right)^2$$

$$2y-1 = 4 + 4\sqrt{y-4} + \left(\sqrt{y-4}\right)^2$$

$$2y-1 = 4 + 4\sqrt{y-4} + y - 4$$

$$2y-1 = y + 4\sqrt{y-4}$$

$$y-1 = 4\sqrt{y-4}$$

$$(y-1)^2 = \left(4\sqrt{y-4}\right)^2$$

$$y^2 - 2y + 1 = 16y - 64$$

$$y^2 - 2y - 16y + 1 + 64 = 0$$

$$y^2 - 18y + 65 = 0$$

$$(y-13)(y-5) = 0$$

$$y-13 = 0 \ \text{ or } \ y-5 = 0$$

$$y = 13 \qquad y = 5$$

Check is left to the student. $\{5,13\}$

Chapter 10: Radicals and Rational Exponents

53)
$$1+\sqrt{3s-2}=\sqrt{2s+5}$$
$$\left(1+\sqrt{3s-2}\right)^2=\left(\sqrt{2s+5}\right)^2$$
$$1+2\sqrt{3s-2}+3s-2=2s+5$$
$$3s-1+2\sqrt{3s-2}=2s+5$$
$$2\sqrt{3s-2}=6-s$$
$$\left(2\sqrt{3s-2}\right)^2=(6-s)^2$$
$$4(3s-2)=36-12s+s^2$$
$$12s-8=s^2-12s+36$$
$$0=s^2-24s+44$$
$$0=(s-2)(s-22)$$
$$s-2=0 \ \text{ or } \ s-22=0$$
$$s=2 \qquad\qquad s=22$$
22 is an extraneous solution. $\{2\}$

55) $\sqrt{5a+19}-\sqrt{a+12}=1$
$$\sqrt{5a+19}=1+\sqrt{a+12}$$
$$\left(\sqrt{5a+19}\right)^2=\left(1+\sqrt{a+12}\right)^2$$
$$5a+19=1+2\sqrt{a+12}+a+12$$
$$5a+19=a+13+2\sqrt{a+12}$$
$$4a+6=2\sqrt{a+12}$$
$$2a+3=\sqrt{a+12}$$
$$\left(2a+3\right)^2=\left(\sqrt{a+12}\right)^2$$
$$4a^2+12a+9=a+12$$
$$4a^2+11a-3=0$$
$$(4a-1)(a+3)=0$$
$$4a-1=0 \ \text{ or } \ a+3=0$$
$$4a=1$$
$$a=\frac{1}{4} \qquad a=-3$$
-3 is an extraneous solution. $\left\{\dfrac{1}{4}\right\}$

57) $\sqrt{3k+1}-\sqrt{k-1}=2$
$$\sqrt{3k+1}=2+\sqrt{k-1}$$
$$\left(\sqrt{3k+1}\right)^2=\left(2+\sqrt{k-1}\right)^2$$
$$3k+1=4+4\sqrt{k-1}+k-1$$
$$3k+1=k+3+4\sqrt{k-1}$$
$$2k-2=4\sqrt{k-1}$$
$$k-1=2\sqrt{k-1}$$
$$(k-1)^2=\left(2\sqrt{k-1}\right)^2$$
$$k^2-2k+1=4(k-1)$$
$$k^2-2k+1=4k-4$$
$$k^2-6k+5=0$$
$$(k-1)(k-5)=0$$
$$k-1=0 \ \text{ or } \ k-5=0$$
$$k=1 \qquad\qquad k=5$$
Check is left to the student. $\{1,5\}$

59)
$$\sqrt{3x+4}-5=\sqrt{3x-11}$$
$$\left(\sqrt{3x+4}-5\right)^2=\left(\sqrt{3x-11}\right)^2$$
$$3x+4-10\sqrt{3x+4}+25=3x-11$$
$$3x+29-10\sqrt{3x+4}=3x-11$$
$$-10\sqrt{3x+4}=-40$$
$$\sqrt{3x+4}=4$$
$$\left(\sqrt{3x+4}\right)^2=(4)^2$$
$$3x+4=16$$
$$3x=12$$
$$x=4$$
4 is an extraneous solution. $\varnothing$

61) $\sqrt{3v+3} - \sqrt{v-2} = 3$

$$\sqrt{3v+3} = \sqrt{v-2} + 3$$

$$\left(\sqrt{3v+3}\right)^2 = \left(\sqrt{v-2} + 3\right)^2$$

$$3v+3 = 9 + 6\sqrt{v-2} + v - 2$$

$$3v+3 = 7 + 6\sqrt{v-2} + v$$

$$2v-4 = 6\sqrt{v-2}$$

$$v-2 = 3\sqrt{v-2}$$

$$(v-2)^2 = \left(3\sqrt{v-2}\right)^2$$

$$v^2 - 4v + 4 = 9(v-2)$$

$$v^2 - 4v + 4 = 9v - 18$$

$$v^2 - 13v + 22 = 0$$

$$(v-2)(v-11) = 0$$

$$v - 2 = 0 \quad \text{or} \quad v - 11 = 0$$

$$v = 2 \qquad v = 11$$

Check is left to the student. $\{2, 11\}$

63) Raise both sides of the equation to the third power.
and the cube root of -27 is -3.

65) $\sqrt[3]{y} = 5$

$$\left(\sqrt[3]{y}\right)^3 = 5^3$$

$$y = 125$$

Check is left to the student. $\{125\}$

67) $\sqrt[3]{m} = -4$

$$\left(\sqrt[3]{m}\right)^3 = (-4)^3$$

$$m = -64$$

Check is left to the student. $\{-64\}$

69) $\sqrt[3]{2x-5} + 3 = 1$

$$\sqrt[3]{2x-5} = -2$$

$$\left(\sqrt[3]{2x-5}\right)^3 = (-2)^3$$

$$2x - 5 = -8$$

$$2x = -3$$

$$x = -\frac{3}{2}$$

Check is left to the student. $\left\{-\dfrac{3}{2}\right\}$

71) $\sqrt[3]{6j-2} = \sqrt[3]{j-7}$

$$\left(\sqrt[3]{6j-2}\right)^3 = \left(\sqrt[3]{j-7}\right)^3$$

$$6j - 2 = j - 7$$

$$5j = -5$$

$$j = -1$$

Check is left to the student.
$\{-1\}$

73) $\sqrt[3]{3y-1} - \sqrt[3]{2y-3} = 0$

$$\left(\sqrt[3]{3y-1}\right)^3 = \left(\sqrt[3]{2y-3}\right)^3$$

$$3y - 1 = 2y - 3$$

$$y = -2$$

Check is left to the student. $\{-2\}$

75) $\sqrt[3]{2n^2} = \sqrt[3]{7n+4}$

$\left(\sqrt[3]{2n^2}\right)^3 = \left(\sqrt[3]{7n+4}\right)^3$

$2n^2 = 7n+4$

$2n^2 - 7n - 4 = 0$

$(2n+1)(n-4) = 0$

$2n+1 = 0 \qquad n-4 = 0$

$n = -\dfrac{1}{2} \qquad n = 4$

Check is left to the student. $\left\{-\dfrac{1}{2}, 4\right\}$

77) $p^{1/2} = 6$

$\left(p^{1/2}\right)^2 = (6)^2$

$p = 36$

Check is left to the student $\{36\}$

79) $7 = (2z-3)^{1/2}$

$(7)^2 = \left((2z-3)^{1/2}\right)^2$

$49 = 2z - 3$

$52 = 2z$

$26 = z$

Check is left to the student. $\{26\}$

81) $(y+4)^{1/3} = 3$

$\left((y+4)^{1/3}\right)^3 = (3)^3$

$y + 4 = 27$

$y = 23$

Check is left to the student. $\{23\}$

83) $\sqrt[4]{n+7} = 2$

$\left(\sqrt[4]{n+7}\right)^4 = (2)^4$

$n + 7 = 16$

$n = 9$

Check is left to the student. $\{9\}$

85) $\sqrt{13+\sqrt{r}} = \sqrt{r+7}$

$\left(\sqrt{13+\sqrt{r}}\right)^2 = \left(\sqrt{r+7}\right)^2$

$13 + \sqrt{r} = r + 7$

$\sqrt{r} = r - 6$

$\left(\sqrt{r}\right)^2 = (r-6)^2$

$r = r^2 - 12r + 36$

$0 = r^2 - 13r + 36$

$0 = (r-9)(r-4)$

$r - 9 = 0 \;$ or $\; r - 4 = 0$

$r = 9 \qquad\qquad r = 4$

4 is an extraneous solution. $\{9\}$

87) $\sqrt{y+\sqrt{y+5}} = \sqrt{y+2}$

$\left(\sqrt{y+\sqrt{y+5}}\right)^2 = \left(\sqrt{y+2}\right)^2$

$y + \sqrt{y+5} = y + 2$

$\sqrt{y+5} = 2$

$\left(\sqrt{y+5}\right)^2 = (2)^2$

$y + 5 = 4$

$y = -1$

Check is left to the student. $\{-1\}$

89) $v = \sqrt{\dfrac{2E}{m}}$

$v^2 = \left(\sqrt{\dfrac{2E}{m}}\right)^2$

$v^2 = \dfrac{2E}{m}$

$mv^2 = 2E$

$\dfrac{mv^2}{2} = E$

91) $c = \sqrt{a^2 + b^2}$

$c^2 = \left(\sqrt{a^2 + b^2}\right)^2$

$c^2 = a^2 + b^2$

$c^2 - a^2 = b^2$

93) $T = \sqrt[4]{\dfrac{E}{\sigma}}$

$T^4 = \left(\sqrt[4]{\dfrac{E}{\sigma}}\right)^4$

$T^4 = \dfrac{E}{\sigma}$

$\sigma T^4 = E$

$\sigma = \dfrac{E}{T^4}$

95) a) Let $T = -17$

$V_s = 20\sqrt{-17 + 273}$

$= 20\sqrt{256}$

$= 20(16)$

$= 320 \qquad\qquad$ 320 m/s

b) Let $T = 16$

$V_s = 20\sqrt{16 + 273}$

$= 20\sqrt{289}$

$= 20(17)$

$= 340 \qquad\qquad$ 340 m/s

c) The speed of sound increases.

d) $\qquad V_s = 20\sqrt{T + 273}$

$V_s^2 = \left(20\sqrt{T + 273}\right)^2$

$V_s^2 = 400(T + 273)$

$\dfrac{V_s^2}{400} = T + 273$

$\dfrac{V_s^2}{400} - 273 = T$

97) a) Let $V = 28\pi$ and $h = 7$, solve for r.

$r = \sqrt{\dfrac{28\pi}{\pi(7)}} = \sqrt{\dfrac{28}{7}} = \sqrt{4} = 2$

2 in.

b) $r = \sqrt{\dfrac{V}{\pi h}}$

$r^2 = \left(\sqrt{\dfrac{V}{\pi h}}\right)^2$

$r^2 = \dfrac{V}{\pi h}$

$\pi r^2 h = V$

99) a) $c = \sqrt{(32)(14400)}$

$c = 678.82$ ft/sec

$c = \dfrac{678.82(3600)}{5280}$

$c = 463$ mph

b) distance $= (\text{rate})(\text{time})$

$60\,\text{miles} = (678.82\ \text{ft/sec})(t)$

$t = \dfrac{(60)(5280)}{678.82}$

$t = 466.69$ seconds

$= \dfrac{466.69}{60} = 7.78$

≈ 8 minutes

101) $D = 1.2\sqrt{h}$

$D^2 = \left(1.2\sqrt{h}\right)^2$

$D^2 = 1.44h$

$\dfrac{D^2}{1.44} = h$

$\dfrac{(4.8)^2}{1.44} = h$

$h = 16\,\text{ft}$

103) $W = 35.74 - 35.75V^{4/25}$

$W - 35.74 = -35.75V^{4/25}$

$\dfrac{W - 35.74}{-35.75} = V^{4/25}$

$V = \left(\dfrac{W - 35.74}{-35.75}\right)^{25/4}$

$V = \left(\dfrac{-10 - 35.74}{-35.75}\right)^{25/4}$

$V = \left(\dfrac{-45.74}{-35.75}\right)^{25/4}$

$V = 4.67\,\text{mph}$

$\approx 5\,\text{mph}$

Section 10.8: Exercises

1) False

3) True

5) $\sqrt{-81} = \sqrt{-1} \cdot \sqrt{81} = i \cdot 9 = 9i$

7) $\sqrt{-25} = \sqrt{-1} \cdot \sqrt{25} = i \cdot 5 = 5i$

9) $\sqrt{-6} = \sqrt{-1} \cdot \sqrt{6} = i\sqrt{6}$

11) $\sqrt{-27} = \sqrt{-1} \cdot \sqrt{27} = i \cdot 3\sqrt{3} = 3i\sqrt{3}$

13) $\sqrt{-60} = \sqrt{-1} \cdot \sqrt{60} = i \cdot 2\sqrt{15} = 2i\sqrt{15}$

15) Write each radical in terms of i before multiplying.

$\sqrt{-5} \cdot \sqrt{-10} = i\sqrt{5} \cdot i\sqrt{10} = i^2\sqrt{50}$

$= -1\sqrt{25} \cdot \sqrt{2} = -5\sqrt{2}$

17) $\sqrt{-1} \cdot \sqrt{-5} = (i)(i\sqrt{5}) = i^2\sqrt{5} = -\sqrt{5}$

19) $\sqrt{-12} \cdot \sqrt{-3} = \left(i\sqrt{12}\right)\left(i\sqrt{3}\right)$

$= i^2\sqrt{36} = -1(6) = -6$

21) $\dfrac{\sqrt{-60}}{\sqrt{-15}} = \dfrac{i\sqrt{60}}{i\sqrt{15}} = \sqrt{\dfrac{60}{15}} = \sqrt{4} = 2$

23) $\left(\sqrt{-13}\right)^2 = \left(i\sqrt{13}\right)^2 = i^2(13)$

$= -1(13) = -13$

25) Add the real parts and add the imaginary parts.

27) -1

29) $(-4 + 9i) + (7 + 2i) = 3 + 11i$

31) $(13 - 8i) - (9 + i) = 4 - 9i$

33) $\left(-\dfrac{3}{4} - \dfrac{1}{6}i\right) - \left(-\dfrac{1}{2} + \dfrac{2}{3}i\right)$

$= \left(-\dfrac{3}{4} + \dfrac{1}{2}\right) + \left(-\dfrac{1}{6}i - \dfrac{2}{3}i\right)$

$= \left(-\dfrac{3}{4} + \dfrac{2}{4}\right) + \left(-\dfrac{1}{6}i - \dfrac{4}{6}i\right) = -\dfrac{1}{4} - \dfrac{5}{6}i$

35) $16i - (3 + 10i) + (3 + i)$

$= 16i - 3 - 10i + 3 + i = 7i$

37) $3(8 - 5i) = 24 - 15i$

39) $\dfrac{2}{3}(-9 + 2i) = -6 + \dfrac{4}{3}i$

41) $6i(5 + 6i) = 30i + 36i^2$

$= 30i + 36(-1) = -36 + 30i$

43) $(2 + 5i)(1 + 6i) = 2 + 12i + 5i + 30i^2$

$= 2 + 17i + 30(-1)$

$= 2 + 17i - 30$

$= -28 + 17i$

45) $(-1+3i)(4-6i) = -4+6i+12i-18i^2$

$= -4+18i-18(-1)$

$= -4+18i+18$

$= 14+18i$

47) $(5-3i)(9-3i) = 45-15i-27i+9i^2$

$= 45-42i+9(-1)$

$= 45-42i-9$

$= 36-42i$

49) $\left(\dfrac{3}{4}+\dfrac{3}{4}i\right)\left(\dfrac{2}{5}+\dfrac{1}{5}i\right)$

$= \dfrac{3}{10}+\dfrac{3}{20}i+\dfrac{3}{10}i+\dfrac{3}{20}i^2$

$= \dfrac{3}{10}+\dfrac{9}{20}i+\dfrac{3}{20}(-1)$

$= \dfrac{3}{10}+\dfrac{9}{20}i-\dfrac{3}{20} = \dfrac{3}{20}+\dfrac{9}{20}i$

51) $(11+4i)(11-4i)$

$= 121-44i+44i-16i^2$

$= 121-16(-1) = 121+16 = 137$

53) $(-3-7i)(-3+7i)$

$= 9-21i+21i-49i^2 = 9-49(-1)$

$= 9+49 = 58$

55) $(-6+4i)(-6-4i)$

$= 36+24i-24i-16i^2$

$= 36-16(-1) = 36+16 = 52$

57) Answers may vary.

59) $\dfrac{4}{2-3i} = \dfrac{4}{2-3i}\cdot\dfrac{2+3i}{2+3i} = \dfrac{8+12i}{2^2+3^2}$

$= \dfrac{8+12i}{4+9} = \dfrac{8+12i}{13} = \dfrac{8}{13}+\dfrac{12}{13}i$

61) $\dfrac{8i}{4+i} = \dfrac{8i}{4+i}\cdot\dfrac{4-i}{4-i} = \dfrac{32i-8i^2}{4^2+1^2}$

$= \dfrac{32i-8(-1)}{16+1} = \dfrac{8}{17}+\dfrac{32}{17}i$

63) $\dfrac{2i}{-3+7i} = \dfrac{2i}{-3+7i}\cdot\dfrac{-3-7i}{-3-7i}$

$= \dfrac{-6i-14i^2}{(-3)^2+7^2} = \dfrac{-6i-14(-1)}{9+49}$

$= \dfrac{-6i+14}{58} = \dfrac{14}{58}-\dfrac{6}{58}i$

$= \dfrac{7}{29}-\dfrac{3}{29}i$

65) $\dfrac{3-8i}{-6+7i} = \dfrac{3-8i}{-6+7i}\cdot\dfrac{-6-7i}{-6-7i}$

$= \dfrac{-18-21i+48i+56i^2}{(-6)^2+7^2}$

$= \dfrac{-18+27i+56(-1)}{36+49}$

$= \dfrac{-74+27i}{85} = -\dfrac{74}{85}+\dfrac{27}{85}i$

67) $\dfrac{2+3i}{5-6i} = \dfrac{2+3i}{5-6i}\cdot\dfrac{5+6i}{5+6i}$

$= \dfrac{20+12i+15i+18i^2}{5^2+6^2}$

$= \dfrac{20+27i+18(-1)}{25+36}$

$= \dfrac{-8+27i}{61} = -\dfrac{8}{61}+\dfrac{27}{61}i$

69) $\dfrac{9}{i} = \dfrac{9}{i}\cdot\dfrac{-i}{-i} = \dfrac{-9i}{1^2} = -9i$

71) $i^{24} = \boxed{\left(i^2\right)^{12}}$ Rewrite i^{24} in terms of i^2 using the power rule

$= \left(-1\right)^{12}$ $\boxed{i^2 = -1}$

$= \boxed{1}$ Simplify

73) $i^{24} = \left(i^2\right)^{12}$

$= \left(-1\right)^{12}$

$= 1$

75) $i^{28} = \left(i^2\right)^{14}$

$= \left(-1\right)^{14}$

$= 1$

77) $i^9 = i^8 \cdot i$

$= \left(i^2\right)^4 \cdot i$

$= \left(-1\right)^4 \cdot i$

$= i$

79) $i^{35} = i^{34} \cdot i$

$= \left(i^2\right)^{17} \cdot i$

$= \left(-1\right)^{17} \cdot i$

$= -i$

81) $i^{23} = i^{22} \cdot i$

$= \left(i^2\right)^{11} \cdot i$

$= \left(-1\right)^{11} \cdot i$

$= -i$

83) $i^{42} = \left(i^2\right)^{21}$

$= \left(-1\right)^{21}$

$= -1$

85) $\left(2i\right)^5 = 2^5 \cdot i^4 \cdot i$

$= 32\left(i^2\right)^2 \cdot i$

$= 32(1) \cdot i$

$= 32i$

87) $\left(-i\right)^{14} = \left(-1\right)^{14} \cdot i^{14}$

$= 1\left(i^2\right)^7$

$= 1\left(-1\right)^7$

$= -1$

89)
$\left(-2+5i\right)^3$

$= \left(-2+5i\right)^2\left(-2+5i\right)$

$= \left(\left(-2\right)^2 + 2\left(-2\right)5i + \left(5i\right)^2\right)\left(-2+5i\right)$

$= \left(4 - 20i + 25i^2\right)\left(-2+5i\right)$

$= \left(4 - 20i - 25\right)\left(-2+5i\right)$

$= \left(-21 - 20i\right)\left(-2+5i\right)$

$= 42 - 105i + 40i - 100i^2$

$= 42 - 65i + 100$

$= 142 - 65i$

91) $1 + \sqrt{-8} = 1 + \sqrt{8(-1)}$

$= 1 + \sqrt{8}\sqrt{-1}$

$= 1 + 2i\sqrt{2}$

93) $8 - \sqrt{-45} = 8 - \sqrt{45(-1)}$

$= 8 - \sqrt{45}\sqrt{-1}$

$= 8 - 3i\sqrt{5}$

95) $\dfrac{-12+\sqrt{-32}}{4} = \dfrac{-12+\sqrt{32(-1)}}{4}$

$= \dfrac{-12+\sqrt{32}\sqrt{-1}}{4}$

$= -\dfrac{12}{4} + \dfrac{4i\sqrt{2}}{4}$

$= -3 + i\sqrt{2}$

97) $Z = Z_1 + Z_2$

$= 3 + 2j + 7 + 4j$

$= 10 + 6j$

99) $Z = Z_1 + Z_2$
$= 5 - 2j + 11 + 6j$
$= 16 + 4j$

Chapter 10 Review

1) $\sqrt{25} = 5$

3) $-\sqrt{81} = -9$

5) $\sqrt[3]{64} = 4$

7) $\sqrt[3]{-1} = -1$

9) $\sqrt[6]{-64}$ is not real

11) $\sqrt{(-13)^2} = \sqrt{169} = 13$

13) $\sqrt{p^2} = |p|$

15) $\sqrt[3]{h^3} = h$

17) Since 34 is between 25 and 36,
$$\sqrt{25} < \sqrt{34} < \sqrt{36}$$
$$5 < \sqrt{34} < 6$$
$$\sqrt{34} \approx 5.8$$

19) The denominator of the fractional exponent becomes the index on the radical. The numerator is the power to which we raise the radical expression. $8^{2/3} = \left(\sqrt[3]{8}\right)^2$

21) $36^{1/2} = \sqrt{36} = 6$

23) $\left(\dfrac{27}{125}\right)^{1/3} = \sqrt[3]{\dfrac{27}{125}} = \dfrac{3}{5}$

25) $32^{3/5} = \left(\sqrt[5]{32}\right)^3 = (2)^3 = 8$

27) $81^{-1/2} = \left(\dfrac{1}{81}\right)^{1/2} = \sqrt{\dfrac{1}{81}} = \dfrac{1}{9}$

29) $81^{-3/4} = \left(\dfrac{1}{81}\right)^{3/4} = \left(\sqrt[4]{\dfrac{1}{81}}\right)^3$
$= \left(\dfrac{1}{3}\right)^3 = \dfrac{1}{27}$

31) $\left(\dfrac{27}{1000}\right)^{-2/3} = \left(\dfrac{1000}{27}\right)^{2/3} = \left(\sqrt[3]{\dfrac{1000}{27}}\right)^2$
$= \left(\dfrac{10}{3}\right)^2 = \dfrac{100}{9}$

33) $3^{6/7} \cdot 3^{8/7} = 3^{6/7+8/7} = 3^{14/7} = 3^2 = 9$

35) $\left(8^{1/5}\right)^{10} = 8^{\frac{1}{5}\cdot 10} = 8^2 = 64$

37) $\dfrac{7^2}{7^{5/3} \cdot 7^{1/3}} = \dfrac{7^2}{7^{5/3+1/3}} = \dfrac{7^2}{7^{6/3}} = \dfrac{7^2}{7^2} = 1$

39) $\left(64a^4b^{12}\right)^{5/6} = 64^{5/6} \cdot \left(a^4\right)^{5/6} \cdot \left(b^{12}\right)^{5/6}$
$= \left(\sqrt[6]{64}\right)^5 \cdot a^{4\cdot\frac{5}{6}} \cdot b^{12\cdot\frac{5}{6}}$
$= 2^5 \cdot a^{10/3} \cdot b^{10}$
$= 32a^{10/3}b^{10}$

41) $\left(\dfrac{81c^{-5}d^9}{16c^{-1}d^2}\right)^{-1/4} = \left(\dfrac{81d^7}{16c^4}\right)^{-1/4}$
$= \left(\dfrac{16c^4}{81d^7}\right)^{1/4} = \dfrac{16^{1/4} \cdot c^{4\cdot\frac{1}{4}}}{81^{1/4} d^{7\cdot\frac{1}{4}}} = \dfrac{2c}{3d^{7/4}}$

43) $\sqrt[12]{27^4} = \left(27^4\right)^{1/12} = 27^{1/3} = \sqrt[3]{27} = 3$

45) $\sqrt[3]{7^3} = 7^{3/3} = 7$

47) $\sqrt[4]{k^{28}} = k^{28/4} = k^7$

49) $\sqrt{w^6} = w^{6/2} = w^3$

51) $\sqrt{1000} = \sqrt{100 \cdot 10} = \sqrt{100} \cdot \sqrt{10}$
$= 10\sqrt{10}$

53) $\sqrt{\dfrac{18}{49}} = \dfrac{\sqrt{18}}{\sqrt{49}} = \dfrac{\sqrt{9\cdot 2}}{7} = \dfrac{3\sqrt{2}}{7}$

55) $\sqrt{k^{12}} = k^{12/2} = k^6$

57) $\sqrt{x^9} = \sqrt{x^8} \cdot \sqrt{x} = x^{8/2} \cdot \sqrt{x} = x^4\sqrt{x}$

59) $\sqrt{45t^2} = \sqrt{45} \cdot \sqrt{t^2} = \sqrt{9} \cdot \sqrt{5} \cdot t^{2/2}$
$= 3\sqrt{5} \cdot t = 3t\sqrt{5}$

61) $\sqrt{72x^7y^{13}}$
$= \sqrt{72} \cdot \sqrt{x^7} \cdot \sqrt{y^{13}}$
$= 6\sqrt{2} \cdot x^3\sqrt{x} \cdot y^6\sqrt{y} = 6x^3y^6\sqrt{2xy}$

63) $\sqrt{5} \cdot \sqrt{3} = \sqrt{5 \cdot 3} = \sqrt{15}$

65) $\sqrt{2} \cdot \sqrt{12} = \sqrt{2 \cdot 12} = \sqrt{24}$

$\qquad = \sqrt{4} \cdot \sqrt{6} = 2\sqrt{6}$

67) $\sqrt{11x^5} \cdot \sqrt{11x^8} = \sqrt{121x^{13}}$

$\qquad = 11 \cdot \sqrt{x^{12}} \cdot \sqrt{x}$

$\qquad = 11x^6 \sqrt{x}$

69) $\dfrac{\sqrt{200k^{21}}}{\sqrt{2k^5}} = \sqrt{\dfrac{200k^{21}}{2k^5}} = \sqrt{100k^{21-5}}$

$\qquad = \sqrt{100k^{16}} = 10k^8$

71) $\sqrt[3]{16} = \sqrt[3]{8} \cdot \sqrt[3]{2} = 2\sqrt[3]{2}$

73) $\sqrt[4]{48} = \sqrt[4]{16} \cdot \sqrt[4]{3} = 2\sqrt[4]{3}$

75) $\sqrt[4]{z^{24}} = z^{24/4} = z^6$

77) $\sqrt[3]{a^{20}} = \sqrt[3]{a^{18}} \cdot \sqrt[3]{a^2} = a^{18/3} \cdot \sqrt[3]{a^2}$

$\qquad = a^6 \sqrt[3]{a^2}$

79) $\sqrt[3]{16z^{15}} = \sqrt[3]{16} \cdot \sqrt[3]{z^{15}} = 2\sqrt[3]{2} \cdot z^{15/3}$

$\qquad = 2\sqrt[3]{2} \cdot z^5 = 2z^5 \sqrt[3]{2}$

81) $\sqrt[4]{\dfrac{h^{12}}{81}} = \dfrac{\sqrt[4]{h^{12}}}{\sqrt[4]{81}} = \dfrac{h^{12/4}}{3} = \dfrac{h^3}{3}$

83) $\sqrt[3]{3} \cdot \sqrt[3]{7} = \sqrt[3]{3 \cdot 7} = \sqrt[3]{21}$

85) $\sqrt[4]{4t^7} \cdot \sqrt[4]{8t^{10}} = \sqrt[4]{32t^{17}} = \sqrt[4]{32} \cdot \sqrt[4]{t^{17}}$

$\qquad = 2\sqrt[4]{2} \cdot t^4 \sqrt[4]{t} = 2t^4 \sqrt[4]{2t}$

87) $\sqrt[3]{n} \cdot \sqrt{n} = n^{1/3} \cdot n^{1/2} = n^{2/6+3/6}$

$\qquad = n^{5/6} = \sqrt[6]{n^5}$

89) $8\sqrt{5} + 3\sqrt{5} = 11\sqrt{5}$

91) $\sqrt{80} - \sqrt{48} + \sqrt{20}$

$\qquad = 4\sqrt{5} - 4\sqrt{3} + 2\sqrt{5} = 6\sqrt{5} - 4\sqrt{3}$

93) $3p\sqrt{p} - 7\sqrt{p^3}$

$\qquad = 3p\sqrt{p} - 7(p\sqrt{p})$

$\qquad = 3p\sqrt{p} - 7p\sqrt{p} = -4p\sqrt{p}$

95) $10d^2\sqrt{8d} - 32d\sqrt{2d^3}$

$\qquad = 10d^2(2\sqrt{2d}) - 32d(d\sqrt{2d})$

$\qquad = 20d^2\sqrt{2d} - 32d^2\sqrt{2d}$

$\qquad = -12d^2\sqrt{2d}$

97) $3\sqrt{k}(\sqrt{20k} + \sqrt{2})$

$\qquad = 3\sqrt{k}(2\sqrt{5k} + \sqrt{2})$

$\qquad = 6\sqrt{5k^2} + 3\sqrt{2k} = 6k\sqrt{5} + 3\sqrt{2k}$

99) $(\sqrt{2r} + 5\sqrt{s})(3\sqrt{s} + 4\sqrt{2r})$

$\qquad = 3\sqrt{2rs} + 4\sqrt{4r^2} + 15\sqrt{s^2} + 20\sqrt{2rs}$

$\qquad = 23\sqrt{2rs} + 4(2r) + 15s$

$\qquad = 23\sqrt{2rs} + 8r + 15s$

101) $\left(1 + \sqrt{y+1}\right)^2$

$\qquad = 1^2 + 2(1)\left(\sqrt{y+1}\right) + \left(\sqrt{y+1}\right)^2$

$\qquad = 1 + 2\sqrt{y+1} + y + 1$

$\qquad = 2 + 2\sqrt{y+1} + y$

103) $\dfrac{14}{\sqrt{3}} = \dfrac{14}{\sqrt{3}} \cdot \dfrac{\sqrt{3}}{\sqrt{3}} = \dfrac{14\sqrt{3}}{3}$

105) $\dfrac{\sqrt{18k}}{\sqrt{n}} = \dfrac{3\sqrt{2k}}{\sqrt{n}} = \dfrac{3\sqrt{2k}}{\sqrt{n}} \cdot \dfrac{\sqrt{n}}{\sqrt{n}}$

$\qquad = \dfrac{3\sqrt{2kn}}{n}$

107) $\dfrac{7}{\sqrt[3]{2}} = \dfrac{7}{\sqrt[3]{2}} \cdot \dfrac{\sqrt[3]{2^2}}{\sqrt[3]{2^2}} = \dfrac{7\sqrt[3]{2^2}}{\sqrt[3]{2^3}} = \dfrac{7\sqrt[3]{4}}{2}$

109) $\dfrac{\sqrt[3]{x^2}}{\sqrt[3]{y}} = \dfrac{\sqrt[3]{x^2}}{\sqrt[3]{y}} \cdot \dfrac{\sqrt[3]{y^2}}{\sqrt[3]{y^2}} = \dfrac{\sqrt[3]{x^2 y^2}}{\sqrt[3]{y^3}}$

$\qquad = \dfrac{\sqrt[3]{x^2 y^2}}{y}$

111) $\dfrac{2}{3+\sqrt{3}} = \dfrac{2}{3+\sqrt{3}} \cdot \dfrac{3-\sqrt{3}}{3-\sqrt{3}}$

$= \dfrac{2\left(3-\sqrt{3}\right)}{\left(3\right)^2 - \left(\sqrt{3}\right)^2}$

$= \dfrac{2\left(3-\sqrt{3}\right)}{9-3}$

$= \dfrac{2\left(3-\sqrt{3}\right)}{6} = \dfrac{3-\sqrt{3}}{3}$

113) $\dfrac{8-24\sqrt{2}}{8} = \dfrac{8\left(1-3\sqrt{2}\right)}{8} = 1-3\sqrt{2}$

115) $\sqrt{x+8} = 3$

$\left(\sqrt{x+8}\right)^2 = 3^2$

$x+8 = 9$

$x = 1$

Check $\sqrt{1+8} = 1$

$\sqrt{9} = 3$ $\quad \{1\}$

117) $\sqrt{3j+4} = -\sqrt{4j-1}$

$\left(\sqrt{3j+4}\right)^2 = \left(-\sqrt{4j-1}\right)^2$

$3j+4 = 4j-1$

$5 = j$

Check $\sqrt{3(5)+4} = -\sqrt{4(5)-1}$

$\sqrt{19} \neq -\sqrt{19}$ $\qquad \varnothing$

119) $a = \sqrt{a+8} - 6$

$a+6 = \sqrt{a+8}$

$\left(a+6\right)^2 = \left(\sqrt{a+8}\right)^2$

$a^2 +12a+36 = a+8$

$a^2 +11a+28 = 0$

$\left(a+7\right)\left(a+4\right) = 0$

$a+7 = 0$ or $a+4 = 0$

$a = -7 \qquad a = -4 \qquad \{-4\}$

-7 is an extraneous solution.

121) $\sqrt{4a+1} - \sqrt{a-2} = 3$

$\sqrt{4a+1} = 3 + \sqrt{a-2}$

$\left(\sqrt{4a+1}\right)^2 = \left(3+\sqrt{a-2}\right)^2$

$4a+1 = 9 + 6\sqrt{a-2} + a - 2$

$4a+1 = 7 + a + 6\sqrt{a-2}$

$3a-6 = 6\sqrt{a-2}$

$a-2 = 2\sqrt{a-2}$

$\left(a-2\right)^2 = \left(2\sqrt{a-2}\right)^2$

$a^2 -4a+4 = 4\left(a-2\right)$

$a^2 -4a+4 = 4a-8$

$a^2 -8a+12 = 0$

$\left(a-6\right)\left(a-2\right) = 0$

$a-6 = 0$ or $a-2 = 0$

$a = 6 \qquad a = 2 \qquad \{2,6\}$

Check is left to the student.

123) $r = \sqrt{\dfrac{3V}{\pi h}}$

$r^2 = \left(\sqrt{\dfrac{3V}{\pi h}}\right)^2$

$r^2 = \dfrac{3V}{\pi h}$

$\pi r^2 h = 3V$

$\dfrac{1}{3}\pi r^2 h = V$

125) $\sqrt{-49} = i\sqrt{49} = 7i$

127) $\sqrt{-2} \cdot \sqrt{-8} = i\sqrt{2} \cdot i\sqrt{8}$

$= i^2\sqrt{16} = -1 \cdot 4 = -4$

129) $\left(2+i\right) + \left(10-4i\right) = 12-3i$

131) $\left(\dfrac{4}{5}-\dfrac{1}{3}i\right)-\left(\dfrac{1}{2}+i\right)$

$=\left(\dfrac{4}{5}-\dfrac{1}{2}\right)+\left(-\dfrac{1}{3}i-i\right)$

$=\left(\dfrac{8}{10}-\dfrac{5}{10}\right)+\left(-\dfrac{1}{3}i-\dfrac{3}{3}i\right)$

$=\dfrac{3}{10}-\dfrac{4}{3}i$

133) $5(-6+7i)=-30+35i$

135) $3i(-7+12i)=-21i+36i^2$

$=-21i+36(-1)$

$=-36-21i$

137) $(4-6i)(3-6i)$

$=12-24i-18i+36i^2$

$=12-42i+36(-1)$

$=12-42i-36=-24-42i$

139) $(2-7i)(2+7i)=(2)^2+(7)^2$

$=4+49=53$

141) $\dfrac{6}{2+5i}=\dfrac{6}{2+5i}\cdot\dfrac{2-5i}{2-5i}$

$=\dfrac{12-30i}{(2)^2+(5)^2}=\dfrac{12-30i}{4+25}$

$=\dfrac{12-30i}{29}=\dfrac{12}{29}-\dfrac{30}{29}i$

143) $\dfrac{8}{i}=\dfrac{8}{i}\cdot\dfrac{-i}{-i}=\dfrac{-8i}{(1)^2}=-8i$

145) $\dfrac{9-4i}{6-i}=\dfrac{9-4i}{6-i}\cdot\dfrac{6+i}{6+i}$

$=\dfrac{54+9i-24i-4i^2}{(6)^2+(1)^2}$

$=\dfrac{54-15i-4(-1)}{36+1}$

$=\dfrac{54-15i+4}{37}=\dfrac{58-15i}{37}$

$=\dfrac{58}{37}-\dfrac{15}{37}i$

147) $i^{10}=\left(i^2\right)^5$

$=(-1)^5$

$=-1$

149) $i^{33}=i^{32}\cdot i$

$=\left(i^2\right)^{16}\cdot i$

$=(-1)^{16}\cdot i$

$=i$

Chapter 10 Test

1) $\sqrt{144}=12$

3) not real

5) $\sqrt[5]{(-19)^5}=-19$

7) $27^{4/3}=\left(27^{1/3}\right)^4=\left(\sqrt[3]{27}\right)^4=3^4=81$

9) $\left(\dfrac{8}{125}\right)^{-2/3}=\left(\dfrac{125}{8}\right)^{2/3}=\left(\sqrt[3]{\dfrac{125}{8}}\right)^2$

$=\left(\dfrac{5}{2}\right)^2=\dfrac{25}{4}$

11) $\dfrac{35a^{1/6}}{14a^{5/6}}=\dfrac{5}{2}a^{1/6-5/6}=\dfrac{5}{2}a^{-4/6}=\dfrac{5}{2a^{2/3}}$

13) $\sqrt{75}=\sqrt{25}\cdot\sqrt{3}=5\sqrt{3}$

15) $\sqrt{\dfrac{24}{2}}=\sqrt{12}=\sqrt{4}\cdot\sqrt{3}=2\sqrt{3}$

17) $\sqrt[4]{p^{24}}=p^{24/4}=p^6$

19) $\sqrt{63m^5n^8}=\sqrt{63}\cdot\sqrt{m^5}\cdot\sqrt{n^8}$

$=3\sqrt{7}\cdot m^2\sqrt{m}\cdot n^4=3m^2n^4\sqrt{7m}$

21) $\sqrt[3]{\dfrac{a^{14}b^7}{27}}=\dfrac{\sqrt[3]{a^{14}}\cdot\sqrt[3]{b^7}}{\sqrt[3]{27}}$

$=\dfrac{a^4\sqrt[3]{a^2}\cdot b^2\sqrt[3]{b}}{3}=\dfrac{a^4b^2\sqrt[3]{a^2b}}{3}$

23) $\sqrt[3]{z^4}\cdot\sqrt[3]{z^6}=\sqrt[3]{z^{10}}=\sqrt[3]{z^9}\cdot\sqrt[3]{z}=z^3\sqrt[3]{z}$

25) $9\sqrt{7}-3\sqrt{7}=6\sqrt{7}$

27) $2h^3\sqrt[4]{h} - 16\sqrt[4]{h^{13}}$

$= 2h^3\sqrt[4]{h} - 16\left(\sqrt[4]{h^{12}} \cdot \sqrt[4]{h}\right)$

$= 2h^3\sqrt[4]{h} - 16h^3\sqrt[4]{h} = -14h^3\sqrt[4]{h}$

29) $\left(3 - 2\sqrt{5}\right)\left(\sqrt{2} + 1\right)$

$= 3\sqrt{2} + 3 - 2\sqrt{10} - 2\sqrt{5}$

31) $\left(\sqrt{2p+1} + 2\right)^2$

$= \left(\sqrt{2p+1}\right)^2 + 2\left(\sqrt{2p+1}\right)(2) + 2^2$

$= 2p + 1 + 4\sqrt{2p+1} + 4$

$= 2p + 5 + 4\sqrt{2p+1}$

33) $\dfrac{2}{\sqrt{5}} = \dfrac{2}{\sqrt{5}} \cdot \dfrac{\sqrt{5}}{\sqrt{5}} = \dfrac{2\sqrt{5}}{5}$

35) $\dfrac{\sqrt{6}}{\sqrt{a}} = \dfrac{\sqrt{6}}{\sqrt{a}} \cdot \dfrac{\sqrt{a}}{\sqrt{a}} = \dfrac{\sqrt{6a}}{a}$

37) $\dfrac{2 - \sqrt{48}}{2} = \dfrac{2 - 4\sqrt{3}}{2} = \dfrac{2\left(1 - 2\sqrt{3}\right)}{2}$

$= 1 - 2\sqrt{3}$

39) $\qquad z = \sqrt{1 - 4z} - 5$

$z + 5 = \sqrt{1 - 4z}$

$(z+5)^2 = \left(\sqrt{1-4z}\right)^2$

$z^2 + 10z + 25 = 1 - 4z$

$z^2 + 14z + 24 = 0$

$(z+2)(z+12) = 0$

$z + 2 = 0 \ \text{ or } \ z + 12 = 0$

$z = -2 \qquad z = -12 \qquad \{-2\}$

-12 is an extraneous solution.

41) $\sqrt{3k+1} - \sqrt{2k-1} = 1$

$\sqrt{3k+1} = 1 + \sqrt{2k-1}$

$\left(\sqrt{3k+1}\right)^2 = \left(1 + \sqrt{2k-1}\right)^2$

$3k+1 = 1 + 2\sqrt{2k-1} + 2k - 1$

$3k+1 = 2k + 2\sqrt{2k-1}$

$k+1 = 2\sqrt{2k-1}$

$(k+1)^2 = \left(2\sqrt{2k-1}\right)^2$

$k^2 + 2k + 1 = 4(2k-1)$

$k^2 + 2k + 1 = 8k - 4$

$k^2 - 6k + 5 = 0$

$(k-5)(k-1) = 0$

$k - 5 = 0 \ \text{ or } \ k - 1 = 0$

$k = 5 \qquad\qquad k = 1 \qquad \{1, 5\}$

Check is left to the student.

43) $\sqrt{-64} = i\sqrt{64} = 8i$

45) $i^{19} = i^{18} \cdot i$

$= \left(i^2\right)^9 \cdot i$

$= (-1)^9 \cdot i$

$= -i$

47) $(2 - 7i)(-1 + 3i)$

$= -2 + 6i + 7i - 21i^2$

$= -2 + 13i - 21(-1)$

$= -2 + 13i + 21 = 19 + 13i$

Cumulative Review: Chapters 1-10

1) $4x - 3y + 9 - \dfrac{2}{3}x + y - 1$

$= \dfrac{12}{3}x - \dfrac{2}{3}x - 3y + y + 9 - 1$

$= \dfrac{10}{3}x - 2y + 8$

3) $3(2c-1)+7=9c+5(c+2)$

$6c-3+7=9c+5c+10$

$6c+4=14c+10$

$-6=8c$

$-\dfrac{6}{8}=c$

$-\dfrac{3}{4}=c$

$\left\{-\dfrac{3}{4}\right\}$

5) $m=\dfrac{-2-3}{1-5}=\dfrac{-5}{-4}=\dfrac{5}{4}$

$y-y_1=m(x-x_1)$

$y-3=\dfrac{5}{4}(x-5)$

$y-3=\dfrac{5}{4}x-\dfrac{25}{4}$

$y=\dfrac{5}{4}x-\dfrac{13}{4}$

7) $(5p^2-2)(3p^2-4p-1)$

$=15p^4-20p^3-5p^2-6p^2+8p+2$

$=15p^4-20p^3-11p^2+8p+2$

9) $4w^2+5w-6=(4w-3)(w+2)$

11) $\qquad 6y^2-4=5y$

$6y^2-5y-4=0$

$6y^2-8y+3y-4=0$

$2y(3y-4)+1(3y-4)=0$

$(2y+1)(3y-4)=0$

$2y+1=0$ or $3y-4=0$

$y=-\dfrac{1}{2}\qquad y=\dfrac{4}{3}$

$\left\{-\dfrac{1}{2},\dfrac{4}{3}\right\}$

13) l = length

$w=l-5$

$84=lw$

$84=l(l-5)$

$84=l^2-5l$

$0=l^2-5l-84$

$0=(l-12)(l+7)$

$l-12=0$ or $l+5=0$

$l=12\qquad l=-5$

$w=12-5$

$w=7$

length $=12$ in., width $=7$ in.

15) $\dfrac{10m^2}{9n}\cdot\dfrac{6n^2}{35m^5}=\dfrac{\overset{2}{\cancel{10m^2}}}{\underset{3}{\cancel{9n}}}\cdot\dfrac{\overset{2n}{\cancel{6n^2}}}{\underset{7m^3}{\cancel{35m^5}}}$

$=\dfrac{4n}{21m^3}$

17) $|6g+1|\geq11$

$6g+1\geq11$ or $6g+1\leq-11$

$6g\geq10\qquad 6g\leq-12$

$g\geq\dfrac{10}{6}\qquad g\leq-2$

$g\geq\dfrac{5}{3}$

$(-\infty,-2]\cup\left[\dfrac{5}{3},\infty\right)$

19) a) $\sqrt{500}=\sqrt{100}\cdot\sqrt5=10\sqrt5$

b) $\sqrt[3]{56}=\sqrt[3]{8}\cdot\sqrt[3]{7}=2\sqrt[3]{7}$

c) $\sqrt{p^{10}q^7}=\sqrt{p^{10}}\cdot\sqrt{q^7}=p^5q^3\sqrt{q}$

d) $\sqrt[4]{32a^{15}}=\sqrt[4]{32}\cdot\sqrt[4]{a^{15}}=2\sqrt[4]{2}\cdot a^3\sqrt[4]{a^3}$

$=2a^3\sqrt[4]{2a^3}$

21) $2\sqrt3(5-\sqrt3)=10\sqrt3-2(3)$

$=10\sqrt3-6$

23) a) $\sqrt{2b-1}+7=6$

$\qquad \sqrt{2b-1}=-1 \quad \varnothing$

b) $\qquad \sqrt{3z+10}=2-\sqrt{z+4}$

$\qquad \left(\sqrt{3z+10}\right)^2 = \left(2-\sqrt{z+4}\right)^2$

$\qquad 3z+10 = 4-4\sqrt{z+4}+z+4$

$\qquad 3z+10 = 8+z-4\sqrt{z+4}$

$\qquad 2z+2 = -4\sqrt{z+4}$

$\qquad z+1 = -2\sqrt{z+4}$

$\qquad \left(z+1\right)^2 = \left(-2\sqrt{z+4}\right)^2$

$\qquad z^2+2z+1 = 4\left(z+4\right)$

$\qquad z^2+2z+1 = 4z+16$

$\qquad z^2-2z-15 = 0$

$\qquad \left(z-5\right)\left(z+3\right)=0$

$\qquad z-5=0 \quad \text{or} \quad z+3=0$

$\qquad z=5 \quad \text{or} \qquad z=-3 \quad \{-3\}$

5 is an extraneous solution.

25) a) $\left(-3+4i\right)+\left(5+3i\right)=2+7i$

b) $\left(3+6i\right)\left(-2+7i\right)$

$\qquad = -6+21i-12i+42i^2$

$\qquad = -6+9i+42\left(-1\right)$

$\qquad = -6+9i-42 = -48+9i$

c) $\dfrac{2-i}{-4+3i} = \dfrac{2-i}{-4+6i} \cdot \dfrac{-4-6i}{-4-6i}$

$\qquad = \dfrac{-8-6i+4i+3i^2}{\left(-4\right)^2+\left(3\right)^2}$

$\qquad = \dfrac{-8-2i+3\left(-1\right)}{16+9}$

$\qquad = \dfrac{-8-2i-3}{25} = \dfrac{-11-2i}{25}$

$\qquad = -\dfrac{11}{25}-\dfrac{2}{25}i$

Section 11.1: Exercises

1) $(t+7)(t-6)=0$

$t+7=0$ or $t-6=0$

$t=-7 \qquad t=6 \qquad \{-7,6\}$

3) $u^2+15u+44=0$

$(u+11)(u+4)=0$

$u+11=0$ or $u+4=0$

$u=-11 \qquad u=-4 \quad \{-11,-4\}$

5) $\qquad x^2=x+56$

$x^2-x-56=0$

$(x-8)(x+7)=0$

$x-8=0$ or $x+7=0$

$x=8 \qquad x=-7 \qquad \{-7,8\}$

7) $\qquad 1-100w^2=0$

$(1+10w)(1-10w)=0$

$1+10w=0$ or $1-10w=0$

$10w=-1 \qquad -10w=-1$

$w=-\dfrac{1}{10} \qquad w=\dfrac{1}{10}$

$\left\{-\dfrac{1}{10},\dfrac{1}{10}\right\}$

9) $\qquad 5m^2+8=22m$

$5m^2-22m+8=0$

$(5m-2)(m-4)=0$

$5m-2=0$ or $m-4=0$

$5m=2 \qquad m=4$

$m=\dfrac{2}{5} \qquad \left\{\dfrac{2}{5},4\right\}$

11) $\qquad 23d=-10-6d^2$

$6d^2+23d+10=0$

$(3d+10)(2d+1)=0$

$3d+10=0$ or $2d+1=0$

$3d=-10 \qquad 2d=-1$

$a=-\dfrac{10}{3} \qquad 2d=-\dfrac{1}{2}$

$\left\{-\dfrac{10}{3},-\dfrac{1}{2}\right\}$

13) $2r=7r^2$

$0=7r^2-2r$

$0=r(7r-2)$

$7r-2=0$ or $r=0$

$7r=2$

$r=\dfrac{2}{7} \qquad \left\{0,\dfrac{2}{7}\right\}$

15) quadratic

17) linear

19) quadratic

21) linear

23) $13c=2c^2+6$

$0=2c^2-13c+6$

$0=(2c-1)(c-6)$

$2c-1=0$ or $c-6=0$

$2c=1 \qquad c=6$

$c=\dfrac{1}{2} \qquad \left\{\dfrac{1}{2},6\right\}$

25) $\quad 2p(p+4)=p^2+5p+10$

$2p^2+8p=p^2+5p+10$

$p^2+3p-10=0$

$(p+5)(p-2)=0$

$p+5=0 \ \text{or} \ p-2=0$

$p=-5 \qquad p=2 \qquad \{-5,2\}$

27) $5(3n-2)-11n=2n-1$

$15n-10-11n=2n-1$

$4n-10=2n-1$

$2n=9$

$n=\dfrac{9}{2} \qquad \left\{\dfrac{9}{2}\right\}$

29) $\quad 3t^3+5t=-8t^2$

$3t^3+8t^2+5t=0$

$t(3t^2+8t+5)=0$

$t(3t+5)(t+1)=0$

$t=0 \ \text{or} \ 3t+5=0 \ \text{or} \ t+1=0$

$3t=-5 \qquad t=-1$

$t=-\dfrac{5}{3}$

$\left\{-\dfrac{5}{3},-1,0\right\}$

31) $\quad 2(r+5)=10-4r^2$

$2r+10=10-4r^2$

$2r=-4r^2$

$4r^2+2r=0$

$2r(2r+1)=0$

$2r+1=0 \ \text{or} \ 2r=0$

$2r=-1 \qquad r=0$

$r=-\dfrac{1}{2} \qquad \left\{-\dfrac{1}{2},0\right\}$

33) $9y-6(y+1)=12-5y$

$9y-6y-6=12-5y$

$3y-6=12-5y$

$8y=18$

$y=\dfrac{18}{8}=\dfrac{9}{4} \qquad \left\{\dfrac{9}{4}\right\}$

35) $\quad \dfrac{1}{16}w^2+\dfrac{1}{8}w=\dfrac{1}{2}$

$16\left(\dfrac{1}{16}w^2+\dfrac{1}{8}w\right)=16\cdot\dfrac{1}{2}$

$w^2+2w=8$

$w^2+2w-8=0$

$(w+4)(w-2)=0$

$w+4=0 \ \text{or} \ w-2=0$

$w=-4 \qquad w=2 \qquad \{-4,2\}$

37) $\quad 12n+3=-12n^2$

$12n^2+12n+3=0$

$4n^2+4n+1=0 \qquad \text{divide by 3}$

$(2n+1)^2=0$

$2n+1=0$

$2n=-1$

$n=-\dfrac{1}{2} \qquad \left\{-\dfrac{1}{2}\right\}$

Chapter 11: Quadratic Equations

39) $3b^2 - b - 7 = 4b(2b+3) - 1$

$3b^2 - b - 7 = 8b^2 + 12b - 1$

$0 = 5b^2 + 13b + 6$

$0 = (5b+3)(b+2)$

$5b+3 = 0$ or $b+2 = 0$

$5b = -3 \qquad b = -2$

$b = -\dfrac{3}{5} \qquad\qquad \left\{ -2, -\dfrac{3}{5} \right\}$

41) $\quad t^3 + 7t^2 - 4t - 28 = 0$

$t^2(t+7) - 4(t+7) = 0$

$(t+7)(t^2 - 4) = 0$

$(t+7)(t+2)(t-2) = 0$

$t+7 = 0$ or $t+2 = 0$ or $t-2 = 0$

$t = -7 \qquad t = -2 \qquad t = 2$

$\{-7, -2, 2\}$

43) $\quad w = \text{width}$

$w + 5 = \text{length}$

$\text{Area} = (\text{width})(\text{length})$

$14 = w(w+5)$

$14 = w^2 + 5w$

$0 = w^2 + 5w - 14$

$0 = (w+7)(w-2)$

$w+7 = 0$ or $w-2 = 0$

$w = -7 \qquad \boxed{w = 2}$

width = 2 in.

length = 7 in.

45) $\quad w = \text{width}$

$2w - 1 = \text{length}$

$\text{Area} = (\text{width})(\text{length})$

$45 = w(2w-1)$

$45 = 2w^2 - w$

$0 = 2w^2 - w - 45$

$0 = (2w+9)(w-5)$

$2w+9 = 0$ or $w-5 = 0$

$2w = -9 \qquad \boxed{w = 5}$

$w = -\dfrac{9}{2}$

width = 5 cm

length = $2(5) - 1 = 9$ cm

47) $A = \dfrac{1}{2}(\text{base})(\text{height})$

$18 = \dfrac{1}{2}(x+6)(x+1)$

$36 = (x+6)(x+1)$

$36 = x^2 + 7x + 6$

$0 = x^2 + 7x - 30$

$0 = (x+10)(x-3)$

$x+10 = 0$ or $x-3 = 0$

$x = -10 \qquad \boxed{x = 3}$

base = $x+6 = 3+6 = 9$ in.

height = $x+1 = 3+1 = 4$ in.

49) $A = \dfrac{1}{2}(\text{base})(\text{height})$

$36 = \dfrac{1}{2}\left(\dfrac{1}{2}x\right)(x)$

$36 = \dfrac{1}{4}x^2$

$144 = x^2$

$0 = x^2 - 144$

$0 = (x+12)(x-12)$

$x + 12 = 0$ or $x - 12 = 0$

$x = -12$ $\boxed{x = 12}$

$\text{base} = \dfrac{1}{2}x = \dfrac{1}{2}(12) = 6$ cm

$\text{height} = x = 12$ cm

51) $x^2 + (x-7)^2 = (x+1)^2$

$x^2 + x^2 - 14x + 49 = x^2 + 2x + 1$

$2x^2 - 14x + 49 = x^2 + 2x + 1$

$x^2 - 16x + 48 = 0$

$(x-12)(x-4) = 0$

$x - 12 = 0$ or $x - 4 = 0$

$\boxed{x = 12}$ $x = 4$

Reject $x = 4$ because if $x = 4$ then the length of the leg labelled $x - 7$ would be -3 units.

one leg $= x = 12$ units

other leg $= x - 7 = 12 - 7 = 5$ units

hypotenuse $= x + 1 = 12 + 1 = 13$ units

53) $(2x)^2 + (x+5)^2 = (3x+1)^2$

$4x^2 + (x^2 + 10x + 25) = 9x^2 + 6x + 1$

$5x^2 + 10x + 25 = 9x^2 + 6x + 1$

$0 = 4x^2 - 4x - 24$

$0 = x^2 - x - 6$

$0 = (x-3)(x+2)$

$x - 3 = 0$ or $x + 2 = 0$

$\boxed{x = 3}$ $x = -2$

one leg $= 2x = 2(3) = 6$ units

other leg $= x + 5 = 3 + 5 = 8$ units

hypotenuse $= 3x + 1 = 3(3) + 1 = 10$ units

Section 11.2: Exercises

1) Factoring:

$y^2 - 16 = 0$

$(y+4)(y-4) = 0$

$y + 4 = 0$ or $y - 4 = 0$

$y = -4$ $y = 4$

Square Root Property:

$y^2 - 16 = 0$

$y^2 = 16$

$y = \pm\sqrt{16}$

$y = \pm 4$ $\{-4, 4\}$

3) $b^2 = 36$

$b = \pm\sqrt{36}$

$b = \pm 6$ $\{-6, 6\}$

5) $r^2 - 27 = 0$

$r = \pm\sqrt{27}$

$= \pm 3\sqrt{3}$ $\{3\sqrt{3}, -3\sqrt{3}\}$

7) $n^2 = \dfrac{4}{9}$

$n = \pm\sqrt{\dfrac{4}{9}} = \pm\dfrac{2}{3}$ $\left\{-\dfrac{2}{3}, \dfrac{2}{3}\right\}$

9) $q^2 = -4$

$= \pm 2\sqrt{-1}$

$= \pm 2i$ $\{-2i, 2i\}$

11) $z^2 + 3 = 0$

$z^2 = -3$

$z = \pm i\sqrt{3}$ $\left\{-i\sqrt{3}, i\sqrt{3}\right\}$

13) $z^2 + 5 = 19$

$z^2 = 14$

$z = \pm\sqrt{14}$ $\left\{-\sqrt{14}, \sqrt{14}\right\}$

15) $2d^2 + 5 = 55$

$2d^2 = 50$

$d^2 = 25$

$d = \pm 5$ $\{-5, 5\}$

17) $5f^2 + 39 = -21$

$5f^2 = -60$

$f^2 = -12$

$f = \pm 2i\sqrt{3}$ $\left\{-2i\sqrt{3}, 2i\sqrt{3}\right\}$

19) $(r+10)^2 = 4$

$r + 10 = \pm 2$

$r = -12, -8$ $\{-12, -8\}$

21) $(q-7)^2 = 1$

$q - 7 = \pm 1$

$q = 6, 8$ $\{6, 8\}$

23) $(p+4)^2 - 18 = 0$

$(p+4)^2 = 18$

$p + 4 = \pm 3\sqrt{2}$

$p = -4 - 3\sqrt{2}, -4 + 3\sqrt{2}$

$\left\{-4 - 3\sqrt{2}, -4 + 3\sqrt{2}\right\}$

25) $(c+3)^2 - 4 = -29$

$(c+3)^2 = -25$

$c + 3 = \pm 5i$

$c = -3 + 5i, -3 - 5i$

$\{-3 - 5i, -3 + 5i\}$

27) $1 = 15 + (k-2)^2$

$(k-2)^2 = -14$

$k - 2 = \pm i\sqrt{14}$

$k = 2 + i\sqrt{14}, 2 - i\sqrt{14}$

$\left\{2 - i\sqrt{14}, 2 + i\sqrt{14}\right\}$

29) $20 = (2w+1)^2$

$2w + 1 = \pm 2\sqrt{5}$

$w = \dfrac{-1 - 2\sqrt{5}}{2}, \dfrac{-1 + 2\sqrt{5}}{2}$

$\left\{\dfrac{-1 - 2\sqrt{5}}{2}, \dfrac{-1 + 2\sqrt{5}}{2}\right\}$

$8 = (3q-10)^2 - 6$

31) $14 = (3q-10)^2$

$\pm\sqrt{14} = 3q - 10$

$10 \pm \sqrt{14} = 3q$

$\dfrac{10 \pm \sqrt{14}}{3} = q$

$\left\{\dfrac{10 - \sqrt{14}}{3}, \dfrac{10 + \sqrt{14}}{3}\right\}$

33) $36 + (4p - 5)^2 = 6$

$(4p - 5)^2 = -30$

$4p - 5 = \pm i\sqrt{30}$

$p = \dfrac{5 - i\sqrt{30}}{4}, \dfrac{5 + i\sqrt{30}}{4}$

$\left\{ \dfrac{5 - i\sqrt{30}}{4}, \dfrac{5 + i\sqrt{30}}{4} \right\}$

35) $(6g + 11)^2 + 50 = 1$

$(6g + 11)^2 = -49$

$6g + 11 = \pm 7i$

$g = \dfrac{-11 - 7i}{6}, \dfrac{-11 + 7i}{6}$

$\left\{ \dfrac{-11 - 7i}{6}, \dfrac{-11 + 7i}{6} \right\}$

37) $\left(\dfrac{3}{4}n - 8 \right)^2 = 4$

$\dfrac{3}{4}n - 8 = \pm\sqrt{4}$

$\dfrac{3}{4}n - 8 = \pm 2$

$\dfrac{3}{4}n - 8 = 2$ or $\dfrac{3}{4}n - 8 = -2$

$\dfrac{3}{4}n = 10 \qquad\qquad \dfrac{3}{4}n = 6$

$n = \dfrac{40}{3} \qquad\qquad n = 8$

$\left\{ 8, \dfrac{40}{3} \right\}$

39) $(5y - 2)^2 + 6 = 22$

$(5y - 2)^2 = 16$

$5y - 2 = \pm 4$

$y = -\dfrac{2}{5}, \dfrac{6}{5} \qquad \left\{ -\dfrac{2}{5}, \dfrac{6}{5} \right\}$

41) $d = \sqrt{(x_2 - x_1)^2 + (y_2 - y_1)^2}$

$d = \sqrt{(3 - 7)^2 + (-2 - 1)^2}$

$= \sqrt{(-4)^2 + (-3)^2}$

$= \sqrt{25}$

$= 5$

43) $d = \sqrt{(x_2 - x_1)^2 + (y_2 - y_1)^2}$

$d = \sqrt{[-2 - (-5)]^2 + [-8 - (-6)]^2}$

$= \sqrt{(3)^2 + (-2)^2}$

$\sqrt{13}$

45) $d = \sqrt{(x_2 - x_1)^2 + (y_2 - y_1)^2}$

$d = \sqrt{(0 - 0)^2 + (7 - 13)^2}$

$= \sqrt{(-6)^2}$

$= 6$

47) $d = \sqrt{(x_2 - x_1)^2 + (y_2 - y_1)^2}$

$d = \sqrt{[2 - (-4)]^2 + (6 - 11)^2}$

$d = \sqrt{6^2 + (-5)^2}$

$d = \sqrt{36 + 25} = \sqrt{61}$

49) $d = \sqrt{(x_2 - x_1)^2 + (y_2 - y_1)^2}$

$d = \sqrt{(5 - 3)^2 + [-7 - (-3)]^2}$

$d = \sqrt{2^2 + (-4)^2}$

$d = \sqrt{4 + 16} = \sqrt{20} = 2\sqrt{5}$

51) It is a trinomial whose factored form is the square of a binomial.

$x^2 - 6x + 9$

53) $w^2 + 8w$

$\frac{1}{2}(8) = 4$ Find half of the coefficient of w

$4^2 = 16$ Square the result

$w^2 + 8w + 16$ Add the constant to the expression

The perfect square trinomial is $w^2 + 8w + 16$

The factored form of the trinomial is $(w+4)^2$

55) 1) $\frac{1}{2}(12) = 6$ 2) $6^2 = 36$

$a^2 + 12a + 36;$ $(a+6)^2$

57) 1) $\frac{1}{2}(-18) = -9$ 2) $(-9)^2 = 81$

$c^2 - 18c + 81;$ $(c-9)^2$

59) 1) $\frac{1}{2}(3) = \frac{3}{2}$ 2) $\left(\frac{3}{2}\right)^2 = \frac{9}{4}$

$r^2 + 3r + \frac{9}{4};$ $\left(r+\frac{3}{2}\right)^2$

61) 1) $\frac{1}{2}(-9) = -\frac{9}{2}$ 2) $\left(-\frac{9}{2}\right)^2 = \frac{81}{4}$

$b^2 - 9b + \frac{81}{4};$ $\left(b-\frac{9}{2}\right)^2$

63) 1) $\frac{1}{2}\left(\frac{1}{3}\right) = \frac{1}{6}$ 2) $\left(\frac{1}{6}\right)^2 = \frac{1}{36}$

$x^2 + \frac{1}{3}x + \frac{1}{36};$ $\left(x+\frac{1}{6}\right)^2$

65) To solve $2p^2 - 7p = 8$ by completing the square, divide both sides of the equation by 2.

67) $x^2 + 6x + 8 = 0$

$x^2 + 6x = -8$

$x^2 + 6x + 9 = -8 + 9$

$(x+3)^2 = 1$

$x + 3 = \pm\sqrt{1}$

$x + 3 = \pm 1$

$x + 3 = 1$ or $x + 3 = -1$

$x = -2$ $x = -4$ $\{-4, -2\}$

69) $k^2 - 8k + 15 = 0$

$k^2 - 8k = -15$

$k^2 - 8k + 16 = -15 + 16$

$(k-4)^2 = 1$

$k - 4 = \pm\sqrt{1}$

$k - 4 = \pm 1$

$k - 4 = 1$ or $k - 4 = -1$

$k = 5$ $k = 3$ $\{3, 5\}$

71) $s^2 + 10 = -10s$

$s^2 + 10s = -10$

$s^2 + 10s + 25 = -10 + 25$

$(s+5)^2 = 15$

$s + 5 = \pm\sqrt{15}$

$s = -5 \pm \sqrt{15}$

$\left\{-5 - \sqrt{15}, -5 + \sqrt{15}\right\}$

73) $t^2 = 2t - 9$

$t^2 - 2t = -9$

$t^2 - 2t + 1 = -9 + 1$

$(t-1)^2 = -8$

$t - 1 = \pm\sqrt{-8}$

$t = 1 \pm 2i\sqrt{2}$

$\left\{1 - 2i\sqrt{2}, 1 + 2i\sqrt{2}\right\}$

75) $v^2 + 4v + 8 = 0$

$v^2 + 4v = -8$

$v^2 + 4v + 4 = -8 + 4$

$(v+2)^2 = -4$

$v + 2 = \pm\sqrt{-4}$

$v = -2 \pm 2i$

$\{-2 - 2i, -2 + 2i\}$

77) $m^2 + 3m - 40 = 0$

$m^2 + 3m = 40$

$m^2 + 3m + \dfrac{9}{4} = 40 + \dfrac{9}{4}$

$\left(m + \dfrac{3}{2}\right)^2 = \dfrac{169}{4}$

$m + \dfrac{3}{2} = \pm\sqrt{\dfrac{169}{4}}$

$m + \dfrac{3}{2} = \pm\dfrac{13}{2}$

$m + \dfrac{3}{2} = \dfrac{13}{2}$ or $m + \dfrac{3}{2} = -\dfrac{13}{2}$

$m = \dfrac{10}{2}$ $\quad$ $m = -\dfrac{16}{2}$

$m = 5$ $\quad$ $m = -8$

$\{-8, 5\}$

79) $x^2 - 7x + 12 = 0$

$x^2 - 7x = -12$

$x^2 - 7x + \dfrac{49}{4} = -12 + \dfrac{49}{4}$

$\left(x - \dfrac{7}{2}\right)^2 = \dfrac{1}{4}$

$x - \dfrac{7}{2} = \pm\sqrt{\dfrac{1}{4}} = \pm\dfrac{1}{2}$

$x - \dfrac{7}{2} = \dfrac{1}{2}$ or $x - \dfrac{7}{2} = -\dfrac{1}{2}$

$x = \dfrac{8}{2}$ $\quad$ $x = \dfrac{6}{2}$

$x = 4$ $\quad$ $x = 3$ $\quad$ $\{3, 4\}$

81) $r^2 - r = 3$

$r^2 - r + \dfrac{1}{4} = 3 + \dfrac{1}{4}$

$\left(r - \dfrac{1}{2}\right)^2 = \dfrac{13}{4}$

$r - \dfrac{1}{2} = \pm\sqrt{\dfrac{13}{4}}$

$r = \dfrac{1}{2} \pm \dfrac{\sqrt{13}}{2}$

$\left\{\dfrac{1}{2} - \dfrac{\sqrt{13}}{2}, \dfrac{1}{2} + \dfrac{\sqrt{13}}{2}\right\}$

83) $c^2 + 5c + 7 = 0$

$$c + 5c + \frac{25}{4} = -7 + \frac{25}{4}$$

$$\left(c + \frac{5}{2}\right)^2 = -\frac{3}{4}$$

$$c + \frac{5}{2} = \pm i\frac{\sqrt{3}}{2}$$

$$c = -\frac{5}{2} \pm i\frac{\sqrt{3}}{2} \quad \left\{-\frac{5}{2} - i\frac{\sqrt{3}}{2}, -\frac{5}{2} + i\frac{\sqrt{3}}{2}\right\}$$

85) $3k^2 - 6k + 12 = 0$

$$k^2 - 2k + 4 = 0$$

$$k^2 - 2k = -4$$

$$k^2 - 2k + 1 = -4 + 1$$

$$(k-1)^2 = -3$$

$$k - 1 = i\sqrt{3}$$

$$k = 1 \pm i\sqrt{3} \quad \left\{1 - i\sqrt{3}, 1 + i\sqrt{3}\right\}$$

87) $4r^2 + 24r = 8$

$$r^2 + 6r = 2$$

$$r^2 + 6r + 9 = 11$$

$$(r+3)^2 = 11$$

$$r + 3 = \pm\sqrt{11}$$

$$r = -3 \pm \sqrt{11} \quad \left\{-3 - \sqrt{11}, -3 + \sqrt{11}\right\}$$

89) $10d = 2d^2 + 12$

$$5d = d^2 + 6$$

$$d^2 - 5d = -6$$

$$d^2 - 5d + \frac{25}{4} = -6 + \frac{25}{4}$$

$$\left(d - \frac{5}{2}\right)^2 = \frac{1}{4}$$

$$d - \frac{5}{2} = \pm\frac{1}{2}$$

$$d = 2, 3 \qquad \{2, 3\}$$

91) $2n^2 + 8 = 5n$

$$2n^2 - 5n = -8$$

$$n^2 - \frac{5}{2}n + \frac{25}{16} = -4 + \frac{25}{16}$$

$$\left(n - \frac{5}{4}\right)^2 = -\frac{39}{16}$$

$$n - \frac{5}{4} = \pm i\frac{\sqrt{39}}{4}$$

$$n = \frac{5}{4} \pm i\frac{\sqrt{39}}{4} \quad \left\{\frac{5}{4} - i\frac{\sqrt{39}}{4}, \frac{5}{4} + i\frac{\sqrt{39}}{4}\right\}$$

93) $4a^2 - 7a + 3 = 0$

$$4a^2 - 7a = -3$$

$$a^2 - \frac{7}{4}a + \frac{49}{64} = -\frac{3}{4} + \frac{49}{64}$$

$$\left(a - \frac{7}{8}\right)^2 = \frac{1}{64}$$

$$a - \frac{7}{8} = \pm\frac{1}{8}$$

$$a = \frac{7}{8} \pm \frac{1}{8} \qquad \left\{\frac{3}{4}, 1\right\}$$

95) $(y+5)(y-3) = 5$

$$y^2 - 3y + 5y - 15 = 5$$

$$y^2 + 2y = 20$$

$$y^2 + 2y + 1 = 20 + 1$$

$$(y+1)^2 = 21$$

$$y + 1 = \pm\sqrt{21}$$

$$y = -1 \pm \sqrt{21} \quad \left\{-1 - \sqrt{21}, -1 + \sqrt{21}\right\}$$

97) $(2m+1)(m-3)=-7$

$2m-6m+m-3=-7$

$2m-5m=-4$

$m-\dfrac{5}{2}m=-2$

$m-\dfrac{5}{2}m+\dfrac{25}{16}=-2+\dfrac{25}{16}$

$\left(m-\dfrac{5}{4}\right)^2=-\dfrac{7}{16}$

$m-\dfrac{5}{4}=\pm i\dfrac{\sqrt{7}}{4}$

$m=\dfrac{5}{4}\pm i\dfrac{\sqrt{7}}{4}$ $\left\{\dfrac{5}{4}-i\dfrac{\sqrt{7}}{4},\dfrac{5}{4}+i\dfrac{\sqrt{7}}{4}\right\}$

99) $a^2+b^2=c^2$

$a^2+8^2=10^2$

$a^2=100-64$

$a^2=36$

$a=6$

101) $a^2+b^2=c^2$

$5^2+2^2=c^2$

$c^2=29$

$c=\sqrt{29}$

103) $\text{Width}=w=4$

$\text{Length}=l$

$\text{Diagonal}=d=2\sqrt{13}$

$l^2=d^2-w^2$

$=\left(2\sqrt{13}\right)^2-4^2$

$=52-16$

$=36$

$d=6$ in.

105) $l=13=$ Ladder Length

$b=5=$ Distance From the wall

$h=$ Height of the wall

$h^2=l^2-b^2$

$h^2=13^2-5^2$

$h^2=169-25=144$

$h=12$ ft

107) $f(x)=49=(x+3)^2$

$x+3=\pm 7$

$x=-3\pm 7$ $\{-10,4\}$

109) $x=$ Width of the Garden

$x+8=$ Length of the Garden

$153=$ Area of the Garden

$x(x+8)=153$

$x^2+8x=153$

$x^2+8x+16=153+16$

$(x+4)^2=169$

$x+4=\pm 13$

$x=9$ or $x=-17$

Width of the garden: 9 ft.;

Length: 17 ft.

Section 11.3: Exercises

1) The fraction bar should also be under $-b$:

$$x=\dfrac{-b\pm\sqrt{b^2-4ac}}{2a}$$

3) You cannot divide only the -2 by 2

$$\dfrac{-2\pm 6\sqrt{11}}{2}=\dfrac{2\left(-1\pm 3\sqrt{11}\right)}{2}$$

$$=-1\pm 3\sqrt{11}$$

5) $x^2 + 4x + 3 = 0$

$a = 1,\ b = 4$ and $c = 3$

$$x = \frac{-4 \pm \sqrt{4^2 - 4(1)(3)}}{2(1)}$$

$$= \frac{-4 \pm \sqrt{16 - 12}}{2}$$

$$= \frac{-4 \pm \sqrt{4}}{2} = \frac{-4 \pm 2}{2}$$

$$\frac{-4 + 2}{2} = \frac{-2}{2} = -1,\ \frac{-4 - 2}{2} = \frac{-6}{2} = -3$$

$$\{-3,\ -1\}$$

7) $3t^2 + t - 10 = 0$

$a = 3,\ b = 1$ and $c = -10$

$$t = \frac{-1 \pm \sqrt{1^2 - 4(3)(-10)}}{2(3)}$$

$$= \frac{-1 \pm \sqrt{1 + 120}}{6}$$

$$= \frac{-1 \pm \sqrt{121}}{6} = \frac{-1 \pm 11}{6}$$

$$\frac{-1 + 11}{6} = \frac{10}{6} = \frac{5}{3},$$

$$\frac{-1 - 11}{6} = \frac{-12}{6} = -2 \qquad \left\{-2,\ \frac{5}{3}\right\}$$

9) $k^2 + 2 = 5k$

$k^2 - 5k + 2 = 0$

$a = 1,\ b = -5$ and $c = 2$

$$k = \frac{-(-5) \pm \sqrt{(-5)^2 - 4(1)(2)}}{2(1)}$$

$$= \frac{5 \pm \sqrt{25 - 8}}{2} = \frac{5 \pm \sqrt{17}}{2}$$

$$\left\{\frac{5 - \sqrt{17}}{2},\ \frac{5 + \sqrt{17}}{2}\right\}$$

11) $y^2 = 8y - 25$

$y^2 - 8y + 25 = 0$

$a = 1,\ b = -8$ and $c = 25$

$$y = \frac{-(-8) \pm \sqrt{(-8)^2 - 4(1)(25)}}{2(1)}$$

$$= \frac{8 \pm \sqrt{64 - 100}}{2} = \frac{8 \pm \sqrt{-36}}{2}$$

$$= \frac{8 \pm 6i}{2} = \frac{8}{2} \pm \frac{6}{2}i = 4 \pm 3i$$

$$\{4 - 3i,\ 4 + 3i\}$$

13) $3 - 2w = -5w^2$

$5w^2 - 2w + 3 = 0$

$a = 5,\ b = -2$ and $c = 3$

$$w = \frac{-(-2) \pm \sqrt{(-2)^2 - 4(5)(3)}}{2(5)}$$

$$= \frac{2 \pm \sqrt{4 - 60}}{10} = \frac{2 \pm \sqrt{-56}}{10}$$

$$= \frac{2 \pm 2i\sqrt{14}}{10} = \frac{2}{10} \pm \frac{2i\sqrt{14}}{10}$$

$$= \frac{1}{5} \pm \frac{\sqrt{14}}{5}i$$

$$\left\{\frac{1}{5} - \frac{\sqrt{14}}{5}i,\ \frac{1}{5} + \frac{\sqrt{14}}{5}i\right\}$$

15) $r^2 + 7r = 0$

$a = 1,\ b = 7$ and $c = 0$

$$r = \frac{-7 \pm \sqrt{7^2 - 4(1)(0)}}{2(1)}$$

$$= \frac{-7 \pm \sqrt{49}}{2} = \frac{-7 \pm 7}{2}$$

$$\frac{-7 + 7}{2} = \frac{0}{2} = 0,$$

$$\frac{-7 - 7}{2} = \frac{-14}{2} = -7 \qquad \{-7,\ 0\}$$

17) $3v(v+3) = 7v+4$

$3v^2 + 9v = 7v + 4$

$3v^2 + 2v - 4 = 0$

$a = 3,\ b = 2$ and $c = -4$

$v = \dfrac{-2 \pm \sqrt{2^2 - 4(3)(-4)}}{2(3)}$

$= \dfrac{-2 \pm \sqrt{4+48}}{6} = \dfrac{-2 \pm \sqrt{52}}{6}$

$= \dfrac{-2 \pm 2\sqrt{13}}{6} = \dfrac{2(-1 \pm \sqrt{13})}{6}$

$= \dfrac{-1 \pm \sqrt{13}}{3}; \left\{ \dfrac{-1 - \sqrt{13}}{3}, \dfrac{-1 + \sqrt{13}}{3} \right\}$

19) $(2c-5)(c-5) = -3$

$2c^2 - 15c + 25 = -3$

$2c^2 - 15c + 28 = 0$

$a = 2,\ b = -15$ and $c = 28$

$c = \dfrac{-(-15) \pm \sqrt{(-15)^2 - 4(2)(28)}}{2(2)}$

$= \dfrac{15 \pm \sqrt{225 - 224}}{4} = \dfrac{15 \pm \sqrt{1}}{4}$

$= \dfrac{15 \pm 1}{4}$

$\dfrac{15+1}{4} = \dfrac{16}{4} = 4,$

$\dfrac{15-1}{4} = \dfrac{14}{4} = \dfrac{7}{2}$ $\left\{ \dfrac{7}{2}, 4 \right\}$

21) $\dfrac{1}{6}u^2 + \dfrac{4}{3}u = \dfrac{5}{2}$

$6\left(\dfrac{1}{6}u^2 + \dfrac{4}{3}u \right) = 6\left(\dfrac{5}{2} \right)$

$u^2 + 8u = 15$

$u^2 + 8u - 15 = 0$

$a = 1,\ b = 8$ and $c = -15$

$u = \dfrac{-8 \pm \sqrt{8^2 - 4(1)(-15)}}{2(1)}$

$= \dfrac{-8 \pm \sqrt{64 + 60}}{2} = \dfrac{-8 \pm \sqrt{124}}{2}$

$= \dfrac{-8 \pm 2\sqrt{31}}{2} = \dfrac{2(-4 \pm \sqrt{31})}{2}$

$= -4 \pm \sqrt{31}$

$\left\{ -4 - \sqrt{31},\ -4 + \sqrt{31} \right\}$

23) $m^2 + \dfrac{4}{3}m + \dfrac{5}{9} = 0$

$9\left(m^2 + \dfrac{4}{3}m + \dfrac{5}{9} \right) = 9(0)$

$9m^2 + 12m + 5 = 0$

$a = 9,\ b = 12$ and $c = 5$

$k = \dfrac{-12 \pm \sqrt{(12)^2 - 4(9)(5)}}{2(9)}$

$= \dfrac{-12 \pm \sqrt{144 - 180}}{18} = \dfrac{-12 \pm \sqrt{-36}}{18}$

$= \dfrac{-12 \pm 6i}{18} = -\dfrac{12}{18} \pm \dfrac{6}{18}i = -\dfrac{2}{3} \pm \dfrac{1}{3}i$

$\left\{ -\dfrac{2}{3} - \dfrac{1}{3}i,\ -\dfrac{2}{3} + \dfrac{1}{3}i \right\}$

$2(p+10) = (p+10)(p-2)$

$2p + 20 = p^2 + 8p - 20$

25) $0 = p^2 + 6p - 40$

$a = 1,\ b = 6$ and $c = -40$

$p = \dfrac{-6 \pm \sqrt{6^2 - 4(1)(-40)}}{2(1)}$

$= \dfrac{-6 \pm \sqrt{36 + 160}}{2} = \dfrac{-6 \pm \sqrt{196}}{2}$

$= \dfrac{-6 \pm 14}{2}$

Chapter 11: Quadratic Equations

$$\frac{-6+14}{2}=\frac{8}{2}=4,$$

$$\frac{-6-14}{2}=\frac{-20}{2}=-10 \qquad \{-10, 4\}$$

27) $4g^2+9=0$

$a=4$, $b=0$ and $c=9$

$$g=\frac{-0\pm\sqrt{0^2-4(4)(9)}}{2(4)}$$

$$=\frac{\pm\sqrt{-144}}{8}=\frac{\pm12i}{8}=\pm\frac{3}{2}i$$

$$\left\{-\frac{3}{2}i, \frac{3}{2}i\right\}$$

29) $x(x+6)=-34$

$x^2+6x+34=0$

$a=1$, $b=6$ and $c=34$

$$x=\frac{-6\pm\sqrt{6^2-4(1)(34)}}{2(1)}$$

$$=\frac{-6\pm\sqrt{36-136}}{2}=\frac{-6\pm\sqrt{-100}}{2}$$

$$=\frac{-6\pm10i}{2}=-\frac{6}{2}\pm\frac{10i}{2}=-3\pm5i$$

$$\{-3-5i, -3+5i\}$$

31) $(2s+3)(s-1)=s^2-s+6$

$2s^2+s-3=s^2-s+6$

$s^2+2s-9=0$

$a=1$, $b=2$ and $c=-9$

$$s=\frac{-2\pm\sqrt{2^2-4(1)(-9)}}{2(1)}$$

$$=\frac{-2\pm\sqrt{4+36}}{2}=\frac{-2\pm\sqrt{40}}{2}$$

$$=\frac{-2\pm2\sqrt{10}}{2}=\frac{2(-1\pm\sqrt{10})}{2}$$

$$=-1\pm\sqrt{10}$$

$$\{-1-\sqrt{10}, -1+\sqrt{10}\}$$

33) $3(3-4y)=-4y^2$

$9-12y=-4y^2$

$4y^2-12y+9=0$

$a=4$, $b=-12$ and $c=9$

$$y=\frac{-(-12)\pm\sqrt{(-12)^2-4(4)(9)}}{2(4)}$$

$$=\frac{12\pm\sqrt{144-144}}{8}=\frac{12\pm\sqrt0}{8}$$

$$=\frac{12}{8}=\frac{3}{2} \qquad \left\{\frac{3}{2}\right\}$$

35) $-\frac{1}{6}=\frac{2}{3}p^2+\frac{1}{2}p$

$$6\left(-\frac{1}{6}\right)=6\left(\frac{2}{3}p^2+\frac{1}{2}p\right)$$

$-1=4p^2+3p$

$0=4p^2+3p+1$

$a=4$, $b=3$ and $c=1$

$$p=\frac{-3\pm\sqrt{3^2-4(4)(1)}}{2(4)}$$

$$=\frac{-3\pm\sqrt{9-16}}{8}=\frac{-3\pm\sqrt{-7}}{8}$$

$$=\frac{-3\pm i\sqrt7}{8}=-\frac{3}{8}\pm\frac{\sqrt7}{8}i$$

$$\left\{-\frac{3}{8}-\frac{\sqrt7}{8}i, -\frac{3}{8}+\frac{\sqrt7}{8}i\right\}$$

37)
$$4q^2+6=20q$$
$$\frac{4q^2}{2}+\frac{6}{2}=\frac{20q}{2}$$
$$2q^2+3=10q$$
$$2q^2-10q+3=0$$
$$a=2,\ b=-10\text{ and }c=3$$
$$q=\frac{-(-10)\pm\sqrt{(-10)^2-4(2)(3)}}{2(2)}$$
$$=\frac{10\pm\sqrt{100-24}}{4}=\frac{10\pm\sqrt{76}}{4}$$
$$=\frac{10\pm2\sqrt{19}}{4}=\frac{2(5\pm\sqrt{19})}{4}$$
$$=\frac{5\pm\sqrt{19}}{2}$$
$$\left\{\frac{5-\sqrt{19}}{2},\frac{5+\sqrt{19}}{2}\right\}$$

39) There is one rational solution

41)
$$a=10,\ b=-9\text{ and }c=3$$
$$b^2-4ac=(-9)^2-4(10)(3)$$
$$=81-120=-39$$
two complex solutions

43)
$$4y^2-49=-28y$$
$$4y^2+28y-49=0$$
$$a=4,\ b=28\text{ and }c=49$$
$$b^2-4ac=28^2-4(4)(49)$$
$$=784-784=0$$
one rational solution

45) $-5=u(u+6)$
$$-5=u^2+6u$$
$$0=u^2+6u+5$$
$$a=1,\ b=6\text{ and }c=5$$
$$b^2-4ac=6^2-4(1)(5)=36-20=16$$
two rational solutions

47)
$$a=2,\ b=-4\text{ and }c=-5$$
$$b^2-4ac=(-4)^2-4(2)(-5)$$
$$=16+40=56$$
two irrational solutions

49)
$$a=1\text{ and }c=16$$
$$b^2-4ac=0$$
$$b^2-4(1)(16)=0$$
$$b^2-64=0$$
$$b^2=64$$
$$b=\pm\sqrt{64}=\pm8$$
-8 or 8

51)
$$a=4\text{ and }b=-12$$
$$b^2-4ac=0$$
$$(-12)^2-4(4)c=0$$
$$144-16c=0$$
$$144=16c$$
$$9=c$$

53)
$$b=12\text{ and }c=9$$
$$b^2-4ac=0$$
$$(12)^2-4a(9)=0$$
$$144-36a=0$$
$$144=36a$$
$$4=a$$

55)
$x=$ length of one leg
$2x+1=$ length of other leg
$\sqrt{29}=$ length of hypotenuse
$$x^2+(2x+1)^2=(\sqrt{29})^2$$
$$x^2+4x^2+4x+1=29$$
$$5x^2+4x-28=0$$
$$(5x+14)(x-2)=0$$

$5x + 14 = 0$ or $x - 2 = 0$

$5x = -14$ $\boxed{x = 2}$

$x = -\dfrac{14}{5}$

$2x + 1 = 2(2) + 1 = 5$

The lengths of the legs are 2 in.
and 5 in.

57) a) Let $h = 8$ and solve for t.

$8 = -16t^2 + 24t + 24$

$0 = -16t^2 + 24t + 16$

$0 = 2t^2 - 3t - 2$

$0 = (2t + 1)(t - 2)$

$2t + 1 = 0$ or $t - 2 = 0$

$2t = -1$ $\boxed{t = 2}$

$t = -\dfrac{1}{2}$

The ball reaches 8 feet after
2 sec.

b) Let $h = 0$ and solve for t.

$0 = -16t^2 + 24t + 24$

$0 = 2t^2 - 3t - 3$

$a = 2,\ b = -3 \text{ and } c = -3$

$t = \dfrac{-(-3) \pm \sqrt{(-3)^2 - 4(2)(-3)}}{2(2)}$

$= \dfrac{3 \pm \sqrt{9 + 24}}{4} = \dfrac{3 \pm \sqrt{33}}{4}$

Reject $t = \dfrac{3 - \sqrt{33}}{4}$ because it is

negative.

The ball will hit the ground after

$\dfrac{3 + \sqrt{33}}{4}$ sec ≈ 2.2 sec.

Chapter 11 Putting It All Togehter

1) $z^2 - 50 = 0$

$z^2 = 50$

$z = \pm\sqrt{50} = \pm 5\sqrt{2}$

$\left\{-5\sqrt{2},\ 5\sqrt{2}\right\}$

3) $a(a + 1) = 20$

$a^2 + a = 20$

$a^2 + a - 20 = 0$

$(a + 5)(a - 4) = 0$

$a + 5 = 0$ or $a - 4 = 0$

$a = -5$ $a = 4$ $\{-5,\ 4\}$

5) $u^2 + 7u + 9 = 0$

$a = 1,\ b = 7 \text{ and } c = 9$

$u = \dfrac{-7 \pm \sqrt{7^2 - 4(1)(9)}}{2(1)}$

$= \dfrac{-7 \pm \sqrt{49 - 36}}{2} = \dfrac{-7 \pm \sqrt{13}}{2}$

$\left\{\dfrac{-7 - \sqrt{13}}{2},\ \dfrac{-7 + \sqrt{13}}{2}\right\}$

7) $2k(2k+7)=3(k+1)$
$4k^2+14k=3k+3$
$4k^2+11k-3=0$
$(4k-1)(k+3)=0$
$4k-1=0$ or $k+3=0$
$4k=1$ $k=-3$
$k=\dfrac{1}{4}$ $\left\{-3,\dfrac{1}{4}\right\}$

9) $m^2+14m+60=0$
$m^2+14m=-60$
$m^2+14m+49=-60+49$
$(m+7)^2=-11$
$m+7=\pm\sqrt{-11}$
$m+7=\pm i\sqrt{11}$
$m=-7\pm i\sqrt{11}$
$\left\{-7-i\sqrt{11},\,-7+i\sqrt{11}\right\}$

11) $10+(3b-1)^2=4$
$(3b-1)^2=-6$
$3b-1=\pm\sqrt{-6}$
$3b-1=\pm i\sqrt{6}$
$3b=1\pm i\sqrt{6}$
$b=\dfrac{1\pm i\sqrt{6}}{3}$
$\left\{\dfrac{1}{3}-\dfrac{\sqrt{6}}{3}i,\,\dfrac{1}{3}+\dfrac{\sqrt{6}}{3}i\right\}$

13) $1=\dfrac{x^2}{12}-\dfrac{x}{3}$
$12(1)=12\left(\dfrac{x^2}{12}-\dfrac{x}{3}\right)$
$12=x^2-4x$
$0=x^2-4x-12$
$0=(x-6)(x+2)$
$x-6=0$ or $x+2=0$
$x=6$ $x=-2$ $\{-2,6\}$

15) $r^2-4r=3$
$r^2-4r+4=3+4$
$(r-2)^2=7$
$r-2=\pm\sqrt{7}$
$r=2\pm\sqrt{7}$
$\left\{2-\sqrt{7},\,2+\sqrt{7}\right\}$

17) $p(p+8)=3(p^2+2)+p$
$p^2+8p=3p^2+6+p$
$0=2p^2-7p+6$
$0=(2p-3)(p-2)$
$2p-3=0$ or $p-2=0$
$2p=3$ $p=2$
$p=\dfrac{3}{2}$ $\left\{\dfrac{3}{2},2\right\}$

19) $\dfrac{10}{z} = 1 + \dfrac{21}{z^2}$

$z^2\left(\dfrac{10}{z}\right) = z^2\left(1 + \dfrac{21}{z^2}\right)$

$10z = z^2 + 21$

$0 = z^2 - 10z + 21$

$0 = (z-7)(z-3)$

$z - 7 = 0 \ \text{ or } \ z - 3 = 0$

$z = 7 \qquad\qquad z = 3 \qquad \{3,\ 7\}$

21) $(3v+4)(v-2) = -9$

$3v^2 - 2v - 8 = -9$

$3v^2 - 2v + 1 = 0$

$a = 3,\ b = -2 \text{ and } c = 1$

$v = \dfrac{-(-2) \pm \sqrt{(-2)^2 - 4(3)(1)}}{2(3)}$

$= \dfrac{2 \pm \sqrt{4-12}}{6} = \dfrac{2 \pm \sqrt{-8}}{6}$

$= \dfrac{2 \pm 2i\sqrt{2}}{6} = \dfrac{1 \pm i\sqrt{2}}{3}$

$\left\{\dfrac{1}{3} - \dfrac{\sqrt{2}}{3}i,\ \dfrac{1}{3} + \dfrac{\sqrt{2}}{3}i\right\}$

23) $(c-5)^2 + 16 = 0$

$(c-5)^2 = -16$

$c - 5 = \pm\sqrt{-16}$

$c - 5 = \pm 4i$

$c = 5 \pm 4i$

$\{5-4i,\ 5+4i\}$

25) $3g = g^2$

$0 = g^2 - 3g$

$0 = g(g-3)$

$g - 3 = 0 \ \text{ or } \ g = 0$

$g = 3 \qquad\qquad \{0,\ 3\}$

27) $4m^3 = 9m$

$4m^3 - 9m = 0$

$m(4m^2 - 9) = 0$

$m(2m+3)(2m-3) = 0$

$2m + 3 = 0 \ \text{ or } \ 2m - 3 = 0 \ \text{ or } \ m = 0$

$2m = -3 \qquad\qquad 2m = 3$

$m = -\dfrac{3}{2} \qquad\qquad m = \dfrac{3}{2}$

$\left\{-\dfrac{3}{2},\ 0,\ \dfrac{3}{2}\right\}$

29) $\dfrac{1}{3}q^2 + \dfrac{5}{6}q + \dfrac{4}{3} = 0$

$6\left(\dfrac{1}{3}q^2 + \dfrac{5}{6}q + \dfrac{4}{3}\right) = 6(0)$

$2q^2 + 5q + 8 = 0$

$a = 2,\ b = 5 \text{ and } c = 8$

$q = \dfrac{-5 \pm \sqrt{5^2 - 4(2)(8)}}{2(2)}$

$= \dfrac{-5 \pm \sqrt{25-64}}{4} = \dfrac{-5 \pm \sqrt{-39}}{4}$

$= \dfrac{-5 \pm i\sqrt{39}}{4}$

Section 11.4: Exercises

1) $t - \dfrac{48}{t} = 8$

$t\left(t - \dfrac{48}{t}\right) = t(8)$

$t^2 - 48 = 8t$

$t^2 - 8t - 48 = 0$

$(t-12)(t+4) = 0$

$t - 12 = 0 \ \text{ or } \ t + 4 = 0$

$t = 12 \qquad t = -4 \qquad \{-4, 12\}$

3)
$$\frac{2}{x}+\frac{6}{x-2}=-\frac{5}{2}$$

$$2x(x-2)\left(\frac{2}{x}+\frac{6}{x-2}\right)=2x(x-2)\left(-\frac{5}{2}\right)$$

$$4(x-2)+12x=-5x(x-2)$$

$$4x-8+12x=-5x^2+10x$$

$$5x^2+6x-8=0$$

$$(5x-4)(x+2)=0$$

$$5x-4=0 \ \text{ or } \ x+2=0$$

$$5x=4 \qquad x=-2$$

$$x=\frac{4}{5} \qquad\qquad \left\{-2,\frac{4}{5}\right\}$$

5)
$$1=\frac{2}{c}+\frac{1}{c-5}$$

$$c(c-5)(1)=c(c-5)\left(\frac{2}{c}+\frac{1}{c-5}\right)$$

$$c^2-5c=2(c-5)+c$$

$$c^2-5c=2c-10+c$$

$$c^2-5c=3c-10$$

$$c^2-8c=-10$$

$$c^2-8c+16=-10+16$$

$$(c-4)^2=6$$

$$c-4=\pm\sqrt{6}$$

$$c=4\pm\sqrt{6}$$

$$\left\{4-\sqrt{6},\ 4+\sqrt{6}\right\}$$

7)
$$\frac{3}{2v+2}+\frac{1}{v}=\frac{3}{2}$$

$$\frac{3}{2(v+1)}+\frac{1}{v}=\frac{3}{2}$$

$$2v(v+1)\left(\frac{3}{2(v+1)}+\frac{1}{v}\right)=2v(v+1)\left(\frac{3}{2}\right)$$

$$3v+2v+2=3v^2+3v$$

$$0=3v^2-2v-2$$

$$v=\frac{-(-2)\pm\sqrt{(-2)^2-4(3)(-2)}}{2(3)}$$

$$=\frac{2\pm\sqrt{4+24}}{6}=\frac{2\pm\sqrt{28}}{6}$$

$$=\frac{2\pm2\sqrt{7}}{6}=\frac{1\pm\sqrt{7}}{3}$$

$$\left\{\frac{1-\sqrt{7}}{3},\ \frac{1+\sqrt{7}}{3}\right\}$$

9)
$$\frac{9}{n^2}=5+\frac{4}{n}$$

$$n^2\left(\frac{9}{n^2}\right)=n^2\left(5+\frac{4}{n}\right)$$

$$9=5n^2+4n$$

$$0=5n^2+4n-9$$

$$0=(5n+9)(n-1)$$

$$5n+9=0 \ \text{ or } \ n-1=0$$

$$5n=-9 \qquad n=1$$

$$n=-\frac{9}{5} \qquad\qquad \left\{-\frac{9}{5},1\right\}$$

357

11)
$$\frac{5}{6r} = 1 - \frac{r}{6r-6}$$

$$6r(r-1)\left(\frac{5}{6r}\right) = 6r(r-1)\left(1 - \frac{r}{6(r-1)}\right)$$

$$5r - 5 = 6r^2 - 6r - r^2$$

$$0 = 5r^2 - 11r + 5$$

$$r = \frac{-(-11) \pm \sqrt{(-11)^2 - 4(5)(5)}}{2(5)}$$

$$= \frac{11 \pm \sqrt{121 - 100}}{10} = \frac{11 \pm \sqrt{21}}{10}$$

$$\left\{\frac{11 - \sqrt{21}}{10}, \frac{11 + \sqrt{21}}{10}\right\}$$

13)
$$g = \sqrt{g + 20}$$

$$g^2 = g + 20$$

$$g^2 - g - 20 = 0$$

$$(g-5)(g+4) = 0$$

$$g - 5 = 0 \text{ or } g + 4 = 0$$

$$g = 5 \qquad g = -4$$

Only one solution satisfies the original equation. $\{5\}$

15)
$$a = \sqrt{\frac{14a - 8}{5}}$$

$$a^2 = \frac{14a - 8}{5}$$

$$5a^2 = 14a - 8$$

$$5a^2 - 14a + 8 = 0$$

$$(5a - 4)(a - 2) = 0$$

$$5a - 4 = 0 \text{ or } a - 2 = 0$$

$$5a = 4 \qquad a = 2$$

$$a = \frac{4}{5} \qquad \qquad \left\{\frac{4}{5}, 2\right\}$$

17)
$$p - \sqrt{p} = 6$$

$$p - 6 = \sqrt{p}$$

$$(p - 6)^2 = \left(\sqrt{p}\right)^2$$

$$p^2 - 12p + 36 = p$$

$$p^2 - 13p + 36 = 0$$

$$(p - 9)(p - 4) = 0$$

$$p - 9 = 0 \text{ or } p - 4 = 0$$

$$p = 9 \qquad p = 4$$

Only one solution satisfies the original equation. $\{9\}$

19)
$$x = 5\sqrt{x} - 4$$

$$x + 4 = 5\sqrt{x}$$

$$(x + 4)^2 = \left(5\sqrt{x}\right)^2$$

$$x^2 + 8x + 16 = 25x$$

$$x^2 - 17x + 16 = 0$$

$$(x - 16)(x - 1) = 0$$

$$x - 16 = 0 \text{ or } x - 1 = 0$$

$$x = 16 \qquad x = 1 \qquad \{1, 16\}$$

21) $2 + \sqrt{2y - 1} = y$

$$\sqrt{2y - 1} = y - 2$$

$$\left(\sqrt{2y - 1}\right)^2 = (y - 2)^2$$

$$2y - 1 = y^2 - 4y + 4$$

$$0 = y^2 - 6y + 5$$

$$0 = (y - 5)(y - 1)$$

$$y - 5 = 0 \text{ or } y - 1 = 0$$

$$y = 5 \qquad y = 1$$

Only one solution satisfies the original equation. $\{5\}$

23)
$$2 = \sqrt{6k+4} - k$$
$$k + 2 = \sqrt{6k+4}$$
$$(k+2)^2 = \left(\sqrt{6k+4}\right)^2$$
$$k^2 + 4k + 4 = 6k + 4$$
$$k^2 - 2k = 0$$
$$k(k-2) = 0$$
$$k - 2 = 0 \text{ or } k = 0$$
$$k = 2 \qquad\qquad \{0, 2\}$$

25) yes

27) yes

29) no

31) yes

33) no

35)
$$x^4 - 10x^2 + 9 = 0$$
$$(x^2 - 9)(x^2 - 1) = 0$$
$$x^2 - 9 = 0 \text{ or } x^2 - 1 = 0$$
$$x^2 = 9 \qquad x^2 = 1$$
$$x = \pm\sqrt{9} \qquad x = \pm\sqrt{1}$$
$$x = \pm 3 \qquad x = \pm 1$$
$$\{-3, -1, 1, 3\}$$

37)
$$p^4 - 11p^2 + 28 = 0$$
$$(p^2 - 7)(p^2 - 4) = 0$$
$$p^2 - 7 = 0 \text{ or } p^2 - 4 = 0$$
$$p^2 = 7 \qquad p^2 = 4$$
$$p = \pm\sqrt{7} \qquad p = \pm 2$$
$$\{-\sqrt{7}, -2, 2, \sqrt{7}\}$$

39)
$$a^4 + 12a^2 = -35$$
$$a^4 + 12a^2 + 35 = 0$$
$$(a^2 + 7)(a^2 + 5) = 0$$
$$a^2 + 7 = 0 \text{ or } a^2 + 5 = 0$$
$$a^2 = -7 \qquad a^2 = -5$$
$$a = \pm i\sqrt{7} \qquad a = \pm i\sqrt{5}$$
$$\{-i\sqrt{7}, -i\sqrt{5}, i\sqrt{5}, i\sqrt{7}\}$$

41)
$$b^{2/3} + 3b^{1/3} + 2 = 0$$
$$(b^{1/3} + 2)(b^{1/3} + 1) = 0$$
$$b^{1/3} + 2 = 0 \quad \text{or} \quad b^{1/3} + 1 = 0$$
$$b^{1/3} = -2 \qquad b^{1/3} = -1$$
$$\left(\sqrt[3]{b}\right)^3 = (-2)^3 \qquad \left(\sqrt[3]{b}\right)^3 = (-1)^3$$
$$b = -8 \qquad\qquad b = -1$$
$$\{-8, -1\}$$

43)
$$t^{2/3} - 6t^{1/3} = 40$$
$$t^{2/3} - 6t^{1/3} - 40 = 0$$
$$(t^{1/3} + 4)(t^{1/3} - 10) = 0$$
$$t^{1/3} + 4 = 0 \quad \text{or} \quad t^{1/3} - 10 = 0$$
$$t^{1/3} = -4 \qquad t^{1/3} = 10$$
$$\left(\sqrt[3]{t}\right)^3 = (-4)^3 \qquad \left(\sqrt[3]{t}\right)^3 = 10^3$$
$$t = -64 \qquad t = 1000$$
$$\{-64, 1000\}$$

45)
$$2n^{2/3} = 7n^{1/3} + 15$$
$$2n^{2/3} - 7n^{1/3} - 15 = 0$$
$$\left(2n^{1/3} + 3\right)\left(n^{1/3} - 5\right) = 0$$
$$2n^{1/3} + 3 = 0 \quad \text{or} \quad n^{1/3} - 5 = 0$$
$$2n^{1/3} = -3 \qquad\qquad n^{1/3} = 5$$
$$n^{1/3} = -\frac{3}{2} \qquad \left(\sqrt[3]{n}\right)^3 = 5^3$$
$$\left(\sqrt[3]{n}\right)^3 = \left(-\frac{3}{2}\right)^3 \qquad n = 125$$
$$n = -\frac{27}{8}$$
$$\left\{-\frac{27}{8}, 125\right\}$$

47)
$$v - 8v^{1/2} + 12 = 0$$
$$\left(v^{1/2} - 2\right)\left(v^{1/2} - 6\right) = 0$$
$$v^{1/2} - 2 = 0 \quad \text{or} \quad v^{1/2} - 6 = 0$$
$$v^{1/2} = 2 \qquad\qquad v^{1/2} = 6$$
$$\left(\sqrt{v}\right)^2 = 2^2 \qquad \left(\sqrt{v}\right)^2 = 6^2$$
$$v = 4 \qquad\qquad v = 36$$
$$\{4, 36\}$$

49) $4h^{1/2} + 21 = h$
$$0 = h - 4h^{1/2} - 21$$
$$0 = \left(h^{1/2} + 3\right)\left(h^{1/2} - 7\right)$$
$$h^{1/2} + 3 = 0 \quad \text{or} \quad h^{1/2} - 7 = 0$$
$$h^{1/2} = -3 \qquad\qquad h^{1/2} = 7$$
$$\left(\sqrt{h}\right)^2 = (-3)^2 \qquad \left(\sqrt{h}\right)^2 = 7^2$$
$$h = 9 \qquad\qquad h = 49$$
Only one solution satisfies
the original equation. $\{49\}$

51)
$$2a - 5a^{1/2} - 12 = 0$$
$$\left(2a^{1/2} + 3\right)\left(a^{1/2} - 4\right) = 0$$
$$2a^{1/2} + 3 = 0 \quad \text{or} \quad a^{1/2} - 4 = 0$$
$$2a^{1/2} = -3 \qquad\qquad a^{1/2} = 4$$
$$a^{1/2} = -\frac{3}{2} \qquad \left(\sqrt{a}\right)^2 = 4^2$$
$$\left(\sqrt{a}\right)^2 = \left(-\frac{3}{2}\right)^2 \qquad a = 16$$
$$a = \frac{9}{4}$$
Only one solution satisfies
the original equation. $\{16\}$

53)
$$9n^4 = -15n^2 - 4$$
$$9n^4 + 15n^2 + 4 = 0$$
$$\left(3x^2 + 4\right)\left(3x^2 + 1\right) = 0$$
$$3x^2 + 4 = 0 \quad \text{or} \quad 3x^2 + 1 = 0$$
$$3x^2 = -4 \qquad\qquad 3x^2 = -1$$
$$x^2 = -\frac{4}{3} \qquad\qquad x^2 = -\frac{1}{3}$$
$$x = \pm\frac{2}{\sqrt{3}}i \qquad\qquad x = \pm\frac{1}{\sqrt{3}}i$$
$$x = \pm\frac{2\sqrt{3}}{3}i \quad \text{or} \quad x = \pm\frac{\sqrt{3}}{3}i$$
$$\left\{-\frac{2\sqrt{3}}{3}i, -\frac{\sqrt{3}}{3}i, \frac{\sqrt{3}}{3}i, \frac{2\sqrt{3}}{3}i\right\}$$

55)
$$z^4 - 2z^2 = 15$$
$$z^4 - 2z^2 - 15 = 0$$
$$\left(z^2 - 5\right)\left(z^2 + 3\right) = 0$$
$$z^2 - 5 = 0 \quad \text{or} \quad z^2 + 3 = 0$$
$$z^2 = 5 \qquad\qquad z^2 = -3$$
$$z = \pm\sqrt{5} \qquad\qquad z = \pm i\sqrt{3}$$
$$\left\{-\sqrt{5}, \sqrt{5}, -i\sqrt{3}, i\sqrt{3}\right\}$$

57) $w^4 - 6w^2 + 2 = 0$

Let $u = w^2$ and $u^2 = w^4$

$u^2 - 6u + 2 = 0$

$$u = \frac{-(-6) \pm \sqrt{(-6)^2 - 4(1)(2)}}{2(1)}$$

$$= \frac{6 \pm \sqrt{28}}{2} = \frac{6 \pm 2\sqrt{7}}{2} = 3 \pm \sqrt{7}$$

$u = 3 + \sqrt{7}$ or $u = 3 - \sqrt{7}$

$u = w^2$ $u = w^2$

$w^2 = 3 + \sqrt{7}$ $w^2 = 3 - \sqrt{7}$

$w = \pm\sqrt{3 + \sqrt{7}}$ $w = \pm\sqrt{3 - \sqrt{7}}$

$\left\{ -\sqrt{3+\sqrt{7}}, \sqrt{3+\sqrt{7}}, -\sqrt{3-\sqrt{7}}, \sqrt{3-\sqrt{7}} \right\}$

59) $2m^4 + 1 = 7m^2$

$2m^4 - 7m^2 + 1 = 0$

Let $u = m^2$ and $u^2 = m^4$

$2u^2 - 7u + 1 = 0$

$$u = \frac{-(-7) \pm \sqrt{(-7)^2 - 4(2)(1)}}{2(2)} = \frac{7 \pm \sqrt{41}}{4}$$

$u = \dfrac{7 + \sqrt{41}}{4}$ or $u = \dfrac{7 - \sqrt{41}}{4}$

$u = m^2$ $u = m^2$

$m^2 = \dfrac{7 + \sqrt{41}}{4}$ $m^2 = \dfrac{7 - \sqrt{41}}{4}$

$m = \pm\dfrac{\sqrt{7 + \sqrt{41}}}{2}$ $m = \pm\dfrac{\sqrt{7 - \sqrt{41}}}{2}$

$\left\{ -\dfrac{\sqrt{7+\sqrt{41}}}{2}, \dfrac{\sqrt{7+\sqrt{41}}}{2}, \right.$
$\left. -\dfrac{\sqrt{7-\sqrt{41}}}{2}, \dfrac{\sqrt{7-\sqrt{41}}}{2} \right\}$

61) $t^{-2} - 4t^{-1} - 12 = 0$

$\left(t^{-1} + 2\right)\left(t^{-1} - 6\right) = 0$

$t^{-1} + 2 = 0$ or $t^{-1} - 6 = 0$

$t^{-1} = -2$ $t^{-1} = 6$

$t = -\dfrac{1}{2}$ $t = \dfrac{1}{6}$ $\left\{ -\dfrac{1}{2}, \dfrac{1}{6} \right\}$

63) $4 = 13y^{-1} - 3y^{-2}$

$3y^{-2} - 13y^{-1} + 4 = 0$

$\left(3y^{-1} - 1\right)\left(y^{-1} - 4\right) = 0$

$3y^{-1} - 1 = 0$ or $y^{-1} - 4 = 0$

$3y^{-1} = 1$ $y^{-1} = 4$

$y^{-1} = \dfrac{1}{3}$ $y = \dfrac{1}{4}$

$y = 3$

$\left\{ \dfrac{1}{4}, 3 \right\}$

65) $(x-2)^2 + 11(x-2) + 24 = 0$

Let $u = x - 2$

$u^2 + 11u + 24 = 0$

$(u+8)(u+3) = 0$

$u + 8 = 0$ or $u + 3 = 0$

$u = -8$ $u = -3$

Solve for x using $u = x - 2$.

$-8 = x - 2$ $-3 = x - 2$

$-6 = x$ $-1 = x$

$\{-6, -1\}$

67) $2(3q+4)^2 - 13(3q+4) + 20 = 0$

Let $u = 3q + 4$

$2u^2 - 13u + 20 = 0$

$(2u - 5)(u - 4) = 0$

$2u - 5 = 0$ or $u - 4 = 0$

$2u = 5$

$u = \dfrac{5}{2}$ $u = 4$

Solve for q using $u = 3q + 4$.

$\dfrac{5}{2} = 3q + 4$ $4 = 3q + 4$

$-\dfrac{3}{2} = 3q$ $0 = 3q$

$-\dfrac{1}{2} = q$ $0 = q$

$\left\{ -\dfrac{1}{2}, 0 \right\}$

69) $(5a - 3)^2 + 6(5a - 3) = -5$

Let $u = 5a - 3$

$u^2 + 6u + 5 = 0$

$(u + 1)(u + 5) = 0$

$u + 1 = 0$ or $u + 5 = 0$

$u = -1$ $u = -5$

Solve for a using $u = 5a - 3$.

$-1 = 5a - 3$ $-5 = 5a - 3$

$2 = 5a$ $-2 = 5a$

$\dfrac{2}{5} = a$ $-\dfrac{2}{5} = a$

$\left\{ -\dfrac{2}{5}, \dfrac{2}{5} \right\}$

71) $3(k+8)^2 + 5(k+8) = 12$

Let $u = k + 8$

$3u^2 + 5u - 12 = 0$

$(3u - 4)(u + 3) = 0$

$3u - 4 = 0$ or $u + 3 = 0$

$3u = 4$ $u = -3$

$u = \dfrac{4}{3}$

Solve for k using $u = k + 8$.

$\dfrac{4}{3} = k + 8$ $-3 = k + 8$

$-\dfrac{20}{3} = k$ $-11 = k$

$\left\{ -11, -\dfrac{20}{3} \right\}$

73) $1 - \dfrac{8}{2w+1} = -\dfrac{16}{(2w+1)^2}$

Let $u = 2w + 1$

$1 - \dfrac{8}{u} = -\dfrac{16}{u^2}$

$u^2 \left(1 - \dfrac{8}{u} \right) = u^2 \left(-\dfrac{16}{u^2} \right)$

$u^2 - 8u = -16$

$u^2 - 8u + 16 = 0$

$(u - 4)^2 = 0$

$u - 4 = 0$

$u = 4$

Solve for w using $u = 2w + 1$.

$4 = 2w + 1$

$3 = 2w$

$\dfrac{3}{2} = w$ $\left\{ \dfrac{3}{2} \right\}$

75) $1+\dfrac{2}{h-3}=\dfrac{1}{(h-3)^2}$

Let $u = h-3$

$1+\dfrac{2}{u}=\dfrac{1}{u^2}$

$u^2\left(1+\dfrac{2}{u}\right)=u^2\left(\dfrac{1}{u^2}\right)$

$u^2+2u=1$

$u^2+2u+1=1+1$

$(u+1)^2=2$

$u+1=\pm\sqrt{2}$

$u=-1\pm\sqrt{2}$

Solve for h using $u = h-3$.

$-1-\sqrt{2}=h-3 \qquad -1+\sqrt{2}=h-3$

$2-\sqrt{2}=h \qquad\qquad 2+\sqrt{2}=h$

$\{2-\sqrt{2}, 2+\sqrt{2}\}$

77) $x=$ Time taken by Walter

$x+3=$ Time taken by Kevin

$2=$ Time taken together

$\dfrac{1}{x}+\dfrac{1}{x+3}=\dfrac{1}{2}$

$\dfrac{(x+3)+x}{x(x+3)}=\dfrac{1}{2}$

$x(x+3)=2(2x+3)$

$x^2+3x=4x+6$

$x^2-x=6$

$x^2-x+\dfrac{1}{4}=6+\dfrac{1}{4}$

$\left(x-\dfrac{1}{2}\right)^2=\dfrac{25}{4}$

$x-\dfrac{1}{2}=\pm\dfrac{5}{2}$

$x=3$ or $x=-2$

Time taken by Walter: 3 hrs.,

by Kevin: 6 hrs

79) $x=$ Speed of Boat in Still water

$3=$ Speed of the current

$\dfrac{9}{x+3}+\dfrac{6}{x-3}=1$

$\dfrac{9(x-3)+6(x+3)}{x^2-9}=1$

$x^2-9=15x-9$

$x^2-15x=0$

$x(x-15)=0$

$x=15$

Speed of boat in still water: 15 mph

81) $t=$ time - large drain emptying

$t+3=$ time - small drain emptying

$2=$ time - both the drains emptying

$\dfrac{1}{t}+\dfrac{1}{t+3}=\dfrac{1}{2}$

$\dfrac{(t+3)+t}{t(t+3)}=\dfrac{1}{2}$

$2(2t+3)=t^2+3t$

$4t+6=t^2+3t$

$t^2-t-6=0$

$(t-3)(t+2)=0$

$t-3=0 \qquad t+2=0$

$\boxed{t=3} \qquad t=-2$

Reject $t=-2$, because time cannot be negative.

large drain emptys the tank in 3 hrs

small drain emptys the tank in 6 hrs

83)
$$s = \text{speed to Boulder}$$
$$s - 10 = \text{speed to home}$$

$$\frac{600}{s} + \frac{600}{s-10} = 22$$

$$\frac{600(s-10)+600s}{s(s-10)} = 22$$

$$1200s - 6000 = 22s(s-10)$$

$$1200s - 6000 = 22s^2 - 220s$$

$$22s^2 - 1420s + 6000 = 0$$

$$11s - 710s + 3000 = 0$$

$$(s-60)(11s-50) = 0$$

$$s - 60 = 0 \ \text{ or } \ 11s - 50 = 0$$

$$s = 60 \qquad s = \frac{50}{11} \approx 4.55$$

Reject $s \approx 4.55$ because the speed to home cannot be negative.

speed to Boulder: 60 mph

speed to home: 50 mph

Section 11.5: Exercises

1)
$$A = \pi r^2$$

$$\frac{A}{\pi} = r^2$$

$$\pm\sqrt{\frac{A}{\pi}} = r$$

$$\pm\frac{\sqrt{A}}{\sqrt{\pi}} \cdot \frac{\sqrt{\pi}}{\sqrt{\pi}} = r$$

$$\frac{\pm\sqrt{A\pi}}{\pi} = r$$

3)
$$a = \frac{v^2}{r}$$

$$ar = v^2$$

$$\pm\sqrt{ar} = v$$

5)
$$E = \frac{I}{d^2}$$

$$d^2 E = I$$

$$d^2 = \frac{I}{E}$$

$$d = \pm\sqrt{\frac{I}{E}}$$

$$d = \pm\frac{\sqrt{I}}{\sqrt{E}} \cdot \frac{\sqrt{E}}{\sqrt{E}} = \frac{\pm\sqrt{IE}}{E}$$

7)
$$F = \frac{kq_1q_2}{r^2}$$

$$r^2 F = kq_1q_2$$

$$r^2 = \frac{kq_1q_2}{F}$$

$$r = \pm\sqrt{\frac{kq_1q_2}{F}}$$

$$r = \frac{\pm\sqrt{kq_1q_2}}{\sqrt{F}} \cdot \frac{\sqrt{F}}{\sqrt{F}} = \frac{\pm\sqrt{kq_1q_2F}}{F}$$

9)
$$d = \sqrt{\frac{4A}{\pi}}$$

$$d^2 = \frac{4A}{\pi}$$

$$\pi d^2 = 4A$$

$$\frac{\pi d^2}{4} = A$$

11)
$$T_p = 2\pi\sqrt{\frac{l}{g}}$$

$$\frac{T_p}{2\pi} = \sqrt{\frac{l}{g}}$$

$$\frac{T_p^2}{4\pi^2} = \frac{l}{g}$$

$$\frac{gT_p^2}{4\pi^2} = l$$

13) $\quad T_p = 2\pi\sqrt{\dfrac{l}{g}}$

$\quad T_p^2 = 4\pi^2\left(\dfrac{l}{g}\right)$

$\quad gT_p^2 = 4\pi^2 l$

$\quad g = \dfrac{4\pi^2 l}{T_p^2}$

15) a) Both are written in the standard form for a quadratic equation,

$ax^2 + bx + c = 0$

b) Use the quadratic formula.

17) $x = \dfrac{-(-5)\pm\sqrt{(-5)^2 - 4rs}}{2r}$

$\quad = \dfrac{5\pm\sqrt{25 - 4rs}}{2r}$

19) $z = \dfrac{-r\pm\sqrt{r^2 - 4p(-q)}}{2p}$

$\quad = \dfrac{-r\pm\sqrt{r^2 + 4pq}}{2p}$

21) $\quad da^2 - ha = k$

$\quad da^2 - ha - k = 0$

$\quad a = \dfrac{-(-h)\pm\sqrt{(-h)^2 - 4d(-k)}}{2d}$

$\quad = \dfrac{h\pm\sqrt{h^2 + 4dk}}{2d}$

23) $\quad s = \dfrac{1}{2}gt^2 + vt$

$\quad 0 = \dfrac{1}{2}gt^2 + vt - s$

$\quad t = \dfrac{-v\pm\sqrt{v^2 + 2\left(\frac{1}{2}g\right)s}}{2\left(\frac{1}{2}g\right)}$

$\quad \dfrac{-v\pm\sqrt{v^2 + gs}}{g}$

25) $\quad x = $ width of sheet metal

$x + 3 = $ length of sheet metal

length of box $= x + 3 - 1 - 1 = x + 1$

width of box $= x - 1 - 1 = x - 2$

height of box $= 1$

Volume $= (\text{length})(\text{width})(\text{height})$

$\quad 70 = (x+1)(x-2)(1)$

$\quad 70 = x^2 - x - 2$

$\quad 0 = x^2 - x - 72$

$\quad 0 = (x+8)(x-9)$

$x + 8 = 0$ or $x - 9 = 0$

$\quad x = -8 \qquad \boxed{x = 9}$

width $= 9$ in.

length $= 9 + 3 = 12$ in.

27) $\quad x = $ width of non-skid surface

$80 + 2x = $ length of pool plus two strips of non-skid surface

$60 + 2x = $ width of pool plus two strips of non-skid surface

Area of Pool plus strips $-$ Area of Pool $=$ Area of Strips

$(80 + 2x)(60 + 2x) \quad - \quad 80(60) \quad = \quad 576$

$$4800 + 280x + 4x^2 - 4800 = 576$$
$$4x^2 + 280x - 576 = 0$$
$$x^2 + 70x - 144 = 0$$
$$(x+72)(x-2) = 0$$
$$x+72 = 0 \text{ or } x-2 = 0$$
$$x = -72 \qquad \boxed{x = 2}$$
The width of non-skid surface is 2 ft.

29) x = base of the sail

$2x - 1$ = height of the sail

$$\text{Area} = \frac{1}{2}(\text{base})(\text{height})$$
$$60 = \frac{1}{2}(x)(2x-1)$$
$$120 = 2x^2 - x$$
$$0 = 2x^2 - x - 120$$
$$0 = (2x+15)(x-8)$$
$$2x+15 = 0 \text{ or } x-8 = 0$$
$$2x = -15 \qquad \boxed{x = 8}$$
$$x = -\frac{15}{2}$$
base = 8 in.

height = $2(8) - 1 = 15$ in.

31) x = height of the ramp

$2x + 4$ = base of the ramp

$3x - 4$ = hypotenuse of the ramp

$$a^2 + b^2 = c^2$$
$$x^2 + (2x+4)^2 = (3x-4)^2$$
$$x^2 + 4x^2 + 16x + 16 = 9x^2 - 24x + 16$$
$$5x^2 + 16x + 16 = 9x^2 - 24x + 16$$
$$0 = 4x^2 - 40x$$
$$0 = 4x(x-10)$$
$$4x = 0 \text{ or } x-10 = 0$$
$$x = 0 \qquad \boxed{x = 10}$$
The height of the ramp is 10 in.

33) a) $h = 40$
$$h = -16t^2 + 60t + 4$$
$$40 = -16t^2 + 60t + 4$$
$$0 = -16t^2 + 60t - 36$$
$$0 = 4t^2 - 15t + 9$$
$$0 = (4t-3)(t-3)$$
$$4t - 3 = 0 \text{ or } t-3 = 0$$
$$4t = 3$$
$$t = \frac{3}{4} \qquad t = 3$$
0.75 sec on the way up,

3 sec on the way down.

b) $h = 0$
$$h = -16t^2 + 60t + 4$$
$$0 = -16t^2 + 60t + 4$$
$$0 = 4t^2 - 15t - 1$$
$$t = \frac{-(-15) \pm \sqrt{(-15)^2 - 4(4)(-1)}}{2(4)}$$
$$= \frac{15 \pm \sqrt{225+16}}{8} = \frac{15 \pm \sqrt{241}}{8}$$

Reject $\dfrac{15 - \sqrt{241}}{8}$ as a solution since this is a negative number.

$\dfrac{15 + \sqrt{241}}{8}$ sec or about 3.8 sec

35) a) $x = 0$

$y = -0.25x^2 + 1.5x + 9.5$

$y = -0.25(0)^2 + 1.5(0) + 9.5$

$y = 9.5$

9.5 million

b) $y = 11.75$

$y = -0.25x^2 + 1.5x + 9.5$

$11.75 = -0.25x^2 + 1.5x + 9.5$

$0 = -0.25x^2 + 1.5x - 2.25$

$0 = 25x^2 - 150x + 225$

$0 = x^2 - 6x + 9$

$0 = (x-3)^2$

$0 = x - 3$

$3 = x$

11.75 million saw a Broadway play in 1999.

37) $D = \dfrac{65}{P}; \ S = 10P + 3$

$D = S$

$\dfrac{65}{P} = 10P + 3$

$65 = P(10P + 3)$

$65 = 10P^2 + 3P$

$0 = 10P^2 + 3P - 65$

$P = \dfrac{-3 \pm \sqrt{3^2 - 4(10)(-65)}}{2(10)}$

$= \dfrac{-3 \pm \sqrt{9 + 2600}}{20} = \dfrac{-3 \pm \sqrt{2609}}{20}$

$= \dfrac{-3 \pm \sqrt{2609}}{20} \approx \dfrac{-3 \pm 51.08}{20}$

$P \approx \dfrac{-3 - 51.08}{20}$ or $P \approx \dfrac{-3 + 51.08}{20}$

$P \approx -2.70$ $\boxed{P \approx \$2.40}$

39) $P = \dfrac{\sqrt{L^2 - d^2}}{2}$

$= \dfrac{\sqrt{12.5^2 - 12^2}}{2}$

$= \dfrac{\sqrt{12.25}}{2}$

$= \dfrac{3.5}{2}$

$= 1.75\text{pt}$

Chapter 11 Review

1) $a^2 - 3a - 54 = 0$

$(a-9)(a+6) = 0$

$a - 9 = 0$ or $a + 6 = 0$

$a = 9$ $a = -6$ $\{-6, 9\}$

3) $\dfrac{2}{3}c^2 = \dfrac{2}{3}c + \dfrac{1}{2}$

$6\left(\dfrac{2}{3}c^2\right) = 6\left(\dfrac{2}{3}c + \dfrac{1}{2}\right)$

$4c^2 = 4c + 3$

$4c^2 - 4c - 3 = 0$

$(2c+1)(2c-3) = 0$

$2c + 1 = 0$ or $2c - 3 = 0$

$2c = -1$ $2c = 3$

$c = -\dfrac{1}{2}$ $c = \dfrac{3}{2}$ $\left\{-\dfrac{1}{2}, \dfrac{3}{2}\right\}$

5) $x^3 + 3x^2 - 16x - 48 = 0$

$x^2(x+3) - 16(x+3) = 0$

$(x^2 - 16)(x+3) = 0$

$(x-4)(x+4)(x+3) = 0$

$x - 4 = 0$ or $x + 4 = 0$ or $x + 3 = 0$

$x = 4$ $x = -4$ $x = -3$

$\{-4, -3, 4\}$

7) $l = \text{length}$

$l - 4 = \text{width}$

$\text{Area} = (\text{width})(\text{length})$

$96 = (l - 4)l$

$96 = l^2 - 4l$

$0 = l^2 - 4l - 96$

$0 = (l - 12)(l + 8)$

$l - 12 = 0 \ \text{ or } \ l + 8 = 0$

$\boxed{l = 12} \qquad l = -8$

$\text{length} = 12 \text{ cm}$

$\text{width} = 12 - 4 = 8 \text{ cm}$

9) $d^2 = 144$

$d = \pm\sqrt{144}$

$d = \pm 12 \qquad \{-12, 12\}$

11) $v^2 + 4 = 0$

$v^2 = -4$

$v = \pm\sqrt{-4}$

$v = \pm 2i \qquad \{-2i, 2i\}$

13) $(b - 3)^2 = 49$

$b - 3 = \pm\sqrt{49}$

$b - 3 = \pm 7$

$b - 3 = 7 \ \text{ or } \ b - 3 = -7$

$b = 10 \qquad b = -4 \qquad \{-4, 10\}$

15) $27k^2 - 30 = 0$

$27k^2 = 30$

$k^2 = \dfrac{30}{27}$

$k^2 = \dfrac{10}{9}$

$k = \pm\sqrt{\dfrac{10}{9}} \quad k = \pm\dfrac{\sqrt{10}}{3}$

$\left\{ -\dfrac{\sqrt{10}}{3}, \dfrac{\sqrt{10}}{3} \right\}$

17) Let $a = $ length of one side

$a^2 + 3^2 = \left(3\sqrt{2}\right)^2$

$a^2 + 9 = 9(2)$

$a^2 + 9 = 18$

$a^2 = 9$

$a = \pm 3$

Reject -3 for the length of the side. The length of the side is 3 units.

19) $d = \sqrt{(x_2 - x_1)^2 + (y_2 - y_1)^2}$

$d = \sqrt{(7 - 2)^2 + (5 - 3)^2}$

$d = \sqrt{5^2 + 2^2} = \sqrt{25 + 4} = \sqrt{29}$

21) $d = \sqrt{(x_2 - x_1)^2 + (y_2 - y_1)^2}$

$d = \sqrt{(0 - 3)^2 + \left[3 - (-1)\right]^2}$

$d = \sqrt{(-3)^2 + 4^2} = \sqrt{9 + 16}$

$d = \sqrt{25} = 5$

23) 1) $\dfrac{1}{2}(10) = 5$ 2) $5^2 = 25$

$r^2 + 10r + 25; \quad (r + 5)^2$

25) 1) $\dfrac{1}{2}(-5) = -\dfrac{5}{2}$ 2) $\left(-\dfrac{5}{2}\right)^2 = \dfrac{25}{4}$

$c^2 - 5c + \dfrac{25}{4}; \quad \left(c - \dfrac{5}{2}\right)^2$

27) 1) $\dfrac{1}{2}\left(\dfrac{2}{3}\right) = \dfrac{1}{3}$ 2) $\left(\dfrac{1}{3}\right)^2 = \dfrac{1}{9}$

$a^2 + \dfrac{2}{3}a + \dfrac{1}{9}; \quad \left(a + \dfrac{1}{3}\right)^2$

29) $p^2 - 6p - 16 = 0$

$p^2 - 6p = 16$

$p^2 - 6p + 9 = 16 + 9$

$(p-3)^2 = 25$

$p - 3 = \pm\sqrt{25}$

$p - 3 = \pm 5$

$p - 3 = 5$ or $p - 3 = -5$

$p = 8 \qquad p = -2 \qquad \{-2, 8\}$

31) $n^2 + 10n = 6$

$n^2 + 10n + 25 = 6 + 25$

$(n+5)^2 = 31$

$n + 5 = \pm\sqrt{31}$

$n = -5 \pm \sqrt{31}$

$\{-5 - \sqrt{31}, -5 + \sqrt{31}\}$

33) $f^2 + 3f + 1 = 0$

$f^2 + 3f = -1$

$f^2 + 3f + \dfrac{9}{4} = -1 + \dfrac{9}{4}$

$\left(f + \dfrac{3}{2}\right)^2 = \dfrac{5}{4}$

$f + \dfrac{3}{2} = \pm\sqrt{\dfrac{5}{4}}$

$f + \dfrac{3}{2} = \pm\dfrac{\sqrt{5}}{2}$

$f = -\dfrac{3}{2} \pm \dfrac{\sqrt{5}}{2}$

$\left\{-\dfrac{3}{2} - \dfrac{\sqrt{5}}{2}, -\dfrac{3}{2} + \dfrac{\sqrt{5}}{2}\right\}$

35) $-3q^2 + 7q = 12$

$q^2 - \dfrac{7}{3}q = -4$

$q^2 - \dfrac{7}{3}q + \dfrac{49}{36} = -4 + \dfrac{49}{36}$

$\left(q - \dfrac{7}{6}\right)^2 = -\dfrac{95}{36}$

$q - \dfrac{7}{6} = \pm\sqrt{-\dfrac{95}{36}}$

$q - \dfrac{7}{6} = \pm\dfrac{\sqrt{95}}{6}i$

$q = \dfrac{7}{6} \pm \dfrac{\sqrt{95}}{6}i$

$\left\{\dfrac{7}{6} - \dfrac{\sqrt{95}}{6}i, \dfrac{7}{6} + \dfrac{\sqrt{95}}{6}i\right\}$

37) $m^2 + 4m - 12 = 0$

$a = 1, b = 4$ and $c = -12$

$m = \dfrac{-4 \pm \sqrt{4^2 - 4(1)(-12)}}{2(1)}$

$= \dfrac{-4 \pm \sqrt{16 + 48}}{2}$

$= \dfrac{-4 \pm \sqrt{64}}{2} = \dfrac{-4 \pm 8}{2}$

$\dfrac{-4+8}{2} = \dfrac{4}{2} = 2,$

$\dfrac{-4-8}{2} = \dfrac{-12}{2} = -6 \qquad \{-6, 2\}$

39) $10g - 5 = 2g^2$

$$0 = 2g^2 - 10g + 5$$
$$a = 2,\ b = -10 \text{ and } c = 5$$
$$g = \frac{-(-10) \pm \sqrt{(-10)^2 - 4(2)(5)}}{2(2)}$$
$$= \frac{10 \pm \sqrt{100 - 40}}{4}$$
$$= \frac{10 \pm \sqrt{60}}{4} = \frac{10 \pm 2\sqrt{15}}{4} = \frac{5 \pm \sqrt{15}}{2}$$
$$\left\{ \frac{5 - \sqrt{15}}{2}, \frac{5 + \sqrt{15}}{2} \right\}$$

41) $\frac{1}{6}t^2 - \frac{1}{3}t + \frac{2}{3} = 0$

$$6\left(\frac{1}{6}t^2 - \frac{1}{3}t + \frac{2}{3} \right) = 6(0)$$
$$t^2 - 2t + 4 = 0$$
$$a = 1,\ b = -2 \text{ and } c = 4$$
$$t = \frac{-(-2) \pm \sqrt{(-2)^2 - 4(1)(4)}}{2(1)}$$
$$= \frac{2 \pm \sqrt{4 - 16}}{2} = \frac{2 \pm \sqrt{-12}}{2}$$
$$= \frac{2 \pm 2i\sqrt{3}}{2} = 1 \pm i\sqrt{3}$$
$$\left\{ 1 - i\sqrt{3}, 1 + i\sqrt{3} \right\}$$

43) $(6r + 1)(r - 4) = -2(12r + 1)$

$$6r^2 - 23r - 4 = -24r - 2$$
$$6r^2 + r - 2 = 0$$
$$a = 6,\ b = 1 \text{ and } c = -2$$
$$r = \frac{-1 \pm \sqrt{1^2 - 4(6)(-2)}}{2(6)}$$
$$= \frac{-1 \pm \sqrt{1 + 48}}{12} = \frac{-1 \pm \sqrt{49}}{12}$$
$$= \frac{-1 \pm 7}{12}$$

$$\frac{-1 + 7}{12} = \frac{6}{12} = \frac{1}{2},$$
$$\frac{-1 - 7}{12} = \frac{-8}{12} = -\frac{2}{3} \qquad \left\{ -\frac{2}{3}, \frac{1}{2} \right\}$$

45) There are two irrational solutions

47) $a = 3,\ b = -2 \text{ and } c = -5$
$$b^2 - 4ac = (-2)^2 - 4(3)(-5)$$
$$= 4 + 60 = 64$$
two rational solutions

49) $t^2 = -3(t + 2)$
$$t^2 = -3t - 6$$
$$t^2 + 3t + 6 = 0$$
$$a = 1,\ b = 3 \text{ and } c = 6$$
$$b^2 - 4ac = 3^2 - 4(1)(6)$$
$$= 9 - 24 = -15$$
two irrational solutions

51) $4k^2 + bk + 9 = 0$
$$a = 4 \text{ and } c = 9$$
$$b^2 - 4ac = 0$$
$$b^2 - 4(4)(9) = 0$$
$$b^2 - 144 = 0$$
$$b^2 = 144$$
$$b = \pm 12$$

53) $z + 2 = \frac{15}{z}$
$$z(z + 2) = z\left(\frac{15}{z} \right)$$
$$z^2 + 2z = 15$$
$$z^2 + 2z - 15 = 0$$
$$(z + 5)(z - 3) = 0$$
$$z + 5 = 0 \text{ or } z - 3 = 0$$
$$z = -5 \qquad z = 3 \qquad \{-5, 3\}$$

55) $\dfrac{10}{m} = 3 + \dfrac{8}{m^2}$

$m^2\left(\dfrac{10}{m}\right) = m^2\left(3 + \dfrac{8}{m^2}\right)$

$10m = 3m^2 + 8$

$0 = 3m^2 - 10m + 8$

$0 = (3m - 4)(m - 2)$

$3m - 4 = 0$ or $m - 2 = 0$

$3m = 4 \qquad\qquad m = 2$

$m = \dfrac{4}{3} \qquad\qquad \left\{\dfrac{4}{3}, 2\right\}$

57) $x - 4\sqrt{x} = 5$

$x - 5 = 4\sqrt{x}$

$(x - 5)^2 = \left(4\sqrt{x}\right)^2$

$x^2 - 10x + 25 = 16x$

$x^2 - 26x + 25 = 0$

$(x - 25)(x - 1) = 0$

$x - 25 = 0$ or $x - 1 = 0$

$x = 25 \qquad\qquad x = 1$

Only one solution satisfies the original equation. $\{25\}$

59) $b^4 + 5b^2 - 14 = 0$

$(b^2 + 7)(b^2 - 2) = 0$

$b^2 + 7 = 0$ or $b^2 - 2 = 0$

$b^2 = -7 \qquad\qquad b^2 = 2$

$b = \pm i\sqrt{7} \qquad\qquad b = \pm\sqrt{2}$

$\left\{-\sqrt{2}, \sqrt{2}, -i\sqrt{7}, i\sqrt{7}\right\}$

61) $y + 2 = 3y^{1/2}$

$y - 3y^{1/2} + 2 = 0$

$(y^{1/2} - 2)(y^{1/2} - 1) = 0$

$y^{1/2} - 2 = 0$ or $y^{1/2} - 1 = 0$

$y^{1/2} = 2 \qquad\qquad y^{1/2} = 1$

$\left(\sqrt{y}\right)^2 = 2^2 \qquad \left(\sqrt{y}\right)^2 = 1^2$

$y = 4 \qquad\qquad y = 1 \qquad \{1, 4\}$

62) $2(v + 2)^2 + (v + 2) - 3 = 0$

Let $u = v + 2$

$2u^2 + u - 3 = 0$

$(2u + 3)(u - 1) = 0$

$2u + 3 = 0$ or $u - 1 = 0$

$2u = -3 \qquad\qquad u = 1$

$u = -\dfrac{3}{2}$

Solve for v using $u = v + 2$.

$-\dfrac{3}{2} = v + 2 \qquad\qquad 1 = v + 2$

$-\dfrac{7}{2} = v \qquad\qquad -1 = v$

$\left\{-\dfrac{7}{2}, -1\right\}$

65) $F = \dfrac{mv^2}{r}$

$Fr = mv^2$

$\dfrac{Fr}{m} = v^2$

$\pm\sqrt{\dfrac{Fr}{m}} = v$

$\pm\dfrac{\sqrt{Fr}}{\sqrt{m}} \cdot \dfrac{\sqrt{m}}{\sqrt{m}} = v$

$\pm\dfrac{\sqrt{Frm}}{m} = v$

67) $r = \sqrt{\dfrac{A}{\pi}}$

$r^2 = \dfrac{A}{\pi}$

$\pi r^2 = A$

69) $kn^2 - ln - m = 0$

$a = k, b = -l,$ and $c = -m$

$n = \dfrac{l \pm \sqrt{l^2 + 4mk}}{2k}$

71) $\quad x = $ width of border

$18 + 2x = $ length of case + borders

$27 + 2x = $ width of case + borders

Area of case plus border $= 792$

$(18 + 2x)(27 + 2x) \quad = 792$

$486 + 90x + 4x^2 = 792$

$4x^2 + 90x - 306 = 0$

$(4x + 102)(x - 3) = 0$

$4x + 102 = 0 \quad$ or $\quad x - 3 = 0$

$4x = -102 \quad \boxed{x = 3}$

$x = -\dfrac{51}{2}$

The width of the border is 3 in.

73) $D = \dfrac{240}{P}; \ S = 4P - 2$

$D = S$

$\dfrac{240}{P} = 4P - 2$

$240 = P(4P - 2)$

$240 = 4P^2 - 2P$

$0 = 4P^2 - 2P - 240$

$0 = 2P^2 - P - 120$

$0 = (2P + 15)(P - 8)$

$2P + 15 = 0 \quad$ or $\quad P - 8 = 0$

$2P = -15 \qquad \boxed{P = \$8.00}$

$P = -\dfrac{15}{2}$

75) $\quad 3k^2 + 4 = 7k$

$3k^2 - 7k + 4 = 0$

$(k - 1)(3k - 4) = 0$

$k - 1 = 0 \qquad$ or $3k - 4 = 0$

$k = 1 \qquad$ or $3k = 4$

$k = \dfrac{4}{3} \qquad \left\{ 1, \dfrac{4}{3} \right\}$

77) $\quad 3 = 15 + (y + 8)^2$

Let $u = y + 8$

$3 = 15 + u^2$

$u^2 + 12 = 0$

$u = \pm\sqrt{-12} = \pm 2\sqrt{-3} = \pm 2i\sqrt{3}$

$y = u - 8$

$= \pm 2i\sqrt{3} - 8 \quad \left\{ -2i\sqrt{3} - 8, 2i\sqrt{3} - 8 \right\}$

79) $\quad c^4 - 26c^2 + 25 = 0$

$(c^2 - 25)(c^2 - 1) = 0$

$c^2 = 25 \ $ or $c^2 = 1$

$c^2 = \pm\sqrt{25} \quad$ or $\quad c^2 = \pm\sqrt{1}$

$c = \pm 5 \quad$ or $\quad c = \pm 1 \qquad \{-5, -1, 1, 5\}$

81) $\quad \dfrac{8}{n - 4} + \dfrac{n}{n - 3} = -\dfrac{12}{n^2 - 7n + 12}$

$\dfrac{8(n - 3) + n(n - 4)}{(n - 4)(n - 3)} = \dfrac{-12}{(n - 4)(n - 3)}$

$8n - 24 + n^2 - 4n = -12$

$n^2 + 4n - 12 = 0$

$(n + 6)(n - 2) = 0$

$n + 6 = 0 \ $ or $\ n - 2 = 0$

$n = -6 \ $ or $n = 2 \qquad \{-6, 2\}$

83) $\dfrac{1}{3}w^2 + w = -\dfrac{5}{6}$

$\dfrac{2w^2 + 6w}{6} = -\dfrac{5}{6}$

$2w^2 + 6w + 5 = 0$

Solving Quadratic Equation,

$w = \dfrac{-6 \pm \sqrt{(-6)^2 - 4(2)(5)}}{2(2)}$

$= \dfrac{-6 \pm \sqrt{36 - 40}}{4} = \dfrac{-6 \pm \sqrt{-4}}{4}$

$= -\dfrac{3}{2} \pm i\dfrac{1}{2}$ $\qquad \left\{ -\dfrac{3}{2} - i\dfrac{1}{2}, -\dfrac{3}{2} + i\dfrac{1}{2} \right\}$

85) $6 + p(p - 10) = 2(4p - 15)$

$6 + p^2 - 10p = 8p - 30$

$p^2 - 18p + 36 = 0$

Solving Quadratic Equation,

$p = \dfrac{-(-18) \pm \sqrt{(-18)^2 - 4(1)(36)}}{2(1)}$

$= \dfrac{18 \pm \sqrt{324 - 180}}{2} = \dfrac{18 \pm \sqrt{180}}{2}$

$= 9 \pm 3\sqrt{5}$ $\qquad \left\{ 9 - 3\sqrt{5}, 9 + 3\sqrt{5} \right\}$

87) $x^3 = x$

$x^3 - x = 0$

$x(x^2 - 1) = 0$

$x = 0$ or $x^2 - 1 = 0$

$x = 0$ or $x^2 = 1$

$x = 0$ or $x = \pm 1$ $\qquad \{-1, 0, 1\}$

89) $(x - 3)^2 + x^2 = 15^2$

$x^2 - 6x + 9 + x^2 = 225$

$2x^2 - 6x - 216 = 0$

$x^2 - 3x - 108 = 0$

$(x - 12)(x + 9) = 0$

$x - 12 = 0$ or $x + 9 = 0$

$x = 12$ or $x = -9$

One Leg $= x - 3 = 12 - 3 = 9$ cm

Other Leg $= x = 12$ cm

91) $\qquad x = $ time taken by Erica

$\qquad x - 25 = $ time taken by Latrice

$\dfrac{1}{x} + \dfrac{1}{x - 25} = \dfrac{1}{30}$

$\dfrac{x - 25 + x}{x(x - 25)} = \dfrac{1}{30}$

$x^2 - 25x = 30(2x - 25)$

$x^2 - 85x + 750 = 0$

$(x - 75)(x - 10) = 0$

$x - 75 = 0$ or $x - 10 = 0$

$x = 75$ or $x = 10$

Reject $x = 10$ because time by Latrice is negative

Time taken by Erica: 75 min.;

$\qquad$ by Latrice: 50 min.

Chapter 11 Test

1) $k^2 - 8k = 48$

$k^2 - 8k - 48 = 0$

$(k - 12)(k + 4) = 0$

$k = 12$ or $k = -4$ $\qquad \{-4, 12\}$

3) $t^2 + 7 = 25$

$t^2 = 18$

$t = \pm 3\sqrt{2}$

5) $d = \sqrt{(x_2 - x_1)^2 + (y_2 - y_1)^2}$

$d = \sqrt{[4 - (-6)]^2 + (3 - 2)^2}$

$d = \sqrt{10^2 + 1^2} = \sqrt{100 + 1} = \sqrt{101}$

Chapter 11: Quadratic Equations

7) $2x^2 - 6x + 14 = 0$

$x^2 - 3x + 7 = 0$

$x^2 - 3x = -7$

$x^2 - 3x + \dfrac{9}{4} = -7 + \dfrac{9}{4}$

$\left(x - \dfrac{3}{2}\right)^2 = -\dfrac{19}{4}$

$x - \dfrac{3}{2} = \pm\sqrt{-\dfrac{19}{4}}$

$x - \dfrac{3}{2} = \pm\dfrac{\sqrt{19}}{2}i$

$x = \dfrac{3}{2} \pm \dfrac{\sqrt{19}}{2}i$

$\left\{\dfrac{3}{2} - \dfrac{\sqrt{19}}{2}i, \dfrac{3}{2} + \dfrac{\sqrt{19}}{2}i\right\}$

9) $(c + 5)^2 + 8 = 2$

$(c + 5)^2 = -6$

$c + 5 = \pm\sqrt{-6}$

$c + 5 = \pm i\sqrt{6}$

$c = -5 \pm i\sqrt{6}$

$\left\{-5 - i\sqrt{6}, -5 + i\sqrt{6}\right\}$

11) $y^2 - \dfrac{4}{25} = 0$

$\left(y + \dfrac{2}{5}\right)\left(y - \dfrac{2}{5}\right) = 0$

$y = -\dfrac{2}{5}$ or $y = \dfrac{2}{5}$ $\qquad \left\{-\dfrac{2}{5}, \dfrac{2}{5}\right\}$

13) $p^4 + p^2 - 72 = 0$

$(p^2 + 9)(p^2 - 8) = 0$

$p^2 + 9 = 0$ or $p^2 - 8 = 0$

$p^2 = -9 \qquad p^2 = 8$

$p = \pm 3i \qquad p = \pm 2\sqrt{2}$

$\left\{-2\sqrt{2}, 2\sqrt{2}, -3i, 3i\right\}$

15) $(2t - 3)(t - 2) = 2$

$2t^2 - 7t + 6 = 2$

$2t^2 - 7t + 4 = 0$

$a = 2, b = -7$ and $c = 4$

$t = \dfrac{-(-7) \pm \sqrt{(-7)^2 - 4(2)(4)}}{2(2)}$

$= \dfrac{7 \pm \sqrt{49 - 32}}{4} = \dfrac{7 \pm \sqrt{17}}{4}$

$\left\{\dfrac{7 - \sqrt{17}}{4}, \dfrac{7 + \sqrt{17}}{4}\right\}$

17) $5z^2 - 6z - 1 = 0$

$a = 5, b = -6$ and $c = -1$

$b^2 - 4ac = (-6)^2 - 4(5)(-1)$

$= 36 + 20 = 56$

two irrational solutions

19) a) Let $h = 40$ and solve for t.

$40 = -16t^2 + 24t + 200$

$0 = -16t^2 + 24t + 160$

$0 = 2t^2 - 3t - 20$

$0 = (2t + 5)(t - 4)$

$2t + 5 = 0$ or $t - 4 = 0$

$2t = -5 \qquad \boxed{t = 4}$

$t = -\dfrac{5}{2}$

Reject $t = -\dfrac{5}{2}$ because it is negative.

after 4 sec

b) Let $h = 0$ and solve for t.

$$0 = -16t^2 + 24t + 200$$

$$0 = 2t^2 - 3t - 25$$

$$a = 2,\ b = -3 \text{ and } c = -25$$

$$t = \dfrac{-(-3) \pm \sqrt{(-3)^2 - 4(2)(-25)}}{2(2)}$$

$$= \dfrac{3 \pm \sqrt{9 + 200}}{4} = \dfrac{3 \pm \sqrt{209}}{4}$$

Reject $t = \dfrac{3 - \sqrt{209}}{4}$ because it is negative.

The ball will hit the ground after $\dfrac{3 + \sqrt{209}}{4}$ sec ≈ 4.4 sec.

21) $\quad rt^2 - st = 6$

$$rt^2 - st - 6 = 0$$

$$a = r,\ b = -s \text{ and } c = -6$$

$$t = \dfrac{-(-s) \pm \sqrt{(-s)^2 - 4(r)(-6)}}{2(r)}$$

$$= \dfrac{s \pm \sqrt{s^2 + 24r}}{2r}$$

23) $x =$ time taken by Justine to type

$x + 20 =$ time taken by Kimora

$24 =$ time taken together

$$\dfrac{1}{x} + \dfrac{1}{x + 20} = \dfrac{1}{24}$$

$$\dfrac{x + 20 + x}{x(x + 20)} = \dfrac{1}{24}$$

$$x^2 + 20x = 48x + 480$$

$$x^2 - 28x - 480 = 0$$

$$(x - 40)(x + 12) = 0$$

$x = 40$ or $x = -12$

Justine: 40 min; Kimora: 60 min.

Cumulative Review: Chapters 1-11

1) $\quad \dfrac{4}{15} + \dfrac{1}{6} - \dfrac{3}{5} = \dfrac{8}{30} + \dfrac{5}{30} - \dfrac{18}{30}$

$$= -\dfrac{5}{30} = -\dfrac{1}{6}$$

3) $\quad A = (20)(21) - (21 - 8)(20 - 6)$

$$= 420 - (13)(14) = 420 - 182$$

$$= 238 \text{ cm}^2$$

$$P = 21 + 20 + 8 + (20 - 6) + (21 - 8) + 6 = 82 \text{ cm}$$

5) $\quad \left(5x^4 y^{-10}\right)\left(3xy^3\right)^2$

$$= 5x^4 y^{-10}\, 3^2 x^2 y^6$$

$$= 5 \cdot 9 \cdot x^4 \cdot x^2 \cdot y^{-10} \cdot y^6$$

$$= 45x^{4+2} y^{-10+6} = 45x^6 y^{-4} = \dfrac{45x^6}{y^4}$$

7) $\quad x =$ cameras sold in December 2009

$$108 = x + x(.20)$$

$$108 = 1.20x$$

$$90 = x$$

9) $\quad 2x - 5y = 8$

x-int: Let $y = 0$, and solve for x.

$$2x - 5(0) = 8$$

$$2x = 8$$

$$x = 4 \qquad (4,\ 0)$$

y-int: Let $x = 0$, and solve for y.

$$x(0) - 5y = 8$$

$$-5y = 8$$

$$y = -\dfrac{8}{5} \qquad \left(0,\ -\dfrac{8}{5}\right)$$

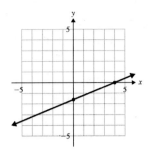

11) $x = $ cost for a bag of chips

$y = $ cost for a can of soda

$2x + 3y = 3.85$

$x + 2y = 2.30$

Use substitution. $x = 2.30 - 2y$

$2(2.30 - 2y) + 3y = 3.85$

$4.60 - 4y + 3y = 3.85$

$-y = -0.75$

$y = 0.75$

$x = 2.30 - 2y; \ y = 0.75$

$x = 2.30 - 2(0.75)$

$x = 2.30 - 1.50$

$x = 0.80$

chips: \$0.80; soda: \$0.75

13) $3(r-5)^2 = 3(r^2 - 10r + 25)$

$= 3r^2 - 30r + 75$

15) $a^3 + 125 = (a+5)(a^2 - 5a + 25)$

17) $\dfrac{2 + \dfrac{6}{c}}{\dfrac{2}{c^2} - \dfrac{8}{c}} = \dfrac{c^2\left(2 + \dfrac{6}{c}\right)}{c^2\left(\dfrac{2}{c^2} - \dfrac{8}{c}\right)} = \dfrac{2c^2 + 6c}{2 - 8c}$

$= \dfrac{2c(c+3)}{2(1-4c)} = \dfrac{c(c+3)}{1-4c}$

19) $\boxed{\text{I}} \quad 4x - 2y + z = -7$

$\boxed{\text{II}} \quad -3x + y - 2z = 5$

$\boxed{\text{III}} \quad 2x + 3y + 5z = 4$

Add: $\boxed{\text{I}} + (2)\boxed{\text{II}}$

$\quad 4x - 2y + z = -7$

$\quad -6x + 2y - 4z = 10$

$\overline{\boxed{\text{A}} \ -2x \quad -3z = 3}$

Add: $\boxed{\text{III}} + (-3)\boxed{\text{II}}$

$\quad 2x + 3y + 5z = 4$

$\quad 9x - 3y + 6z = -15$

$\overline{\quad 11x \quad +11z = -11}$

$\boxed{\text{B}} \ x + z = -1; \ x = -z - 1$

Substitute $x = -z - 1$ into $\boxed{\text{A}}$.

$-2(-z - 1) - 3z = 3$

$2z + 2 - 3z = 3$

$-z = 1$

$z = -1$

Substitute $z = -1$ into $\boxed{\text{A}}$.

$-2x - 3(-1) = 3$

$-2x + 3 = 3$

$-2x = 0$

$x = 0$

Substitute $x = 0$ and $z = -1$ into $\boxed{\text{II}}$.

$-3(0) + y - 2(-1) = 5$

$y + 2 = 5$

$y = 3$

$(0, 3, -1)$

21) $\sqrt[3]{40} = \sqrt[3]{8} \cdot \sqrt[3]{5}$

$= 2\sqrt[3]{5}$

23) $64^{2/3} = \left(\sqrt[3]{64}\right)^2 = (4)^2 = 16$

25) $(10+3i)(1-8i) = 10-80i+3i-24i^2$

$$= 10-77i-24(-1)$$
$$= 10-77i+24$$
$$= 34-77i$$

27) $\dfrac{3}{5}x^2 + \dfrac{1}{5} = \dfrac{1}{5}x$

$$5\left(\dfrac{3}{5}x^2 + \dfrac{1}{5}\right) = 5\left(\dfrac{1}{5}x\right)$$

$$3x^2 + 1 = x$$

$$3x^2 - x + 1 = 0$$

$$a = 3,\ b = -1\ \text{and}\ c = 1$$

$$x = \dfrac{-(-1) \pm \sqrt{(-1)^2 - 4(3)(1)}}{2(3)}$$

$$= \dfrac{1 \pm \sqrt{1-12}}{6} = \dfrac{1 \pm \sqrt{-11}}{6}$$

$$= \dfrac{1 \pm i\sqrt{11}}{6} = \dfrac{1}{6} \pm \dfrac{\sqrt{11}}{6}i$$

$$\left\{\dfrac{1}{6} - \dfrac{\sqrt{11}}{6}i,\ \dfrac{1}{6} + \dfrac{\sqrt{11}}{6}i\right\}$$

29) $p^2 + 6p = 27$

$$p^2 + 6p - 27 = 0$$

$$(p+9)(p-3) = 0$$

$$p+9=0\ \text{ or }\ p-3=0$$

$$p = -9 \qquad p = 3$$

$$\{-9, 3\}$$

Section 12.1: Exercises

1) It is a special type of relation in which each element of the domain corresponds to exactly one element in the range.

3) Domain: $\{5,6,14\}$

Range: $\{-3,0,1,3\}$

Not a function

5) Domain: $\{-2,2,5,8\}$

Range: $\{4,25,64\}$

Is a function

7) Domain: $(-\infty,\infty)$ Range: $[-4,\infty)$

Is a function

9) yes

11) yes

13) no

15) no

17) False. It is read as "f of x."

19)

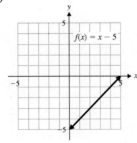

21)

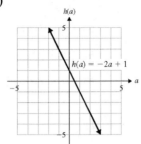

23)

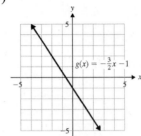

25)

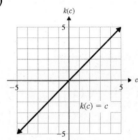

27) $f(6)=3(6)-7=18-7=11$

29) $g(3)=(3)^2-4(3)-9$
$$=9-12-9=-12$$

31) $f(a)=3(a)-7=3a-7$

33) $g(d)=d^2-4d-9$

35) $f(c+4)=3(c+4)-7$
$$=3c+12-7=3c+5$$

37) $g(t+2)=(t+2)^2-4(t+2)-9$
$$=t^2+4t+4-4t-8-9$$
$$=t^2-13$$

39) $g(h-1)=(h-1)^2-4(h-1)-9$
$$=h^2-2h+1-4h+4-9$$
$$=h^2-6h-4$$

41) $f(4) = -5(4) + 2 = -20 + 2 = -18$

43) $g(-6) = (-6)^2 + 7(-6) + 2$
$= 36 - 42 + 2 = -4$

45) $f(-3k) = -5(-3k) + 2 = 15k + 2$

47) $g(5t) = (5t)^2 + 7(5t) + 2$
$= 25t^2 + 35t + 2$

49) $f(b+1) = -5(b+1) + 2$
$= -5b - 5 + 2 = -5b - 3$

51) $g(r+4) = (r+4)^2 + 7(r+4) + 2$
$= r^2 + 8r + 16 + 7r + 28 + 2$
$= r^2 + 15r + 46$

53) $f(x) = 4x + 3$
$23 = 4x + 3$
$20 = 4x$
$5 = x$

55) $h(x) = -2x - 5$
$0 = -2x - 5$
$5 = -2x$
$-\dfrac{5}{2} = x$

57) $p(x) = x^2 - 6x - 16$
$0 = x^2 - 6x - 16$
$0 = (x+2)(x-8)$
$x + 2 = 0 \ \text{ or } \ x - 8 = 0$
$x = -2 \qquad x = 8$

59) $(-\infty, \infty)$

61) Set the denominator equal to 0 and solve for the variable. The domain consists of all real numbers except the values that make the denominator equal to 0.

63) $(-\infty, \infty)$

65) $(-\infty, \infty)$

67) Solve: $x + 8 = 0$
$x = -8$
$(-\infty, -8) \cup (-8, \infty)$

69) Solve: $x = 0$
$(-\infty, 0) \cup (0, \infty)$

71) Solve: $2c - 1 = 0$
$2c = 1$
$c = \dfrac{1}{2}$
$\left(-\infty, \dfrac{1}{2}\right) \cup \left(\dfrac{1}{2}, \infty\right)$

73) Solve: $7t + 3 = 0$
$7t = -3$
$t = -\dfrac{3}{7}$
$\left(-\infty, -\dfrac{3}{7}\right) \cup \left(-\dfrac{3}{7}, \infty\right)$

75) $(-\infty, \infty)$

77) Solve: $x^2 + 11x + 24 = 0$
$(x+8)(x+3) = 0$
$x + 8 = 0 \ \text{ or } \ x + 3 = 0$
$x = -8 \qquad x = -3$
$(-\infty, -8) \cup (-8, -3) \cup (-3, \infty)$

79) Solve: $c^2 - 5c - 36 = 0$
$(c-9)(c+4) = 0$
$c - 9 = 0 \ \text{ or } \ c + 4 = 0$
$c = 9 \qquad c = -4$
$(-\infty, -4) \cup (-4, 9) \cup (9, \infty)$

81) Solve: $x \geq 0$
$[0, \infty)$

83) Solve: $n + 2 \geq 0$

$n \geq -2$

$[-2, \infty)$

85) Solve: $a - 8 \geq 0$

$a \geq 8$

$[8, \infty)$

87) Solve: $2x - 5 \geq 0$

$2x \geq 5$

$x \geq \dfrac{5}{2}$

$\left[\dfrac{5}{2}, \infty\right)$

89) Solve: $-t \geq 0$

$t \leq 0$

$(-\infty, 0]$

91) Solve: $9 - a \geq 0$

$-a \geq -9$

$a \leq 9$

$(-\infty, 9]$

93) $(-\infty, \infty)$

95) a) $C(20) = 22(20) = 440$

$\$440$

b) $C(56) = 22(56) = 1232$

$\$1232$

c) $C(y) = 770$

$770 = 22y$

$35 = y \qquad 35$ sq. yards

d)

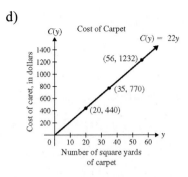

97) a) $L(1) = 50(1) + 40 = 50 + 40 = 90$

The labor charge for a 1-hour repair job is $\$90$.

b) $L(1.5) = 50(1.5) + 40$

$= 75 + 40 = 115$

The labor charge for a 1.5-hour repair job is $\$115$.

c) $L(h) = 165$

$165 = 50h + 40$

$125 = 50h$

$2.5 = h$

If the labor charge is $\$165$, the repair job took 2.5 hours.

99) a) $A(r) = \pi r^2$

b) $A(3) = \pi(3)^2 = 9\pi$

When the radius of the circle is 3 cm, the area of the circle is 9π sq. cm.

c) $A(5) = \pi(5)^2 = 25\pi$

When the radius of the circle is 5 in, the area of the circle is 25π sq. in.

d) $A(r) = 64\pi$

$64\pi = \pi r^2$

$64 = r^2$

$\pm 8 = r \qquad r = 8$ inches

Section 12.2: Exercises

1) Domain: $(-\infty, \infty)$; Range: $[3, \infty)$

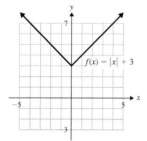

3) Domain: $(-\infty, \infty)$; Range: $[0, \infty)$

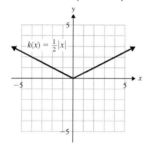

5) Domain: $(-\infty, \infty)$; Range: $[-4, \infty)$

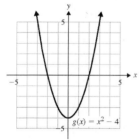

7) Domain: $(-\infty, \infty)$; Range: $(-\infty, -1]$

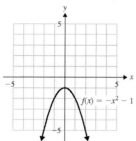

9) Domain: $[-3, \infty)$; Range: $[0, \infty)$

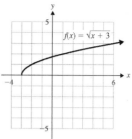

11) Domain: $[0, \infty)$; Range: $[0, \infty)$

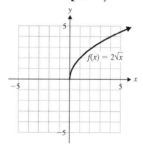

13) The graph of $g(x)$ is the same shape as $f(x)$, but g is shifted down 2 units.

15) The graph of $g(x)$ is the same shape as $f(x)$, but g is shifted left 2 units.

17) The graph of $g(x)$ is the reflection of $f(x)$ about the x-axis.

19)

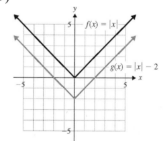

381

21)

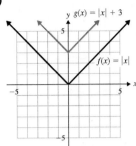

31)

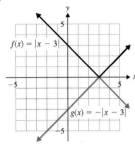

23)

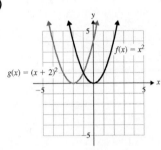

33)

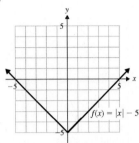

25)

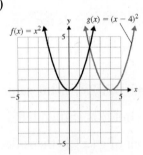

35)

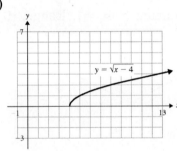

27)

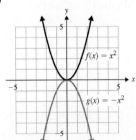

37)

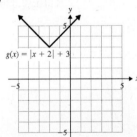

29)

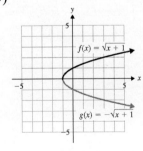

39)

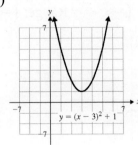

41)

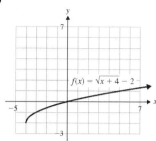

43)

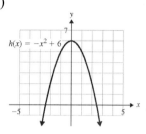

45)

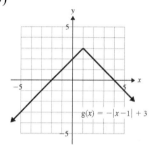

47)

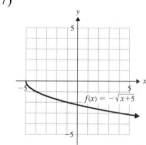

49) a) $h(x)$ b) $f(x)$

c) $g(x)$ d) $k(x)$

51) $g(x) = \sqrt{x+5}$

53) $g(x) = |x+2| - 1$

55) $g(x) = (x+3)^2 + \dfrac{1}{2}$

57) $g(x) = -x^2$

59)

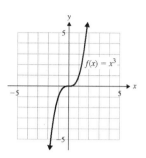

a)

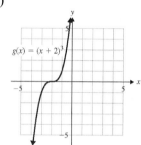

b)

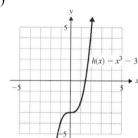

c)

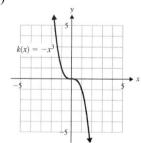

d)

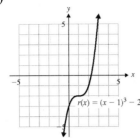

$r(x) = (x-1)^3 - 2$

69)

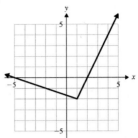

61)

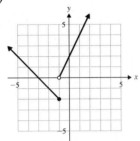

71) $f\left(3\dfrac{1}{4}\right) = \left[\!\left[3\dfrac{1}{4}\right]\!\right] = 3$

73) $f(9.2) = [\![9.2]\!] = 9$

75) $f(8) = [\![8]\!] = 8$

77) $f\left(-6\dfrac{2}{5}\right) = \left[\!\left[-6\dfrac{2}{5}\right]\!\right] = -7$

63)

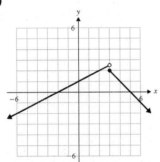

79) $f(-8.1) = [\![-8.1]\!] = -9$

81)

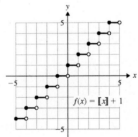

$f(x) = [\![x]\!] + 1$

65)

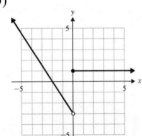

83)

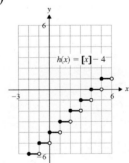

$h(x) = [\![x]\!] - 4$

67)

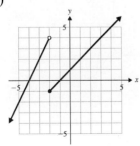

85)

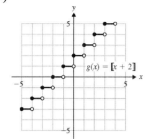

87)

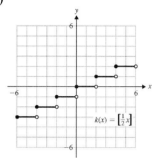

89)

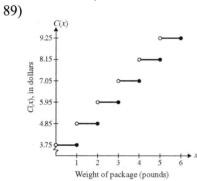

91)

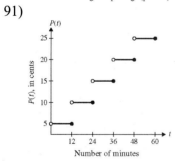

Section 12.3: Exercises

1) (h, k)

3) a is positive

5) $|a| > 1$

7) $V(-1, -4); x = -1$

x-int $\quad 0 = (x+1)^2 - 4$

$\qquad 4 = (x+1)^2$

$\qquad \pm\sqrt{4} = x + 1$

$\qquad \pm 2 = x + 1$

$\qquad -1 \pm 2 = x$

$\qquad x = -3 \text{ or } x = 1 \qquad (-3, 0), (1, 0)$

y-int $\quad y = (0+1)^2 - 4$

$\qquad y = (1)^2 - 4$

$\qquad y = 1 - 4 = -3 \qquad (0, -3)$

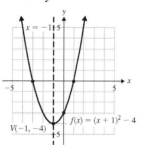

9) $V(2, 3); x = 2$

x-int $\quad 0 = (x-2)^2 + 3$

$\qquad -3 = (x-2)^2$

$\qquad \pm\sqrt{-3} = x - 2$

$\qquad \pm i\sqrt{3} = x - 2$

$\qquad 2 \pm i\sqrt{3} = x \qquad\qquad \text{none}$

y-int $\quad y = (0-2)^2 + 3$

$\qquad y = (-2)^2 + 3$

$\qquad y = 4 + 3 = 7 \qquad (0, 7)$

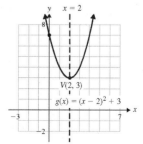

11) $V(4,-2); x=4$

x-int $\quad 0=(x-4)^2-2$

$\qquad 2=(x-4)^2$

$\qquad \pm\sqrt{2}=x-4$

$\qquad 4\pm\sqrt{2}=x$

$\qquad (4-\sqrt{2},0),(4+\sqrt{2},0)$

y-int $\quad y=(0-4)^2-2$

$\qquad y=(-4)^2-2$

$\qquad y=16-2=14 \qquad (0,14)$

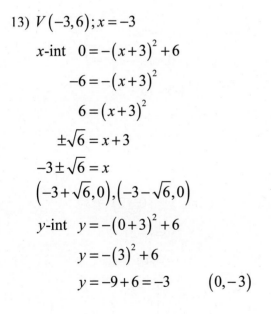

13) $V(-3,6); x=-3$

x-int $\quad 0=-(x+3)^2+6$

$\qquad -6=-(x+3)^2$

$\qquad 6=(x+3)^2$

$\qquad \pm\sqrt{6}=x+3$

$\qquad -3\pm\sqrt{6}=x$

$\qquad (-3+\sqrt{6},0),(-3-\sqrt{6},0)$

y-int $\quad y=-(0+3)^2+6$

$\qquad y=-(3)^2+6$

$\qquad y=-9+6=-3 \qquad (0,-3)$

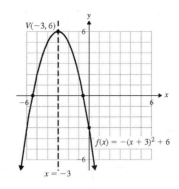

15) $V(-1,-5); x=-1$

x-int $\quad 0=-(x+1)^2-5$

$\qquad 5=-(x+1)^2$

$\qquad -5=(x+1)^2$

$\qquad \pm\sqrt{-5}=x+1$

$\qquad -1\pm i\sqrt{5}=x \qquad\qquad$ none

y-int $\quad y=-(0+1)^2-5$

$\qquad y=-(1)^2-5$

$\qquad y=-1-5=-6 \qquad (0,-6)$

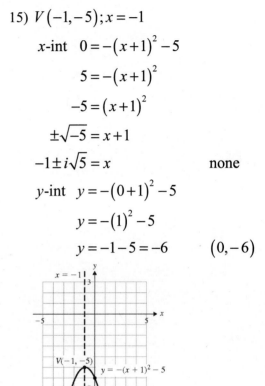

17) $V(1,-8); x=1$

x-int $\quad 0=2(x-1)^2-8$

$\qquad 8=2(x-1)^2$

$\qquad 4=(x-1)^2$

$\qquad \pm2=x-1$

$\qquad 1\pm2=x$

$\qquad x=-1 \text{ or } x=3 \qquad (-1,0),(3,0)$

y-int $y = 2(0-1)^2 - 8$

$\qquad y = 2(-1)^2 - 8$

$\qquad y = 2(1) - 8$

$\qquad y = 2 - 8 = -6 \qquad (0,-6)$

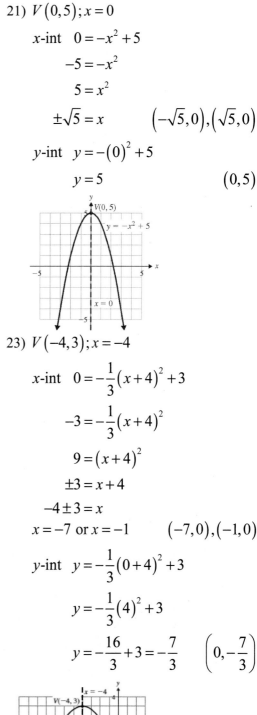

21) $V(0,5); x = 0$

x-int $\quad 0 = -x^2 + 5$

$\qquad -5 = -x^2$

$\qquad 5 = x^2$

$\qquad \pm\sqrt{5} = x \qquad \left(-\sqrt{5},0\right),\left(\sqrt{5},0\right)$

y-int $\quad y = -(0)^2 + 5$

$\qquad y = 5 \qquad\qquad (0,5)$

19) $V(-4,0); x = -4$

x-int $\quad 0 = \dfrac{1}{2}(x+4)^2$

$\qquad 0 = (x+4)^2$

$\qquad 0 = x+4$

$\qquad -4 = x \qquad\qquad (-4,0)$

y-int $\quad y = \dfrac{1}{2}(0+4)^2$

$\qquad y = \dfrac{1}{2}(4)^2$

$\qquad y = \dfrac{1}{2}(16) = 8 \qquad (0,8)$

23) $V(-4,3); x = -4$

x-int $\quad 0 = -\dfrac{1}{3}(x+4)^2 + 3$

$\qquad -3 = -\dfrac{1}{3}(x+4)^2$

$\qquad 9 = (x+4)^2$

$\qquad \pm 3 = x+4$

$\qquad -4 \pm 3 = x$

$\qquad x = -7 \text{ or } x = -1 \qquad (-7,0),(-1,0)$

y-int $\quad y = -\dfrac{1}{3}(0+4)^2 + 3$

$\qquad y = -\dfrac{1}{3}(4)^2 + 3$

$\qquad y = -\dfrac{16}{3} + 3 = -\dfrac{7}{3} \qquad \left(0,-\dfrac{7}{3}\right)$

25) $V(-2,5); x = -2$

x-int $\quad 0 = 3(x+2)^2 + 5$

$\qquad -5 = 3(x+2)^2$

$\qquad -\dfrac{5}{3} = (x+2)^2$

$\qquad \pm\sqrt{-\dfrac{5}{3}} = x+2$

$\qquad -2 \pm i\sqrt{\dfrac{5}{3}} = x$ $\qquad\qquad$ none

y-int $\quad y = 3(0+2)^2 + 5$

$\qquad y = 3(2)^2 + 5$

$\qquad y = 3(4) + 2 = 17 \qquad (0,17)$

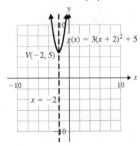

27) $\boxed{f(x) = (x^2 + 8x) + 11}$ Group the variable terms together using parentheses.

$\boxed{\dfrac{1}{2}(8) = 4;\ 4^2 = 16}$ Find the number that completes the square in the parentheses

$f(x) = (x^2 + 8x + 16) + 11 - 16$ $\boxed{\begin{array}{l}\text{Add and subtract the number above} \\ \text{to the same side of the equation}\end{array}}$

$\boxed{f(x) = (x+4)^2 - 5}$ $\qquad$ Factor and simplify.

29) $f(x) = x^2 - 2x - 3$

$f(x) = (x^2 - 2x + 1) - 3 - 1$

$f(x) = (x-1)^2 - 4$

x-int $\quad 0 = (x-1)^2 - 4$

$\qquad 4 = (x-1)^2$

$\qquad \pm\sqrt{4} = x - 1$

$\qquad \pm 2 = x - 1$

$\qquad 1 \pm 2 = x$

$x = 3$ or $x = -1 \qquad (3,0), (-1,0)$

y-int $\quad f(0) = 0^2 - 2(0) - 3$

$\qquad\qquad = -3 \qquad\qquad (0,-3)$

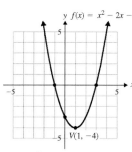

31) $y = x^2 + 6x + 7$

$y = (x^2 + 6x + 9) + 7 - 9$

$y = (x+3)^2 - 2$

x-int $\quad 0 = (x+3)^2 - 2$

$\qquad 2 = (x+3)^2$

$\qquad \pm\sqrt{2} = x + 3$

$\qquad -3 \pm \sqrt{2} = x$

$\left(-3-\sqrt{2},0\right), \left(-3-\sqrt{2},0\right)$

y-int $\quad y = 0^2 + 6(0) + 7$

$\qquad\qquad = 7 \qquad\qquad (0,7)$

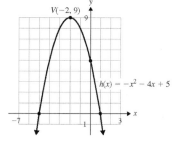

33) $g(x) = x^2 + 4x$

$g(x) = (x^2 + 4x + 4) - 4$

$g(x) = (x+2)^2 - 4$

x-int $\quad 0 = (x+2)^2 - 4$

$\qquad 4 = (x+2)^2$

$\qquad \pm 2 = x + 2$

$\qquad -2 \pm 2 = x$

$x = -4$ or $x = 0 \qquad (-4,0), (0,0)$

y-int $\quad g(0) = 0^2 + 4(0) \qquad (0,0)$

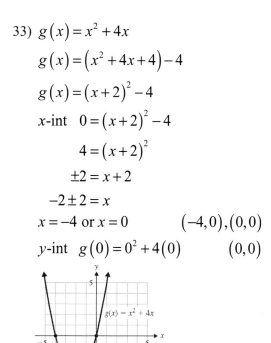

35) $h(x) = -x^2 - 4x + 5$

$h(x) = -(x^2 + 4x) + 5$

$h(x) = -(x^2 + 4x + 4) + 5 + 4$

$h(x) = -(x+2)^2 + 9$

x-int $\quad 0 = -(x+2)^2 + 9$

$\qquad -9 = -(x+2)^2$

$\qquad 9 = (x+2)^2$

$\qquad \pm 3 = x + 2$

$\qquad -2 \pm 3 = x$

$x = -5$ or $x = 1 \qquad (-5,0), (1,0)$

y-int $\quad h(0) = -(0)^2 - 4(0) + 5 = 5 \quad (0,5)$

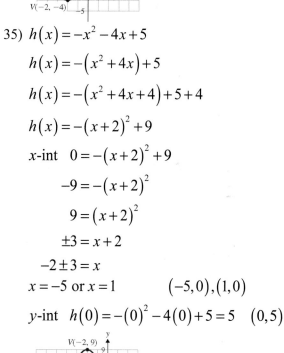

389

37) $y = -x^2 + 6x - 10$

$y = -(x^2 - 6x) - 10$

$y = -(x^2 - 6x + 9) - 10 + 9$

$y = -(x-3)^2 - 1$

x-int $\quad 0 = -(x-3)^2 - 1$

$\qquad 1 = -(x-3)^2$

$\qquad -1 = (x-3)^2$

$\qquad \pm\sqrt{-1} = x - 3$

$\qquad 3 \pm i = x \qquad\qquad$ none

y-int $\quad y = -(0)^2 + 6(0) - 10$

$\qquad\quad = -10 \qquad\qquad (0, -10)$

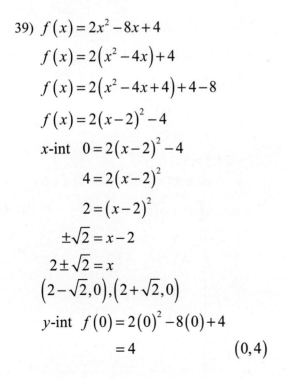

39) $f(x) = 2x^2 - 8x + 4$

$f(x) = 2(x^2 - 4x) + 4$

$f(x) = 2(x^2 - 4x + 4) + 4 - 8$

$f(x) = 2(x-2)^2 - 4$

x-int $\quad 0 = 2(x-2)^2 - 4$

$\qquad 4 = 2(x-2)^2$

$\qquad 2 = (x-2)^2$

$\qquad \pm\sqrt{2} = x - 2$

$\qquad 2 \pm \sqrt{2} = x$

$\left(2 - \sqrt{2}, 0\right), \left(2 + \sqrt{2}, 0\right)$

y-int $\quad f(0) = 2(0)^2 - 8(0) + 4$

$\qquad\qquad = 4 \qquad\qquad (0, 4)$

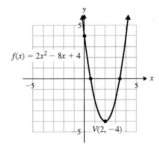

$f(x) = 2x^2 - 8x + 4$

$V(2, -4)$

41) $g(x) = -\dfrac{1}{3}x^2 - 2x - 9$

$g(x) = -\dfrac{1}{3}(x^2 + 6x) - 9$

$g(x) = -\dfrac{1}{3}(x^2 + 6x + 9) - 9 + 3$

$g(x) = -\dfrac{1}{3}(x+3)^2 - 6$

x-int $\quad 0 = -\dfrac{1}{3}(x+3)^2 - 6$

$\qquad 6 = -\dfrac{1}{3}(x+3)^2$

$\qquad -18 = (x+3)^2$

$\qquad \pm\sqrt{-18} = x + 3$

$\qquad -3 \pm 3i\sqrt{2} = x \qquad\qquad$ none

y-int $\quad g(0) = -\dfrac{1}{3}(0)^2 - 2(0) - 9$

$\qquad\qquad = -9$

$(0, -9)$

$V(-3, -6)$

$g(x) = -\dfrac{1}{3}x^2 - 2x - 9$

43) $y = x^2 - 3x + 2$

$$y = \left(x^2 - 3x + \frac{9}{4}\right) + 2 - \frac{9}{4}$$

$$y = \left(x - \frac{3}{2}\right)^2 - \frac{1}{4}$$

x-int $\quad 0 = \left(x - \frac{3}{2}\right)^2 - \frac{1}{4}$

$$\frac{1}{4} = \left(x - \frac{3}{2}\right)^2$$

$$\pm\frac{1}{2} = x - \frac{3}{2}$$

$$\frac{3}{2} \pm \frac{1}{2} = x$$

$x = 1$ or $x = 2$ $\qquad (1,0),(2,0)$

y-int $\quad y = (0)^2 - 3(0) + 2$

$\qquad\qquad = 2 \qquad\qquad (0,2)$

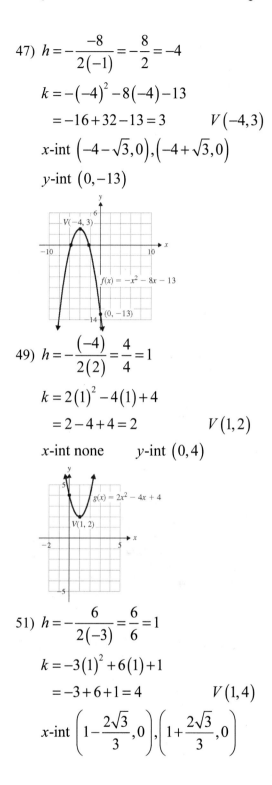

45) $h = -\dfrac{2}{2(1)} = -1$

$k = (-1)^2 + 2(-1) - 3$

$\quad = 1 - 2 - 3 = -4 \qquad V(-1,-4)$

x-int $(-3,0),(1,0) \qquad y$-int $(0,-3)$

47) $h = -\dfrac{-8}{2(-1)} = -\dfrac{8}{2} = -4$

$k = -(-4)^2 - 8(-4) - 13$

$\quad = -16 + 32 - 13 = 3 \qquad V(-4,3)$

x-int $\left(-4 - \sqrt{3}, 0\right), \left(-4 + \sqrt{3}, 0\right)$

y-int $(0,-13)$

49) $h = -\dfrac{(-4)}{2(2)} = \dfrac{4}{4} = 1$

$k = 2(1)^2 - 4(1) + 4$

$\quad = 2 - 4 + 4 = 2 \qquad V(1,2)$

x-int none $\qquad y$-int $(0,4)$

51) $h = -\dfrac{6}{2(-3)} = \dfrac{6}{6} = 1$

$k = -3(1)^2 + 6(1) + 1$

$\quad = -3 + 6 + 1 = 4 \qquad V(1,4)$

x-int $\left(1 - \dfrac{2\sqrt{3}}{3}, 0\right), \left(1 + \dfrac{2\sqrt{3}}{3}, 0\right)$

y-int $(0,1)$

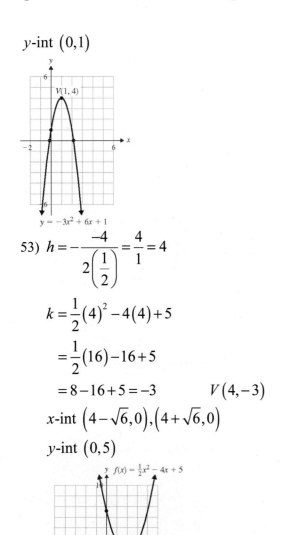

$y = -3x^2 + 6x + 1$

53) $h = -\dfrac{-4}{2\left(\dfrac{1}{2}\right)} = \dfrac{4}{1} = 4$

$k = \dfrac{1}{2}(4)^2 - 4(4) + 5$

$ = \dfrac{1}{2}(16) - 16 + 5$

$ = 8 - 16 + 5 = -3 \qquad V(4,-3)$

x-int $\left(4 - \sqrt{6}, 0\right), \left(4 + \sqrt{6}, 0\right)$

y-int $(0,5)$

$y \quad f(x) = \frac{1}{2}x^2 - 4x + 5$

$V(4,-3)$

55) $h = -\dfrac{-2}{2\left(-\dfrac{1}{3}\right)} = -\dfrac{1}{\dfrac{1}{3}} = -3$

$k = -\dfrac{1}{3}(-3)^2 - 2(-3) - 5$

$ = -\dfrac{1}{3}(9) + 6 - 5$

$ = -3 + 6 - 5 = -2 \qquad V(-3,-2)$

x-int none $\qquad y$-int $(0,-5)$

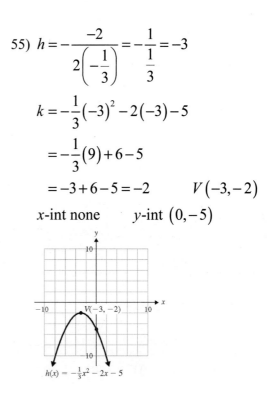

$h(x) = -\frac{1}{3}x^2 - 2x - 5$

Section 12.4: Exercises

1) maximum

3) neither

5) minimum

7) If a is positive the graph opens upward, so the y-coordinate of the vertex is the minimum value of the function. If a is negative the graph opens downward, so the y-coordinate of the vertex is the maximum value of the function.

9) a) minimum

b) $h = -\dfrac{6}{2(1)} = -3$

$k = (-3)^2 + 6(-3) + 9$

$ = 9 - 18 + 9 = 0 \qquad V(-3,0)$

c) 0

d)

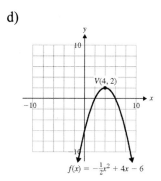

$f(x) = x^2 + 6x + 9$

$V(-3, 0)$

11) a) maximum

d)

$V(4, 2)$

$f(x) = -\frac{1}{2}x^2 + 4x - 6$

13) a) The max height occurs at the
t-coordinate of the vertex.

$$t = -\frac{320}{2(-16)} = \frac{320}{32} = 10$$

10 seconds

b) Since the object reaches max
height $t = 10$, find $h(10)$.

$$h(10) = -16(10)^2 + 320(10)$$
$$= -16(100) + 3200 = 1600$$

1600 ft

c) The object hits the ground

b) $h = -\dfrac{4}{2\left(-\dfrac{1}{2}\right)} = \dfrac{4}{1} = 4$

$$k = -\frac{1}{2}(4)^2 + 4(4) - 6$$
$$= -\frac{1}{2}(16) + 16 - 6$$
$$= -8 + 16 - 6 = 2 \qquad V(4, 2)$$

c) 2
when $h(t) = 0$.

$$0 = -16t^2 + 320t$$
$$0 = t^2 - 20t$$
$$0 = t(t - 20)$$
$$t = 0 \ \text{ or } \ t - 20 = 0$$
$$t = 20$$

The object will hit the
ground after 20 sec.

15) The x-coordinate of the vertex
represents the number of months
after January that had the greatest
number of guests.

$$x = -\frac{120}{2(-10)} = \frac{120}{20} = 6$$

The inn had the greatest number
of guests in July.
The number of guests at the inn
during July is $N(6)$.

$$N(6) = -10(6)^2 + 120(6) + 120$$
$$= -10(36) + 720 + 120 = 480$$

480 guests stayed at the inn during
the month of July.

17) The t-coordinate of the vertex represents the number of years after 1989 in which the greatest number of babies was born to teen mothers.

$$t = -\frac{2.75}{2(-0.721)} \approx 2$$

Greatest number born in 1991. The number of babies born in (in thousands) to teen mothers is $N(2)$.

$$N(2) = -0.721(2)^2 + 2.75(2) + 528$$
$$= -2.884 + 5.5 + 528 \approx 531$$

Approximately 531,000 babies were born to teen mothers in 1991.

19) w = width of ice rink
l = length of ice rink
A = area of the rink
Maximize: $A = lw$
Constraint: $2l + 2w = 100$
$2l + 2w = 100$
$$2l = 100 - 2w$$
$$l = 50 - w$$
$$A = lw$$
$$A(w) = (50 - w)w$$
$$A(w) = -w^2 + 50w$$

Find the w-coordinate of the vertex, the value that maximizes the area.

$$w = -\frac{50}{2(-1)} = 25$$

Find $A(25)$

$$A(25) = -(25)^2 + 50(25)$$
$$= -625 + 1250 = 625 \text{ ft}^2$$

21) w = width of dog pen
l = length of dog pen
A = area of the dog pen
Maximize: $A = lw$
Since the barn is 1 side of the pen, the fence is used for only 3 sides.
Constraint: $l + 2w = 48$
$l + 2w = 48$
$$l = 48 - 2w$$
$$A = lw$$
$$A(w) = (48 - 2w)w$$
$$A(w) = -2w^2 + 48w$$

Find the w-coordinate of the vertex, the value that maximizes the area.

$$w = -\frac{48}{2(-2)} = 12$$

Use $l + 2w = 48$ with $w = 12$ to find the corresponding length.
$$l + 2(12) = 48$$
$$l + 24 = 48$$
$$l = 24$$

The dog pen will be 12 ft × 24 ft.

23) x = one integer, y = other integer,
P = product
Maximize: $P = xy$
Constraint: $x + y = 18$
$x + y = 18$
$$y = 18 - x$$
$$P = xy$$
$$P(x) = x(18 - x)$$
$$P(x) = -x^2 + 18x$$

$$x = -\frac{18}{2(-1)} = 9$$

$$x + y = 18$$

$$9 + y = 18$$

$$y = 9$$

9 and 9

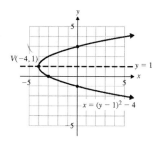

25) $x =$ one integer, $y =$ other integer,

$P =$ product

Maximize: $P = xy$

Constraint: $x - y = 12$

$$x - y = 12$$

$$x - 12 = y$$

$$P = xy$$

$$P(x) = x(x - 12)$$

$$P(x) = x^2 - 12x$$

$$x = -\frac{-12}{2(1)} = 6$$

$$x - y = 12$$

$$6 - y = 12$$

$$-y = 6$$

$$y = -6$$

−6 and 6

27) (h, k)

29) to the left

31) $V(-4, 1); y = 1$

x-int $\quad x = (0 - 1)^2 - 4$

$$x = 1 - 4$$

$$x = -3 \qquad\qquad (-3, 0)$$

y-int $\quad 0 = (y - 1)^2 - 4$

$$4 = (y - 1)^2$$

$$\pm\sqrt{4} = y - 1$$

$$1 \pm 2 = y$$

$y = -1$ or $y = 3 \qquad (0, -1), (0, 3)$

33) $V(2, 0); y = 0$

x-int $\quad x = 0^2 + 2$

$$x = 2 \qquad\qquad (2, 0)$$

y-int $\quad 0 = y^2 + 2$

$$-2 = y^2$$

$$\pm i\sqrt{2} = y \qquad\qquad \text{none}$$

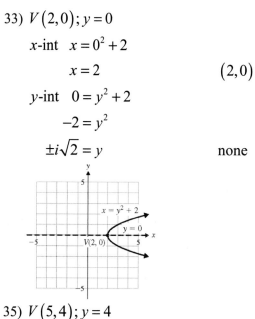

35) $V(5, 4); y = 4$

x-int $\quad x = -(0 - 4)^2 + 5$

$$x = -(-4)^2 + 5$$

$$x = -16 + 5 \qquad (-11, 0)$$

y-int $\quad 0 = -(y - 4)^2 + 5$

$$-5 = -(y - 4)^2$$

$$5 = (y - 4)^2$$

$$\pm\sqrt{5} = y - 4$$

$$4 \pm \sqrt{5} = y$$

$$\left(0, 4 - \sqrt{5}\right), \left(0, 4 + \sqrt{5}\right)$$

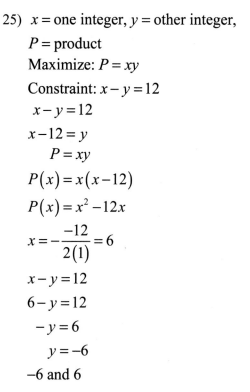

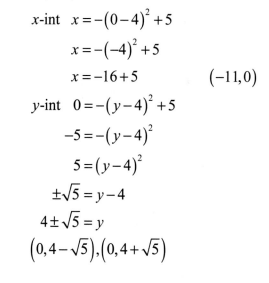

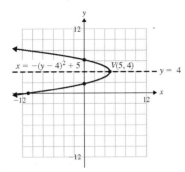

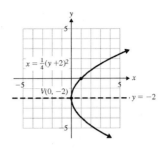

37) $V(-9,2); y=2$

 x-int $x=-2(0-2)^2-9$

    $x=-2(4)-9$

    $x=-8-9=-17$ $(-17,0)$

 y-int $0=-2(y-2)^2-9$

    $9=-2(y-2)^2$

    $-\dfrac{9}{2}=(y-2)^2$

    $\pm i\dfrac{3}{\sqrt{2}}=y-2$

    $2\pm i\dfrac{3}{\sqrt{2}}=y$ none

41) $x=y^2-4y+5$

 $x=(y^2-4y+4)+5-4$

 $x=(y-2)^2+1$

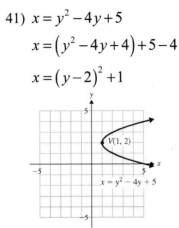

43) $x=-y^2+6y+6$

 $x=-(y^2-6y)+6$

 $x=-(y^2-6y+9)+6+9$

 $x=-(y-3)^2+15$

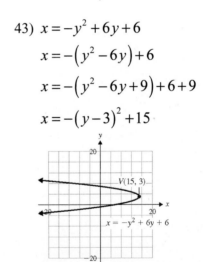

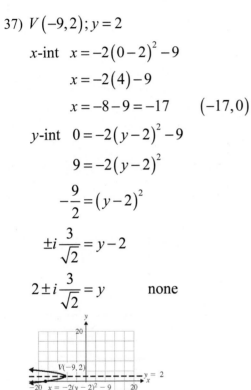

39) $V(0,-2); y=-2$

 x-int $x=\dfrac{1}{4}(0+2)^2$

    $x=\dfrac{1}{4}(4)=1$ $(1,0)$

 y-int $0=\dfrac{1}{4}(y+2)^2$

    $0=(y+2)^2$

    $0=y+2$

    $-2=y$ $(0,-2)$

45) $x = \dfrac{1}{3}y^2 + \dfrac{8}{3}y - \dfrac{5}{3}$

$3x = y^2 + 8y - 5$

$3x = \left(y^2 + 8y + 16\right) - 5 - 16$

$3x = (y+4)^2 - 21$

$x = \dfrac{1}{3}(y+4)^2 - 7$

y-int $(0,1),(0,3)$

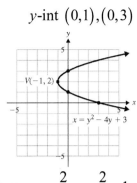

51) $y = -\dfrac{2}{2(-1)} = \dfrac{2}{2} = 1$

$x = -1^2 + 2(1) + 2$

$\quad = -1 + 2 + 2 = 3 \qquad V(3,1)$

x-int $(2,0)$

y-int $\left(0,1-\sqrt{3}\right),\left(0,1+\sqrt{3}\right)$

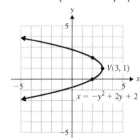

47) $x = -4y^2 - 8y - 10$

$x = -4\left(y^2 + 2y\right) - 10$

$x = -4\left(y^2 + 2y + 1\right) - 10 + 4$

$x = -4(y+1)^2 - 6$

53) $y = -\dfrac{4}{2(-2)} = \dfrac{4}{4} = 1$

$x = -2(1)^2 + 4(1) - 6$

$\quad = -2 + 4 - 6 = -4 \qquad V(-4,1)$

x-int $(-6,0)$

y-int none

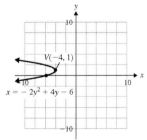

49) $y = -\dfrac{-4}{2(1)} = \dfrac{4}{2} = 2$

$x = 2^2 - 4(2) + 3$

$\quad = 4 - 8 + 3 = -1 \qquad V(-1,2)$

x-int $(3,0)$

55) $y = -\dfrac{-16}{2(4)} = \dfrac{16}{8} = 2$

$x = 4(2)^2 - 16(2) + 13$

$\quad = 16 - 32 + 13 = -3 \qquad V(-3, 2)$

x-int $(13, 0)$

y-int $\left(0, 2 - \dfrac{\sqrt{3}}{2}\right), \left(0, 2 + \dfrac{\sqrt{3}}{2}\right)$

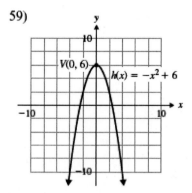

57) $y = -\dfrac{-\dfrac{1}{2}}{2\left(\dfrac{1}{4}\right)} = \dfrac{\dfrac{1}{2}}{\dfrac{1}{2}} = 1$

$x = \dfrac{1}{4}(1)^2 - \dfrac{1}{2}(1) + \dfrac{25}{4}$

$\quad = \dfrac{1}{4} - \dfrac{2}{4} + \dfrac{25}{4} = 6 \qquad V(6, 1)$

x-int $\left(\dfrac{25}{4}, 0\right)$

y-int none

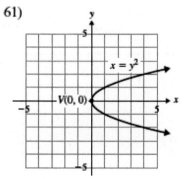

59)

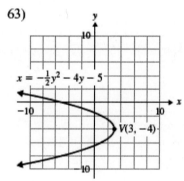

61)

63)

65)

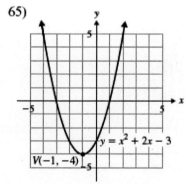

67)

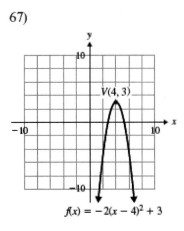

$f(x) = -2(x-4)^2 + 3$

Section 12.5: Exercises

1) a) $(f+g)(x) = f(x)+g(x)$
$$= (-3x+1)+(2x-11)$$
$$= -x-10$$

b) $(f+g)(5) = -5-10 = -15$

c) $(f-g)(x) = f(x)-g(x)$
$$= (-3x+1)-(2x-11)$$
$$= -3x+1-2x+11$$
$$= -5x+12$$

d) $(f-g)(2) = -5(2)+12$
$$= -10+12 = 2$$

3) a) $(f+g)(x)$
$$= f(x)+g(x)$$
$$= (4x^2 -7x-1)+(x^2 +3x-6)$$
$$= 5x^2 -4x-7$$

b) $(f+g)(5) = 5(5)^2 -4(5)-7$
$$= 5(25)-20-7$$
$$= 125-20-7$$
$$= 98$$

c) $(f-g)(x)$
$$= f(x)-g(x)$$
$$= (4x^2 -7x-1)-(x^2 +3x-6)$$
$$= 4x^2 -7x-1-x^2 -3x+6$$
$$= 3x^2 -10x+5$$

d) $(f-g)(2) = 3(2)^2 -10(2)+5$
$$= 12-20+5 = -3$$

5) a) $(fg)(x) = f(x)\cdot g(x)$
$$= (x)(-x+5)$$
$$= -x^2 +5x$$

b) $(fg)(-3) = -(-3)^2 +5(-3)$
$$= -9-15 = -24$$

7) a) $(fg)(x) = f(x)\cdot g(x)$
$$= (2x+3)(3x+1)$$
$$= 6x^2 +11x+3$$

b) $(fg)(-3) = 6(-3)^2 +11(-3)+3$
$$= 54-33+3 = 24$$

9) a) $\left(\dfrac{f}{g}\right)(x) = \dfrac{f(x)}{g(x)} = \dfrac{6x+9}{x+4}, \; x \neq -4$

b) $\left(\dfrac{f}{g}\right)(-2) = \dfrac{6(-2)+9}{-2+4}$
$$= \dfrac{-12+9}{2} = -\dfrac{3}{2}$$

11) a) $\left(\dfrac{f}{g}\right)(x) = \dfrac{f(x)}{g(x)}$

$= \dfrac{x^2 - 5x - 24}{x - 8}$

$= \dfrac{(x-8)(x+3)}{x-8}$

$= x + 3, \qquad x \neq 8$

b) $\left(\dfrac{f}{g}\right)(-2) = -2 + 3 = 1$

13) a) $\left(\dfrac{f}{g}\right)(x) = \dfrac{f(x)}{g(x)}$

$= \dfrac{3x^2 + 14x + 8}{3x + 2}$

$= \dfrac{(3x+2)(x+4)}{3x+2}$

$= x + 4, \qquad x \neq -\dfrac{2}{3}$

b) $\left(\dfrac{f}{g}\right)(-2) = -2 + 4 = 2$

15) Answers may vary.

17) $(fg)(x) = f(x) \cdot g(x) = 8x^2 - 22x + 15$

$g(x) = \dfrac{(fg)(x)}{f(x)} = \dfrac{8x^2 - 22x + 15}{4x - 5}$

$= \dfrac{(2x-3)\,\cancel{(4x-5)}}{\cancel{(4x-5)}}$

$= 2x - 3$

19) a) $P(x) = R(x) - C(x)$

$= 12x - (8x + 2000)$

$= 4x - 2000$

b) $P(1500) = 4(1500) - 2000$

$= 6000 - 2000 = \$4000$

21) a) $P(x) = R(x) - C(x)$

$= 18x - (15x + 2400)$

$= 3x - 2400$

b) $P(800) = 3(800) - 2400$

$= 2400 - 2400 = 0$

23) a) $P(x) = R(x) - C(x)$

$= -0.2x^2 + 23x - (4x + 9)$

$= -0.2x^2 + 23x - 4x - 9$

$= -0.2x^2 + 19x - 9$

b) $P(200) = -0.2(20)^2 + 19(20) - 9$

$= -80 + 380 - 9 = \$291$

Profit is $291,000

25) $(f \circ g)(x) = f(g(x))$

So, substitute the function $g(x)$
into the function $f(x)$ and simplify.

27) a) $g(4) = 2(4) - 9$

$= 8 - 9$

$= -1$

b) $f \circ g(4) = 3(-1) + 1$

$= -3 + 1$

$= -2$

c) $(f \circ g)(x) = f(g(x))$

$\qquad = 3(2x-9)+1$

$\qquad = 6x-27+1$

$\qquad = 6x-26$

d) $(f \circ g)(4) = f(g(4))$

$\qquad = 6(4)-26$

$\qquad = -2$

29) a) $(f \circ g)(x) = f(g(x))$

$\qquad = 5(x+7)-4$

$\qquad = 5x+35-4$

$\qquad = 5x+31$

b) $(g \circ f)(x) = g(f(x))$

$\qquad = (5x-4)+7$

$\qquad = 5x+3$

c) $(f \circ g)(3) = 5(3)+31$

$\qquad = 15+31 = 46$

31) a) $(h \circ g)(x) = h(g(x))$

$\qquad = (x^2-6x+11)-4$

$\qquad = x^2-6x+7$

b) $(g \circ h)(x)$

$\quad = g(h(x))$

$\quad = (x-4)^2-6(x-4)+11$

$\quad = x^2-8x+16-6x+24+11$

$\quad = x^2-14x+51$

c) $(g \circ h)(4) = (4)^2-14(4)+51$

$\qquad = 16-56+51 = 11$

33) a) $(f \circ g)(x)$

$\quad = f(g(x))$

$\quad = 2(3x-5)^2+3(3x-5)-10$

$\quad = 2(9x^2-30x+25)+9x-15-10$

$\quad = 18x^2-60x+50+9x-25$

$\quad = 18x^2-51x+25$

b) $(g \circ f)(x) = g(f(x))$

$\qquad = 3(2x^2+3x-10)-5$

$\qquad = 6x^2+9x-30-5$

$\qquad = 6x^2+9x-35$

c) $(f \circ g)(1) = 18(1)^2-51(1)+25$

$\qquad = 18-51+25 = -8$

35) a) $(n \circ m)(x)$

$\quad = n(m(x))$

$\quad = -(x+8)^2+3(x+8)-8$

$\quad = -(x^2+16x+64)+3x+24-8$

$\quad = -x^2-16x-64+3x+16$

$\quad = -x^2-13x-48$

b) $(m \circ n)(x) = m(n(x))$

$\qquad = (-x^2+3x-8)+8$

$\qquad = -x^2+3x$

c) $(m \circ n)(0) = -(0)^2+3(0) = 0$

37) a) $(f \circ g)(x)$

$\quad = f(g(x))$

$\quad = \sqrt{(x^2-6)+10}$

$\quad = \sqrt{x^2+4}$

b) $(g \circ f)(x) = g(f(x))$

$\qquad = \left(\sqrt{x+10}\right)^2 - 6$

$\qquad = x + 10 - 6$

$\qquad = x + 4$

c) $(f \circ g)(-3) = \sqrt{(-3)^2 + 4}$

$\qquad = \sqrt{13}$

39) a) $(P \circ Q)(t) = P(Q(t))$

$\qquad = \dfrac{1}{t^2 + 8}$

b) $(Q \circ P)(t) = Q(P(t))$

$\qquad = \dfrac{1}{(t+8)^2}$

c) $(Q \circ P)(-5) = \dfrac{1}{(-5+8)^2} = \dfrac{1}{9}$

41) a) $r(5) = 4(5) = 20$

The radius of the spill 5 min. after the ship started leaking was 20 ft.

b) $A(20) = \pi(20)^2 = 400\pi$

The area of the oil slick is 400π ft^2 when its radius is 20 ft.

c) $A(r(t)) = \pi(4t)^2 = 16\pi t^2$

This is the area of the oil slick in terms of t, the number of minutes after the leak began.

d) $A(r(5)) = 16\pi(5)^2$

$\qquad = 16\pi(25) = 400\pi$

The area of the oil slick 5 minutes after the ship began leaking was 400π ft^2.

43) a)

$s(40) = 0.8(40) = 32$

When the regular price of the item is $40, the sale price is $32.

b)

$f(32) = 1.07(32) = 34.24$

When the regular price of the item is $32, the sale price is $34.24.

c)

$(f \circ s)(x) = f(s(x)) = 1.07(0.8x) = 0.856x$

This is the final cost of the item after the discount and sales tax.

d)

$(f \circ s)(40) = f(s(40)) = 0.856(40) = 34.24$

When the original cost of an item is $40, the final cost of the item after the discount and sales tax is $34.24

45) $h(x) = (f \circ g)(x) = \sqrt{x^2 + 13}$

$f(x) = \sqrt{x}$, $g(x) = x^2 + 13$;

answers may vary

47) $h(x) = (f \circ g)(x) = (8x - 3)^2$

$f(x) = x^2$, $g(x) = 8x - 3$;

answers may vary

49) $h(x) = (f \circ g)(x) = \dfrac{1}{6x+5}$

$f(x) = \dfrac{1}{x}, \ g(x) = 6x+5;$

answers may vary

Section 12.6: Exercises

1) increases

3) direct

5) inverse

7) combined

9) $M = kn$

11) $h = \dfrac{k}{j}$

13) $T = \dfrac{k}{c^2}$

15) $s = krt$

17) $Q = \dfrac{k\sqrt{z}}{m}$

19) a) $z = kx$

$63 = k(7)$

$9 = k$

b) $z = 9x$

c) $z = 9(6) = 54$

21) a) $N = \dfrac{k}{y}$

$4 = \dfrac{k}{12}$

$48 = k$

b) $N = \dfrac{48}{y}$

c) $N = \dfrac{48}{3} = 16$

23) a) $Q = \dfrac{kr^2}{w}$

$25 = \dfrac{k(10)^2}{20}$

$500 = k(100)$

$5 = k$

b) $Q = \dfrac{5r^2}{w}$

c) $Q = \dfrac{5(6)^2}{4} = \dfrac{5(36)}{4} = 5(9) = 45$

25) $B = kR$

$35 = k(5)$

$7 = k$

$B = 7R = 7(8) = 56$

27) $L = \dfrac{k}{h^2}$

$8 = \dfrac{k}{3^2}$

$72 = k$

$L = \dfrac{72}{h^2} = \dfrac{72}{2^2} = 18$

29) $y = kxz$

$60 = k(4)(3)$

$60 = 12k$

$5 = k$

$y = 5xz = 5(7)(2) = 70$

31) $E = kh$

$437.50 = k(35)$

$12.50 = k$

$E = 12.50h = 12.50(40) = \$500$

33) $t = \dfrac{d}{r}$

$14 = \dfrac{d}{60}$

$840 = d$

$t = \dfrac{840}{r} = \dfrac{840}{70} = 12$ hours

35) $P = kIR^2$

$100 = k(4)(5)^2$

$100 = k(100)$

$1 = k$

$P = (1)IR^2 = (5)(6)^2 = 180$ watts

37) $E_k = kmv^2$

$112,500 = k(1000)(15)^2$

$112,500 = 225,000k$

$0.5 = k$

$E_k = 0.5mv^2 = 0.5(1000)(18)^2$

$= 500(324) = 162,000$ J

39) $f = \dfrac{k}{L}$

$100 = \dfrac{k}{5}$

$500 = k$

$f = \dfrac{500}{L} = \dfrac{500}{2.5} = 200$ cycles/sec

41) $R = \dfrac{kl}{A}$

$2 = \dfrac{k(40)}{0.05}$

$0.1 = 40k$

$0.0025 = k$

$R = \dfrac{0.0025l}{A} = \dfrac{0.0025(60)}{0.05}$

$= \dfrac{0.15}{0.05} = 3$ ohms

43) $F = kx$

$200 = k5$

$40 = k$

$F = 40x = 40(8) = 320$ lb

Chapter 12 Review

1) Domain: $\{-7, -5, 2, 4\}$

Range: $\{-4, -1, 3, 5, 9\}$

Not a function

3) Domain: $(-\infty, \infty)$ Range: $[0, \infty)$

Is a function

5) yes

7) no

9) yes

11) yes

13)

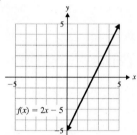

15)

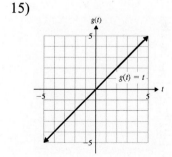

17) a) $f(5) = -8(5) + 3 = -40 + 3 = -37$

b) $f(-4) = -8(-4) + 3 = 32 + 3 = 35$

c) $g(-2)=(-2)^2+7(-2)-12$

$\quad = 4-14-12=-22$

d) $g(3)=(3)^2+7(3)-12$

$\quad = 9+21-12=18$

e) $f(c)=-8(c)+3=-8c+3$

f) $g(r)=r^2+7r-12$

g) $f(p-3)=-8(p-3)+3$

$\quad = -8p+24+3$

$\quad = -8p+27$

h) $g(t+4)$

$\quad = (t+4)^2+7(t+4)-12$

$\quad = t^2+8t+16+7t+28-12$

$\quad = t^2+15t+32$

19) $k(x)=-\dfrac{2}{3}x+8$

$\quad 0=-\dfrac{2}{3}x+8$

$\quad -8=-\dfrac{2}{3}x$

$\quad 12=x$

21) $p(x)=x^2-8x+15$

$\quad 3=x^2-8x+15$

$\quad 0=x^2-8x+12$

$\quad 0=(x-2)(x-4)$

$\quad x-2=0$ or $x-6=0$

$\quad x=2 \qquad x=6$

23) Solve: $x-5=0$

$\qquad x=5$

$\qquad (-\infty,5)\cup(5,\infty)$

25) $(-\infty,\infty)$

27) Solve: $5t-7\geq 0$

$\qquad 5t\geq 7$

$\qquad t\geq \dfrac{7}{5} \qquad \left[\dfrac{7}{5},\infty\right)$

29) Solve: $x=0 \qquad (-\infty,0)\cup(0,\infty)$

31) $(-\infty,\infty)$

33) Solve: $\quad a^2-7a-8=0$

$\qquad (a+1)(a-8)=0$

$\qquad a+1=0$ or $a-8=0$

$\qquad a=-1 \qquad a=8$

$\qquad (-\infty,-1)\cup(-1,8)\cup(8,\infty)$

35) a) $C(30)=0.20(30)+26=32$

$\qquad$ \$32

b) $C(100)=0.20(100)+26=46$

$\qquad$ \$46

c) $C(m)=0.20m+26$

$\qquad 56=0.20m+26$

$\qquad 30=0.20m$

$\qquad 150=m \qquad\qquad$ 150 miles

d) $C(m)=0.20m+26$

$\qquad 42=0.20m+26$

$\qquad 16=0.20m$

$\qquad 80=m \qquad\qquad$ 80 miles

37) Domain: $[0,\infty)$; Range: $[0,\infty)$

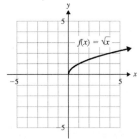

39) Domain: $(-\infty,\infty)$; Range: $[0,\infty)$

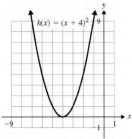

41) Domain: $(-\infty,\infty)$; Range: $(-\infty,5]$

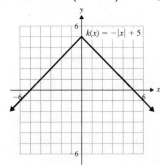

43) Domain: $[2,\infty)$; Range: $[-1,\infty)$

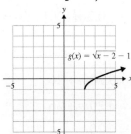

45)

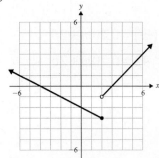

47) $f\left(7\dfrac{2}{3}\right)=\left[\!\!\left[7\dfrac{2}{3}\right]\!\!\right]=7$

49) $f\left(-8\dfrac{1}{2}\right)=\left[\!\!\left[-8\dfrac{1}{2}\right]\!\!\right]=-9$

51) $f\left(\dfrac{3}{8}\right)=\left[\!\!\left[\dfrac{3}{8}\right]\!\!\right]=0$

53)

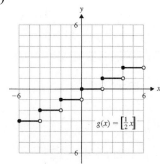

55) $g(x)=|x-5|$

57) a) (h,k) b) $x=h$

c) If a is positive, the parabola opens upward. If a is negative, the parabola opens downward.

59) a) (h,k) b) $y=k$

c) If a is positive, the parabola opens to the right. If a is negative, the parabola opens to the left.

61) $V(-2,-1); x=-2$

x-int $\quad 0=(x+2)^2-1$

$\qquad\qquad 1=(x+2)^2$

$\qquad\qquad \pm 1=x+2$

$\qquad\qquad -2\pm 1=x$

$x=-3$ or $x=-1 \qquad (-3,0),(-1,0)$

y-int $\quad y = (0+2)^2 - 1$

$\qquad\quad y = (2)^2 - 1$

$\qquad\quad y = 4 - 1 = 3 \qquad\qquad (0,3)$

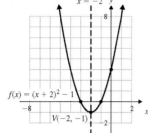

63) $V(-1,0); y = 0$

$\quad x$-int $\quad x = -0^2 - 1$

$\qquad\qquad x = -1 \qquad\qquad (-1,0)$

$\quad y$-int $\quad 0 = -y^2 - 1$

$\qquad\qquad 1 = -y^2$

$\qquad\qquad -1 = y^2$

$\qquad\qquad \pm i = y \qquad\qquad$ none

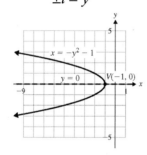

65) $V(11,3); y = 3$

$\quad x$-int $\quad x = -(0-3)^2 + 11$

$\qquad\qquad x = -(-3)^2 + 11$

$\qquad\qquad x = -9 + 11 = 2 \qquad\qquad (2,0)$

y-int $\quad 0 = -(y-3)^2 + 11$

$\qquad\quad -11 = -(y-3)^2$

$\qquad\quad 11 = (y-3)^2$

$\qquad\quad \pm\sqrt{11} = y - 3$

$\qquad\quad 3 \pm \sqrt{11} = y$

$\qquad\quad \left(0, 3-\sqrt{11}\right), \left(0, 3+\sqrt{11}\right)$

67) $x = y^2 + 8y + 7$

$\quad x = \left(y^2 + 8y + 16\right) + 7 - 16$

$\quad x = (y+4)^2 - 9$

$\quad x$-int $\quad x = (0)^2 + 8(0) + 7 = 7 \qquad (7,0)$

$\quad y$-int $\quad 0 = (y+4)^2 - 9$

$\qquad\qquad 9 = (y+4)^2$

$\qquad\qquad \pm 3 = y + 4$

$\qquad\qquad -4 \pm 3 = y$

$\quad y = -1$ or $y = -7 \qquad (0,-1), (0,-7)$

69) $y = \dfrac{1}{2}x^2 - 4x + 9$

$y = \dfrac{1}{2}(x^2 - 8x) + 9$

$y = \dfrac{1}{2}(x^2 - 8x + 16) + 9 - 8$

$y = \dfrac{1}{2}(x - 4)^2 + 1$

x-int $\quad 0 = \dfrac{1}{2}(x - 4)^2 + 1$

$-1 = \dfrac{1}{2}(x - 4)^2$

$-2 = (x - 4)^2$

$\pm i\sqrt{2} = x - 4$

$4 \pm i\sqrt{2} = x \qquad\qquad$ none

y-int $\quad y = \dfrac{1}{2}(0)^2 - 4(0) + 9$

$= 9 \qquad\qquad (0, 9)$

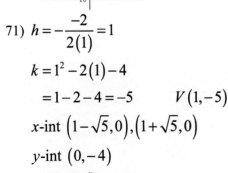

71) $h = -\dfrac{-2}{2(1)} = 1$

$k = 1^2 - 2(1) - 4$

$= 1 - 2 - 4 = -5 \qquad V(1, -5)$

x-int $\left(1 - \sqrt{5}, 0\right), \left(1 + \sqrt{5}, 0\right)$

y-int $(0, -4)$

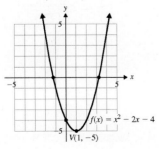

73) $k = -\dfrac{-3}{2\left(-\dfrac{1}{2}\right)} = -\dfrac{3}{1} = -3$

$h = -\dfrac{1}{2}(-3)^2 - 3(-3) - \dfrac{5}{2}$

$= -\dfrac{9}{2} + 9 - \dfrac{5}{2} = 2 \qquad V(2, -3)$

x-int $\left(-\dfrac{5}{2}, 0\right)$

y-int $(0, -5), (0, -1)$

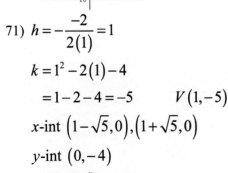

75) a) The max height occurs at the
    t-coordinate of the vertex.

$t = -\dfrac{32}{2(-16)} = \dfrac{32}{32} = 1$

1 second

b) Since the object reaches max
    height $t = 1$, find $h(1)$.

$h(1) = -16(1)^2 + 32(1) + 240$

$= -16 + 32 + 240 = 256$

256 ft

c) The object hits ground
when $h(t) = 0$.

$0 = -16t^2 + 32t + 240$

$0 = t^2 - 2t - 15$

$0 = (t-5)(t+3)$

$t - 5 = 0$ or $t + 3 = 0$

$\boxed{t = 5}$ $t = -3$

The object will hit the
ground after 5 sec.

77) $(f+g)(x) = f(x) + g(x)$

$= (5x+2) + (-x+4)$

$= 4x + 6$

79) $(g-h)(2) = g(2) - h(2)$

$= (-2+4) - (3(2)^2 - 7)$

$= 2 - (12 - 7)$

$= 2 - 5 = -3$

81) $(fg)(x) = f(x) \cdot g(x)$

$= (5x+2)(-x+4)$

$= -5x^2 + 18x + 8$

83) a) $\left(\dfrac{f}{g}\right)(x) = \dfrac{f(x)}{g(x)} = \dfrac{6x-5}{x+4}, x \neq 4$

b) $\left(\dfrac{f}{g}\right)(3) = \dfrac{6(3)-5}{3+4} = \dfrac{13}{7}$

85) a) $P(x) = R(x) - C(x)$

$= 20x - (14x + 400)$

$= 6x - 400$

b) $P(200) = 6(200) - 400$

$= 1200 - 400 = \$800$

87) a) $(k \circ h)(x)$

$= k(h(x))$

$= (2x-1)^2 + 5(2x-1) - 4$

$= 4x^2 - 4x + 1 + 10x - 5 - 4$

$= 4x^2 + 6x - 8$

b) $(h \circ k)(x) = h(k(x))$

$= 2(x^2 + 5x - 4) - 1$

$= 2x^2 + 10x - 8 - 1$

$= 2x^2 + 10x - 9$

c) $(h \circ k)(-3) = 2(-3)^2 + 10(-3) - 9$

$= 2(9) - 30 - 9$

$= 18 - 39 = -21$

89) a) $(N \circ G)(h) = N(G(x))$

$= 0.8(12h) = 9.6h$

This is Antoine's net pay in terms
of how many hours he has worked.

b) $(N \circ G)(30) = 9.6(30) = 288$

When Antoine works 30 hours,
his net pay is \$288.

c) $(N \circ G)(40) = 9.6(40) = \$384$

91) $A = ktr$

$15 = k\left(\dfrac{1}{2}\right)(5)$

$6 = k$

$A = 6tr = 6(3)(4) = 72$

93) $w = kr^3$

$0.96 = k(2)^3$

$0.12 = k$

$w = 0.12r^3 = 0.12(3)^3 = 3.24$ lb

Chapter 12 Test

1) It is a special type of relation in which each element of the domain corresponds to exactly one element of the range.

3) a) $\left[-\dfrac{7}{3}, \infty\right)$ b) yes

5) $\left(-\infty, \dfrac{8}{7}\right) \cup \left(\dfrac{8}{7}, \infty\right)$

7) $f(c) = 4c + 3$

9) $g(k+5)$

$= (k+5)^2 - 6(k+5) + 10$

$= k^2 + 10k + 25 - 6k - 30 + 10$

$= k^2 + 4k + 5$

11) a) $C(3) = 50(3) + 60$

$= 150 + 60 = 210$

The cost of delivering 3 cubic yards of cedar mulch is $210.

b) $C(m) = 50m + 60$

$360 = 50m + 60$

$300 = 50m$

$6 = m$ 6 cubic yards

13) Domain: $[-3, \infty)$; Range: $[0, \infty)$

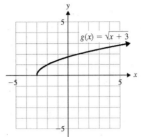

15)

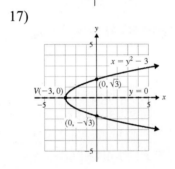

17)

19)

21) $(g-f)(x) = g(x) - f(x)$

$= (x^2 + 5x - 3) - (2x + 7)$

$= x^2 + 3x - 10$

23) $(f \circ g)(x) = f(g(x))$

$= 2(x^2 + 5x - 3) + 7$

$= 2x^2 + 10x - 6 + 7$

$= 2x^2 + 10x + 1$

25) $n = krs^2$

$72 = k(2)(3)^2$

$72 = 18k$

$4 = k$

$n = 4rs^2 = 4(3)(5)^2 = 12(25) = 300$

Cumulative Review: Chapters 1-12

1) inverse

3) $\left(\dfrac{1}{2}\right)^5 = \dfrac{1^5}{2} = \dfrac{1}{32}$

5) $10^0 = 1$

7) $x + 8 \le 6$ or $1 - 2x \le -5$

$x \le -2 \qquad -2x \le -6$

$x \ge 3$

$(-\infty, -2] \cup [3, \infty)$

9) $x - \dfrac{1}{4}y = \dfrac{5}{2}$

$x = \dfrac{1}{4}y + \dfrac{5}{2}$

$6\left(\dfrac{1}{2}x + \dfrac{1}{3}y\right) = 6\left(\dfrac{13}{6}\right)$

$3x + 2y = 13$

Use substitution.

$3\left(\dfrac{1}{4}y + \dfrac{5}{2}\right) + 2y = 13$

$4\left(\dfrac{3}{4}y + \dfrac{15}{2} + 2y\right) = 4(13)$

$3y + 30 + 8y = 52$

$11y = 22$

$y = 2$

$x - \dfrac{1}{4}(2) = \dfrac{5}{2}$

$x - \dfrac{1}{2} = \dfrac{5}{2}$

$x = \dfrac{6}{2} = 3 \qquad (3, 2)$

11) $\dfrac{12r - 40r^2 + 6r^3 + 4}{4r^2}$

$= \dfrac{6r^3}{4r^2} - \dfrac{40r^2}{4r^2} + \dfrac{12r}{4r^2} + \dfrac{4}{4r^2}$

$= \dfrac{3}{2}r - 10 + \dfrac{3}{r} + \dfrac{1}{r^2}$

13) $100 - 9m^2 = (10 + 3m)(10 - 3m)$

411

15) $\dfrac{c-8}{2c^2-5c-12} \div \dfrac{3c-24}{c^2-16}$

$= \dfrac{c-8}{2c^2-5c-12} \cdot \dfrac{c^2-16}{3c-24}$

$= \dfrac{\cancel{c-8}}{(2c+3)\cancel{(c-4)}} \cdot \dfrac{(c+4)\cancel{(c-4)}}{3\cancel{(c-8)}}$

$= \dfrac{c+4}{3(2c+3)}$

17) $|7y+6| \le -8$

$\varnothing$

19) $\sqrt{60} = \sqrt{4}\cdot\sqrt{15} = 2\sqrt{15}$

21) $\sqrt{18c^6 d^{11}} = 3\sqrt{2}\cdot c^3 \cdot d^5 \sqrt{d}$

$\qquad = 3c^3 d^5 \sqrt{2d}$

23) $\sqrt{12}+\sqrt{3}+\sqrt{48} = 2\sqrt{3}+\sqrt{3}+4\sqrt{3}$

$\qquad\qquad\qquad = 7\sqrt{3}$

25) $\dfrac{4-2i}{2+3i} = \dfrac{4-2i}{2+3i}\cdot\dfrac{2-3i}{2-3i}$

$\qquad = \dfrac{8-12i-4i+6i^2}{2^2+3^2}$

$\qquad = \dfrac{8-16i+6(-1)}{4+9}$

$\qquad = \dfrac{2-16i}{13} = \dfrac{2}{13}-\dfrac{16}{13}i$

27) $\qquad 4(y^2+2y)=5$

$\qquad\quad 4y^2+8y=5$

$\qquad 4y^2+8y-5=0$

$\qquad (2y+5)(2y-1)=0$

$\qquad 2y+5=0 \ \text{ or } \ 2y-1=0$

$\qquad\quad 2y=-5 \qquad\quad 2y=1$

$\qquad\qquad y=-\dfrac{5}{2} \qquad\quad y=\dfrac{1}{2}$

$\qquad \left\{-\dfrac{5}{2},\dfrac{1}{2}\right\}$

29) a) $g(7)=7+1=8$

b) $\left(\dfrac{h}{g}\right)(x)=\dfrac{h(x)}{g(x)}$

$\qquad = \dfrac{x^2+4x+3}{x+1}$

$\qquad = \dfrac{(x+3)(x+1)}{x+1}$

$\qquad = x+3, \qquad x\ne -1$

c) $g(x)=x+1$

$\qquad 5=x+1$

$\qquad 4=x$

d) $(g\circ h)(x)=g(h(x))$

$\qquad = (x^2+4x+3)+1$

$\qquad = x^2+4x+4$

31)

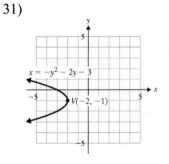

Section 13.1: Exercises

1) no

3) yes; $h^{-1} = \{(-16,-5),(-4,-1),(8,3)\}$

5) yes;

$g^{-1} = \{(1,2),(2,5),(14,7),(19,10)\}$

7) yes

9) No; only one-to-one functions have inverses.

11) False; it is read "f inverse of x."

13) True

15) False; they are symmetric with respect to $y = x$

17) a) yes

b)

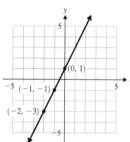

19) no

21) a) yes

b)

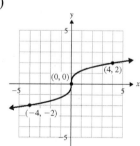

23) $f(x) = 2x - 10$

$\quad y = 2x - 10$ Replace $f(x)$ with y

$\quad x = 2y - 10$ Interchange x and y

$\quad 2y = x + 10$ Solve for y

$\quad y = \dfrac{1}{2}x + 5$ Divide by 2 and simplify

$\quad f^{-1}(x) = \dfrac{1}{2}x + 5$ Replace y with $f^{-1}(x)$

25) $y = x - 6$

$\quad\quad x = y - 6$

$\quad\quad x + 6 = y$

$\quad\quad g^{-1}(x) = x + 6$

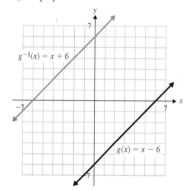

27) $y = -2x + 5$

$\quad\quad x = -2y + 5$

$\quad\quad x - 5 = -2y$

$\quad\quad -\dfrac{1}{2}x + \dfrac{5}{2} = y$

$\quad\quad f^{-1}(x) = -\dfrac{1}{2}x + \dfrac{5}{2}$

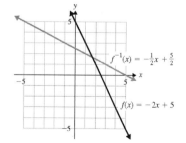

29) $y = \dfrac{1}{2}x$

$x = \dfrac{1}{2}y$

$2x = y$

$g^{-1}(x) = 2x$

31) $y = x^3$

$x = y^3$

$\sqrt[3]{x} = y$

$f^{-1}(x) = \sqrt[3]{x}$

33) $y = 2x - 6$

$x = 2y - 6$

$x + 6 = 2y$

$\dfrac{1}{2}x + 3 = y$

$f^{-1}(x) = \dfrac{1}{2}x + 3$

35) $y = -\dfrac{3}{2}x + 4$

$x = -\dfrac{3}{2}y + 4$

$x - 4 = -\dfrac{3}{2}y$

$-\dfrac{2}{3}x + \dfrac{8}{3} = y$

$h^{-1}(x) = -\dfrac{2}{3}x + \dfrac{8}{3}$

37) $y = \sqrt[3]{x+2}$

$x = \sqrt[3]{y+2}$

$x^3 = y + 2$

$x^3 - 2 = y$

$g^{-1}(x) = x^3 - 2$

39) $y = \sqrt{x}$

$x = \sqrt{y}$

$x^2 = y$

$f^{-1}(x) = x^2, \ x \geq 0$

41) a) $f(1) = 5(1) - 2 = 3$

b) $f^{-1}(3) = 1$

43) a) $f(9) = -\dfrac{1}{3}(9) + 5 = -3 + 5 = 2$

b) $f^{-1}(2) = 9$

45) a) $f(-7) = -(-7) + 3 = 7 + 3 = 10$

b) $f^{-1}(10) = -7$

47) a) $f(3) = 2^3 = 8$

b) $f^{-1}(8) = 3$

49) $\left(f^{-1} \circ f\right)(x) = f^{-1}\left(f(x)\right)$

$= f^{-1}(x+9)$

$= (x+9)-9$

$= x$

$\left(f \circ f^{-1}\right)(x) = f\left(f^{-1}(x)\right)$

$= f(x-9)$

$= (x-9)+9 = x$

51) $\left(f^{-1} \circ f\right)(x) = f^{-1}\left(f(x)\right)$

$= f^{-1}(-6x+4)$

$= -\dfrac{1}{6}(-6x+4)+\dfrac{2}{3}$

$= x-\dfrac{4}{6}+\dfrac{2}{3} = x$

$\left(f \circ f^{-1}\right)(x) = f\left(f^{-1}(x)\right)$

$= f\left(-\dfrac{1}{6}x+\dfrac{2}{3}\right)$

$= -6\left(-\dfrac{1}{6}x+\dfrac{2}{3}\right)+4$

$= x-4+4 = x$

53) $\left(f^{-1} \circ f\right)(x) = f^{-1}\left(f(x)\right)$

$= f^{-1}\left(\dfrac{3}{2}x-9\right)$

$= \dfrac{2}{3}\left(\dfrac{3}{2}x-9\right)+6$

$= x-6+6 = x$

$\left(f \circ f^{-1}\right)(x) = f\left(f^{-1}(x)\right)$

$= f\left(\dfrac{2}{3}x+6\right)$

$= \dfrac{3}{2}\left(\dfrac{2}{3}x+6\right)-9$

$= x+9-9 = x$

55) $\left(f^{-1} \circ f\right)(x) = f^{-1}\left(f(x)\right)$

$= f^{-1}\left(\sqrt[3]{x-10}\right)$

$= \left(\sqrt[3]{x-10}\right)^{3}+10$

$= x-10+10 = x$

$\left(f \circ f^{-1}\right)(x) = f\left(f^{-1}(x)\right)$

$= f\left(x^{3}+10\right)$

$= \left(\sqrt[3]{\left(x^{3}+10\right)}\right)^{3}-10$

$= x+10-10 = x$

Section 13.2: Exercises

1) Choose values for the variable that will give positive numbers, negative numbers, and zero in the exponent.

3) Domain: $(-\infty,\infty)$; Range: $(0,\infty)$

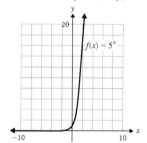

5) Domain: $(-\infty,\infty)$; Range: $(0,\infty)$

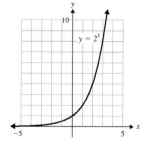

7) Domain: $(-\infty, \infty)$; Range: $(0, \infty)$

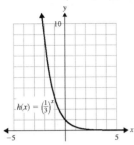

9) $(-\infty, \infty)$

11) Domain: $(-\infty, \infty)$; Range: $(0, \infty)$

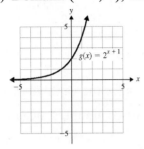

13) Domain: $(-\infty, \infty)$; Range: $(0, \infty)$

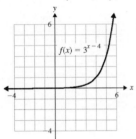

15) Domain: $(-\infty, \infty)$; Range: $(0, \infty)$

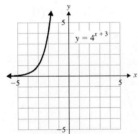

17) Domain: $(-\infty, \infty)$; Range: $(0, \infty)$

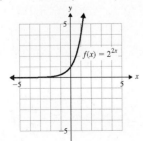

19) Domain: $(-\infty, \infty)$; Range: $(1, \infty)$

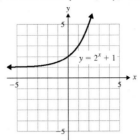

21) Domain: $(-\infty, \infty)$; Range: $(-2, \infty)$

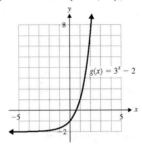

23) Domain: $(-\infty, \infty)$; Range: $(-\infty, 0)$

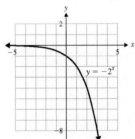

25) $g(x) = 2^x$ would grow faster for values of $x > 2$, $2^x > 2x$

27) Shift the graph of $f(x)$ down 2 units.

29) 2.7183

31) B

33) D

35) Domain: $(-\infty, \infty)$; Range: $(-2, \infty)$

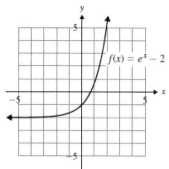

$f(x) = e^x - 2$

37) Domain: $(-\infty, \infty)$; Range: $(0, \infty)$

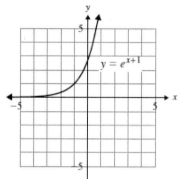

$y = e^{x+1}$

39) Domain: $(-\infty, \infty)$; Range: $(0, \infty)$

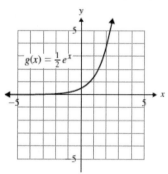

$g(x) = \frac{1}{2}e^x$

41) Domain: $(-\infty, \infty)$; Range: $(-\infty, 0)$

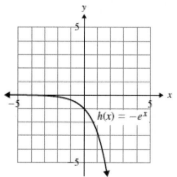

$h(x) = -e^x$

43) They are symmetric with respect to the x-axis

45) $6^{3n} = 36^{n-4}$

$6^{3n} = \left(6^2\right)^{n-4}$

Express both sides with same base

$6^{3n} = 6^{2(n-4)}$

$3n = 2(n-4)$ Set the exponents equal

$3n = 2n - 8$

$n = -8$ $\qquad \{-8\}$

47) $9^x = 81$

$9^x = 9^2$

$x = 2$ $\qquad \{2\}$

49) $5^{4d} = 125$

$5^{4d} = 5^3$

$4d = 3$

$d = \frac{3}{4}$ $\qquad \left\{\frac{3}{4}\right\}$

51) $16^{m-2} = 2^{3m}$

$\left(2^4\right)^{m-2} = 2^{3m}$

$2^{4m-8} = 2^{3m}$

$4m - 8 = 3m$

$m = 8$ $\qquad \{8\}$

53) $7^{2k-6} = 49^{3k+1}$

$7^{2k-6} = \left(7^2\right)^{3k+1}$

$7^{2k-6} = 7^{6k+2}$

$2k - 6 = 6k + 2$

$-8 = 4k$

$-2 = k$ $\qquad \{-2\}$

55) $32^{3c} = 8^{c+4}$

$\left(2^5\right)^{3c} = \left(2^3\right)^{c+4}$

$2^{15c} = 2^{3c+12}$

$15c = 3c + 12$

$12c = 12$

$c = 1$ $\qquad \{1\}$

57) $100^{5z-1} = (1000)^{2z+7}$

$(10^2)^{5z-1} = (10^3)^{2z+7}$

$10^{10z-2} = 10^{6z+21}$

$10z - 2 = 6z + 21$

$4z = 23$

$z = \dfrac{23}{4}$ $\left\{\dfrac{23}{4}\right\}$

59) $81^{3n+9} = 27^{2n+6}$

$(3^4)^{3n+9} = (3^3)^{2n+6}$

$3^{12n+36} = 3^{6n+18}$

$12n + 36 = 6n + 18$

$6n = -18$

$n = -3$ $\{-3\}$

61) $6^x = \dfrac{1}{36}$

$6^x = \left(\dfrac{1}{6}\right)^2$

$6^x = 6^{-2}$

$x = -2$ $\{-2\}$

63) $2^a = \dfrac{1}{8}$

$2^a = \left(\dfrac{1}{2}\right)^3$

$2^a = 2^{-3}$

$a = -3$ $\{-3\}$

65) $9^r = \dfrac{1}{27}$

$(3^2)^r = \left(\dfrac{1}{3}\right)^3$

$3^{2r} = 3^{-3}$

$2r = -3$

$r = -\dfrac{3}{2}$ $\left\{-\dfrac{3}{2}\right\}$

67) $\left(\dfrac{3}{4}\right)^{5k} = \left(\dfrac{27}{64}\right)^{k+1}$

$\left(\dfrac{3}{4}\right)^{5k} = \left[\left(\dfrac{3}{4}\right)^3\right]^{k+1}$

$\left(\dfrac{3}{4}\right)^{5k} = \left(\dfrac{3}{4}\right)^{3k+3}$

$5k = 3k + 3$

$2k = 3$

$k = \dfrac{3}{2}$ $\left\{\dfrac{3}{2}\right\}$

69) $\left(\dfrac{5}{6}\right)^{3x+7} = \left(\dfrac{36}{25}\right)^{2x}$

$\left(\dfrac{5}{6}\right)^{3x+7} = \left[\left(\dfrac{6}{5}\right)^2\right]^{2x}$

$\left(\dfrac{5}{6}\right)^{3x+7} = \left[\left(\dfrac{5}{6}\right)^{-2}\right]^{2x}$

$\left(\dfrac{5}{6}\right)^{3x+7} = \left(\dfrac{5}{6}\right)^{-4x}$

$3x + 7 = -4x$

$7 = -7x$

$-1 = x$ $\{-1\}$

71)a) When the SUV was purchased $t = 0$. Find $V(0)$.

$V(0) = 32,700(0.812)^0$

$V(0) = 32,700$

$\$32,700$

b) Find $V(3)$.

$V(3) = 32,700(0.812)^3$

$V(3) \approx 17,507.17$

$\$17,507.17$

73) a) When the minivan was purchased $t = 0$. Find $V(0)$.

$V(0) = 16,800(0.803)^0$

$V(0) = 16,800$

$16,800

b) Find $V(6)$.

$V(6) = 16,800(0.803)^6$

$V(6) \approx 4504.04$

$4504.04

75) a) The value of the house in 1995 was $V(0)$.

$V(0) = 185,200(1.03)^0$

$V(0) = 185,200$

$185,200

b) Find $V(7)$.

$V(7) = 185,200(1.03)^7$

$V(7) \approx 21,973.12$

$227,772.64

77) $c = 2000$, $t = 18$, $r = 0.09$

$$A = 2000 \left[\frac{(1+0.09)^{18} - 1}{0.09} \right](1+0.09)$$

$$= 2000 \left[\frac{(1.09)^{18} - 1}{0.09} \right](1.09)$$

$$\approx \$90,036.92$$

79) $c = 4000$, $t = 10$, $r = 0.07$

$$A = 4000 \left[\frac{(1+0.07)^{10} - 1}{0.07} \right](1+0.07)$$

$$= 4000 \left[\frac{(1.07)^{10} - 1}{0.07} \right](1.07)$$

$$\approx \$59,134.40$$

81) $A(6) = 1000e^{-0.5332(6)}$

$$\approx 40.8 \text{ mg}$$

Section 13.3: Exercises

1) a must be a positive real number that is not equal to 1.

3) 10

5) $7^2 = 49$

7) $2^3 = 8$

9) $9^{-2} = \dfrac{1}{81}$

11) $10^6 = 1,000,000$

13) $25^{1/2} = 5$

15) $13^1 = 13$

17) $\log_9 81 = 2$

19) $\log 100 = 2$

21) $\log_3 \dfrac{1}{81} = -4$

23) $\log_{10} 1 = 0$

25) $\log_{169} 13 = \dfrac{1}{2}$

27) $\sqrt{9} = 3$

$9^{1/2} = 3$

$\log_9 3 = \dfrac{1}{2}$

29) $\sqrt[3]{64} = 4$

$64^{1/3} = 4$

$\log_{64} 4 = \dfrac{1}{3}$

31) Write the equation in exponential form, then solve for the variable.

33) $\log_2 x = 6$

$2^6 = x$ Rewrite in exponential form

$64 = x$ Solve for x

The solution set is $\{64\}$

35) $\log_{11} x = 2$

$11^2 = x$

$121 = x$ $\{121\}$

37) $\log_4 r = 3$

$4^3 = r$

$64 = k$ $\{64\}$

39) $\log p = 5$

$10^5 = p$

$100,000 = p$ $\{100,000\}$

41) $\log_m 49 = 2$

$m^2 = 49$

$m = \pm 7$

the base must be positive $\{7\}$

43) $\log_6 h = -2$

$6^{-2} = h$

$\dfrac{1}{36} = h$ $\left\{\dfrac{1}{36}\right\}$

45) $\log_2 (a+2) = 4$

$2^4 = a + 2$

$16 = a + 2$

$14 = a$ $\{14\}$

47) $\log_3 (4t - 3) = 3$

$3^3 = 4t - 3$

$27 = 4t - 3$

$30 = 4t$

$\dfrac{30}{4} = t$

$\dfrac{15}{2} = t$ $\left\{\dfrac{15}{2}\right\}$

49) $\log_{81} \sqrt[4]{9} = x$

$81^x = \sqrt[4]{9}$

$\left(9^2\right)^x = 9^{1/4}$

$9^{2x} = 9^{1/4}$

$2x = \dfrac{1}{4}$

$x = \dfrac{1}{8}$ $\left\{\dfrac{1}{8}\right\}$

51) $\log_{125} \sqrt{5} = c$

$125^c = \sqrt{5}$

$\left(5^3\right)^c = 5^{1/2}$

$5^{3c} = 5^{1/2}$

$3c = \dfrac{1}{2}$

$c = \dfrac{1}{6}$ $\left\{\dfrac{1}{6}\right\}$

53) $\log_{144} w = \dfrac{1}{2}$

$144^{1/2} = w$

$12 = w$ $\{12\}$

55) $\log_8 x = \dfrac{2}{3}$

$\quad 8^{2/3} = x$

$\quad \left(\sqrt[3]{8}\right)^2 = x$

$\quad 2^2 = x$

$\quad 4 = x \qquad\qquad \{4\}$

57) $\log_{(3m-1)} 25 = 2$

$\quad (3m-1)^2 = 25$

$\quad 3m-1 = \pm 5$

$\quad 3m-1 = 5 \ \text{ or } \ 3m-1 = -5$

$\qquad 3m = 6 \qquad\quad 3m = -4$

$\qquad m = 2 \qquad\qquad m = -\dfrac{4}{3}$

the base must be positive $\qquad \{2\}$

59) $\log_5 25 = 2$ since $5^{\boxed{2}} = 25$

61) $\log_2 32 = 5$ since $2^{\boxed{5}} = 32$

63) $\log 100 = 2$ since $10^{\boxed{2}} = 100$

65) Let $\log_{49} 7 = x$

$\qquad 49^x = 7$

$\qquad \left(7^2\right)^x = 7^1$

$\qquad 2x = 1$

$\qquad x = \dfrac{1}{2}$

$\quad \log_{49} 7 = \dfrac{1}{2}$

67) Let $\log_8 \dfrac{1}{8} = x$

$\qquad 8^x = \dfrac{1}{8}$

$\qquad 8^x = 8^{-1}$

$\qquad x = -1$

$\quad \log_8 \dfrac{1}{8} = -1$

69) $\log_5 5 = 1$

71) Let $\log_{1/4} 16 = x$

$\qquad \left(\dfrac{1}{4}\right)^x = 16$

$\qquad \left(4^{-1}\right)^x = 4^2$

$\qquad -x = 2$

$\qquad x = -2$

$\quad \log_{1/4} 16 = -2$

$\quad \log_{1/3} 27 = -3$

73) Replace $f(x)$ with y, write $y = \log_a x$ in exponential form, make a table of values, then plot the points and draw the curve.

75)

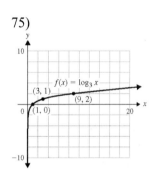

77)

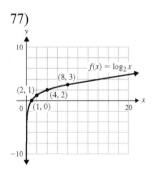

79)

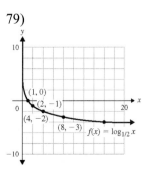

421

81)

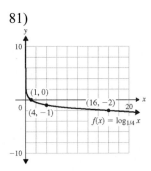

83) $y = 3^x$

$x = 3^y$

$y = \log_3 x$

$f^{-1}(x) = \log_3 x$

85) $y = \log_2 x$

$x = \log_2 y$

$y = 2^x$

$f^{-1}(x) = 2^x$

87)a) Find $L(0)$.

$L(0) = 1800 + 68\log_3(0+3)$

$= 1800 + 68\log_3 3$

$= 1800 + 68(1) = 1868$

b) Find $L(24)$.

$L(24) = 1800 + 68\log_3(24+3)$

$= 1800 + 68\log_3 27$

$= 1800 + 68(3)$

$= 1800 + 204 = 2004$

c) Let $L(t) = 2072$ and solve for t.

$2072 = 1800 + 68\log_3(t+3)$

$272 = 68\log_3(t+3)$

$4 = \log_3(t+3)$

$3^4 = t+3$

$81 = t+3$

$78 = t$

78 years after 1980 is the
year 2058.

89)a) Find $S(1)$.

$S(1) = 14\log_3 \big[2(1)+1\big]$

$= 14\log_3 3$

$= 14(1) = 14$ \qquad 14,000

b) Find $S(4)$.

$S(4) = 14\log_3 \big[2(4)+1\big]$

$= 14\log_3 9$

$= 14(2) = 28$ \qquad 28,000

c) Find $S(13)$ and compare this
value to the actual value.

$S(13) = 14\log_3 \big[2(13)+1\big]$

$= 14\log_3 27$

$= 14(3) = 42$ \qquad 42,000

The actual number sold was
1000 more boxes than what was
predicted by the formula.

Section 13.4: Exercises

1) True
3) False
5) False
7) True
9) $\log_5 25y = \log_5 25 + \log_5 y$
$\qquad = 2 + \log_5 y$
11) $\log_8(3 \cdot 10) = \log_8 3 + \log_8 10$

13) $\log_7 5d = \log_7 5 + \log_7 d$

15) $\log_9 \dfrac{4}{7} = \log_9 4 - \log_9 7$

17) $\log_5 2^3 = 3\log_5 2$

19) $\log p^8 = 8\log p$

21) $\log_3 \sqrt{7} = \log_3 7^{1/2} = \dfrac{1}{2}\log_3 7$

23) $\log_5 25t = \log_5 25 + \log_5 t$
$\qquad = 2 + \log_5 t$

25) $\log_2 \dfrac{8}{k} = \log_2 8 - \log_2 k = 3 - \log_2 k$

27) $\log_7 49^3 = 3\log_7 49 = 3(2) = 6$

29) $\log 1000b = \log 1000 + \log b$
$\qquad = 3 + \log b$

31) $\log_2 32^7 = 7\log_2 32 = 7(5) = 35$

33) $\log_5 \sqrt{5} = \log_5 5^{1/2} = \dfrac{1}{2}\log_5 5 = \dfrac{1}{2}$

35) $\log \sqrt[3]{100} = \log 100^{1/3} = \dfrac{1}{3}\log 100$
$\qquad = \dfrac{1}{3}(2) = \dfrac{2}{3}$

37) $\log_6 w^4 z^3 = \log_6 w^4 + \log_6 z^3$
$\qquad = 4\log_6 w + 3\log_6 z$

39) $\log_7 \dfrac{a^2}{b^5} = \log_7 a^2 - \log_7 b^5$
$\qquad = 2\log_7 a - 5\log_7 b$

41) $\log \dfrac{\sqrt[5]{11}}{y^2} = \log \sqrt[5]{11} - \log y^2$
$\qquad = \log 11^{1/5} - 2\log y$
$\qquad = \dfrac{1}{5}\log 11 - 2\log y$

43) $\log_2 \dfrac{4\sqrt{n}}{m^3}$
$\qquad = \log_2 4\sqrt{n} - \log_2 m^3$
$\qquad = \log_2 4 + \log_2 \sqrt{n} - 3\log_2 m$
$\qquad = 2 + \log_2 n^{1/2} - 3\log_2 m$
$\qquad = 2 + \dfrac{1}{2}\log_2 n - 3\log_2 m$

45) $\log_4 \dfrac{x^3}{yz^2} = \log_4 x^3 - \log_4 yz^2$
$\qquad = 3\log_4 x - \left(\log_4 y + \log_4 z^2\right)$
$\qquad = 3\log_4 x - \left(\log_4 y + 2\log_4 z\right)$
$\qquad = 3\log_4 x - \log_4 y - 2\log_4 z$

47) $\log_5 \sqrt{5c} = \log_5 (5c)^{1/2} = \dfrac{1}{2}\log_5 5c$
$\qquad = \dfrac{1}{2}\left(\log_5 5 + \log_5 c\right)$
$\qquad = \dfrac{1}{2}\left(1 + \log_5 c\right)$
$\qquad = \dfrac{1}{2} + \dfrac{1}{2}\log_5 c$

49) $\log k\,(k-6) = \log k + \log(k-6)$

51) $2\log_6 x + \log_6 y = \log_6 x^2 + \log_6 y$
$\qquad = \log_6 x^2 y$

53) $\log_a m + \log_a n = \log_a mn$

55) $\log_7 d - \log_7 3 = \log_7 \dfrac{d}{3}$

57) $4\log_3 f + \log_3 g = \log_3 f^4 + \log_3 g$
$\qquad = \log_3 f^4 g$

Chapter 13: Exponential, and Logarithmic Functions

59) $\log_8 t + 2\log_8 u - 3\log_8 v$

$= \log_8 t + \log_8 u^2 - \log_8 v^3$

$= \log_8 tu^2 - \log_8 v^3 = \log_8 \dfrac{tu^2}{v^3}$

61) $\log(r^2+3) - 2\log(r^2-3)$

$= \log(r^2+3) - \log(r^2-3)^2$

$= \log \dfrac{r^2+3}{(r^2-3)^2}$

63) $3\log_n 2 + \dfrac{1}{2}\log_n k$

$= \log_n 2^3 + \log_n k^{1/2}$

$= \log_n 8 + \log_n \sqrt{k} = \log_n 8\sqrt{k}$

65) $\dfrac{1}{3}\log_d 5 - 2\log_d z$

$= \log_d 5^{1/3} - \log_d z^2$

$= \log_d \sqrt[3]{5} - \log_d z^2 = \log_d \dfrac{\sqrt[3]{5}}{z^2}$

67) $\log_6 y - \log_6 3 - 3\log_6 z$

$= \log_6 y - \log_6 3 - \log_6 z^3$

$= \log_6 y - (\log_6 3 + \log_6 z^3)$

$= \log_6 y - \log_6 3z^3 = \log_6 \dfrac{y}{3z^3}$

69) $4\log_3 t - 2\log_3 6 - 2\log_3 u$

$= \log_3 t^4 - \log_3 6^2 - \log_3 u^2$

$= \log_3 t^4 - (\log_3 36 + \log_3 u^2)$

$= \log_3 t^4 - (\log_3 36u^2) = \log_3 \dfrac{t^4}{36u^2}$

71) $\dfrac{1}{2}\log_b(c+4) - 2\log_b(c+3)$

$= \log_b(c+4)^{1/2} - \log_b(c+3)^2$

$= \log_b \sqrt{c+4} - \log_b(c+3)^2$

$= \log_b \dfrac{\sqrt{c+4}}{(c+3)^2}$

73) $\log(a^2+b^2) - \log(a^4-b^4)$

$= \log\left(\dfrac{a^2+b^2}{a^4-b^4}\right)$

$= \log\left[\dfrac{a^2+b^2}{(a^2-b^2)(a^2+b^2)}\right]$

$= \log\dfrac{1}{(a^2-b^2)} = -\log(a^2-b^2)$

75) $\log 45 = \log(5\cdot9) = \log 5 + \log 9$

$= 0.6990 + 0.9542 = 1.6532$

77) $\log 81 = \log 9^2 = 2\log 9$

$= 2(0.9542) = 1.9084$

79) $\log \dfrac{5}{9} = \log 5 - \log 9$

$= 0.6990 - 0.9542 = -0.2552$

81) $\log 3 = \log\sqrt{9} = \log 9^{1/2} = \dfrac{1}{2}\log 9$

$= \dfrac{1}{2}(0.9542) = 0.4771$

83) $\log\dfrac{1}{5} = \log 5^{-1} = -1\log 5$

$= -(0.6990) = -0.6990$

85) $\log\dfrac{1}{81} = \log 1 - \log 81$

$= 0 - \log 9^2 = -2\log 9$

$= -2(0.9542) = -1.9084$

87) $\log 50 = \log(10 \cdot 5) = \log 10 + \log 5$

$\qquad = 1 + 0.6990 = 1.6990$

89) No. $\text{Log}_a xy$ is defined only if

$\qquad x$ and y are positive

Section 13.5: Exercises

1) e

3) 2

5) $\log \dfrac{1}{1000} = \log 10^{-3}$

$\qquad = -3\log 10 = -3 \cdot 1 = -3$

7) $\log 0.1 = \log \dfrac{1}{10} = \log 10^{-1}$

$\qquad = -1\log 10 = -1 \cdot 1 = -1$

9) $\log 10^9 = 9\log 10 = 9 \cdot 1 = 9$

11) $\log \sqrt[4]{10} = \log 10^{1/4}$

$\qquad = \dfrac{1}{4}\log 10 = \dfrac{1}{4} \cdot 1 = \dfrac{1}{4}$

13) $\ln e^6 = 6\ln e = 6 \cdot 1 = 6$

15) $\ln \sqrt{e} = \ln e^{1/2} = \dfrac{1}{2}\ln e = \dfrac{1}{2} \cdot 1 = \dfrac{1}{2}$

17) $\ln \dfrac{1}{e^5} = \ln e^{-5} = -5\ln e = -5 \cdot 1 = -5$

19) $\ln 1 = 0$

21) $\log 16 \approx 1.2041$

23) $\log 0.5 = -0.3010$

25) $\ln 3 \approx 1.0986$

27) $\ln 1.31 \approx 0.2700$

29) $\log x = 3$

$\qquad 10^3 = x$

$\qquad 1000 = x \qquad \{1000\}$

31) $\log k = -1$

$\qquad 10^{-1} = k$

$\qquad \dfrac{1}{10} = k \qquad\qquad \left\{\dfrac{1}{10}\right\}$

33) $\log(4a) = 2$

$\qquad 10^2 = 4a$

$\qquad 100 = 4a$

$\qquad 25 = a \qquad\qquad \{25\}$

35) $\log(3t + 4) = 1$

$\qquad 10^1 = 3t + 4$

$\qquad 10 = 3t + 4$

$\qquad 6 = 3t$

$\qquad 2 = t \qquad\qquad \{2\}$

37) $\log a = 1.5$

$\qquad 10^{1.5} = a \qquad \{10^{1.5}\};\{31.6228\}$

39) $\log r = 0.8$

$\qquad 10^{0.8} = r \qquad \{10^{0.8}\};\{6.3096\}$

41) $\ln x = 1.6$

$\qquad e^{1.6} = x \qquad \{e^{1.6}\};\{4.9530\}$

43) $\ln t = -2$

$\qquad e^{-2} = t$

$\qquad \dfrac{1}{e^2} = t \qquad\qquad \left\{\dfrac{1}{e^2}\right\};\{0.1353\}$

45) $\ln(3q) = 2.1$

$\qquad e^{2.1} = 3q$

$\qquad \dfrac{e^{2.1}}{3} = q \qquad\qquad \left\{\dfrac{e^{2.1}}{3}\right\};\{2.7221\}$

47) $\log\left(\dfrac{1}{2}c\right) = 0.47$

$\qquad 10^{0.47} = \dfrac{1}{2}c$

$\qquad 2(10)^{0.47} = c$

$\qquad \{2(10)^{0.47}\};\{5.9024\}$

49) $\log(5y-3) = 3.8$

$$10^{3.8} = 5y - 3$$

$$3 + 10^{3.8} = 5y$$

$$\frac{3 + 10^{3.8}}{5} = y$$

$$\left\{\frac{3 + 10^{3.8}}{5}\right\}; \{1262.5147\}$$

51) $\ln(10w+19) = 1.85$

$$e^{1.85} = 10w + 19$$

$$e^{1.85} - 19 = 10w$$

$$\frac{e^{1.85} - 19}{10} = w$$

$$\left\{\frac{e^{1.85} - 19}{10}\right\}; \{-1.2640\}$$

53) $\ln(2d-5) = 0$

$$e^0 = 2d - 5$$

$$5 + 1 = 2d$$

$$6 = 2d$$

$$3 = d \qquad\qquad \{3\}$$

55)

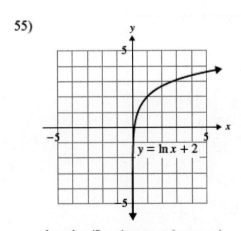

domain: $(0, \infty)$; range: $(-\infty, \infty)$

57)

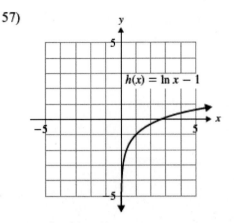

domain: $(0, \infty)$; range: $(-\infty, \infty)$

59)

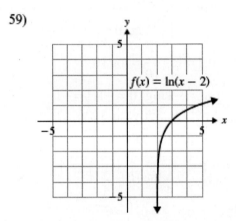

domain: $(2, \infty)$; range: $(-\infty, \infty)$

61)

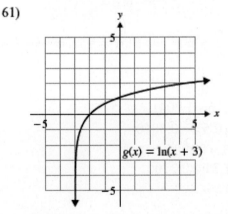

domain: $(-3, \infty)$; range: $(-\infty, \infty)$

63)

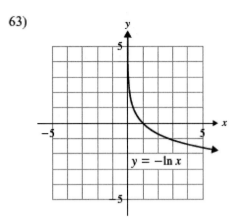

domain: $(0, \infty)$; range: $(-\infty, \infty)$

65)

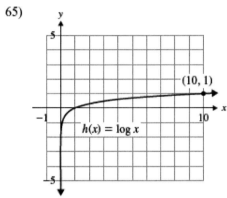

domain: $(0, \infty)$; range: $(-\infty, \infty)$

67) Shift the graph of $f(x)$ left 5 units

69) $\log_2 13 = \dfrac{\ln 13}{\ln 2} \approx 3.7004$

71) $\log_9 70 = \dfrac{\ln 70}{\ln 9} \approx 1.9336$

73) $\log_{1/3} 16 = \dfrac{\ln 16}{\ln \dfrac{1}{3}} \approx -2.5237$

75) $\log_5 3 = \dfrac{\ln 3}{\ln 5} \approx 0.6826$

77) $L(0.1) = 10 \log \dfrac{0.1}{10^{-12}}$

$= 10 \log \dfrac{10^{-1}}{10^{-12}}$

$= 10 \log 10^{11}$

$= 110 \log 10 = 110$ dB

79) $L(0.00000001) = 10 \log \dfrac{0.00000001}{10^{-12}}$

$= 10 \log \dfrac{10^{-8}}{10^{-12}}$

$= 10 \log 10^4$

$= 40 \log 10$

$= 40$ dB

81) $A = 3000\left(1 + \dfrac{.05}{12}\right)^{12(3)}$

$\approx 3000(1.004)^{36} \approx \$3484.42$

83) $A = 4000\left(1 + \dfrac{.065}{4}\right)^{4(5)}$

$\approx 4000(1.0163)^{20} \approx \$5521.68$

85) $A = 3000e^{.05(3)} = 3000e^{.15} \approx \$3485.50$

87) $A = 10{,}000e^{.075(6)} = 10{,}000e^{.45}$

$\approx \$15{,}683.12$

89) a) $N(0) = 5000e^{0.0617(0)}$

$= 5000e^0$

$= 5000(1) = 5000$ bacteria

b) $N(8) = 5000e^{0.0617(8)}$

$= 5000e^{0.4936}$

≈ 8191 bacteria

91) $N(24) = 10{,}000e^{0.0492(24)}$

$= 10{,}000e^{1.1808}$

$= 32{,}570$ bacteria

93) $\text{pH} = -\log\left[2\times10^{-3}\right]$

$\quad = -\log(0.002) \approx 2.7; \text{ acidic}$

95) $\text{pH} = -\log\left[6\times10^{-12}\right]$

$\quad = -\log(0.000000000006)$

$\quad \approx 11.2; \text{ basic}$

97) $y = \ln x$

$\quad x = \ln y$

$\quad y = e^x$

Section 13.6: Exercises

1) $7^x = 49$

$\quad 7^x = 7^2$

$\quad x = 2 \qquad \qquad \{2\}$

3) $7^n = 15$

$\quad \ln 7^n = \ln 15$

$\quad n \ln 7 = \ln 15$

$\quad n = \dfrac{\ln 15}{\ln 7}$

$\left\{\dfrac{\ln 15}{\ln 7}\right\}; \{1.3917\}$

5) $8^z = 3$

$\quad \ln 8^z = \ln 3$

$\quad z \ln 8 = \ln 3$

$\quad z = \dfrac{\ln 3}{\ln 8}$

$\left\{\dfrac{\ln 3}{\ln 8}\right\}; \{0.5283\}$

7) $6^{5p} = 36$

$\quad 6^{5p} = 6^2$

$\quad 5p = 2$

$\quad p = \dfrac{2}{5} \qquad \qquad \left\{\dfrac{2}{5}\right\}$

9) $\quad 4^{6k} = 2.7$

$\quad \ln 4^{6k} = \ln 2.7$

$\quad 6k \ln 4 = \ln 2.7$

$\quad k = \dfrac{\ln 2.7}{6 \ln 4}$

$\left\{\dfrac{\ln 2.7}{6 \ln 4}\right\}; \{0.1194\}$

11) $\quad 2^{4n+1} = 5$

$\quad \ln 2^{4n+1} = \ln 5$

$\quad (4n+1)\ln 2 = \ln 5$

$\quad 4n\ln 2 + \ln 2 = \ln 5$

$\quad 4n \ln 2 = \ln 5 - \ln 2$

$\quad n = \dfrac{\ln 5 - \ln 2}{4 \ln 2}$

$\left\{\dfrac{\ln 5 - \ln 2}{4 \ln 2}\right\}, \{0.3305\}$

13) $\quad 5^{3a-2} = 8$

$\quad \ln 5^{3a-2} = \ln 8$

$\quad (3a-2)\ln 5 = \ln 8$

$\quad 3a\ln 5 - 2\ln 5 = \ln 8$

$\quad 3a\ln 5 = \ln 8 + 2\ln 5$

$\quad a = \dfrac{\ln 8 + 2\ln 5}{3 \ln 5}$

$\left\{\dfrac{\ln 8 + 2\ln 5}{3 \ln 5}\right\}, \{1.0973\}$

15) $4^{2c+7} = 64^{3c-1}$

$\quad 4^{2c+7} = \left(4^3\right)^{3c-1}$

$\quad 4^{2c+7} = 4^{9c-3}$

$\quad 10 = 7c$

$\quad \dfrac{10}{7} = c \qquad \qquad \left\{\dfrac{10}{7}\right\}$

17)
$$9^{5d-2} = 4^{3d}$$
$$\ln 9^{5d-2} = \ln 4^{3d}$$
$$(5d-2)\ln 9 = 3d \ln 4$$
$$5d \ln 9 - 2 \ln 9 = 3d \ln 4$$
$$5d \ln 9 - 3d \ln 4 = 2 \ln 9$$
$$d(5 \ln 9 - 3 \ln 4) = 2 \ln 9$$
$$d = \frac{2 \ln 9}{5 \ln 9 - 3 \ln 4}$$
$$\left\{\frac{2 \ln 9}{5 \ln 9 - 3 \ln 4}\right\}; \{0.6437\}$$

19)
$$e^y = 12.5$$
$$\ln e^y = \ln 12.5$$
$$y \ln e = \ln 12.5$$
$$y(1) = \ln 12.5$$
$$y = \ln 12.5$$
$$\{\ln 12.5\}; \{2.5257\}$$

21)
$$e^{-4x} = 9$$
$$\ln e^{-4x} = \ln 9$$
$$-4x \ln e = \ln 9$$
$$-4x(1) = \ln 9$$
$$x = -\frac{\ln 9}{4}$$
$$\left\{-\frac{\ln 9}{4}\right\}; \{-0.5493\}$$

23)
$$e^{0.01r} = 2$$
$$\ln e^{0.01r} = \ln 2$$
$$0.01r \ln e = \ln 2$$
$$0.01r(1) = \ln 2$$
$$r = \frac{\ln 2}{0.01}$$
$$\left\{\frac{\ln 2}{0.01}\right\}, \{69.3147\}$$

25)
$$e^{0.006t} = 3$$
$$\ln e^{0.006t} = \ln 3$$
$$0.006t \ln e = \ln 3$$
$$0.006t(1) = \ln 3$$
$$t = \frac{\ln 3}{0.006}$$
$$\left\{\frac{\ln 3}{0.006}\right\}, \{183.1021\}$$

27)
$$e^{-0.4y} = 5$$
$$\ln e^{-0.4y} = \ln 5$$
$$-0.4y \ln e = \ln 5$$
$$-0.4y(1) = \ln 5$$
$$y = -\frac{\ln 5}{0.4}$$
$$\left\{-\frac{\ln 5}{0.4}\right\}, \{-4.0236\}$$

29) $\log_6 (k+9) = \log_6 11$
$$k + 9 = 11$$
$$k = 2 \qquad \{2\}$$

31) $\log_7 (3p-1) = \log_7 9$
$$3p - 1 = 9$$
$$3p = 10$$
$$p = \frac{10}{3} \qquad \left\{\frac{10}{3}\right\}$$

33) $\log x + \log (x-2) = \log 15$
$$\log x(x-2) = \log 15$$
$$x(x-2) = 15$$
$$x^2 - 2x = 15$$
$$x^2 - 2x - 15 = 0$$
$$(x-5)(x+3) = 0$$
$$x - 5 = 0 \ \text{ or } \ x + 3 = 0$$
$$x = 5 \qquad x = -3$$

Only one solution satisfies the original equation. $\{5\}$

35) $\log_3 n + \log_3 (12-n) = \log_3 20$

$\log_3 n(12-n) = \log_3 20$

$n(12-n) = 20$

$12n - n^2 = 20$

$n^2 - 12n + 20 = 0$

$(n-10)(n-2) = 0$

$n-10 = 0 \text{ or } n-2 = 0$

$n = 10 \qquad n = 2 \qquad \{2,10\}$

37) $\log_2 (-z) + \log_2 (z-8) = \log_2 15$

$\log_2 \left[-z(z-8)\right] = \log_2 15$

$-z(z-8) = 15$

$-z^2 + 8z = 15$

$z^2 - 8z + 15 = 0$

$(z-5)(z-3) = 0$

$z-5 = 0 \text{ or } z-3 = 0$

$z = 5 \qquad z = 3$

Neither solution satisfies the original equation. $\varnothing$

39) $\log_6 (5b-4) = 2$

$6^2 = 5b - 4$

$36 = 5b - 4$

$40 = 5b$

$8 = b \qquad \{8\}$

41) $\log (3p+4) = 1$

$10^1 = 3p + 4$

$6 = 3p$

$2 = p \qquad \{2\}$

43) $\log_3 y + \log_3 (y-8) = 2$

$\log_3 y(y-8) = 2$

$3^2 = y(y-8)$

$9 = y^2 - 8y$

$0 = y^2 - 8y - 9$

$0 = (y-9)(y+1)$

$y-9 = 0 \text{ or } y+1 = 0$

$y = 9 \qquad y = -1$

Only one solution satisfies the original equation. $\{9\}$

45) $\log_2 r + \log_2 (r+2) = 3$

$\log_2 r(r+2) = 3$

$2^3 = r(r+2)$

$8 = r^2 + 2r$

$0 = r^2 + 2r - 8$

$0 = (r+4)(r-2)$

$r+4 = 0 \text{ or } r-2 = 0$

$r = -4 \qquad r = 2$

Only one solution satisfies the original equation. $\{2\}$

47) $\log_4 20c - \log_4 (c+1) = 2$

$\log_4 \dfrac{20c}{c+1} = 2$

$4^2 = \dfrac{20c}{c+1}$

$16 = \dfrac{20c}{c+1}$

$16(c+1) = 20c$

$16c + 16 = 20c$

$16 = 4c$

$4 = c \qquad \{4\}$

49) $\log_2 8d - \log_2(2d-1) = 4$

$$\log_2 \frac{8d}{2d-1} = 4$$

$$2^4 = \frac{8d}{2d-1}$$

$$16 = \frac{8d}{2d-1}$$

$$16(2d-1) = 8d$$

$$32d - 16 = 8d$$

$$-16 = -24d$$

$$\frac{-16}{-24} = d$$

$$\frac{2}{3} = d \qquad \left\{\frac{2}{3}\right\}$$

51) a) $2500 = 2000e^{0.06t}$

$$\frac{5}{4} = e^{0.06t}$$

$$\ln \frac{5}{4} = \ln e^{0.06t}$$

$$\ln \frac{5}{4} = 0.06t \ln e$$

$$\ln \frac{5}{4} = 0.06t(1)$$

$$\frac{\ln \frac{5}{4}}{0.06} = t$$

$$3.72 \text{ yr} \approx t$$

b) $4000 = 2000e^{0.06t}$

$$2 = e^{0.06t}$$

$$\ln 2 = \ln e^{0.06t}$$

$$\ln 2 = 0.06t \ln e$$

$$\ln 2 = 0.06t(1)$$

$$\frac{\ln 2}{0.06} = t$$

$$11.55 \text{ yr} \approx t$$

53) $\quad 7800 = 7000e^{0.075t}$

$$\frac{7800}{7000} = e^{0.075t}$$

$$\ln \frac{39}{35} = \ln e^{0.075t}$$

$$\ln \frac{39}{35} = 0.075t \ln e$$

$$\ln \frac{39}{35} = 0.075t(1)$$

$$\frac{\ln \frac{39}{35}}{0.075} = t$$

$$1.44 \text{ yr} \approx t$$

55) $\quad 5000 = Pe^{0.08(10)}$

$$5000 = Pe^{0.80}$$

$$\frac{5000}{e^{0.40}} = P$$

$$\$2246.64 \approx P$$

57) $4000 = 3000e^{r(4)}$

$$\frac{4}{3} = e^{4r}$$

$$\ln \frac{4}{3} = \ln e^{4r}$$

$$\ln \frac{4}{3} = 4r \ln e$$

$$\ln \frac{4}{3} = 4r(1)$$

$$\frac{\ln \frac{4}{3}}{4} = r$$

$$0.072 \approx r \qquad 7.2\%$$

Chapter 13: Exponential, and Logarithmic Functions

59) a) $5000 = 4000e^{0.0374t}$

$\dfrac{5}{4} = e^{0.0374t}$

$\ln \dfrac{5}{4} = \ln e^{0.0374t}$

$\ln \dfrac{5}{4} = 0.0374t \ln e$

$\ln \dfrac{5}{4} = 0.0374t(1)$

$\dfrac{\ln \dfrac{5}{4}}{0.0374} = t$

$6 \text{ hr} \approx t$

b) $8000 = 4000e^{0.0374t}$

$2 = e^{0.0374t}$

$\ln 2 = \ln e^{0.0374t}$

$\ln 2 = 0.0374t \ln e$

$\ln 2 = 0.0374t(1)$

$\dfrac{\ln 2}{0.0374} = t$

$18.5 \text{ hr} \approx t$

61) Let $t = 8$, $y_0 = 21,000$

$y = 21,000e^{0.036(8)}$

$y = 21,000e^{0.288}$

$y \approx 28,009 \text{ people}$

63) a) Let $t = 15$, $y_0 = 2470$

$y = 2470e^{-0.013(15)}$

$y = 2470e^{-0.195}$

$y \approx 2032 \text{ people}$

b) Let $y = 1600$, $y_0 = 2470$

$1600 = 2470e^{-0.013t}$

$\dfrac{1600}{2470} = e^{-0.013t}$

$\ln \dfrac{1600}{2470} = \ln e^{-0.013t}$

$\ln \dfrac{1600}{2470} = -0.013t \ln e$

$\ln \dfrac{1600}{2470} = -0.013t(1)$

$\dfrac{\ln \dfrac{1600}{2470}}{-0.013} = t$

$33 \approx t \qquad\qquad 2023$

65) a) Let $t = 2000$, $y_0 = 15$

$y = 15e^{-0.000121(2000)}$

$y = 15e^{-0.242}$

$y \approx 11.78 \text{ g}$

b) Let $y = 10$, $y_0 = 15$

$10 = 15e^{-0.000121t}$

$\dfrac{2}{3} = e^{-0.000121t}$

$\ln \dfrac{2}{3} = \ln e^{-0.000121t}$

$\ln \dfrac{2}{3} = -0.000121t \ln e$

$\ln \dfrac{2}{3} = -0.000121t(1)$

$\dfrac{\ln \dfrac{2}{3}}{-0.000121} = t$

$3351 \text{ yr} \approx t$

c) Let $y = 7.5, y_0 = 15$

$$7.5 = 15e^{-0.000121t}$$

$$\frac{1}{2} = e^{-0.000121t}$$

$$\ln\frac{1}{2} = \ln e^{-0.000121t}$$

$$\ln\frac{1}{2} = -0.000121t \ln e$$

$$\ln\frac{1}{2} = -0.000121t(1)$$

$$\frac{\ln\frac{1}{2}}{-0.000121} = t$$

$$5728 \text{ yr} \approx t$$

67) a) $y = 0.4e^{-0.086(0)} = 0.4e^0 = 0.4$ units

b) $y = 0.4e^{-0.086(7)} = 0.4e^{-0.602}$
$$= 0.22 \text{ units}$$

69) $\log_2(\log_2 x) = 2$

$$\log_2 x = 2^2 = 4$$

$$x = 2^4 = 16 \qquad \{16\}$$

71) $\log_3 \sqrt{n^2 + 5} = 1$

$$\sqrt{n^2 + 5} = 3^1 = 3$$

$$n^2 + 5 = 9$$

$$n^2 = 4$$

$$n = \pm 2 \qquad \{-2, 2\}$$

73) $e^{|t|} = 13$

$$\ln e^{|t|} = \ln 13$$

$$t = -\ln 13, \ln 13$$

$$t \approx -2.5649, 2.5649 \qquad \{-2.5649, 2.5649\}$$

75) $e^{2y} + 3e^y - 4 = 0$

$$(e^y + 4)(e^y - 1) = 0$$

$$e^y = 1 \text{ No negative value for } e^y$$

$$\ln(e^y) = \ln 1 = 0$$

$$y = 0 \qquad \{0\}$$

77) $5^{2c} - 4 \cdot 5^c - 21 = 0$

$$(5^c - 7)(5^c + 3) = 0$$

$$5^c = 7$$

$$\log 5^c = \log 7$$

$$c = \frac{\log 7}{\log 5} \approx 1.2091 \qquad \{1.2091\}$$

79) $(\log x)^2 = \log x^3$

$$(\log x)^2 - 3\log x = 0$$

$$\log x(\log x - 3) = 0$$

$$\log_{10} x = 0 \text{ or } \log_{10} x = 3$$

$$x = 1 \text{ or } x = 1000 \qquad \{1, 1000\}$$

Chapter 13 Review

1) yes; $\{(-4, -7), (1, -2), (5, 1), (11, 6)\}$

3) yes

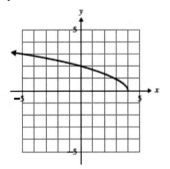

5) $\quad y = x + 4$

$\qquad x = y + 4$

$\qquad x - 4 = y$

$\qquad f^{-1}(x) = x - 4$

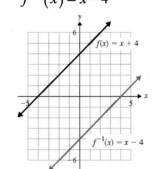

11)

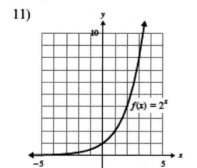

domain: $(-\infty, \infty)$;
range: $(0, \infty)$

7) $\quad y = \dfrac{1}{3}x - 1$

$\qquad x = \dfrac{1}{3}y - 1$

$\qquad x + 1 = \dfrac{1}{3}y$

$\qquad 3x + 3 = y$

$\qquad h^{-1}(x) = 3x + 3$

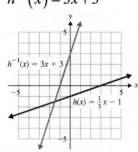

13)

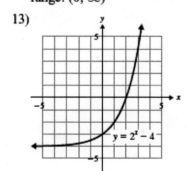

domain: $(-\infty, \infty)$;
range: $(-4, \infty)$

15)

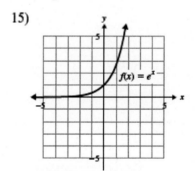

domain: $(-\infty, \infty)$;
range: $(0, \infty)$

9) a) $f(2) = 6(2) - 1 = 12 - 1 = 11$

b) $f^{-1}(11) = 2$

17) $2^c = 64$

$\qquad 2^c = 2^6$

$\qquad c = 6 \qquad\qquad \{6\}$

19) $16^{3z} = 32^{2z-1}$

$\left(2^4\right)^{3z} = \left(2^5\right)^{2z-1}$

$2^{12z} = 2^{10z-5}$

$12z = 10z - 5$

$2z = -5$

$z = -\dfrac{5}{2}$ $\qquad \left\{-\dfrac{5}{2}\right\}$

21) $\left(\dfrac{3}{2}\right)^{x+4} = \left(\dfrac{4}{9}\right)^{x-3}$

$\left(\dfrac{3}{2}\right)^{x+4} = \left[\left(\dfrac{3}{2}\right)^{-2}\right]^{x-3}$

$\left(\dfrac{3}{2}\right)^{x+4} = \left(\dfrac{3}{2}\right)^{-2x+6}$

$x + 4 = -2x + 6$

$3x = 2$

$x = \dfrac{2}{3}$ $\qquad \left\{\dfrac{2}{3}\right\}$

23) x must be a positive number.

25) $5^3 = 125$

27) $10^2 = 100$

29) $\log_3 81 = 4$

31) $\log 1000 = 3$

33) $\log_2 x = 3$

$2^3 = x$

$8 = x$ $\qquad \{8\}$

35) $\log_{32} 16 = x$

$32^x = 16$

$\left(2^5\right)^x = 2^4$

$2^{5x} = 2^4$

$5x = 4$

$x = \dfrac{4}{5}$ $\qquad \left\{\dfrac{4}{5}\right\}$

37) Let $\log_8 64 = x$

$8^x = 64$

$8^x = 8^2$

$x = 2$

$\log_8 64 = 2$

39) Let $\log 1000 = x$

$10^x = 1000$

$10^x = 10^3$

$x = 3$

$\log 1000 = 3$

41) Let $\log_{1/2} 16 = x$

$\left(\dfrac{1}{2}\right)^x = 16$

$\left(\dfrac{1}{2}\right)^x = \left(\dfrac{1}{2}\right)^{-4}$

$x = -4$

$\log_{1/2} 16 = -4$

43)

45)

47) $y = 5^x$

$x = 5^y$

$\log_5 x = y$

$f^{-1}(x) = \log_5 x$

49) $y = \log_6 x$

$x = \log_6 y$

$y = 6^x$

$h^{-1}(x) = 6^x$

51) False

53) $\log_8 (3z) = \log_8 3 + \log_8 z$

55) $\log_4 \sqrt{64} = \log_4 64^{1/2} = \frac{1}{2}\log_4 64$

$= \frac{1}{2}(3) = \frac{3}{2}$

57) $\log_5 c^4 d^3 = \log_5 c^4 + \log_5 d^3$

$= 4\log_5 c + 3\log_5 d$

59) $\log_a \frac{xy}{z^3} = \log_a xy - \log_a z^3$

$= \log_a x + \log_a y - 3\log_a z$

61) $\log p(p+8) = \log p + \log(p+8)$

63) $\log c + \log d = \log(cd)$

65) $9\log_2 a + 3\log_2 b = \log_2 a^9 + \log_2 b^3$

$= \log_2 a^9 b^3$

67) $\log_3 5 + 4\log_3 m - 2\log_3 n$

$= \log_3 5 + \log_3 m^4 - \log_3 n^2$

$= \log_3 5m^4 - \log_3 n^2 = \log_3 \frac{5m^4}{n^2}$

69) $3\log_5 c - \log_5 d - 2\log_5 f$

$= \log_5 c^3 - (\log_5 d + \log_5 f^2)$

$= \log_5 c^3 - \log_5 df^2 = \log_5 \frac{c^3}{df^2}$

71) $\log 49 = \log 7^2 = 2\log 7$

$\approx 2(0.8451) \approx 1.6902$

73) $\log \frac{7}{9} = \log 7 - \log 9$

$\approx 0.8451 - 0.9542 \approx -0.1091$

75) e

77) $\log 10 = 1$ since $10^1 = 10$

79) $\log \sqrt{10} = \frac{1}{2}\log 10 = \frac{1}{2}\cdot 1 = \frac{1}{2}$

81) $\log 0.001 = \log 10^{-3}$

$= -3\log 10 = -3\cdot 1 = -3$

83) $\ln 1 = 0$

85) $\log 8 \approx 0.9031$

87) $\ln 1.75 \approx 0.5596$

89) $\log p = 2$

$10^2 = p$

$100 = p$ \qquad $\{100\}$

91) $\log\left(\frac{1}{2}c\right) = -1$

$10^{-1} = \frac{1}{2}c$

$\frac{1}{10} = \frac{1}{2}c$

$\frac{1}{5} = c$ \qquad $\left\{\frac{1}{5}\right\}$

93) $\log x = 2.1$

$10^{2.1} = x$ \qquad $\{10^{2.1}\};\{125.8925\}$

95) $\ln y = 2$

$e^2 = y$ \qquad $\{e^2\};\{7.3891\}$

97) $\log(4t) = 1.75$

$10^{1.75} = 4t$

$\frac{10^{1.75}}{4} = t$

$\left\{\frac{10^{1.75}}{4}\right\};\{14.0585\}$

99)

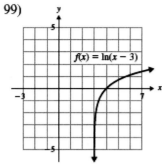

domain: $(3, \infty)$;
range: $(-\infty, \infty)$

101) $\quad \log_4 19 = \dfrac{\log 19}{\log 4} \approx 2.1240$

103) $\quad \log_{1/2} 38 = \dfrac{\log 38}{\log \dfrac{1}{2}} \approx -5.2479$

105) $\quad L(0.1) = 10 \log \dfrac{0.1}{10^{-12}} = 10 \log \dfrac{10^{-1}}{10^{-12}}$

$\qquad = 10 \log 10^{11} = 110 \log 10$

$\qquad = 110(1) = 110 \text{ dB}$

107) $\quad A = 2500 \left(1 + \dfrac{.06}{4}\right)^{4(5)}$

$\qquad = 2500(1.015)^{20} \approx \$3367.14$

109) $\quad A = 9000 e^{.062(4)}$

$\qquad = 9000 e^{.248} \approx \$11,533.14$

111) a)
$N(0) = 6000 e^{0.0514(0)}$

$\qquad = 6000(1) = 6000 \text{ bacteria}$

b) $\quad N(12) = 6000 e^{0.0514(12)}$

$\qquad = 6000 e^{0.6168}$

$\qquad \approx 11,118 \text{ bacteria}$

113) $\quad 2^y = 16$

$\qquad 2^y = 2^4$

$\qquad y = 4 \qquad\qquad \{4\}$

115) $\qquad 9^{4k} = 2$

$\qquad \ln 9^{4k} = \ln 2$

$\qquad 4k \ln 9 = \ln 2$

$\qquad k = \dfrac{\ln 2}{4 \ln 9}$

$\qquad \left\{ \dfrac{\ln 2}{4 \ln 9} \right\}; \{0.0789\}$

117) $\qquad 6^{2c} = 8^{c-5}$

$\qquad \ln 6^{2c} = \ln 8^{c-5}$

$\qquad 2c \ln 6 = (c-5) \ln 8$

$\qquad 2c \ln 6 = c \ln 8 - 5 \ln 8$

$\qquad 5 \ln 8 = c \ln 8 - 2c \ln 6$

$\qquad 5 \ln 8 = c(\ln 8 - 2 \ln 6)$

$\qquad \dfrac{5 \ln 8}{\ln 8 - 2 \ln 6} = c$

$\qquad \left\{ \dfrac{5 \ln 8}{\ln 8 - 2 \ln 6} \right\}; \{-6.9127\}$

119) $\qquad e^{5p} = 8$

$\qquad \ln e^{5p} = \ln 8$

$\qquad 5p \ln e = \ln 8$

$\qquad 5p = \ln 8$

$\qquad p = \dfrac{\ln 8}{5} \qquad \left\{ \dfrac{\ln 8}{5} \right\}; \{0.4159\}$

121) $\quad \log_3 (5w + 3) = 2$

$\qquad 3^2 = 5w + 3$

$\qquad 9 = 5w + 3$

$\qquad 6 = 5w$

$\qquad \dfrac{6}{5} = w \qquad \left\{ \dfrac{6}{5} \right\}$

123) $\log_2 x + \log_2 (x+2) = \log_2 24$

$\log_2 x(x+2) = \log_2 24$

$x(x+2) = 24$

$x^2 + 2x = 24$

$x^2 + 2x - 24 = 0$

$(x+6)(x-4) = 0$

$x+6 = 0$ or $x-4 = 0$

$x = -6 \qquad x = 4$

Only one solution satisfies the original equation. $\{4\}$

125) $\log_4 k + \log_4 (k-12) = 3$

$\log_4 k(k-12) = 3$

$4^3 = k(k-12)$

$64 = k^2 - 12k$

$0 = k^2 - 12k - 64$

$0 = (k-16)(k+4)$

$k-16 = 0$ or $k+4 = 0$

$k = 16 \qquad k = -4$

Only one solution satisfies the original equation. $\{16\}$

127) $10,000 = Pe^{.065(6)}$

$10,000 = Pe^{0.39}$

$\dfrac{10,000}{e^{0.39}} = P$

$\$6770.57 \approx P$

129) a) $y = 16,410e^{0.016(5)}$

$= 16,410e^{0.08} \approx 17,777$ people

b) $23,000 = 16,410e^{0.016t}$

$\dfrac{23,000}{16,410} = e^{0.016t}$

$\ln \dfrac{23,000}{16,410} = \ln e^{0.016t}$

$\ln \dfrac{23,000}{16,410} = 0.016t \ln e$

$\dfrac{\ln \dfrac{23,000}{16,410}}{0.016} = t$

$21 \approx t$

The year 2011

Chapter 13 Test

1) no

3) yes

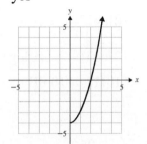

5)

7) a) $(0, \infty)$

b) $(-\infty, \infty)$

9) $\log_3 \dfrac{1}{9} = -2$

438

11) $125^{2c} = 25^{c-4}$

$\left(5^3\right)^{2c} = \left(5^2\right)^{c-4}$

$5^{6c} = 5^{2c-8}$

$6c = 2c - 8$

$4c = -8$

$c = -2 \qquad \{-2\}$

13) $\log(3r+13) = 2$

$10^2 = 3r + 13$

$100 = 3r + 13$

$87 = 3r$

$29 = r \qquad \{29\}$

15) a) let $\log_2 16 = x$

$2^x = 16$

$2^x = 2^4$

$x = 4$

b) let $\log_7 \sqrt{7} = x$

$7^x = \sqrt{7}$

$7^x = 7^{1/2}$

$x = \dfrac{1}{2}$

17) $\log_8(5n) = \log_8 5 + \log_8 n$

19) $2\log x - 3\log(x+1)$

$= \log x^2 - \log(x+1)^3 = \log \dfrac{x^2}{(x+1)^3}$

21) $e^{0.3t} = 5$

$\ln e^{0.3t} = \ln 5$

$0.3t \ln e = \ln 5$

$t = \dfrac{\ln 5}{0.3} \qquad \left\{\dfrac{\ln 5}{0.3}\right\};\{5.3648\}$

23) $4^{4a+3} = 9$

$\ln 4^{4a+3} = \ln 9$

$(4a+3)\ln 4 = \ln 9$

$4a \ln 4 + 3 \ln 4 = \ln 9$

$4a \ln 4 = \ln 9 - 3 \ln 4$

$a = \dfrac{\ln 9 - 3\ln 4}{4\ln 4}$

$\left\{\dfrac{\ln 9 - 3\ln 4}{4\ln 4}\right\};\{-0.3538\}$

25)

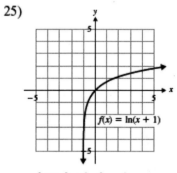

domain: $(-1, \infty)$;
range: $(-\infty, \infty)$

27) $A = 6000e^{0.074(5)} = 6000e^{0.37}$

$\approx \$8686.41$

Cumulative Review: Chapters 1-13

1) $40 + 8 \div 2 - 3^2 = 40 + 8 \div 2 - 9$

$= 40 + 4 - 9$

$= 44 - 9 = 35$

3) $\left(-5a^2\right)\left(3a^4\right) = -15a^{2+4} = -15a^6$

5) $\left(\dfrac{2c^{10}}{d^3}\right)^{-3} = \left(\dfrac{d^3}{2c^{10}}\right)^3 = \dfrac{d^9}{8c^{30}}$

7) $x =$ regular price of watch

$38.40 = x - .20x$

$3840 = 100x - 20x$

$3840 = 80x$

$48 = x$ $48.00

9) $x + 4y = -2$

$2(x + 4y) = 2(-2)$

$2x + 8y = -4$

Add the equations.

$2x + 8y = -4$

$+ -2x + 3y = 15$

—————————

$11y = 11$

$y = 1$

Substitute $y = 1$ into

$x + 4y = -2$

$x + 4(1) = -2$

$x + 4 = -2$

$x = -6$ $(-6, 1)$

11) $m = \dfrac{-1-5}{2-(-2)} = \dfrac{-6}{4} = -\dfrac{3}{2}$

$(x_1, y_1) = (-2, 5)$

$y - y_1 = m(x - x_1)$

$y - 5 = -\dfrac{3}{2}(x - (-2))$

$y - 5 = -\dfrac{3}{2}(x + 2)$

$y - 5 = -\dfrac{3}{2}x - 3$

$y = -\dfrac{3}{2}x + 2$

13) $4w^2 + w - 18 = (4w + 9)(w - 2)$

15) $y^2 - 6y + 9 = (y - 3)^2$

17) $\dfrac{r}{r^2 - 49} - \dfrac{3}{r^2 - 2r - 63}$

$= \dfrac{r}{(r-7)(r+7)} - \dfrac{3}{(r+7)(r-9)}$

$= \dfrac{r(r-9) - 3(r-7)}{(r-7)(r+7)(r+9)}$

$= \dfrac{r^2 - 9r - 3r + 21}{(r-7)(r+7)(r+9)}$

$= \dfrac{r^2 - 12r + 21}{(r-7)(r+7)(r+9)}$

19)

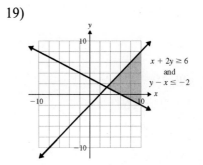

21) $\sqrt{45t^9} = 3\sqrt{5} \cdot t^4 \sqrt{t} = 3t^4 \sqrt{5t}$

23) $(27)^{2/3} = \left(\sqrt[3]{27}\right)^2 = 3^2 = 9$

25) $(2 - 7i)(3 + i) = 6 + 2i - 21i - 7i^2$

$= 6 - 19i - 7(-1)$

$= 13 - 19i$

27) $r^2 + 5r = -2$

$r^2 + 5r + \dfrac{25}{4} = -2 + \dfrac{25}{4}$

$\left(r + \dfrac{5}{2}\right)^2 = \dfrac{17}{4}$

$r + \dfrac{5}{2} = \pm\sqrt{\dfrac{17}{4}}$

$r = -\dfrac{5}{2} \pm \sqrt{\dfrac{17}{4}}$

$\left\{ -\dfrac{5}{2} - \dfrac{\sqrt{17}}{2}, -\dfrac{5}{2} + \dfrac{\sqrt{17}}{2} \right\}$

29) $4m^4 + 4 = 17m^2$

$4m^4 - 17m^2 + 4 = 0$

$(4m^2 - 1)(m^2 - 4) = 0$

$4m^2 - 1 = 0$ or $m^2 - 4 = 0$

$4m^2 = 1$ $\qquad m^2 = 4$

$m^2 = \dfrac{1}{4}$

$m = \pm\dfrac{1}{2}$ or $m = \pm 2$

$\left\{-2, -\dfrac{1}{2}, \dfrac{1}{2}, 2\right\}$

31) Domain: $(-\infty, \infty)$ Range: $[-4, \infty)$

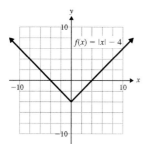

33) a) $f(-1) = (-1)^2 - 6(-1) + 2$

$= 1 + 6 + 2 = 9$

b) $(f \circ g)(x)$

$= f(g(x)) = f(x-3)$

$= (x-3)^2 - 6(x-3) + 2$

$= x^2 - 6x + 9 - 6x + 18 + 2$

$= x^2 - 12x + 29$

c) $g(x) = -7$

$-7 = x - 3$

$-4 = x$

35) $16^y = \dfrac{1}{64}$

$\left(4^2\right)^y = 4^{-3}$

$4^{2y} = 4^{-3}$

$2y = -3$

$y = -\dfrac{3}{2}$ $\qquad \left\{-\dfrac{3}{2}\right\}$

37) $\log a + 2\log b - 5\log c$

$= \log a + \log b^2 - \log c^5$

$= \log ab^2 - \log c^5 = \log \dfrac{ab^2}{c^5}$

39) $e^{-0.04t} = 6$

$\ln e^{-0.04t} = \ln 6$

$-0.04t \ln e = \ln 6$

$t = -\dfrac{\ln 6}{0.04}$

$\left\{-\dfrac{\ln 6}{0.04}\right\}; \{-44.7940\}$

Section 14.1: Exercises

1) $(x_1, y_1)\ (x_2, y_2)$
    $(1,3)$ and $(7,9)$

    Mid point $= \left(\dfrac{x_1 + x_2}{2}, \dfrac{y_1 + y_2}{2} \right)$

    $= \left(\dfrac{1+7}{2}, \dfrac{3+9}{2} \right) = (4,6)$

3) $(x_1, y_1)\ (x_2, y_2)$
    $(-5,2)$ and $(-1,-8)$

    Mid point $= \left(\dfrac{x_1 + x_2}{2}, \dfrac{y_1 + y_2}{2} \right)$

    $= \left(\dfrac{-5+(-1)}{2}, \dfrac{2+(-8)}{2} \right)$

    $= (-3,-3)$

5) $(x_1, y_1)\ (x_2, y_2)$
    $(-3,-7)$ and $(1,-2)$

    Mid point $= \left(\dfrac{x_1 + x_2}{2}, \dfrac{y_1 + y_2}{2} \right)$

    $= \left(\dfrac{-3+1}{2}, \dfrac{-7+(-2)}{2} \right)$

    $= \left(-1, -\dfrac{9}{2} \right)$

7) $(x_1, y_1)\ (x_2, y_2)$
    $(4,0)$ and $(-3,-5)$

    Mid point $= \left(\dfrac{x_1 + x_2}{2}, \dfrac{y_1 + y_2}{2} \right)$

    $= \left(\dfrac{4+(-3)}{2}, \dfrac{0+(-5)}{2} \right) = \left(\dfrac{1}{2}, -\dfrac{5}{2} \right)$

9) $(x_1, y_1)\ (x_2, y_2)$

    $\left(\dfrac{3}{2}, -1 \right)$ and $\left(\dfrac{5}{2}, \dfrac{7}{2} \right)$

    Mid point $= \left(\dfrac{x_1 + x_2}{2}, \dfrac{y_1 + y_2}{2} \right)$

    $= \left(\dfrac{\frac{3}{2}+\frac{5}{2}}{2}, \dfrac{-1+\frac{7}{2}}{2} \right) = \left(\dfrac{\frac{8}{2}}{2}, \dfrac{\frac{5}{2}}{2} \right)$

    $= \left(2, \dfrac{5}{4} \right)$

11) $(x_1, y_1)\ (x_2, y_2)$
    $(-6.2, 1.5)$ and $(4.8, 5.7)$

    Mid point $= \left(\dfrac{x_1 + x_2}{2}, \dfrac{y_1 + y_2}{2} \right)$

    $= \left(\dfrac{-6.2+4.8}{2}, \dfrac{1.5+5.7}{2} \right)$

    $= \left(-\dfrac{1.4}{2}, \dfrac{7.2}{2} \right)$

    $= (-0.7, 3.6)$

13) No; there are values in the domain that give more than one value in the range. The graph fails the vertical line test.

15) Center: $(-2,4)$; $r = \sqrt{9} = 3$

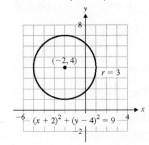

442

17) Center: $(5,3)$; $r = \sqrt{1} = 1$

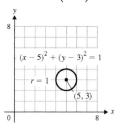

19) Center: $(-3,0)$; $r = \sqrt{4} = 2$

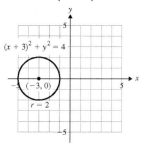

21) Center: $(6,-3)$; $r = \sqrt{16} = 4$

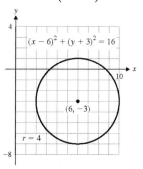

23) Center: $(0,0)$; $r = \sqrt{36} = 6$

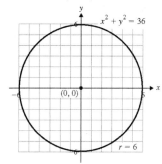

25) Center: $(0,0)$; $r = \sqrt{9} = 3$

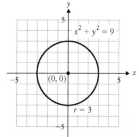

27) Center: $(0,1)$; $r = \sqrt{25} = 5$

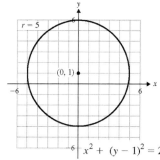

29) $(x-4)^2 + (y-1)^2 = 5^2$

$(x-4)^2 + (y-1)^2 = 25$

31) $(x-(-3))^2 + (y-2)^2 = 1^2$

$(x+3)^2 + (y-2)^2 = 1$

33) $(x-(-1))^2 + (y-(-5))^2 = (\sqrt{3})^2$

$(x+1)^2 + (y+5)^2 = 3$

35) $(x-0)^2 + (y-0)^2 = (\sqrt{10})^2$

$x^2 + y^2 = 10$

37) $(x-6)^2 + (y-0)^2 = (4)^2$

$(x-6)^2 + y^2 = 16$

39) $(x-0)^2 + (y-(-4))^2 = (2\sqrt{2})^2$

$x^2 + (y+4)^2 = 8$

41) $x^2 + y^2 - 8x + 2y + 8 = 0$

$x^2 - 8x + y^2 + 2y = -8$ Group x and y terms separately.

$(x^2 - 8x + 16) + (y^2 + 2y + 1) = -8 + 16 + 1$ Complete the square.

$(x-4)^2 + (y+1)^2 = 9$ Factor.

43) $x^2 + y^2 + 2x + 10y + 17 = 0$

$(x^2 + 2x) + (y^2 + 10y) = -17$

$(x^2 + 2x + 1) + (y^2 + 10y + 25) = -17 + 1 + 25$

$(x+1)^2 + (y+5)^2 = 9$

Center: $(-1, -5)$; $r = \sqrt{9} = 3$

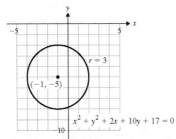

47) $x^2 + y^2 - 10x - 14y + 73 = 0$

$(x^2 - 10x) + (y^2 - 14y) = -73$

$(x^2 - 10x + 25) + (y^2 - 14y + 49) = -73 + 25 + 49$

$(x-5)^2 + (y-7)^2 = 1$

Center: $(5, 7)$; $r = \sqrt{1} = 1$

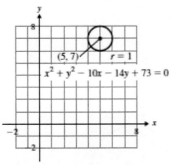

45) $x^2 + y^2 + 8x - 2y - 8 = 0$

$(x^2 + 8x) + (y^2 - 2y) = 8$

$(x^2 + 8x + 16) + (y^2 - 2y + 1) = 8 + 16 + 1$

$(x+4)^2 + (y-1)^2 = 25$

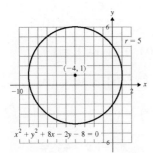

Center: $(-4, 1)$; $r = \sqrt{25} = 5$

49) $x^2 + y^2 + 6y + 5 = 0$

$x^2 + (y^2 + 6y) = -5$

$x^2 + (y^2 + 6y + 9) = -5 + 9$

$x^2 + (y+3)^2 = 4$

Center: $(0, -3)$; $r = \sqrt{4} = 2$

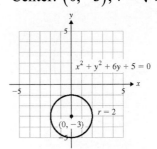

51) $x^2 + y^2 - 4x - 1 = 0$

$\left(x^2 - 4x\right) + y^2 = 1$

$\left(x^2 - 4x + 4\right) + y^2 = 1 + 4$

$(x - 2)^2 + y^2 = 5$

Center: $(2,0)$; $r = \sqrt{5}$

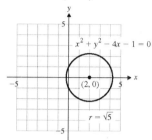

53) $x^2 + y^2 - 8x + 8y - 4 = 0$

$\left(x^2 - 8x\right) + \left(y^2 + 8y\right) = 4$

$\left(x - 8x + 16\right)^2 + \left(y^2 + 8y + 16\right) = 4 + 16 + 16$

$(x - 4)^2 + (y + 4)^2 = 36$

Center: $(4, -4)$; $r = \sqrt{36} = 6$

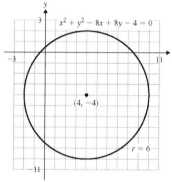

55) $4x^2 + 4y^2 - 12x - 4y - 6 = 0$

$x^2 + y^2 - 3x - y - \dfrac{3}{2} = 0$

$\left(x^2 - 3x\right) + \left(y^2 - y\right) = \dfrac{3}{2}$

$\left(x^2 - 3x + \dfrac{9}{4}\right) + \left(y^2 - y + \dfrac{1}{4}\right) = \dfrac{3}{2} + \dfrac{9}{4} + \dfrac{1}{4}$

$\left(x - \dfrac{3}{2}\right)^2 + \left(y - \dfrac{1}{2}\right)^2 = 4$

Center: $\left(\dfrac{3}{2}, \dfrac{1}{2}\right)$; $r = \sqrt{4} = 2$

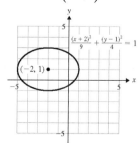

57) a) $128\,\text{m}$

   b) $64\,\text{m}$

   c) $(0, 71)$

   d) $x^2 + (y - 71)^2 = 4096$

59) $A = \pi\left(r_2^2 - r_1^2\right)$

$\approx 3.14(3600 - 56.25)$

$= 3.14(3543.75)$

$\approx 11{,}127\,\text{mm}^2$

Section 14.2: Exercises

1) ellipse

3) hyperbola

5) hyperbola

7) ellipse

9) Center: $(-2, 1)$

11) Center: $(3,-2)$

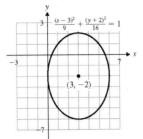

13) Center: $(0,0)$

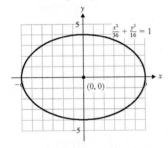

15) Center: $(0,0)$

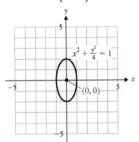

17) Center: $(0,-4)$

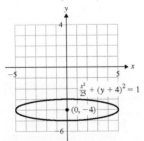

19) Center: $(-1,-3)$

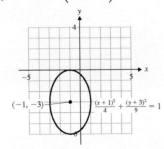

21) $4x^2 + 9y^2 = 36$

$$\frac{4x^2}{36} + \frac{9y^2}{36} = \frac{36}{36}$$

$$\frac{x^2}{9} + \frac{y^2}{4} = 1$$

Center: $(0,0)$

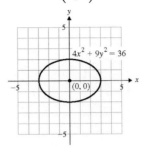

23) $25x^2 + y^2 = 25$

$$\frac{25x^2}{25} + \frac{y^2}{25} = \frac{25}{25}$$

$$x^2 + \frac{y^2}{25} = 1$$

Center: $(0,0)$

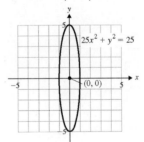

25) Center: $(0,0)$

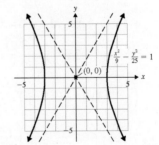

27) Center: $(0,0)$

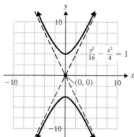

29) Center: $(2,-3)$

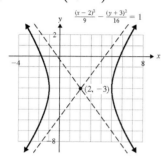

31) Center: $(-4,-1)$

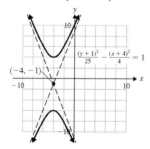

33) Center: $(1,0)$

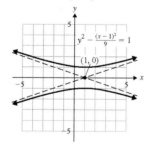

35) Center: $(1,2)$

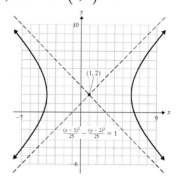

37) $9x^2 - y^2 = 36$

$$\frac{9x^2}{36} - \frac{y^2}{36} = \frac{36}{36}$$

$$\frac{x^2}{4} - \frac{y^2}{36} = 1$$

Center: $(0,0)$

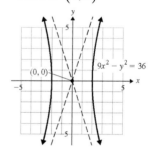

39) Center: $(0,0)$

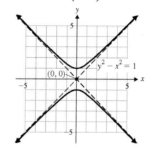

41) domain: $[-3, 3]$; range: $[0, 3]$

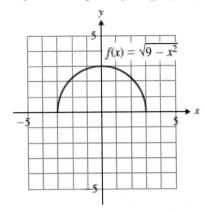

43) domain: $[-1, 1]$; range: $[-1, 0]$

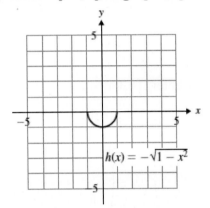

45) domain: $[-3, 3]$; range: $[-2, 0]$

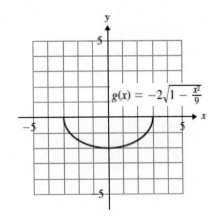

47) domain: $(-\infty, -2] \cup [2, \infty)$; range: $(-\infty, 0]$

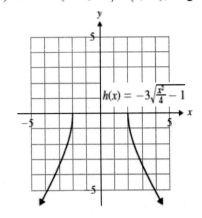

49)

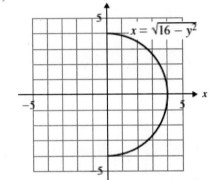

51)

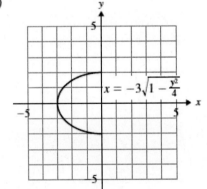

53) $\dfrac{36}{2} = 18$; $\dfrac{29}{2} = 14.5$

$$\dfrac{x^2}{18^2} + \dfrac{y^2}{14.5^2} = 1$$

$$\dfrac{x^2}{324} + \dfrac{y^2}{210.25} = 1$$

Chapter 14: Putting It All Together:

1) parabola

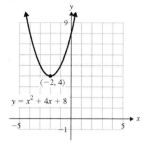

$y = x^2 + 4x + 8$

3) hyperbola

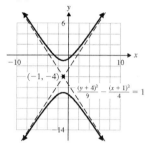

$\frac{(y+4)^2}{9} - \frac{(x+1)^2}{4} = 1$

5) $16x^2 + 9y^2 = 144$

$$\frac{16x^2}{144} + \frac{9y^2}{144} = \frac{144}{144}$$

$$\frac{x^2}{9} + \frac{y^2}{16} = 1$$

ellipse

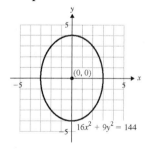

$16x^2 + 9y^2 = 144$

7) $x^2 + y^2 + 8x - 6y - 11 = 0$

$$\left(x^2 + 8x\right) + \left(y^2 - 6y\right) = 11$$

$$\left(x^2 + 8x + 16\right) + \left(y^2 - 6y + 9\right) = 11 + 16 + 9$$

$$\left(x + 4\right)^2 + \left(y - 3\right)^2 = 36$$

circle

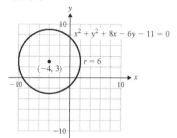

$x^2 + y^2 + 8x - 6y - 11 = 0$

$r = 6$

9) ellipse

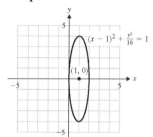

$(x - 1)^2 + \frac{y^2}{16} = 1$

11) parabola

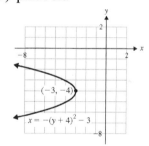

$x = -(y + 4)^2 - 3$

13) $25x^2 - 4y^2 = 100$

$$\frac{25x^2}{100} - \frac{4y^2}{100} = \frac{100}{100}$$

$$\frac{x^2}{4} - \frac{y^2}{25} = 1$$

hyperbola

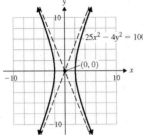

15) circle

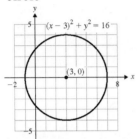

17) parabola

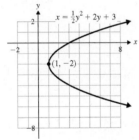

19) $(x-2)^2 - (y+1)^2 = 9$

$$\frac{(x-2)^2}{9} - \frac{(y+1)^2}{9} = 1$$

hyperbola

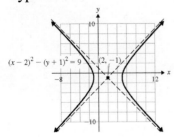

Section 14.3: Exercises

1) a)

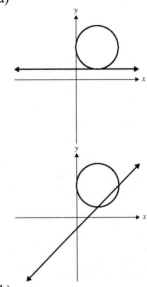

b)

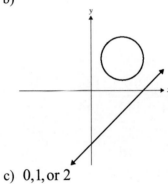

c) 0, 1, or 2

3) a)

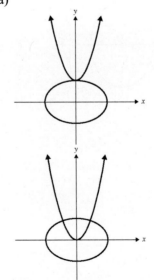

450

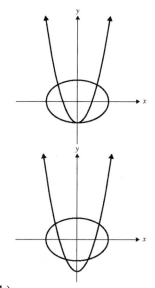

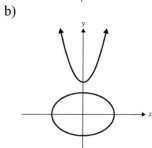

b)

c) $0, 1, 2, 3,$ or 4

5) a)

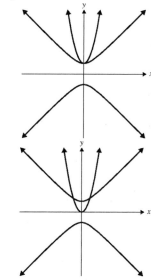

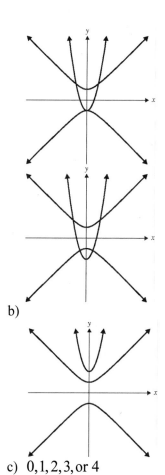

b)

c) $0, 1, 2, 3,$ or 4

7) $x^2 + 4y = 8$ (1)

    $x + 2y = -8$ (2)

Substitute $x = -2y - 8$ into (1).

$$(-2y - 8)^2 + 4y = 8$$

$$4y^2 + 32y + 64 + 4y = 8$$

$$4y^2 + 36y + 56 = 0$$

$$y^2 + 9y + 14 = 0$$

$$(y + 7)(y + 2) = 0$$

$$y + 7 = 0 \text{ or } y + 2 = 0$$

$$y = -7 \qquad y = -2$$

$$y = -7 : x + 2(-7) = -8$$

$$x - 14 = -8$$

$$x = 6$$

$y = -2 : x + 2(-2) = -8$

$x - 4 = -8$

$x = -4$

Verify by substituting into (1).

$\{(-4, -2), (6, -7)\}$

9) $x + 2y = 5 \qquad (1)$

$x^2 + y^2 = 10 \quad (2)$

Substitute $x = 5 - 2y$ into (2).

$(5 - 2y)^2 + y^2 = 10$

$4y^2 - 20y + 25 + y^2 = 10$

$5y^2 - 20y + 15 = 0$

$y^2 - 4y + 3 = 0$

$(y - 3)(y - 1) = 0$

$y - 3 = 0 \ \text{ or } \ y - 1 = 0$

$y = 3 \qquad\qquad y = 1$

$y = 3 : x + 2(3) = 5$

$x + 6 = 5$

$x = -1$

$y = 1 : x + 2(1) = 5$

$x + 2 = 5$

$x = 3$

Verify by substituting into (2).

$\{(-1, 3), (3, 1)\}$

11) $y = x^2 - 6x + 10 \qquad (1)$

$y = 2x - 6 \qquad\qquad (2)$

Substitute (2) into (1).

$2x - 6 = x^2 - 6x + 10$

$0 = x^2 - 8x + 16$

$0 = (x - 4)^2$

$0 = x - 4$

$4 = x$

$x = 4 : y = 2(4) - 6 = 8 - 6 = 2$

Verify by substituting into (1).

$\{(4, 2)\}$

13) $x^2 + 2y^2 = 11 \qquad (1)$

$x^2 - y^2 = 8 \qquad (2)$

Solve using elimination.

$x^2 + 2y^2 = 11 \qquad (1)$

$+ \quad -x^2 + y^2 = -8 \qquad -1 \cdot (2)$

$\overline{ 3y^2 = 3}$

$y^2 = 1$

$y = \pm\sqrt{1} = \pm 1$

$y = 1 : x^2 - 1^2 = 8$

$x^2 - 1 = 8$

$x^2 = 9$

$x = \pm 3$

$y = -1 : x^2 - (-1)^2 = 8$

$x^2 - 1 = 8$

$x^2 = 9$

$x = \pm 3$

Verify by substituting into (1).

$\{(3, 1), (3, -1), (-3, 1), (-3, -1)\}$

15) $x^2 + y^2 = 6 \qquad (1)$

$2x^2 + 5y^2 = 18 \qquad (2)$

Solve using elimination.

$-2x^2 - 2y^2 = -12 \qquad -2 \cdot (1)$

$+ \quad 2x^2 + 5y^2 = 18 \qquad (2)$

$\overline{ 3y^2 = 6}$

$y^2 = 2$

$y = \pm\sqrt{2}$

$y = \sqrt{2} : x^2 + (\sqrt{2})^2 = 6$

$x^2 + 2 = 6$

$x^2 = 4$

$x = \pm 2$

$y = -\sqrt{2} : x^2 + \left(-\sqrt{2}\right)^2 = 6$

$$x^2 + 2 = 6$$

$$x^2 = 4$$

$$x = \pm 2$$

Verify by substituting into (2).

$$\left\{\left(2, \sqrt{2}\right), \left(2, -\sqrt{2}\right), \left(-2, \sqrt{2}\right), \left(-2, -\sqrt{2}\right)\right\}$$

17) $3x^2 + 4y = -1$ (1)

    $x^2 + 3y = -12$ (2)

    Solve using elimination.

$$3x^2 + 4y = -1 \qquad (1)$$
$$+ \quad -3x^2 - 9y = 36 \qquad -3 \cdot (2)$$
$$\overline{\qquad\qquad -5y = 35}$$
$$y = -7$$

$y = -7 : x^2 + 3(-7) = -12$

$$x^2 - 21 = -12$$

$$x^2 = 9$$

$$x = \pm 3$$

Verify by substituting into (2).

$$\left\{(3, -7), (-3, -7)\right\}$$

19) $y = 6x^2 - 1$ (1)

    $2x^2 + 5y = -5$ (2)

    Substitute (1) into (2).

$$2x^2 + 5\left(6x^2 - 1\right) = -5$$

$$2x^2 + 30x^2 - 5 = -5$$

$$32x^2 = 0$$

$$x^2 = 0$$

$$x = 0$$

$x = 0 : y = 6(0)^2 - 1 = -1$

Verify by substituting into (2).

$$\left\{(0, -1)\right\}$$

21) $x^2 + y^2 = 4$ (1)

    $-2x^2 + 3y = 6$ (2)

    Solve using elimination.

$$2x^2 + 2y^2 = 8 \qquad 2 \cdot (1)$$
$$+ \quad -2x^2 + 3y = 6 \qquad (2)$$
$$\overline{\qquad\qquad 2y^2 + 3y = 14}$$

$$2y^2 + 3y - 14 = 0$$

$$(2y + 7)(y - 2) = 0$$

$$2y + 7 = 0 \text{ or } y - 2 = 0$$

$$2y = -7 \qquad y = 2$$

$$y = -\frac{7}{2}$$

$y = -\frac{7}{2} : x^2 + \left(-\frac{7}{2}\right)^2 = 4$

$$x^2 + \frac{49}{4} = 4$$

$$x^2 = -\frac{33}{4}$$

$$x = \pm \frac{\sqrt{33}}{2} i$$

does not give real number solutions.

$y = 2 : x^2 + 2^2 = 4$

$$x^2 + 4 = 4$$

$$x^2 = 0$$

$$x = 0$$

Verify by substituting into (2).

$$\left\{(0, 2)\right\}$$

23) $x^2 + y^2 = 3$ (1)

    $x + y = 4$ (2)

    Substitute $y = 4 - x$ into (1).

$$x^2 + (4 - x)^2 = 3$$

$$x^2 + 16 - 8x + x^2 = 3$$

$$2x^2 - 8x + 13 = 0$$

$$x = \frac{-(-8) \pm \sqrt{(-8)^2 - 4(2)(13)}}{2(2)}$$

$$= \frac{8 \pm \sqrt{64 - 104}}{4} = \frac{8 \pm \sqrt{-40}}{4}$$

No real number values for x. $\quad \varnothing$

25) $\qquad x = \sqrt{y} \qquad$ (1)

$x^2 - 9y^2 = 9 \qquad$ (2)

Substitute (1) into (2).

$$\left(\sqrt{y}\right)^2 - 9y^2 = 9$$

$$y - 9y^2 = 9$$

$$0 = 9y^2 - y + 9$$

$$y = \frac{-(-1) \pm \sqrt{(-1)^2 - 4(9)(9)}}{2(9)}$$

$$= \frac{1 \pm \sqrt{1 - 324}}{18} = \frac{1 \pm \sqrt{-323}}{18}$$

No real number values for y. $\quad \varnothing$

27) $9x^2 + y^2 = 9 \qquad$ (1)

$x^2 + y^2 = 5 \qquad$ (2)

Solve using elimination.

$$\begin{array}{ll} 9x^2 + y^2 = 9 & (1) \\ + \quad -x^2 - y^2 = -5 & -1 \cdot (2) \\ \hline 8x^2 = 4 \end{array}$$

$$x^2 = \frac{1}{2}$$

$$x = \pm\sqrt{\frac{1}{2}} = \pm\frac{\sqrt{2}}{2}$$

$$x = \frac{\sqrt{2}}{2} : \left(\frac{\sqrt{2}}{2}\right)^2 + y^2 = 5$$

$$\frac{1}{2} + y^2 = 5$$

$$y^2 = \frac{9}{2}$$

$$y = \pm\frac{3\sqrt{2}}{2}$$

$$x = -\frac{\sqrt{2}}{2} : \left(-\frac{\sqrt{2}}{2}\right)^2 + y^2 = 5$$

$$\frac{1}{2} + y^2 = 5$$

$$y^2 = \frac{9}{2}$$

$$y = \pm\frac{3\sqrt{2}}{2}$$

Verify by substituting into (1).

$$\left\{ \left(\frac{\sqrt{2}}{2}, \frac{3\sqrt{2}}{2}\right), \left(\frac{\sqrt{2}}{2}, -\frac{3\sqrt{2}}{2}\right), \left(-\frac{\sqrt{2}}{2}, \frac{3\sqrt{2}}{2}\right), \left(-\frac{\sqrt{2}}{2}, -\frac{3\sqrt{2}}{2}\right) \right\}$$

29) $\qquad y = -x^2 - 2 \qquad$ (1)

$x^2 + y^2 = 4 \qquad$ (2)

Substitute (1) into (2).

$$x^2 + \left(-x^2 - 2\right)^2 = 4$$

$$x^2 + x^4 + 4x^2 + 4 = 4$$

$$x^4 + 5x^2 = 0$$

$$x^2\left(x^2 + 5\right) = 0$$

$$x^2 + 5 = 0 \ \text{ or } \ x^2 = 0$$

$$x^2 = -5 \qquad \boxed{x = 0}$$

$$x = \pm i\sqrt{5}$$

$x = \pm i\sqrt{5}$ does not give

real number solutions.

$x = 0: y = -(0)^2 - 2 = -2$

Verify by substituting into (2).

$\{(0, -2)\}$

31) x = one number

y = other number

$xy = 40 \qquad (1)$

$x + y = 13 \qquad (2)$

Substitute $y = 13 - x$ into (1).

$x(13 - x) = 40$

$13x - x^2 = 40$

$0 = x^2 - 13x + 40$

$0 = (x - 8)(x - 5)$

$x - 8 = 0 \quad$ or $\quad x - 5 = 0$

$x = 8 \qquad\qquad x = 5$

$x = 8: y = 13 - 8 = 5$

$x = 5: y = 13 - 5 = 8$

The numbers are 8 and 5.

33) l = length of screen

w = width of screen

$2l + 2w = 38 \qquad (1)$

$lw = 88 \qquad\qquad (2)$

Solve (1) for l.

$2l = 38 - 2w$

$l = 19 - w \qquad (3)$

Substitute (3) into (2).

$(19 - w)w = 88$

$19w - w^2 = 88$

$0 = w^2 - 19w + 88$

$0 = (w - 11)(w - 8)$

$w - 11 = 0 \quad$ or $\quad w - 8 = 0$

$w = 11 \qquad\qquad w = 8$

$w = 11: l = 19 - 11 = 8$

$w = 8: l = 19 - 8 = 11$

The dimensions of the screen
are 8 in $\times$ 11 in.

35) $\qquad\qquad 15x^2 = 6x^2 + 33x + 12$

$9x^2 - 33x - 12 = 0$

$3x^2 - 11x - 4 = 0$

$(3x + 1)(x - 4) = 0$

$3x + 1 = 0 \quad$ or $\quad x - 4 = 0$

$3x = -1 \qquad\qquad \boxed{x = 4}$

$x = -\dfrac{1}{3}$

$x = 4: y = 15(4)^2 = 240$

The break-even point is 4000
basketballs and \$240.

Section 14.4: Exercises

1) The endpoints are included when the
inequality symbol is $\leq$ or $\geq$. The
endpoints are not included when the
symbol is $<$ or $>$.

3) a) $[-5, 1]$ $\qquad$ b) $(-\infty, -5) \cup (1, \infty)$

5) a) $[-1, 3]$ $\qquad$ b) $(-\infty, -1) \cup (3, \infty)$

7) $\qquad x^2 + 6x - 7 \geq 0$

$(x + 7)(x - 1) \geq 0$

$(x + 7)(x - 1) = 0$

$x + 7 = 0 \quad$ or $\quad x - 1 = 0$

$x = -7 \qquad\qquad x = 1$

Interval A: $(-\infty, -7) \quad$ Positive

Interval B: $(-7, 1) \qquad$ Negative

Interval C: $(1, \infty) \qquad$ Positive

$(-\infty, -7] \cup [1, \infty)$

9) $\qquad c^2 + 5c < 36$

$c^2 + 5c - 36 < 0$

$(c+9)(c-4) < 0$

$(c+9)(c-4) = 0$

$c + 9 = 0$ or $c - 4 = 0$

$\qquad c = -9 \qquad\quad c = 4$

Interval A: $(-\infty, -9)$ Positive

Interval B: $(-9, 4)$ Negative

Interval C: $(4, \infty)$ Positive

$(-9, 4)$

11) $\qquad r^2 - 13r > -42$

$r^2 - 13r + 42 > 0$

$(r-6)(r-7) > 0$

$(r-6)(r-7) = 0$

$r - 6 = 0$ or $r - 7 = 0$

$\qquad r = 6 \qquad\quad r = 7$

Interval A: $(-\infty, 6)$ Positive

Interval B: $(6, 7)$ Negative

Interval C: $(7, \infty)$ Positive

$(-\infty, 6) \cup (7, \infty)$

13) $3z^2 + 14z - 24 \le 0$

$(3z - 4)(z + 6) \le 0$

$(3z - 4)(z + 6) = 0$

$3z - 4 = 0$ or $z + 6 = 0$

$\qquad 3z = 4 \qquad\quad z = -6$

$\qquad z = \dfrac{4}{3}$

Interval A: $(-\infty, -6)$ Positive

Interval B: $\left(-6, \dfrac{4}{3}\right)$ Negative

Interval C: $\left(\dfrac{4}{3}, \infty\right)$ Positive

$\left[-6, \dfrac{4}{3}\right]$

15) $\qquad 7p^2 - 4 > 12p$

$7p^2 - 12p - 4 > 0$

$(7p + 2)(p - 2) > 0$

$(7p + 2)(p - 2) = 0$

$7p + 2 = 0$ or $p - 2 = 0$

$\qquad 7p = -2 \qquad\quad p = 2$

$\qquad p = -\dfrac{2}{7}$

Interval A: $\left(-\infty, -\dfrac{2}{7}\right)$ Positive

Interval B: $\left(-\dfrac{2}{7}, 2\right)$ Negative

Interval C: $(2, \infty)$ Positive

$\left(-\infty, -\dfrac{2}{7}\right) \cup (2, \infty)$

17) $b^2 - 9b > 0$

$b(b - 9) > 0$

$b(b - 9) = 0$

$b - 9 = 0$ or $b = 0$

$\qquad b = 9$

Interval A: $(-\infty, 0)$ Positive

Interval B: $(0, 9)$ Negative

Interval C: $(9, \infty)$ Positive

$(-\infty, 0) \cup (9, \infty)$

19) $\qquad 4y^2 \le -5y$

$4y^2 + 5y \le 0$

$y(4y+5) \le 0$

$y(4y+5) = 0$

$4y+5 = 0$ or $y = 0$

$4y = -5$

$y = -\dfrac{5}{4}$

Interval A: $\left(-\infty, -\dfrac{5}{4}\right)$ Positive

Interval B: $\left(-\dfrac{5}{4}, 0\right)$ Negative

Interval C: $(0, \infty)$ Positive

$\left[-\dfrac{5}{4}, 0\right]$

21) $\qquad m^2 - 64 < 0$

$(m+8)(m-8) < 0$

$(m+8)(m-8) = 0$

$m+8 = 0$ or $m+8 = 0$

$\quad m = -8 \qquad m = 8$

Interval A: $(-\infty, -8)$ Positive

Interval B: $(-8, 8)$ Negative

Interval C: $(8, \infty)$ Positive

$(-8, 8)$

23) $\qquad 121 - h^2 \le 0$

$(11+h)(11-h) \le 0$

$(11+h)(11-h) = 0$

$11+h = 0$ or $11-h = 0$

$\quad h = -11 \qquad h = 11$

Interval A: $(-\infty, -11)$ Negative

Interval B: $(-11, 11)$ Positive

Interval C: $(11, \infty)$ Negative

$(-\infty, -11] \cup [11, \infty)$

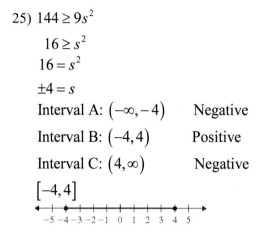

25) $144 \ge 9s^2$

$16 \ge s^2$

$16 = s^2$

$\pm 4 = s$

Interval A: $(-\infty, -4)$ Negative

Interval B: $(-4, 4)$ Positive

Interval C: $(4, \infty)$ Negative

$[-4, 4]$

27) $(-\infty, \infty)$

29) $(-\infty, \infty)$

31) $\varnothing$

33) $\varnothing$

35) $(r+2)(r-5)(r-1) \le 0$

$(r+2)(r-5)(r-1) = 0$

$r+2 = 0$ or $r-5 = 0$ or $r-1 = 0$

$\quad r = -2 \qquad r = 5 \qquad r = 1$

Interval A: $(-\infty, -2)$ Negative

Interval B: $(-2, 1)$ Positive

Interval C: $(1, 5)$ Negative

Interval D: $(5, \infty)$ Positive

$(-\infty, -2] \cup [1, 5]$

37) $(j-7)(j-5)(j+9) \geq 0$

$(j-7)(j-5)(j+9) = 0$

$j-7=0$ or $j-5=0$ or $j+9=0$

$j=7 \qquad j=5 \qquad j=-9$

Interval A: $(-\infty,-9)$ Negative

Interval B: $(-9,5)$ Positive

Interval C: $(5,7)$ Negative

Interval D: $(7,\infty)$ Positive

$[-9,5] \cup [7,\infty)$

39) $(6c+1)(c+7)(4c-3) < 0$

$(6c+1)(c+7)(4c-3) = 0$

$6c+1=0$ or $c+7=0$ or $4c-3=0$

$6c=-1 \qquad c=-7 \qquad 4c=3$

$c=-\dfrac{1}{6} \qquad\qquad c=\dfrac{3}{4}$

Interval A: $(-\infty,-7)$ Negative

Interval B: $\left(-7,-\dfrac{1}{6}\right)$ Positive

Interval C: $\left(-\dfrac{1}{6},\dfrac{3}{4}\right)$ Negative

Interval D: $\left(\dfrac{3}{4},\infty\right)$ Positive

$(-\infty,-7) \cup \left(-\dfrac{1}{6},\dfrac{3}{4}\right)$

41) $\dfrac{7}{p+6} > 0$

Set the numerator and denominator

equal to zero and solve for p.

$7 \neq 0 \quad p+6=0$

$p=-6$

Interval A: $(-\infty,-6)$ Negative

Interval B: $(-6,\infty)$ Positive

$(-6,\infty)$

43) $\dfrac{5}{z+3} \leq 0$

Set the numerator and denominator

equal to zero and solve for z.

$5 \neq 0 \quad z+3=0$

$z=-3$

Interval A: $(-\infty,-3)$ Negative

Interval B: $(-3,\infty)$ Positive

$z \neq -3$ because it makes the

denominator equal to zero.

$(-\infty,-3)$

45) $\dfrac{x-4}{x-3} > 0$

Set the numerator and denominator

equal to zero and solve for x.

$x-4=0 \qquad x-3=0$

$x=4 \qquad\qquad x=3$

Interval A: $(-\infty,3)$ Positive

Interval B: $(3,4)$ Negative

Interval C: $(4,\infty)$ Positive

$(-\infty,3) \cup (4,\infty)$

47) $\dfrac{h-9}{3h+1} \leq 0$

Set the numerator and denominator

equal to zero and solve for h.

$h-9=0 \qquad 3h+1=0$

$h=9 \qquad\qquad 3h=-1$

$h=-\dfrac{1}{3}$

Interval A: $\left(-\infty, -\dfrac{1}{3}\right)$ Positive

Interval B: $\left(-\dfrac{1}{3}, 9\right)$ Negative

Interval C: $(9, \infty)$ Positive

$h \neq -\dfrac{1}{3}$ because it makes the denominator equal to zero.

$\left(-\dfrac{1}{3}, 9\right]$

49) $\dfrac{k}{k+3} \leq 0$

Set the numerator and denominator equal to zero and solve for k.

$k = 0 \quad k + 3 = 0$
$\qquad\qquad k = -3$

Interval A: $(-\infty, -3)$ Positive

Interval B: $(-3, 0)$ Negative

Interval C: $(0, \infty)$ Positive

$k \neq -3$ because it makes the denominator equal to zero.

$(-3, 0]$

51) $\dfrac{7}{t+6} < 3$

$\dfrac{7}{t+6} - 3 < 0$

$\dfrac{7}{t+6} - \dfrac{3(t+6)}{t+6} < 0$

$\dfrac{7 - 3(t+6)}{t+6} < 0$

$\dfrac{7 - 3t - 18}{t+6} < 0$

$\dfrac{-3t - 11}{t+6} < 0$

Set the numerator and denominator equal to zero and solve for t.

$-3t - 11 = 0 \qquad\qquad t + 6 = 0$
$-3t = 11 \qquad\qquad\qquad t = -6$
$t = -\dfrac{11}{3}$

Interval A: $(-\infty, -6)$ Negative

Interval B: $\left(-6, -\dfrac{11}{3}\right)$ Positive

Interval C: $\left(-\dfrac{11}{3}, \infty\right)$ Negative

$(-\infty, -6) \cup \left(-\dfrac{11}{3}, \infty\right)$

53) $\dfrac{3}{a+7} \geq 1$

$\dfrac{3}{a+7} - 1 \geq 0$

$\dfrac{3}{a+7} - \dfrac{(a+7)}{a+7} \geq 0$

$\dfrac{3 - (a+7)}{a+7} \geq 0$

$\dfrac{-a - 4}{a+7} \geq 0$

Set the numerator and denominator equal to zero and solve for a.

$$-a-4=0 \qquad a+7=0$$
$$-a=4 \qquad a=-7$$
$$a=-4$$

Interval A: $(-\infty,-7)$ Negative

Interval B: $(-7,-4)$ Positive

Interval C: $(-4,\infty)$ Negative

$a \neq -7$ because it makes the denominator equal to zero.

$$(-7,-4]$$

55) $$\frac{2y}{y-6} \leq -3$$

$$\frac{2y}{y-6}+3 \leq 0$$

$$\frac{2y}{y-6}+\frac{3(y-6)}{y-6} \leq 0$$

$$\frac{2y+3(y-6)}{y-6} \leq 0$$

$$\frac{2y+3y-18}{y-6} \leq 0$$

$$\frac{5y-18}{y-6} \leq 0$$

Set the numerator and denominator equal to zero and solve for y.

$$5y-18=0 \qquad y-6=0$$
$$5y=18 \qquad y=6$$
$$y=\frac{18}{5}$$

Interval A: $\left(-\infty,\frac{18}{5}\right)$ Positive

Interval B: $\left(\frac{18}{5},6\right)$ Negative

Interval C: $(6,\infty)$ Positive

$y \neq 6$ because it makes the denominator equal to zero.

$$\left[\frac{18}{5},6\right)$$

57) $$\frac{3w}{w+2} > -4$$

$$\frac{3w}{w+2}+4 > 0$$

$$\frac{3w}{w+2}+\frac{4(w+2)}{w+2} > 0$$

$$\frac{3w+4(w+2)}{w+2} > 0$$

$$\frac{3w+4w+8}{w+2} > 0$$

$$\frac{7w+8}{w+2} > 0$$

Set the numerator and denominator equal to zero and solve for w.

$$7w+8=0 \qquad w+2=0$$
$$7w=-8 \qquad w=-2$$
$$w=-\frac{8}{7}$$

Interval A: $(-\infty,-2)$ Positive

Interval B: $\left(-2,-\frac{8}{7}\right)$ Negative

Interval C: $\left(-\frac{8}{7},\infty\right)$ Positive

$$(-\infty,-2)\cup\left(-\frac{8}{7},\infty\right)$$

59) $\dfrac{(6d+1)^2}{d-2} \le 0$

Numerator can not be negative.
Set the denominator equal to zero
and solve for d.
$d-2=0$
$d=2$
Interval A: $(-\infty, 2)$ Negative
Interval B: $(2, \infty)$ Positive
$d \ne 2$ because it makes the
denominator equal to zero.
$(-\infty, 2)$

61) $\dfrac{(4t-3)^2}{t-5} > 0$

Numerator can not be negative.
Set the denominator equal to zero
and solve for t.
$t-5=0$
$t=5$
Interval A: $(-\infty, 5)$ Negative
Interval B: $(5, \infty)$ Positive
$(5, \infty)$

63) $\dfrac{n+6}{n^2+4} < 0$

Denominator will always be
positive. Set the numerator equal
to zero and solve for n.
$n+6=0$
$n=-6$
Interval A: $(-\infty, -6)$ Negative
Interval B: $(-6, \infty)$ Positive
$(-\infty, -6)$

65) $\dfrac{m+1}{m^2+3} \ge 0$

Denominator will always be
positive. Set the numerator equal
to zero and solve for m.
$m+1=0$
$m=-1$
Interval A: $(-\infty, -1)$ Negative
Interval B: $(-1, \infty)$ Positive
$[-1, \infty)$

67) $\dfrac{s^2+2}{s-4} \le 0$

Numerator will always be positive.
Set the denominator equal to zero
and solve for s.
$s-4=0$
$s=4$
Interval A: $(-\infty, 4)$ Negative
Interval B: $(4, \infty)$ Positive
$s \ne 4$ because it makes the
denominator equal to zero.
$(-\infty, 4)$

69) a) $P(x) = -2x^2 + 32x - 96 > 0$
$-2(x^2 - 16x + 48) > 0$
$-2(x-4)(x-12) > 0$
$(x-4)(x-12) < 0$
$(x-4)(x-12) = 0$
$x-4=0$ or $x-12=0$
$x=4$ $x=12$
Interval A: $(-\infty, 4)$ Positive
Interval B: $(4, 12)$ Negative
Interval C: $(12, \infty)$ Positive

461

$(4,12)$

Between 4000 and 12000 units.

b) $(-\infty,4)\cup(12,\infty)$

When they produce less than 4000 units or more than 12000 units.

71) $\overline{C}(x)=\dfrac{10x+100,000}{x}\le 20$

$10x+100,000\le 20x$

$-10x+100,000\le 0$

$-10x\le -100,000$

$x\ge 10,000$

$10,000$ or more.

Chapter 14 Review

1) $\qquad (x_1,y_2)\ (x_2,y_2)$

$(3,8)$ and $(5,2)$

Mid point $=\left(\dfrac{x_1+x_2}{2},\dfrac{y_1+y_2}{2}\right)$

$=\left(\dfrac{3+5}{2},\dfrac{8+2}{2}\right)=(4,5)$

3) $\qquad (x_1,y_2)(x_2,y_2)$

$(7,-3)$ and $(6,-4)$

Mid point $=\left(\dfrac{x_1+x_2}{2},\dfrac{y_1+y_2}{2}\right)$

$=\left(\dfrac{7+6}{2},\dfrac{-3+(-4)}{2}\right)=\left(\dfrac{13}{2},-\dfrac{7}{2}\right)$

5) Center: $(-3,5)$; $r=\sqrt{36}=6$

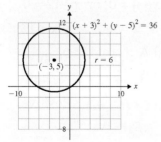

7) $x^2+y^2-10x-4y+13=0$

$\left(x^2-10x\right)+\left(y^2-4y\right)=-13$

$\left(x^2-10x+25\right)+\left(y^2-4y+4\right)=-13+25+4$

$(x-5)^2+(y-2)^2=16$

Center: $(5,2)$; $r=\sqrt{16}=4$

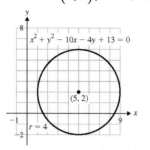

9) $(x-3)^2+(y-0)^2=4^2$

$(x-3)^2+y^2=16$

11) The equation of an ellipse contains the sum of two squares, but the equation of hyperbola contains the difference two squares.

13) Center: $(0,0)$

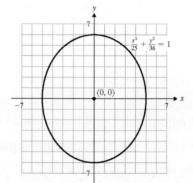

462

15) Center: $(4, 2)$

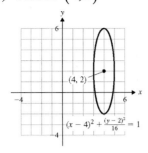

17) Center: $(0, 0)$

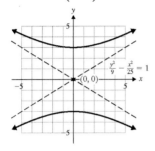

19) Center: $(-1, -2)$

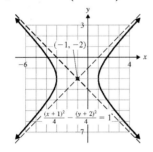

21) $x^2 + 9y^2 = 9$

$$\frac{x^2}{9} + \frac{9y^2}{9} = \frac{9}{9}$$

$$\frac{x^2}{9} + y^2 = 1$$

ellipse

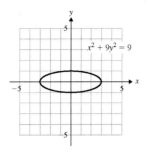

23) parabola

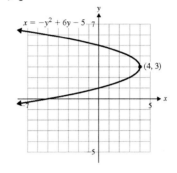

25) hyperbola

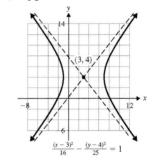

27) $\qquad x^2 + y^2 - 2x + 2y - 2 = 0$

$$\left(x^2 - 2x\right) + \left(y^2 + 2y\right) = 2$$

$$\left(x^2 - 2x + 1\right) + \left(y^2 + 2y + 1\right) = 2 + 1 + 1$$

$$\left(x - 1\right)^2 + \left(y + 1\right)^2 = 4$$

circle

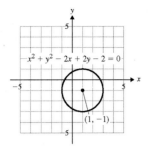

29) parabola

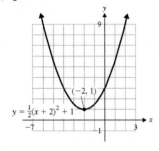

31) Domain:$[-3,3]$; Range:$[0,2]$

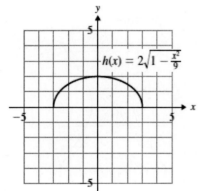

33) $0,1,2,3,$ or 4

35) $-4x^2+3y^2=3$ (1)

     $7x^2-5y^2=7$ (2)

$$-20x^2+15y^2=15 \quad 5\cdot(1)$$
$$\underline{+ \quad 21x^2-15y^2=21 \quad 3\cdot(2)}$$
$$x^2=36$$
$$x=\pm6$$

$$x=6: -4(6)^2+3y^2=3$$
$$-4(36)+3y^2=3$$
$$-144+3y^2=3$$
$$3y^2=147$$
$$y^2=49$$
$$y=\pm7$$
$$x=-6: -4(-6)^2+3y^2=3$$
$$-4(36)+3y^2=3$$
$$-144+3y^2=3$$
$$3y^2=147$$
$$y^2=49$$
$$y=\pm7$$

Verify by substituting into (2).

$$\{(6,7),(6,-7),(-6,7),(-6,-7)\}$$

37) $y=3-x^2$ (1)

       $x-y=-1$ (2)

Substitute (1) into (2).

$$x-(3-x^2)=-1$$
$$x^2+x-3=-1$$
$$x^2+x-2=0$$
$$(x-1)(x+2)=0$$
$$x-1=0 \ \text{ or } \ x+2=0$$
$$x=1 \qquad x=-2$$
$$x=1: y=3-1^2=3-1=2$$
$$x=-2: y=3-(-2)^2=3-4=-1$$

Verify by substituting into (2).

$$\{(1,2),(-2,-1)\}$$

464

39) $4x^2 + 9y^2 = 36$ (1)

$$y = \frac{1}{3}x - 5 \quad (2)$$

Substitute (2) into (1).

$$4x^2 + 9\left(\frac{1}{3}x - 5\right)^2 = 36$$

$$4x^2 + 9\left(\frac{1}{9}x^2 - \frac{10}{3}x + 25\right) = 36$$

$$4x^2 + x^2 - 30x + 225 = 36$$

$$5x^2 - 30x + 189 = 0$$

$$x = \frac{-(-30) \pm \sqrt{(-30)^2 - 4(5)(189)}}{2(5)}$$

$$= \frac{30 \pm \sqrt{900 - 3780}}{10}$$

$$= \frac{30 \pm \sqrt{-2880}}{10}$$

does not give real number solutions.

$\varnothing$

41) $x =$ one number

$y =$ other number

$$xy = 36$$

$$x + y = 13$$

Substitute $y = 13 - x$ into (1).

$$x(13 - x) = 36$$

$$13x - x^2 = 36$$

$$0 = x^2 - 13x + 36$$

$$0 = (x - 9)(x - 4)$$

$x - 9 = 0$ or $x - 4 = 0$

$x = 9$ $x = 4$

$x = 9: y = 13 - 9 = 4$

$x = 4: y = 13 - 4 = 9$

The numbers are 9 and 4.

43) $a^2 + 2a - 3 < 0$

$$(a + 3)(a - 1) < 0$$

$$(a + 3)(a - 1) = 0$$

$a + 3 = 0$ or $a - 1 = 0$

$a = -3$ $a = 1$

Interval A: $(-\infty, -3)$ Positive

Interval B: $(-3, 1)$ Negative

Interval C: $(1, \infty)$ Positive

$(-3, 1)$

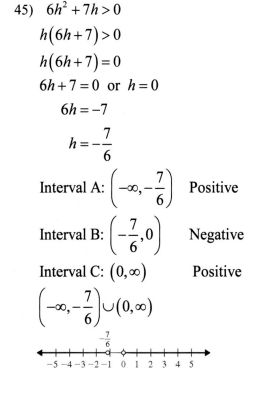

45) $6h^2 + 7h > 0$

$$h(6h + 7) > 0$$

$$h(6h + 7) = 0$$

$6h + 7 = 0$ or $h = 0$

$$6h = -7$$

$$h = -\frac{7}{6}$$

Interval A: $\left(-\infty, -\frac{7}{6}\right)$ Positive

Interval B: $\left(-\frac{7}{6}, 0\right)$ Negative

Interval C: $(0, \infty)$ Positive

$$\left(-\infty, -\frac{7}{6}\right] \cup (0, \infty)$$

47) $\qquad 36 - r^2 > 0$

$(6-r)(6+r) > 0$

$(6-r)(6+r) = 0$

$6 - r = 0$ or $6 + r = 0$

$\qquad -r = -6 \qquad\qquad r = -6$

$\qquad r = 6$

Interval A: $(-\infty, -6)$ Negative

Interval B: $(-6, 6)$ Positive

Interval C: $(6, \infty)$ Negative

$(-6, 6)$

49) $(-\infty, \infty)$

51) $\dfrac{t+7}{2t-3} > 0$

Set the numerator and denominator

equal to zero and solve for t.

$t + 7 = 0 \qquad\qquad 2t - 3 = 0$

$\qquad t = -7 \qquad\qquad 2t = 3$

$\qquad\qquad\qquad\qquad\qquad t = \dfrac{3}{2}$

Interval A: $(-\infty, -7)$ Positive

Interval B: $\left(-7, \dfrac{3}{2}\right)$ Negative

Interval C: $\left(\dfrac{3}{2}, \infty\right)$ Positive

$(-\infty, -7) \cup \left(\dfrac{3}{2}, \infty\right)$

53) $\dfrac{4w+3}{5w-6} \le 0$

Set the numerator and denominator

equal to zero and solve for w.

$4w + 3 = 0 \qquad\quad 5w - 6 = 0$

$\qquad 4w = -3 \qquad\qquad 5w = 6$

$\qquad w = -\dfrac{3}{4} \qquad\qquad w = \dfrac{6}{5}$

Interval A: $\left(-\infty, -\dfrac{3}{4}\right)$ Positive

Interval B: $\left(-\dfrac{3}{4}, \dfrac{6}{5}\right)$ Negative

Interval C: $\left(\dfrac{6}{5}, \infty\right)$ Positive

$w \ne \dfrac{6}{5}$ because it makes the

denominator equal to zero.

$\left[-\dfrac{3}{4}, \dfrac{6}{5}\right)$

55) $\qquad\qquad\qquad \dfrac{1}{n-4} > -3$

$\qquad\qquad \dfrac{1}{n-4} + 3 > 0$

$\dfrac{1}{n-4} + \dfrac{3(n-4)}{n-4} > 0$

$\qquad \dfrac{1 + 3(n-4)}{n-4} > 0$

$\qquad \dfrac{1 + 3n - 12}{n-4} > 0$

$\qquad\qquad \dfrac{3n - 11}{n-4} > 0$

Set the numerator and denominator

equal to zero and solve for n.

$$3n - 11 = 0 \qquad n - 4 = 0$$
$$3n = 11 \qquad\qquad n = 4$$
$$n = \frac{11}{3}$$

Interval A: $\left(-\infty, \dfrac{11}{3}\right)$ Positive

Interval B: $\left(\dfrac{11}{3}, 4\right)$ Negative

Interval C: $(4, \infty)$ Positive

$$\left(-\infty, \frac{11}{3}\right) \cup (4, \infty)$$

57) $\dfrac{r^2 + 4}{r - 7} \geq 0$

Numerator will always be positive.
Set the denominator equal to zero
and solve for r.
$$r - 7 = 0$$
$$r = 7$$

Interval A: $(-\infty, 7)$ Negative

Interval B: $(7, \infty)$ Positive

$r \neq 7$ because it makes the
denominator equal to zero.

$(7, \infty)$

Chapter 14 Test

1) $(x_1, y_1)\ (x_2, y_2)$
       $(2,1)$ and $(10, -6)$

$$\text{Mid point} = \left(\frac{x_1 + x_2}{2}, \frac{y_1 + y_2}{2}\right)$$

$$= \left(\frac{2 + 10}{2}, \frac{1 + (-6)}{2}\right) = \left(6, -\frac{5}{2}\right)$$

3)   parabola

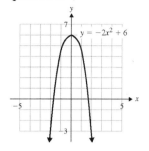

5)   circle

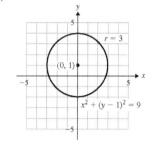

7) $(x - 5)^2 + (y - 2)^2 = \left(\sqrt{11}\right)^2$
     $(x - 5)^2 + (y - 2)^2 = 11$

9)   domain: $[-5, 5]$; range: $[-5, 0]$

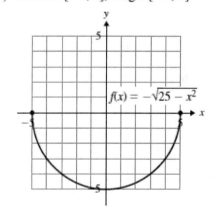

11) $x - 2y^2 = -1$ (1)

    $x + 4y = -1$ (2)

Substitute $x = -4y - 1$ into (1).

$-4y - 1 - 2y^2 = -1$

    $-2y^2 - 4y = 0$

    $-2y(y + 2) = 0$

  $-2y = 0$ or $y + 2 = 0$

    $y = 0$ $y = -2$

$y = 0 : x + 4(0) = -1$

            $x = -1$

$y = -2 : x + 4(-2) = -1$

            $x - 8 = -1$

               $x = 7$

Verify by substituing into (1).

$\{(-1, 0), (7, -2)\}$

13) $x^2 + y^2 = 7$ (1)

   $3x - 2y^2 = 0$ (2)

Solve using elimination.

    $2x^2 + 2y^2 = 14$ $2 \cdot (1)$

  $+ \quad 3x - 2y^2 = 0$ (2)

      $\overline{2x^2 + 3x = 14}$

    $2x^2 + 3x - 14 = 0$

  $(2x + 7)(x - 2) = 0$

  $2x + 7 = 0$ or $x - 2 = 0$

    $2x = -7$ $x = 2$

      $x = -\dfrac{7}{2}$

$x = -\dfrac{7}{2} : 3\left(-\dfrac{7}{2}\right) - 2y^2 = 0$

$-\dfrac{21}{2} - 2y^2 = 0$

   $-\dfrac{21}{2} = 2y^2$

   $-\dfrac{21}{4} = y^2$

   $\pm \dfrac{\sqrt{21}}{2} i = y$

does not give real number solutions.

$x = 2 : 3(2) - 2y^2 = 0$

      $6 - 2y^2 = 0$

         $6 = 2y^2$

         $3 = y^2$

     $\pm\sqrt{3} = y$

Verify by substituting into (1).

$\left\{\left(2, \sqrt{3}\right), \left(2, -\sqrt{3}\right)\right\}$

15) $y^2 + 4y - 45 \geq 0$

  $(y + 9)(y - 5) \geq 0$

  $(y + 9)(y - 5) = 0$

  $y + 9 = 0$ or $y - 5 = 0$

    $y = -9$ $y = 5$

Interval A: $(-\infty, -9)$ Positive

Interval B: $(-9, 5)$ Negative

Interval C: $(5, \infty)$ Positive

$(-\infty, -9] \cup [5, \infty)$

-12 -11 -10 -9 -8 -7 -6 -5 -4 -3 -2 -1 0 1 2 3 4 5 6 7 8

17) $$49 - 9p^2 \le 0$$

$$(7 - 3p)(7 + 3p) \le 0$$

$$(7 - 3p)(7 + 3p) = 0$$

$$7 - 3p = 0 \quad \text{or} \quad 7 + 3p = 0$$

$$-3p = -7 \qquad\qquad 3p = -7$$

$$p = \frac{7}{3} \qquad\qquad p = -\frac{7}{3}$$

Interval A: $\left(-\infty, -\frac{7}{3}\right)$ Negative

Interval B: $\left(-\frac{7}{3}, \frac{7}{3}\right)$ Positive

Interval C: $\left(\frac{7}{3}, \infty\right)$ Negative

$$\left(-\infty, -\frac{7}{3}\right] \cup \left[\frac{7}{3}, \infty\right)$$

19) $$\frac{6}{n-2} > 2$$

$$\frac{6}{n-2} - 2 > 0$$

$$\frac{6}{n-2} - \frac{2(n-2)}{n-2} > 0$$

$$\frac{6 - 2(n-2)}{n-2} > 0$$

$$\frac{-2n + 10}{n-2} > 0$$

Set the numerator and denominator equal to zero and solve for n.

$$-2n + 10 = 0 \qquad\qquad n - 2 = 0$$

$$-2n = -10 \qquad\qquad n = 2$$

$$n = 5$$

Interval A: $(-\infty, 2)$ Negative

Interval B: $(2, 5)$ Positive

Interval C: $(5, \infty)$ Negative

$(2, 5)$

Cumulative Review: Chapters 1-14

1) $\dfrac{1}{6} - \dfrac{11}{12} = \dfrac{2}{12} - \dfrac{11}{12} = -\dfrac{9}{12} = -\dfrac{3}{4}$

3) $A = \dfrac{1}{2}(6)(5) = 3(5) = 15$ sq cm

$P = 7 + 6 + 5.5 = 18.5$ cm

5) $(-1)^5 = -1$

7) $\left(\dfrac{2a^8 b}{a^2 b^{-4}}\right)^{-3} = \left(\dfrac{a^2 b^{-4}}{2a^8 b}\right)^{3}$

$= \left(\dfrac{1}{2} a^{2-8} b^{-4-1}\right)^{3}$

$= \left(\dfrac{1}{2} a^{-6} b^{-5}\right)^{3}$

$= \dfrac{1}{8} a^{-18} b^{-15} = \dfrac{1}{8a^{18} b^{15}}$

9) $an + z = c$

$an = c - z$

$n = \dfrac{c - z}{a}$

11) $n + (n+2) + (n+4) = 2(n+4) + 13$

$3n + 6 = 2n + 8 + 13$

$3n + 6 = 2n + 21$

$n = 15$

$n = 15; n + 2 = 17; n + 4 = 19$

$15, 17, 19$

13)   0

15) $m = \dfrac{1-7}{4-(-4)} = \dfrac{-6}{4+4} = \dfrac{-6}{8} = -\dfrac{3}{4}$

$(y-1) = -\dfrac{3}{4}(x-4)$

$y-1 = -\dfrac{3}{4}x+3$

$y = -\dfrac{3}{4}x+4$

17) x = milliliters of 8% solution
$20-x$ = milliliters of 16% solution
$0.08x + 0.16(20-x) = 0.14(20)$
$100(0.08x + 0.16(20-x)) = 100(0.14(20))$
$8x + 16(20-x) = 14(20)$
$8x + 320 - 16x = 280$
$-8x = -40$
$x = 5$
$20 - x = 15$
8% solution: 5 ml; 16% solution: 15 ml

19) $(4w-3)(2w^2+9w-5)$

$= 8w^3 + 36w^2 - 20w - 6w^2 - 27w + 15$

$= 8w^3 + 30w^2 - 47w + 15$

21) $6c^2 - 14c + 8 = 2(3c^2 - 7c + 4)$

$\qquad\qquad\quad = 2(3c-4)(c-1)$

23) $(x+1)(x+2) = 2(x+7) + 5x$

$x^2 + 3x + 2 = 2x + 14 + 5x$

$x^2 + 3x + 2 = 7x + 14$

$x^2 - 4x - 12 = 0$

$(x-6)(x+2) = 0$

$x - 6 = 0 \text{ or } x + 2 = 0$

$x = 6 \qquad x = -2 \qquad \{-2,6\}$

25) $\dfrac{\dfrac{t^2-9}{4}}{\dfrac{t-3}{24}} = \dfrac{\dfrac{(t+3)(t-3)}{4}}{\dfrac{t-3}{24}}$

$= \dfrac{(t+3)(t-3)}{4} \div \dfrac{t-3}{24}$

$= \dfrac{(t+3)\,(t\!-\!3)}{\cancel{4}} \cdot \dfrac{\cancel{24}^{\,6}}{\cancel{t\!-\!3}}$

$= 6(t+3), \quad t \neq 3$

27) $|5r+3| > 12$

$5r + 3 > 12 \text{ or } 5r + 3 < -12$

$5r > 9 \qquad\qquad 5r < -15$

$r > \dfrac{9}{5} \qquad\qquad r < -3$

$(-\infty, -3) \cup \left(\dfrac{9}{5}, \infty\right)$

29) $\sqrt[3]{48} = \sqrt[3]{8} \cdot \sqrt[3]{6} = 2\sqrt[3]{6}$

31) $(16)^{-3/4} = \left(\dfrac{1}{16}\right)^{3/4} = \left(\sqrt[4]{\dfrac{1}{16}}\right)^3$

$= \left(\dfrac{1}{2}\right)^3 = \dfrac{1}{8}$

33) $\dfrac{5}{\sqrt{3}+4} = \dfrac{5}{\sqrt{3}+4} \cdot \dfrac{\sqrt{3}-4}{\sqrt{3}-4}$

$= \dfrac{5(\sqrt{3}-4)}{(\sqrt{3})^2 - 4^2} = \dfrac{5\sqrt{3}-20}{3-16}$

$= \dfrac{5\sqrt{3}-20}{-13} = \dfrac{20-5\sqrt{3}}{13}$

35)
$$y^2 = -7y - 3$$
$$y^2 + 7y = -3$$
$$y^2 + 7y + \frac{49}{4} = -3 + \frac{49}{4}$$
$$\left(y + \frac{7}{2}\right)^2 = \frac{37}{4}$$
$$y + \frac{7}{2} = \pm\frac{\sqrt{37}}{2}$$
$$y = -\frac{7}{2} \pm \frac{\sqrt{37}}{2}$$
$$\left\{ -\frac{7}{2} - \frac{\sqrt{37}}{2}, -\frac{7}{2} + \frac{\sqrt{37}}{2} \right\}$$

37)

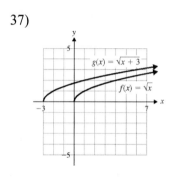

39) a) no b) no

41)
$$8^{5t} = 4^{t-3}$$
$$\left(2^3\right)^{5t} = \left(2^2\right)^{t-3}$$
$$2^{15t} = 2^{2t-6}$$
$$15t = 2t - 6$$
$$13t = -6$$
$$t = -\frac{6}{13} \qquad \left\{ -\frac{6}{13} \right\}$$

43) $\log 100 = \log 10^2 = 2\log 10 = 2(1) = 2$

45)
$$e^{3k} = 8$$
$$\ln e^{3k} = \ln 8$$
$$3k \ln e = \ln 8$$
$$k = \frac{\ln 8}{3} \qquad \left\{\frac{\ln 8}{3}\right\}; \{0.6931\}$$

47)

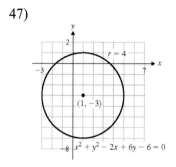

49)
$$25p^2 \le 144$$
$$25p^2 - 144 \le 0$$
$$(5p + 12)(5p - 12) \le 0$$
$$(5p + 12)(5p - 12) = 0$$
$$p = \pm\frac{12}{5}$$

Interval A: $\left(-\infty, -\frac{12}{5}\right)$ Positive

Interval B: $\left(-\frac{12}{5}, \frac{12}{5}\right)$ Negative

Interval C: $\left(\frac{12}{5}, \infty\right)$ Positive

$$\left[-\frac{12}{5}, \frac{12}{5}\right]$$

Section 15.1: Exercises

1) $a_1 = 1+2 = 3$

$a_2 = 2+2 = 4$

$a_3 = 3+2 = 5$

$a_4 = 4+2 = 6$

$a_5 = 5+2 = 7$

3) $a_1 = 3(1)-4 = 3-4 = -1$

$a_2 = 3(2)-4 = 6-4 = 2$

$a_3 = 3(3)-4 = 9-4 = 5$

$a_4 = 3(4)-4 = 12-4 = 8$

$a_5 = 3(5)-4 = 15-4 = 11$

5) $a_1 = 2(1)^2 - 1 = 2(1)-1 = 1$

$a_2 = 2(2)^2 - 1 = 2(4)-1 = 8-1 = 7$

$a_3 = 2(3)^2 - 1 = 2(9)-1 = 18-1 = 17$

$a_4 = 2(4)^2 - 1 = 2(16)-1$

$\quad = 32-1 = 31$

$a_5 = 2(5)^2 - 1 = 2(25)-1 = 50-1 = 49$

7) $a_1 = 3^{1-1} = 3^0 = 1$

$a_2 = 3^{2-1} = 3^1 = 3$

$a_3 = 3^{3-1} = 3^2 = 9$

$a_4 = 3^{4-1} = 3^3 = 27$

$a_5 = 3^{5-1} = 3^4 = 81$

9) $a_1 = 5 \cdot \left(\dfrac{1}{2}\right)^1 = 5 \cdot \dfrac{1}{2} = \dfrac{5}{2}$

$a_2 = 5 \cdot \left(\dfrac{1}{2}\right)^2 = 5 \cdot \dfrac{1}{4} = \dfrac{5}{4}$

$a_3 = 5 \cdot \left(\dfrac{1}{2}\right)^3 = 5 \cdot \dfrac{1}{8} = \dfrac{5}{8}$

$a_4 = 5 \cdot \left(\dfrac{1}{2}\right)^4 = 5 \cdot \dfrac{1}{16} = \dfrac{5}{16}$

$a_5 = 5 \cdot \left(\dfrac{1}{2}\right)^5 = 5 \cdot \dfrac{1}{32} = \dfrac{5}{32}$

11) $a_1 = (-1)^{1+1} \cdot 7(1) = 1 \cdot 7 = 7$

$a_2 = (-1)^{2+1} \cdot 7(2) = -1 \cdot 14 = -14$

$a_3 = (-1)^{3+1} \cdot 7(3) = 1 \cdot 21 = 21$

$a_4 = (-1)^{4+1} \cdot 7(4) = -1 \cdot 28 = -28$

$a_5 = (-1)^{5+1} \cdot 7(5) = 1 \cdot 35 = 35$

13) $a_1 = \dfrac{1-4}{1+3} = \dfrac{-3}{4} = -\dfrac{3}{4}$

$a_2 = \dfrac{2-4}{2+3} = \dfrac{-2}{5} = -\dfrac{2}{5}$

$a_3 = \dfrac{3-4}{3+3} = \dfrac{-1}{6} = -\dfrac{1}{6}$

$a_4 = \dfrac{4-4}{4+3} = \dfrac{0}{7} = 0$

$a_5 = \dfrac{5-4}{5+3} = \dfrac{1}{8}$

15) a) $a_1 = 3(1)+2 = 5$

b) $a_5 = 3(5)+2 = 15+2 = 17$

c) $a_{28} = 3(28)+2 = 84+2 = 86$

17) a) $a_1 = \dfrac{1-4}{1+6} = \dfrac{-3}{7} = -\dfrac{3}{7}$

    b) $a_2 = \dfrac{2-4}{2+6} = \dfrac{-2}{8} = -\dfrac{1}{4}$

    c) $a_{16} = \dfrac{16-4}{16+6} = \dfrac{12}{22} = \dfrac{6}{11}$

19) a) $a_1 = 10 - 1^2 = 10 - 1 = 9$

    b) $a_6 = 10 - 6^2 = 10 - 36 = -26$

    c) $a_{20} = 10 - 20^2 = 10 - 400 = -390$

21) $a_n = 2n$

23) $a_n = n^2$

25) $a_n = \left(\dfrac{1}{3}\right)^n$

27) $a_n = \dfrac{n}{n+1}$

29) $a_n = (-1)^{n+1} \cdot (5n)$

31) $a_n = (-1)^n \cdot \left(\dfrac{1}{2}\right)^n$

33) $2592 - \dfrac{1}{3}(2592) = 2592 - 864 = 1728$

    $1728 - \dfrac{1}{3}(1728) = 1728 - 576 = 1152$

    $1152 - \dfrac{1}{3}(1152) = 1152 - 384 = 768$

    $768 - \dfrac{1}{3}(768) = 768 - 256 = 512$

    $1728, \$1152, \$768, \$512$

35) $100 + 10 = 110$

    $110 + 10 = 120$

    $120 + 10 = 130$

    $130 + 10 = 140$

    $140 + 10 = 150$

    $150 + 10 = 160$ 160 lb

37) A sequence is a list of terms in a certain order, and a series is a sum of the terms of a sequence.

39) $\displaystyle\sum_{i=1}^{6}(2i+1) = (2(1)+1) + (2(2)+1)$

    $+ (2(3)+1) + (2(4)+1)$

    $+ (2(5)+1) + (2(6)+1)$

    $= 3 + 5 + 7 + 9 + 11 + 13$

    $= 48$

41) $\displaystyle\sum_{i=1}^{5}(i-8) = (1-8) + (2-8) + (3-8)$

    $+ (4-8) + (5-8)$

    $= -7 - 6 - 5 - 4 - 3 = -25$

43) $\displaystyle\sum_{i=1}^{4}\left(4i^2-2i\right)$

$=\left(4(1)^2-2(1)\right)+\left(4(2)^2-2(2)\right)$

$+\left(4(3)^2-2(3)\right)+\left(4(4)^2-2(4)\right)$

$=(4-2)+(16-4)$

$+(36-6)+(64-8)$

$=2+12+30+56=100$

45) $\displaystyle\sum_{i=1}^{6}\frac{i}{2}=\frac{1}{2}+\frac{2}{2}+\frac{3}{2}+\frac{4}{2}+\frac{5}{2}+\frac{6}{2}=\frac{21}{2}$

47) $\displaystyle\sum_{i=1}^{5}(-1)^{i+1}\cdot(i)=(-1)^{1+1}\cdot 1+(-1)^{2+1}\cdot 2$

$+(-1)^{3+1}\cdot 3+(-1)^{4+1}\cdot 4$

$+(-1)^{5+1}\cdot 5$

$=1-2+3-4+5=3$

49) $\displaystyle\sum_{i=5}^{9}(i-2)=(5-2)+(6-2)+(7-2)$

$+(8-2)+(9-2)$

$=3+4+5+6+7=25$

51) $\displaystyle\sum_{i=3}^{6}\left(i^2\right)=3^2+4^2+5^2+6^2$

$=9+16+25+36=86$

53) $\displaystyle\sum_{i=1}^{5}\frac{1}{i}$ 55) $\displaystyle\sum_{i=1}^{4}(3i)$

57) $\displaystyle\sum_{i=1}^{6}(i+4)$

59) $\displaystyle\sum_{i=1}^{7}(-1)^{i}\cdot(i)$

60) $\displaystyle\sum_{i=1}^{4}(-1)^{i+1}\cdot\left(3^i\right)$

63) $\bar{x}=\dfrac{19+24+20+17+23+17}{6}=20$

65) $\bar{x}=\dfrac{8+7+11+9+12}{5}=9.4$

67) $\bar{x}=\dfrac{1431.60+1117.82+985.43+1076.22+900.00+813.47}{6}$

$=\$1054.09$

Section 15.2: Exercises

1) It is a list of numbers in a specific order such that the difference between any two successive terms is the same. 1, 4, 7, 10,...

3) yes, $d=11-3=8$

5) yes, $d=6-10=-4$

7) no

9) yes, $d=-14-(-17)=3$

11) $a_1=7$

$a_2=7+2=9$

$a_3=9+2=11$

$a_4=11+2=13$

$a_5=13+2=15$

7, 9, 11, 13, 15

13) $a_1 = 15$
$a_2 = 15 - 8 = 7$
$a_3 = 7 - 8 = -1$
$a_4 = -1 - 8 = -9$
$a_5 = -9 - 8 = -17$
15, 7, −1, −9, −17

15) $a_1 = -10$
$a_2 = -10 + 3 = -7$
$a_3 = -7 + 3 = -4$
$a_4 = -4 + 3 = -1$
$a_5 = -1 + 3 = 2$
−10, −7, −4, −1, 2

17) $a_1 = 6(1) + 7 = 6 + 7 = 13$
$a_2 = 6(2) + 7 = 12 + 7 = 19$
$a_3 = 6(3) + 7 = 18 + 7 = 25$
$a_4 = 6(4) + 7 = 24 + 7 = 31$
$a_5 = 6(5) + 7 = 30 + 7 = 37$
13, 19, 25, 31, 37

19) $a_1 = 5 - 1 = 4$
$a_2 = 5 - 2 = 3$
$a_3 = 5 - 3 = 2$
$a_4 = 5 - 4 = 1$
$a_5 = 5 - 5 = 0$
4, 3, 2, 1, 0

21) a) $a_1 = 4; d = 7 - 4 = 3$
b) $a_n = 4 + (n-1)3$
$a_n = 4 + 3n - 3$
$a_n = 3n + 1$
c) $a_{35} = 3(35) + 1 = 105 + 1 = 106$

23) a) $a_1 = 4; d = -1 - 4 = -5$
b) $a_n = 4 + (n-1)(-5)$
$a_n = 4 - 5n + 5$
$a_n = -5n + 9$
c) $a_{19} = -5(19) + 9 = -95 + 9 = -86$

25)
$a_n = -7 + (n-1)(2)$
$a_n = -7 + 2n - 2$
$a_n = 2n - 9$
$a_{25} = 2(25) - 9 = 50 - 9 = 41$

27) $a_n = 1 + (n-1)\left(\frac{1}{2}\right)$
$a_n = 1 + \frac{1}{2}n - \frac{1}{2}$
$a_n = \frac{1}{2}n + \frac{1}{2}$
$a_{18} = \frac{1}{2}(18) + \frac{1}{2} = \frac{18}{2} + \frac{1}{2} = \frac{19}{2}$

29) $a_n = 0 + (n-1)(-5)$
$a_n = -5n + 5$
$a_{23} = -5(23) + 5 = -115 + 5 = -110$

31) $a_{16} = -5 + (16-1)(4)$
$= -5 + (15)(4) = -5 + 60 = 55$

33) $a_{21} = -7 + (21-1)(-5)$

$\quad = -7 + (20)(-5)$

$\quad = -7 - 100 = -107$

35) $a_3 = a_1 + (3-1)d$

$\quad 11 = a_1 + (3-1)d$

$\quad 11 = a_1 + 2d \qquad (1)$

$\quad a_7 = a_1 + (7-1)d$

$\quad 19 = a_1 + (7-1)d$

$\quad 19 = a_1 + 6d \qquad (2)$

Solve using elimination.

$\quad -11 = -a_1 - 2d \qquad -1 \cdot (1)$

$\underline{+ \quad 19 = a_1 + 6d \qquad \quad (2)}$

$\qquad 8 = 4d$

$\qquad 2 = d$

$d = 2 : 11 = a_1 + 2(2)$

$\qquad 11 = a_1 + 4$

$\qquad 7 = a_1$

$a_n = 7 + (n-1)2$

$\quad = 7 + 2n - 2 = 2n + 5$

$a_{11} = 2(11) + 5 = 22 + 5 = 27$

37) $a_2 = a_1 + (2-1)d$

$\quad 7 = a_1 + (2-1)d$

$\quad 7 = a_1 + d \qquad (1)$

$\quad a_6 = a_1 + (6-1)d$

$\quad -13 = a_1 + (6-1)d$

$\quad -13 = a_1 + 5d \qquad (2)$

Solve using elimination.

$\quad -7 = -a_1 - d \qquad -1 \cdot (1)$

$\underline{+ \quad -13 = a_1 + 5d \qquad \quad (2)}$

$\quad -20 = 4d$

$\quad -5 = d$

$d = -5 : \ 7 = a_1 + (-5)$

$\qquad 12 = a_1$

$\quad a_n = 12 + (n-1)(-5)$

$\qquad = 12 - 5n + 5 = -5n + 17$

$\quad a_{14} = -5(14) + 17 = -70 + 17 = -53$

39) $a_4 = a_1 + (4-1)d$

$\quad -5 = a_1 + (4-1)d$

$\quad -5 = a_1 + 3d \qquad (1)$

$\quad a_{11} = a_1 + (11-1)d$

$\quad 16 = a_1 + (11-1)d$

$\quad 16 = a_1 + 10d \qquad (2)$

Solve using elimination.

$\quad 5 = -a_1 - 3d \qquad -1 \cdot (1)$

$\underline{+ \quad 16 = a_1 + 10d \qquad \quad (2)}$

$\quad 21 = 7d$

$\quad 3 = d$

$d = 3 : \ -5 = a_1 + 3(3)$

$\qquad -5 = a_1 + 9$

$\qquad -14 = a_1$

$\quad a_n = -14 + (n-1)(3)$

$\qquad = -14 + 3n - 3 = 3n - 17$

$\quad a_{18} = 3(18) - 17 = 54 - 17 = 37$

41) $a_n = 63, \ a_1 = 8, \ d = 13 - 8 = 5$

$\quad 63 = 8 + (n-1)(5)$

$\quad 63 = 8 + 5n - 5$

$\quad 63 = 3 + 5n$

$\quad 60 = 5n$

$\quad 12 = n$

43) $a_n = -27,\ a_1 = 9,\ d = 7-9 = -2$

$-27 = 9 + (n-1)(-2)$

$-27 = 9 - 2n + 2$

$-27 = 11 - 2n$

$-38 = -2n$

$19 = n$

45) S_{15} is the sum of the first 15 terms of the sequence.

47) $S_{10} = \dfrac{10}{2}(14+68) = 5(82) = 410$

49) $S_7 = \dfrac{7}{2}(3+(-9)) = \dfrac{7}{2}(-6) = -21$

51) $S_8 = \dfrac{8}{2}(-1+(-29)) = 4(-30) = -120$

53) $S_8 = \dfrac{8}{2}[2(3)+(8-1)5]$

$= 4(6+35) = 4(41) = 164$

55) $S_8 = \dfrac{8}{2}[2(10)+(8-1)(-6)]$

$= 4(20-42) = 4(-22) = -88$

57) $a_1 = -4(1)-1 = -4-1 = -5$

$a_8 = -4(8)-1 = -32-1 = -33$

$S_8 = \dfrac{8}{2}(-5+(-33)) = 4(-38) = -152$

59) $a_1 = 3(1)+4 = 3+4 = 7$

$a_8 = 3(8)+4 = 24+4 = 28$

$S_8 = \dfrac{8}{2}(7+28) = 4(35) = 140$

61) a) $\displaystyle\sum_{i=1}^{10}(2i+7)$

$= (2(1)+7)+(2(2)+7)$

$+(2(3)+7)+(2(4)+7)$

$+(2(5)+7)+(2(6)+7)$

$+(2(7)+7)+(2(8)+7)$

$+(2(9)+7)+(2(10)+7)$

$= 9+11+13+15+17+$

$19+21+23+25+27 = 180$

b) $a_1 = 2(1)+7 = 2+7 = 9$

$a_{10} = 2(10)+7 = 20+7 = 27$

$S_{10} = \dfrac{10}{2}(9+27) = 5(36) = 180$

c) Answers may vary.

63) $a_1 = 8(1)-5 = 8-5 = 3$

$a_5 = 8(5)-5 = 40-5 = 35$

$S_5 = \dfrac{5}{2}(3+35) = \dfrac{5}{2}(38) = 95$

65) $a_1 = -2(1)+7 = -2+7 = 5$

$a_7 = -2(7)+7 = -14+7 = -7$

$S_7 = \dfrac{7}{2}(5+(-7)) = \dfrac{7}{2}(-2) = -7$

67) $a_1 = 3(1) - 11 = 3 - 11 = -8$

$a_{18} = 3(18) - 11 = 54 - 11 = 43$

$S_{18} = \dfrac{18}{2}(-8 + 43) = 9(35) = 315$

69) $a_1 = 1$

$a_{500} = 500$

$S_{500} = \dfrac{500}{2}(1 + 500)$

$= 250(501) = 125,250$

71) $a_1 = \$1500, \; d = \$100$

$a_n = 1500 + (n-1)100$

$a_n = 1500 + 100n - 100$

$a_n = 100n + 1400$

$a_9 = 100(9) + 1400$

$= 900 + 1400 = \$2300$

73) $a_1 = \$1, \; d = \$1$

$S_{24} = \dfrac{24}{2}\left[2(1) + (24-1)(1)\right]$

$= 12(2 + 23) = 12(25) = \$300$

75) a) $a_1 = 12, \; d = -1$

$a_n = 12 + (n-1)(-1)$

$a_n = 12 - n + 1$

$a_n = 13 - n$

$a_8 = 13 - 8 = 5$ logs

b) $1 = 13 - n$

$n = 12$

The stack has 12 rows of logs.

$S_{12} = \dfrac{12}{2}(12 + 1) = 6(13) = 78$ logs

77) $a_1 = 12, \; d = 2$

$a_n = 12 + (n-1)(2)$

$a_n = 12 + 2n - 2$

$a_n = 10 + 2n$

$a_{14} = 10 + 2(14)$

$= 10 + 28 = 38$ seats in last row

$S_{14} = \dfrac{14}{2}(12 + 38)$

$= 7(50) = 350$ seats

79) $S_n = 860, \; a_1 = 24, \; a_n = 62$

$860 = \dfrac{n}{2}(24 + 62)$

$860 = \dfrac{n}{2}(86)$

$860 = 43n$

$20 = n$

Section 15.3: Exercises

1) A sequence is arithmetic if the difference between consecutive terms is constant, but a sequence is geometric if each term after the first is obtained by multiplying the preceding term by a common ratio.

3) $r = \dfrac{2}{1} = 2$

5) $r = \dfrac{3}{9} = \dfrac{1}{3}$

7) $r = \dfrac{\frac{1}{2}}{-2} = -\dfrac{1}{4}$

9) $a_1 = 2$

$a_2 = 2(5) = 10$

$a_3 = 10(5) = 50$

$a_4 = 50(5) = 250$

$a_5 = 250(5) = 1250$

11) $a_1 = \dfrac{1}{4}$

$a_2 = \dfrac{1}{4}(-2) = -\dfrac{1}{2}$

$a_3 = -\dfrac{1}{2}(-2) = 1$

$a_4 = 1(-2) = -2$

$a_5 = -2(-2) = 4$

13) $a_1 = 72$

$a_2 = 72\left(\dfrac{2}{3}\right) = 48$

$a_3 = 48\left(\dfrac{2}{3}\right) = 32$

$a_4 = 32\left(\dfrac{2}{3}\right) = \dfrac{64}{3}$

$a_5 = \dfrac{64}{3}\left(\dfrac{2}{3}\right) = \dfrac{128}{9}$

15) $a_n = 4(7)^{n-1}$

$a_3 = 4(7)^{3-1} = 4(7)^2 = 4(49) = 196$

17) $a_n = -1(3)^{n-1}$

$a_5 = -1(3)^{5-1} = -1(81) = -81$

19) $a_n = 2\left(\dfrac{1}{5}\right)^{n-1}$

$a_4 = 2\left(\dfrac{1}{5}\right)^{4-1} = 2\left(\dfrac{1}{5}\right)^3$

$= 2\left(\dfrac{1}{125}\right) = \dfrac{2}{125}$

21) $a_n = -\dfrac{1}{2}\left(-\dfrac{3}{2}\right)^{n-1}$

$a_4 = -\dfrac{1}{2}\left(-\dfrac{3}{2}\right)^{4-1} = -\dfrac{1}{2}\left(-\dfrac{3}{2}\right)^3$

$= -\dfrac{1}{2}\left(-\dfrac{27}{8}\right) = \dfrac{27}{16}$

23) $a_1 = 5, \; r = \dfrac{10}{5} = 2$

$a_n = 5(2)^{n-1}$

25) $a_1 = -3, \; r = \dfrac{-\dfrac{3}{5}}{-3} = \dfrac{1}{5}$

$a_n = -3\left(\dfrac{1}{5}\right)^{n-1}$

27) $a_1 = 3, \; r = \dfrac{-6}{3} = -2$

$a_n = 3(-2)^{n-1}$

29) $a_1 = \dfrac{1}{3}, \; r = \dfrac{\dfrac{1}{12}}{\dfrac{1}{3}} = \dfrac{1}{4}$

$a_n = \dfrac{1}{3}\left(\dfrac{1}{4}\right)^{n-1}$

Chapter 15: Sequences and Series

31) $a_1 = 1, r = \dfrac{2}{1} = 2$

$a_n = 1(2)^{n-1} = 2^{n-1}$

$a_{12} = 2^{12-1} = 2^{11} = 2048$

33) $a_1 = 27, r = \dfrac{-9}{27} = -\dfrac{1}{3}$

$a_n = 27\left(-\dfrac{1}{3}\right)^{n-1}$

$a_8 = 27\left(-\dfrac{1}{3}\right)^{8-1} = 27\left(-\dfrac{1}{3}\right)^7$

$= 3^3 \cdot 3^{-7} \cdot (-1)^7 = -3^{-4} = -\dfrac{1}{81}$

35) $a_1 = -\dfrac{1}{64}, r = \dfrac{-\dfrac{1}{32}}{-\dfrac{1}{64}} = 2$

$a_n = -\dfrac{1}{64}(2)^{n-1}$

$a_{12} = -\dfrac{1}{64}(2)^{12-1} = -2^{-6} \cdot 2^{11}$

$= -2^5 = -32$

37) arithmetic

$a_n = 15 + (n-1)(9)$

$a_n = 15 + 9n - 9$

$a_n = 9n + 6$

39) geometric

$a_n = -2(-3)^{n-1}$

41) geometric

$a_n = \dfrac{1}{9}\left(\dfrac{1}{2}\right)^{n-1}$

43) arithmetic

$a_n = -31 + (n-1)(7)$

$a_n = -31 + 7n - 7$

$a_n = 7n - 38$

45) a) $a_1 = 40,000, r = 1 - 0.20 = 0.80$

$a_n = 40,000(0.80)^{n-1}$

b) $a_5 = 40,000(0.80)^{5-1}$

$= 40,000(0.80)^4 = \$16,384$

47) a) $a_1 = 500,000, r = 1 - 0.10 = 0.90$

$a_n = 500,000(0.90)^{n-1}$

b) $a_4 = 500,000(0.90)^{4-1}$

$= 500,000(0.90)^3 = \$364,500$

49) a) $a_1 = 160,000, r = 1 + 0.04 = 1.04$

$a_n = 160,000(1.04)^{n-1}$

b) $a_6 = 160,000(1.04)^{6-1}$

$= 160,000(1.04)^5 \approx \$194,664$

51) $S_6 = \dfrac{9(1-2^6)}{1-2} = \dfrac{9(1-64)}{-1}$

$= -9(-63) = 567$

53) $a_1 = 7$, $r = \dfrac{28}{7} = 4$, $n = 7$

$$S_7 = \frac{7(1-4^7)}{1-4} = \frac{7(1-16,384)}{-3}$$

$$= \frac{7(-16,383)}{-3} = 7(5461)$$

$$= 38,227$$

55) $a_1 = -\dfrac{1}{4}$, $r = \dfrac{-\dfrac{1}{2}}{-\dfrac{1}{4}} = 2$, $n = 6$

$$S_6 = \frac{-\dfrac{1}{4}(1-2^6)}{1-2} = \frac{-\dfrac{1}{4}(1-64)}{-1}$$

$$= \frac{1}{4}(-63) = -\frac{63}{4}$$

57) $a_1 = 1$, $r = \dfrac{\dfrac{1}{3}}{1} = \dfrac{1}{3}$, $n = 5$

$$S_5 = \frac{1\left(1-\left(\dfrac{1}{3}\right)^5\right)}{1-\dfrac{1}{3}} = \frac{\left(1-\dfrac{1}{243}\right)}{\dfrac{2}{3}}$$

$$= \frac{3}{2}\left(\frac{242}{243}\right) = \frac{121}{81}$$

59) $a_1 = 18$, $r = 2$, $n = 7$

$$S_7 = \frac{18(1-2^7)}{1-2} = \frac{18(1-128)}{-1}$$

$$= -18(-127) = 2286$$

61) $a_1 = -12$, $r = 3$, $n = 5$

$$S_5 = \frac{-12(1-3^5)}{1-3} = \frac{-12(1-243)}{-2}$$

$$= 6(-242) = -1452$$

63) $a_1 = -\dfrac{3}{2}$, $r = -\dfrac{1}{2}$, $n = 6$

$$S_6 = \frac{-\dfrac{3}{2}\left(1-\left(-\dfrac{1}{2}\right)^6\right)}{1-\left(-\dfrac{1}{2}\right)}$$

$$= \frac{-\dfrac{3}{2}\left(1-\dfrac{1}{64}\right)}{\dfrac{3}{2}} = -\left(\frac{63}{64}\right) = -\frac{63}{64}$$

65) $a_1 = 12$, $r = -\dfrac{2}{3}$, $n = 4$

$$S_4 = \frac{12\left(1-\left(-\dfrac{2}{3}\right)^4\right)}{1-\left(-\dfrac{2}{3}\right)} = \frac{12\left(1-\dfrac{16}{81}\right)}{\dfrac{5}{3}}$$

$$= 12\left(\frac{65}{81}\right)\frac{3}{5} = \frac{52}{9}$$

67) a) $a_1 = 1$, $r = 2$

$$a_n = 1(2)^{n-1}$$

$$a_{10} = 1(2)^{10-1} = 2^9 = 512$$

b) $S_{10} = \dfrac{1(1-2^{10})}{1-2} = \dfrac{(1-1024)}{-1}$

$$= -(-1023) = 1023 = \$10.23$$

69) $S = \dfrac{8}{1-\dfrac{1}{4}} = \dfrac{8}{\dfrac{3}{4}} = 8 \cdot \dfrac{4}{3} = \dfrac{32}{3}$

71) $S = \dfrac{5}{1-\left(-\dfrac{4}{5}\right)} = \dfrac{5}{\dfrac{9}{5}} = 5 \cdot \dfrac{5}{9} = \dfrac{25}{9}$

73) $\left|\dfrac{5}{3}\right| > 1$. The sum does not exist.

75) $a_1 = 8,\ r = \dfrac{\dfrac{16}{3}}{8} = \dfrac{2}{3}$

$S = \dfrac{8}{1-\dfrac{2}{3}} = \dfrac{8}{\dfrac{1}{3}} = 8 \cdot 3 = 24$

77) $a_1 = -\dfrac{15}{2},\ r = \dfrac{\dfrac{15}{4}}{-\dfrac{15}{2}} = -\dfrac{1}{2}$

$S = \dfrac{-\dfrac{15}{2}}{1-\left(-\dfrac{1}{2}\right)} = \dfrac{-\dfrac{15}{2}}{\dfrac{3}{2}}$

$= -\dfrac{15}{2} \cdot \dfrac{2}{3} = -5$

79) $a_1 = \dfrac{1}{25},\ r = \dfrac{\dfrac{1}{5}}{\dfrac{1}{25}} = 5$

$|5| > 1$. The sum does not exist.

81) $a_1 = -40,\ r = \dfrac{-30}{-40} = \dfrac{3}{4}$

$S = \dfrac{-40}{1-\dfrac{3}{4}} = -\dfrac{40}{\dfrac{1}{4}} = -40 \cdot 4 = -160$

83) $a_1 = 3,\ r = 0.75$

$S = \dfrac{3}{1-0.75} = \dfrac{3}{0.25} = 12$ ft

85) a) $a_1 = 27,\ r = \dfrac{2}{3}$

$a_n = 27\left(\dfrac{2}{3}\right)^{n-1}$

$a_5 = 27\left(\dfrac{2}{3}\right)^{6-1} = 27\left(\dfrac{2}{3}\right)^{5}$

$= 27 \cdot \dfrac{32}{243} = 3\dfrac{5}{9}$ ft

b) $S = \dfrac{27}{1-\dfrac{2}{3}} = \dfrac{27}{\dfrac{1}{3}} = 81$

The ball travels this distance twice minus the inital height.

$d = 2 \cdot 81 - 27 = 162 - 27 = 135$ ft

Section 15.4: Exercises

1) Answers may vary.

3) $(r+s)^3 = r^3 + 3r^2 s + 3rs^2 + s^3$

5) $(y+z)^5 = y^5 + 5y^4 z + 10y^3 z^2$
$+ 10y^2 z^3 + 5yz^4 + z^5$

7) $(x+5)^4 = x^4 + 4x^3(5) + 6x^2(5)^2$
$\qquad + 4x(5)^3 + 5^4$
$\qquad = x^4 + 20x^3 + 150x^2$
$\qquad\quad + 500x + 625$

9) Answers may vary.

11) $2! = 2 \cdot 1 = 2$

13) $5! = 5 \cdot 4 \cdot 3 \cdot 2 \cdot 1 = 120$

15) $\binom{5}{2} = \dfrac{5!}{2!(5-2)!} = \dfrac{5!}{2!3!} = \dfrac{5 \cdot 4 \cdot \cancel{3!}}{(2 \cdot 1)\cancel{3!}}$
$\qquad = \dfrac{20}{2} = 10$

17) $\binom{7}{3} = \dfrac{7!}{3!(7-3)!} = \dfrac{7!}{3!4!} = \dfrac{7 \cdot 6 \cdot 5 \cdot \cancel{4!}}{(3 \cdot 2 \cdot 1)\cancel{4!}}$
$\qquad = \dfrac{210}{6} = 35$

19) $\binom{10}{4} = \dfrac{10!}{4!(10-4)!} = \dfrac{10!}{4!6!}$
$\qquad = \dfrac{10 \cdot 9 \cdot 8 \cdot 7 \cdot \cancel{6!}}{(4 \cdot 3 \cdot 2 \cdot 1)\cancel{6!}} = \dfrac{5040}{24} = 210$

21) $\binom{9}{7} = \dfrac{9!}{7!(9-7)!} = \dfrac{9!}{7!2!}$
$\qquad = \dfrac{9 \cdot 8 \cdot \cancel{7!}}{\cancel{7!}(2 \cdot 1)} = \dfrac{72}{2} = 36$

23) $\binom{4}{4} = 1$

25) $\binom{6}{1} = \dfrac{6!}{1!(6-1)!} = \dfrac{6!}{1!5!} = \dfrac{6 \cdot \cancel{5!}}{1 \cdot \cancel{5!}} = 6$

27) $\binom{5}{0} = 1$

29) 10

31) $(f+g)^3 = f^3 + \binom{3}{1}f^{3-1}g + \binom{3}{2}f^{3-2}g^2 + g^3$
$\qquad = f^3 + 3f^2g + 3fg^2 + g^3$

33) $(w+2)^4 = w^4 + \binom{4}{1}w^{4-1}(2) + \binom{4}{2}w^{4-2}$
$\qquad (2)^2 + \binom{4}{3}w^{4-3}(2)^3 + 2^4$
$\qquad = w^4 + 4w^3(2) + 6w^2(4) + 4w(8)$
$\qquad + 16$
$\qquad = w^4 + 8w^3 + 24w^2 + 32w + 16$

35) $(b+3)^5 = b^5 + \binom{5}{1}b^{5-1}(3) + \binom{5}{2}b^{5-2}(3)^2$
$\qquad + \binom{5}{3}b^{5-3}(3)^3 + \binom{5}{4}b^{5-4}(3)^4 + 3^5$
$\qquad = b^5 + 5b^4(3) + 10b^3(9) + 10b^2(27)$
$\qquad + 5b(81) + 243$
$\qquad = b^5 + 15b^4 + 90b^3 + 270b^2 + 405b + 243$

37) $\left[a+(-3)\right]^4 = a^4 + \binom{4}{1}a^{4-1}(-3) + \binom{4}{2}a^{4-2}(-3)^2 + \binom{4}{3}a^{4-3}(-3)^3 + (-3)^4$

$$= a^4 + 4a^3(-3) + 6a^2(9) + 4a(-27) + 81 = a^4 - 12a^3 + 54a^2 - 108a + 81$$

39) $\left[u+(-v)\right]^3 = u^3 + \binom{3}{1}u^{3-1}(-v) + \binom{3}{2}u^{3-2}(-v)^2 + (-v)^3$

$$= u^3 + 3u^2(-v) + 3uv^2 - v^3 = u^3 - 3u^2v + 3uv^2 - v^3$$

41) $(3m+2)^4 = (3m)^4 + \binom{4}{1}(3m)^{4-1}(2) + \binom{4}{2}(3m)^{4-2}(2)^2 + \binom{4}{3}(3m)^{4-3}(2)^3 + 2^4$

$$= 81m^4 + 4(3m)^3(2) + 6(3m)^2(4) + 4(3m)(8) + 16$$
$$= 81m^4 + 4(27m^3)(2) + 6(9m^2)(4) + 4(3m)(8) + 16$$
$$= 81m^4 + 216m^3 + 216m^2 + 96m + 16$$

43) $\left[3a+(-2b)\right]^5$

$$= (3a)^5 + \binom{5}{1}(3a)^{5-1}(-2b) + \binom{5}{2}(3a)^{5-2}(-2b)^2 + \binom{5}{3}(3a)^{5-3}(-2b)^3 + \binom{5}{4}(3a)^{5-4}(-2b)^4 + (-2b)^5$

$$= 243a^5 + 5(3a)^4(-2b) + 10(3a)^3(4b^2) + 10(3a)^2(-8b^3) + 5(3a)(16b^4) - 32b^5$$
$$= 243a^5 + 5(81a^4)(-2b) + 10(27a^3)(4b^2) + 10(9a^2)(-8b^3) + 5(3a)(16b^4) - 32b^5$$
$$= 243a^5 - 810a^4b + 1080a^3b^2 - 720a^2b^3 + 240ab^4 - 32b^5$$

45) $(x^2+1)^3 = (x^2)^3 + \binom{3}{1}(x^2)^{3-1}(1) + \binom{3}{2}(x^2)^{3-2}(1)^2 + 1^3$

$$= x^6 + 3(x^2)^2 + 3(x^2) + 1 = x^6 + 3x^4 + 3x^2 + 1$$

47) $\left[\frac{1}{2}m+(-3n)\right]^4$

$$= \left(\frac{1}{2}m\right)^4 + \binom{4}{1}\left(\frac{1}{2}m\right)^{4-1}(-3n) + \binom{4}{2}\left(\frac{1}{2}m\right)^{4-2}(-3n)^2 + \binom{4}{3}\left(\frac{1}{2}m\right)^{4-3}(-3n)^3 + (-3n)^4$

$$= \frac{1}{16}m^4 + 4\left(\frac{1}{2}m\right)^3(-3n) + 6\left(\frac{1}{2}m\right)^2(9n^2) + 4\left(\frac{1}{2}m\right)(-27n^3) + 81n^4$$
$$= \frac{1}{16}m^4 + 4\left(\frac{1}{8}m^3\right)(-3n) + 6\left(\frac{1}{4}m^2\right)(9n^2) + 4\left(\frac{1}{2}m\right)(-27n^3) + 81n^4$$
$$= \frac{1}{16}m^4 - \frac{3}{2}m^3n + \frac{27}{2}m^2n^2 - 54mn^3 + 81n^4$$

49) $\left(\frac{1}{3}y+2z^2\right)^3 = \left(\frac{1}{3}y\right)^3 + \binom{3}{1}\left(\frac{1}{3}y\right)^{3-1}\left(2z^2\right) = \frac{1}{27}y^3 + 3\left(\frac{1}{3}y\right)^2\left(2z^2\right)$

$+ \binom{3}{2}\left(\frac{1}{3}y\right)^{3-2}\left(2z^2\right)^2 + \left(2z^2\right)^3 + 3\left(\frac{1}{3}y\right)\left(4z^4\right) + 8z^6$

$= \frac{1}{27}y^3 + \frac{2}{3}y^2z^2 + 4yz^4 + 8z^6$

51) $a=k, b=5, n=8, r=3$

$\frac{8!}{(8-3+1)!(3-1)!}k^{8-3+1}5^{3-1} = \frac{8!}{6!2!}k^6 5^2 = 28k^6(25) = 700k^6$

53) $a=w, b=1, n=15, r=10$

$\frac{15!}{(15-10+1)!(10-1)!}w^{15-10+1}1^{10-1} = \frac{15!}{6!9!}w^6 1^9 = 5005w^6$

55) $a=q, b=-3, n=9, r=2$

$\frac{9!}{(9-2+1)!(2-1)!}q^{9-2+1}(-3)^{2-1} = \frac{9!}{8!1!}q^8(-3)^1 = 9q^8(-3) = -27q^8$

57) $a=3x, b=-2, n=6, r=5$

$\frac{6!}{(6-5+1)!(5-1)!}(3x)^{6-5+1}(-2)^{5-1} = \frac{6!}{2!4!}(3x)^2(-2)^4 = 15(9x^2)(16) = 2160x^2$

59) $a=2y^2, b=z, n=10, r=8$

$\frac{10!}{(10-8+1)!(8-1)!}(2y^2)^{10-8+1}(z)^{8-1} = \frac{10!}{3!7!}(2y^2)^3(z)^7 = 120(8y^6)(z^7) = 960y^6z^7$

61) $a=c^3, b=-3d^2, n=7, r=3$

$\frac{7!}{(7-3+1)!(3-1)!}(c^3)^{7-3+1}(-3d^2)^{3-1} = \frac{7!}{5!2!}(c^3)^5(-3d^2)^2 = 21c^{15}(9d^4) = 189c^{15}d^4$

63) v^{33}

65) $\dbinom{n}{n} = \dfrac{n!}{n!(n-n)!} = \dfrac{1}{0!} = \dfrac{1}{1} = 1$

Chapter 15 Review

1) $a_1 = 7(1)+1 = 7+1 = 8$
$a_2 = 7(2)+1 = 14+1 = 15$
$a_3 = 7(3)+1 = 21+1 = 22$
$a_4 = 7(4)+1 = 28+1 = 29$
$a_5 = 7(5)+1 = 35+1 = 36$

3) $a_1 = (-1)^{1+1} \cdot \left(\dfrac{1}{1^2}\right) = (-1)^2 \cdot (1) = 1 \cdot 1 = 1$

$a_2 = (-1)^{2+1} \cdot \left(\dfrac{1}{2^2}\right) = (-1)^3 \cdot \left(\dfrac{1}{4}\right)$
$= -1 \cdot \dfrac{1}{4} = -\dfrac{1}{4}$

$a_3 = (-1)^{3+1} \cdot \left(\dfrac{1}{3^2}\right) = (-1)^4 \cdot \left(\dfrac{1}{9}\right)$
$= 1 \cdot \dfrac{1}{9} = \dfrac{1}{9}$

$a_4 = (-1)^{4+1} \cdot \left(\dfrac{1}{4^2}\right) = (-1)^5 \cdot \left(\dfrac{1}{16}\right)$
$= -1 \cdot \dfrac{1}{16} = -\dfrac{1}{16}$

$a_5 = (-1)^{5+1} \cdot \left(\dfrac{1}{5^2}\right) = (-1)^6 \cdot \left(\dfrac{1}{25}\right)$
$= 1 \cdot \dfrac{1}{25} = \dfrac{1}{25}$

5) $a_n = 5n$

7) $a_n = -\dfrac{n+1}{n}$

9) $8.25+0.25 = 8.50$
$8.50+0.25 = 8.75$
$8.75+0.25 = 9.00$
$\$8.25, \$8.50, \$8.75, \$9.00$

11) A sequence is a list of terms in a certain order, and a series is a sum of the terms of a sequence.

13) $\displaystyle\sum_{i=1}^{5}(2i^2+1) = \left(2(1)^2+1\right)+\left(2(2)^2+1\right)$
$+\left(2(3)^2+1\right)+\left(2(4)^2+1\right)$
$+\left(2(5)^2+1\right)$
$= 3+9+19+33+51 = 115$

15) $\displaystyle\sum_{i=1}^{4}\dfrac{13}{i}$

17) $\bar{x} = \dfrac{18+25+26+20+22}{5} = 22.2$

19) $a_1 = 11$
$a_2 = 11+7 = 18$
$a_3 = 18+7 = 25$
$a_4 = 25+7 = 32$
$a_5 = 32+7 = 39$
11, 18, 25, 32, 39

21) $a_1 = -58$
$a_2 = -58+8 = -50$
$a_3 = -50+8 = -42$
$a_4 = -42+8 = -34$
$a_5 = -34+8 = -26$
$-58, -50, -42, -34, -26$

23) a) $a_1 = 6;\ d = 10 - 6 = 4$

b) $a_n = 6 + (n-1)4$

$a_n = 6 + 4n - 4$

$a_n = 4n + 2$

c) $a_{20} = 4(20) + 2 = 80 + 2 = 82$

25) a) $a_1 = -8;\ d = -13 - (-8) = -5$

b) $a_n = -8 + (n-1)(-5)$

$a_n = -8 - 5n + 5$

$a_n = -5n - 3$

c) $a_{20} = -5(20) - 3$

$= -100 - 3 = -103$

27) $a_1 = -15,\ d = -9 - (-15) = 6$

$a_n = -15 + (n-1)(6)$

$a_n = -15 + 6n - 6$

$a_n = 6n - 21$

$a_{15} = 6(15) - 21 = 90 - 21 = 69$

29) $a_n = -4 + (n-1)\left(-\dfrac{3}{2}\right)$

$a_n = -4 - \dfrac{3}{2}n + \dfrac{3}{2}$

$a_n = -\dfrac{3}{2}n - \dfrac{5}{2}$

$a_{21} = -\dfrac{3}{2}(21) - \dfrac{5}{2} = -\dfrac{63}{2} - \dfrac{5}{2} = -34$

31) $a_6 = a_1 + (6-1)d$

$24 = a_1 + (6-1)d$

$24 = a_1 + 5d \qquad (1)$

$a_9 = a_1 + (9-1)d$

$36 = a_1 + (9-1)d$

$36 = a_1 + 8d \qquad (2)$

Solve using elimination.

$-24 = -a_1 - 5d \qquad -1 \cdot (1)$

$+ \quad 36 = a_1 + 8d \qquad (2)$

$\overline{\qquad\qquad\qquad\qquad}$

$12 = 3d$

$4 = d$

$d = 4:\ 24 = a_1 + 5(4)$

$24 = a_1 + 20$

$4 = a_1$

$a_n = 4 + (n-1)4$

$= 4 + 4n - 4 = 4n$

$a_{12} = 4(12) = 48$

33) $a_5 = a_1 + (5-1)d$

$-5 = a_1 + (5-1)d$

$-5 = a_1 + 4d \qquad (1)$

$a_{10} = a_1 + (10-1)d$

$-15 = a_1 + (10-1)d$

$-15 = a_1 + 9d \qquad (2)$

Solve using elimination.

$5 = -a_1 - 4d \qquad -1 \cdot (1)$

$+ \ -15 = a_1 + 9d \qquad (2)$

$\overline{\qquad\qquad\qquad\qquad}$

$-10 = 5d$

$-2 = d$

$d = -2:\ -5 = a_1 + 4(-2)$

$-5 = a_1 - 8$

$3 = a_1$

$a_n = 3 + (n-1)(-2)$

$= 3 - 2n + 2 = -2n + 5$

$a_{17} = -2(17) + 5 = -34 + 5 = -29$

35) $a_n = 34$, $a_1 = -4$, $d = -2 - (-4) = 2$

$34 = -4 + (n-1)(2)$

$34 = -4 + 2n - 2$

$34 = -6 + 2n$

$40 = 2n$

$20 = n$

37) $a_n = 43$, $a_1 = -8$, $d = -7 - (-8) = 1$

$43 = -8 + (n-1)(1)$

$43 = -8 + n - 1$

$43 = -9 + n$

$52 = n$

39) $S_8 = \dfrac{8}{2}(5 + (-27)) = 4(-22) = -88$

41) $S_{10} = \dfrac{10}{2}[2(-6) + (10-1)(7)]$

$= 5(-12 + 63) = 5(51) = 255$

43) $S_{10} = \dfrac{10}{2}(13 + (-59))$

$= 5(-46) = -230$

45) $a_1 = 2(1) - 5 = 2 - 5 = -3$

$a_{10} = 2(10) - 5 = 20 - 5 = 15$

$S_{10} = \dfrac{10}{2}(-3 + 15) = 5(12) = 60$

47) $a_1 = -7(1) + 16 = -7 + 16 = 9$

$a_{10} = -7(10) + 16 = -70 + 16 = -54$

$S_{10} = \dfrac{10}{2}(9 + (-54))$

$= 5(-45) = -225$

49) $a_1 = -11(1) - 2 = -11 - 2 = -13$

$a_4 = -11(4) - 2 = -44 - 2 = -46$

$S_4 = \dfrac{4}{2}(-13 + (-46))$

$= 2(-59) = -118$

51) $a_1 = 3(1) + 4 = 3 + 4 = 7$

$a_{13} = 3(13) + 4 = 39 + 4 = 43$

$_{13} = \dfrac{13}{2}(7 + 43) = \dfrac{13}{2}(50) = 325$

53) $a_1 = -4(1) + 2 = -4 + 2 = -2$

$a_{11} = -4(11) + 2 = -44 + 2 = -42$

$S_{11} = \dfrac{11}{2}(-2 + (-42))$

$= \dfrac{11}{2}(-44) = -242$

55) $a_1 = 15$, $d = 2$

$a_n = 15 + (n-1)(2)$

$a_n = 15 + 2n - 2$

$a_n = 13 + 2n$

$a_{20} = 13 + 2(20)$

$= 13 + 40 = 53$ seats

$S_{20} = \dfrac{20}{2}(15 + 53) = 10(68)$

$= 680$ seats

57) $a_1 = \$2$, $d = \$2$

$S_{30} = \dfrac{30}{2}[2(2) + (30-1)(2)]$

$= 15(4 + 58) = 15(62) = \$930$

59) $r = \dfrac{20}{4} = 5$

61) $a_1 = 7$

$a_2 = 7(2) = 14$

$a_3 = 14(2) = 28$

$a_4 = 28(2) = 56$

$a_5 = 56(2) = 112$

63) $a_1 = 48$

$a_2 = 48\left(\dfrac{1}{4}\right) = 12$

$a_3 = 12\left(\dfrac{1}{4}\right) = 3$

$a_4 = 3\left(\dfrac{1}{4}\right) = \dfrac{3}{4}$

$a_5 = \dfrac{3}{4}\left(\dfrac{1}{4}\right) = \dfrac{3}{16}$

65) $a_n = 3(2)^{n-1}$

$a_6 = 3(2)^{6-1} = 3(2)^5 = 3(32) = 96$

67) $a_n = 8\left(\dfrac{1}{3}\right)^{n-1}$

$a_4 = 8\left(\dfrac{1}{3}\right)^{4-1} = 8\left(\dfrac{1}{3}\right)^3 = 8\left(\dfrac{1}{27}\right) = \dfrac{8}{27}$

69) $a_1 = 7,\ r = \dfrac{42}{7} = 6$

$a_n = 7(6)^{n-1}$

71) $a_1 = -15,\ r = \dfrac{45}{-15} = -3$

$a_n = (-15)(-3)^{n-1}$

73) $a_1 = 1,\ r = \dfrac{3}{1} = 3$

$a_n = (1)(3)^{n-1} = 3^{n-1}$

$a_8 = 3^{8-1} = 3^7 = 2187$

75) $a_1 = 8,\ r = 3,\ n = 5$

$S_5 = \dfrac{8(1-3^5)}{1-3} = \dfrac{8(1-243)}{-2}$

$= -4(-242) = 968$

77) $a_1 = 8,\ r = \dfrac{40}{8} = 5,\ n = 5$

$S_5 = \dfrac{8(1-5^5)}{1-5} = \dfrac{8(1-3125)}{-4}$

$= -2(-3124) = 6248$

79) $a_1 = 8,\ r = \dfrac{4}{8} = \dfrac{1}{2},\ n = 6$

$S_6 = \dfrac{8\left(1-\left(\dfrac{1}{2}\right)^6\right)}{1-\dfrac{1}{2}} = \dfrac{8\left(1-\dfrac{1}{64}\right)}{\dfrac{1}{2}}$

$= 16\left(\dfrac{63}{64}\right) = \dfrac{63}{4}$

81) $a_1 = \dfrac{7}{2},\ r = \dfrac{1}{2},\ n = 5$

$S_5 = \dfrac{\dfrac{7}{2}\left(1-\left(\dfrac{1}{2}\right)^5\right)}{1-\dfrac{1}{2}} = \dfrac{\dfrac{7}{2}\left(1-\dfrac{1}{32}\right)}{\dfrac{1}{2}}$

$= 7\left(\dfrac{31}{32}\right) = \dfrac{217}{32}$

83) a) $a_1 = 28,000$, $r = 1 - 0.20 = 0.80$

$$a_n = 28,000(0.80)^{n-1}$$

b) $a_3 = 28,000(0.80)^{3-1}$

$$= 28,000(0.80)^2 = \$17,920$$

85) a) arithmetic

b) $a_1 = 7$, $d = 9 - 7 = 2$

$$a_n = 7 + (n-1)2$$
$$= 7 + 2n - 2 = 2n + 5$$

c) $S_8 = \dfrac{8}{2}\big[2(7) + (8-1)(2)\big]$

$$= 4(14 + 14) = 4(28) = 112$$

87) a) geometric

b) $a_1 = 9$, $r = \dfrac{\frac{9}{2}}{9} = \dfrac{1}{2}$

$$a_n = 9\left(\dfrac{1}{2}\right)^{n-1}$$

c) $S_8 = \dfrac{9\left(1 - \left(\frac{1}{2}\right)^8\right)}{1 - \frac{1}{2}} = \dfrac{9\left(1 - \frac{1}{256}\right)}{\frac{1}{2}}$

$$= 18\left(\dfrac{255}{256}\right) = \dfrac{2295}{128}$$

89) a) geometric

b) $a_1 = -1$, $r = \dfrac{-3}{-1} = 3$

$$a_n = (-1)(3)^{n-1}$$

c) $S_8 = \dfrac{-1(1 - 3^8)}{1 - 3} = \dfrac{-(1 - 6561)}{-2}$

$$= \dfrac{-6560}{2} = -3280$$

91) $a_1 = 2000$, $r = 1.5$

$$a_n = 2000(1.5)^{n-1}$$

$$a_5 = 2000(1.5)^{5-1} = 2000(1.5)^4$$

$$= 10,125 \text{ bacteria}$$

93) when $|r| < 1$

95) $S = \dfrac{-3}{1 - \frac{1}{8}} = \dfrac{-3}{\frac{7}{8}} = -3 \cdot \dfrac{8}{7} = -\dfrac{24}{7}$

97) $a_1 = -15$, $r = \dfrac{10}{-15} = -\dfrac{2}{3}$

$$S = \dfrac{-15}{1 - \left(-\frac{2}{3}\right)} = \dfrac{-15}{\frac{5}{3}} = -15 \cdot \dfrac{3}{5} = -9$$

99) $a_1 = -4$, $r = \dfrac{12}{-4} = -3$

$|-3| > 1$. The sum does not exist.

101) $(y + z)^4 = y^4 + 4y^3 z + 6y^2 z^2$
$$+ 4yz^3 + z^4$$

103) $6! = 6 \cdot 5 \cdot 4 \cdot 3 \cdot 2 \cdot 1 = 720$

105) $\dbinom{5}{3} = \dfrac{5!}{3!(5-3)!} = \dfrac{5!}{3!2!} = \dfrac{5 \cdot 4 \cdot \cancel{3!}}{\cancel{3!}(2 \cdot 1)}$

$$= \dfrac{20}{2} = 10$$

107) $\dbinom{9}{1} = \dfrac{9!}{1!(9-1)!} = \dfrac{9!}{1!8!} = \dfrac{9 \cdot \cancel{8!}}{1 \cdot \cancel{8!}}$

$$= \dfrac{9}{1} = 9$$

109) $(m+n)^4 = m^4 + \binom{4}{1}m^{4-1}n + \binom{4}{2}m^{4-2}n^2 + \binom{4}{3}m^{4-3}n^3 + n^4 = m^4 + 4m^3n + 6m^2n^2 + 4mn^3 + n^4$

111) $\left[h+(-9)\right]^3 = h^3 + \binom{3}{1}h^{3-1}(-9) + \binom{3}{2}h^{3-2}(-9)^2 + (-9)^3$

$= h^3 + 3h^2(-9) + 3h(81) - 729$

$= h^3 - 27h^2 + 243h - 729$

113) $\left[2p^2 + (-3r)\right]^5$

$= (2p^2)^5 + \binom{5}{1}(2p^2)^{5-1}(-3r) + \binom{5}{2}(2p^2)^{5-2}(-3r)^2 + \binom{5}{3}(2p^2)^{5-3}(-3r)^3$

$+ \binom{5}{4}(2p^2)^{5-4}(-3r)^4 + (-3r)^5$

$= 32p^{10} + 5(2p^2)^4(-3r) + 10(2p^2)^3(9r^2) + 10(2p^2)^2(-27r^3) + 5(2p^2)(81r^4) - 243r^5$

$= 32p^{10} + 5(16p^8)(-3r) + 10(8p^6)(9r^2) + 10(4p^4)(-27r^3) + 5(2p^2)(81r^4) - 243r^5$

$= 32p^{10} - 240p^8 r + 720p^6 r^2 - 1080p^4 r^3 + 810p^2 r^4 - 243r^5$

115) $a = z,\ b = 4,\ n = 8,\ r = 5$

$\dfrac{8!}{(8-5+1)!(5-1)!}z^{8-5+1}4^{5-1} = \dfrac{8!}{4!4!}z^4 4^4$

$= 70z^4(256) = 17{,}920z^4$

117) $a = 2k,\ b = -1,\ n = 13,\ r = 11$

$\dfrac{13!}{(13-11+1)!(11-1)!}(2k)^{13-11+1}(-1)^{11-1}$

$= \dfrac{13!}{3!10!}(2k)^3(-1)^{10} = 286(8k^3) = 2288k^3$

Chapter 15 Test

1) $a_1 = 2(1) - 3 = 2 - 3 = -1$

$a_2 = 2(2) - 3 = 4 - 3 = 1$

$a_3 = 2(3) - 3 = 6 - 3 = 3$

$a_4 = 2(4) - 3 = 8 - 3 = 5$

$a_5 = 2(5) - 3 = 10 - 7 = 7$

3) An arithmetic sequence is obtained by adding the common difference to each term to obtain the next term, while a geometric sequence is obtained by muliplying a term by the common ratio to get the next term

5) $d = -11 - (-17) = 6$

7) arithmetic

$a_1 = 5$, $d = 2 - 5 = -3$

$a_n = 5 + (n-1)(-3) = 5 - 3n + 3 = -3n + 8$

9) $\displaystyle\sum_{i=1}^{4} (5i^2 + 6) = (5(1)^2 + 6) + (5(2)^2 + 6) + (5(3)^2 + 6) + (5(4)^2 + 6) = 11 + 26 + 51 + 86 = 174$

11) $S_{11} = \dfrac{11}{2}[2(5) + (11-1)(3)] = \dfrac{11}{2}(10 + 30) = \dfrac{11}{2}(40) = 220$

13) $S = \dfrac{7}{1 - \dfrac{3}{10}} = \dfrac{7}{\dfrac{7}{10}} = 7 \cdot \dfrac{10}{7} = 10$

15) $a_1 = 11,000$, $d = 1 + 0.10 = 1.10$

$a_n = 11,000(1.10)^{n-1}$

$a_4 = 11,000(1.10)^{4-1} = 11,000(1.10)^3 \approx 14,641$

17) $5! = 5 \cdot 4 \cdot 3 \cdot 2 \cdot 1 = 120$

19) $(3x+1)^4 = (3x)^4 + \dbinom{4}{1}(3x)^{4-1}(1) + \dbinom{4}{2}(3x)^{4-2}(1)^2 + \dbinom{4}{3}(3x)^{4-3}(1)^3 + (1)^4$

$= (3x)^4 + 4(3x)^3 + 6(3x)^2 + 4(3x) + 1$

$= 81x^4 + 4(27x^3) + 6(9x^2) + 12x + 1$

$= 8x^4 + 108x^3 + 54x^2 + 12x + 1$

1) $\dfrac{5}{8}+\dfrac{1}{6}+\dfrac{3}{4}=\dfrac{15}{24}+\dfrac{4}{24}+\dfrac{18}{24}=\dfrac{37}{24}$

3) a) $\left(-9k^4\right)^2=(-9)^2k^{4\cdot2}=81k^8$

b) $\left(-7z^3\right)\left(8z^{-9}\right)=-56z^{3+(-9)}$

$=-56z^{-6}=-\dfrac{56}{z^6}$

c) $\left(\dfrac{40a^{-7}b^3}{8ab^{-2}}\right)^{-3}=\left(\dfrac{8ab^{-2}}{40a^{-7}b^3}\right)^3$

$=\left(\dfrac{1}{5}a^{1-(-7)}b^{-2-3}\right)^3$

$=\left(\dfrac{a^8}{5b^5}\right)^3$

$=\dfrac{a^{8\cdot3}}{5^3b^{5\cdot3}}=\dfrac{a^{24}}{125b^{15}}$

5) a) $\dfrac{4}{3}y-7=13$

$4y-21=39$

$4y=60$

$y=15 \qquad \{15\}$

b) $3(n-4)+11=5n-2(3n+8)$

$3n-12+11=5n-6n-16$

$3n-1=-n-16$

$4n=-15$

$n=-\dfrac{15}{4}$

$\left\{-\dfrac{15}{4}\right\}$

7) $x=$ amt invested at 4%

$y=$ amt invested at 6%

$x+y=9000;\ x=9000-y$

$0.04x+0.06y=480$

$0.04(9000-y)+0.06y=480$

$4(9000-y)+6y=48{,}000$

$36{,}000-4y+6y=48{,}000$

$2y=12{,}000$

$y=6{,}000$

$x=9000-6000=3000$

$\$3000@4\%$ and $\$6000@6\%$

9) $2c+9<3\ $ or $\ c-7>-2$

$2c<-6 \qquad c>5$

$c<-3$

$(-\infty,-3)\cup(5,\infty)$

11) $m=\dfrac{2-5}{6-(-3)}=\dfrac{-3}{9}=-\dfrac{1}{3}$

$y-2=-\dfrac{1}{3}(x-6)$

$y-2=-\dfrac{1}{3}x+2$

$y=-\dfrac{1}{3}x+4$

13) $3x+5y=12 \qquad (1)$

$2x-3y=8 \qquad (2)$

Solve using elimination.

$-6x-10y=-24 \qquad -2\cdot(1)$

$+\quad 6x-9y=24 \qquad 3\cdot(2)$

$-19y=0$

$y=0$

$y=0:3x+5(0)=12$

$3x=12$

$x=4 \qquad (4,0)$

15) a) $\left(9m^3 - 7m^2 + 3m + 2\right)$

$\qquad -\left(4m^3 + 11m^2 - 7m - 1\right)$

$\qquad = 9m^3 - 7m^2 + 3m + 2$

$\qquad \quad -4m^3 - 11m^2 + 7m + 1$

$\qquad = 5m^3 - 18m^2 + 10m + 3$

b) $5(2p+3) + \dfrac{2}{3}(4p-9)$

$\qquad = 10p + 15 + \dfrac{8}{3}p - 6 = \dfrac{38}{3}p + 9$

17) a) $3x-2\overline{)\,12x^3 + 7x^2 - 37x + 18}$ quotient $4x^2 + 5x - 9$

$\qquad \underline{-\;(12x^3 - 8x^2)}$

$\qquad \qquad 15x^2 - 37x$

$\qquad \qquad \underline{-(15x^2 - 10x)}$

$\qquad \qquad \qquad -27x + 18$

$\qquad \qquad \qquad \underline{-\;(-27x + 18)}$

$\qquad \qquad \qquad \qquad 0$

b) $\dfrac{20a^3b^3 - 45a^2b + 10ab + 60}{10ab}$

$\qquad = \dfrac{20a^3b^3}{10ab} - \dfrac{45a^2b}{10ab} + \dfrac{10ab}{10ab} + \dfrac{60}{10ab}$

$\qquad = 2a^2b^2 - \dfrac{9a}{2} + 1 + \dfrac{6}{ab}$

19) a) $m^2 - 15m + 54 = 0$

$\qquad (m-6)(m-9) = 0$

$\qquad m - 6 = 0 \text{ or } m - 9 = 0$

$\qquad \quad m = 6 \qquad \quad m = 9 \quad \{6, 9\}$

b) $\qquad \qquad x^3 = 3x^2 + 28x$

$\qquad x^3 - 3x^2 - 28x = 0$

$\qquad x\left(x^2 - 3x - 28\right) = 0$

$\qquad x(x-7)(x+4) = 0$

$x - 7 = 0 \text{ or } x + 4 = 0 \text{ or } x = 0$

$\qquad x = 7 \qquad \qquad x = -4$

$\qquad \{-4, 0, 7\}$

21) $\dfrac{k}{k^2 - 11k + 18} + \dfrac{k+2}{2k^2 - 17k - 9}$

$\qquad = \dfrac{k}{(k-9)(k-2)} + \dfrac{k+2}{(2k+1)(k-9)}$

$\qquad = \dfrac{k(2k+1) + (k+2)(k-2)}{(2k+1)(k-2)(k-9)}$

$\qquad = \dfrac{2k^2 + k + k^2 - 4}{(2k+1)(k-2)(k-9)}$

$\qquad = \dfrac{3k^2 + k - 4}{(2k+1)(k-2)(k-9)}$

$\qquad = \dfrac{(3k+4)(k-1)}{(2k+1)(k-2)(k-9)}$

23) $\qquad \dfrac{1}{a+1} - \dfrac{a}{6} = \dfrac{a-4}{a+1}$ Multiply by $6(a+1)$

$\qquad 6 - a(a+1) = 6(a-4)$

$\qquad 6 - a^2 - a = 6a - 24$

$\qquad 0 = a^2 + 7a - 30$

$\qquad 0 = (a+10)(a-3)$

$\qquad a + 10 = 0 \text{ or } a - 3 = 0$

$\qquad a = -10 \qquad a = 3 \qquad \{-10, 3\}$

25) $\left|\dfrac{1}{4}t - 5\right| \le 2$

$\qquad -2 \le \dfrac{1}{4}t - 5 \le 2$

$\qquad 3 \le \dfrac{1}{4}t \le 7$

$\qquad 12 \le t \le 28$

$\qquad [12, 28]$

27) $\begin{bmatrix} -1 & 3 & 3 & | & 5 \\ 3 & -2 & -1 & | & -9 \\ -3 & 1 & 3 & | & 5 \end{bmatrix} \xrightarrow[\substack{3R_1+R_2\to R_2 \\ -3R_1+R_3\to R_3}]{} \begin{bmatrix} -1 & 3 & 3 & | & 5 \\ 0 & 7 & 8 & | & 6 \\ 0 & -8 & -6 & | & -10 \end{bmatrix} \xrightarrow[\substack{-R_1\to R_1 \\ -\frac{1}{2}R_3\to R_3}]{}$

$\begin{bmatrix} 1 & -3 & -3 & | & -5 \\ 0 & 7 & 8 & | & 6 \\ 0 & 4 & 3 & | & 5 \end{bmatrix} \xrightarrow[-4R_2+7R_3\to R_3]{} \begin{bmatrix} 1 & -3 & -3 & | & -5 \\ 0 & 7 & 8 & | & 6 \\ 0 & 0 & -11 & | & 11 \end{bmatrix} \xrightarrow[\substack{\frac{1}{7}R_2\to R_2 \\ -\frac{1}{11}R_3\to R_3}]{} \begin{bmatrix} 1 & -3 & -3 & | & -5 \\ 0 & 1 & \frac{8}{7} & | & \frac{6}{7} \\ 0 & 0 & 1 & | & -1 \end{bmatrix}$

$x - 3y - 3z = -5$

$y + \frac{8}{7}z = \frac{6}{7}$

$z = -1$

$y + \frac{8}{7}z = \frac{6}{7}$

$y + \frac{8}{7}(-1) = \frac{6}{7}$

$y - \frac{8}{7} = \frac{6}{7}$

$y = 2$

$x - 3y - 3z = -5$

$x - 3(2) - 3(-1) = -5$

$x - 6 + 3 = -5$

$x - 3 = -5$

$x = -2$

$(-2, 2, -1)$

29) a) $\sqrt{63} = \sqrt{9} \cdot \sqrt{7} = 3\sqrt{7}$

b) $\sqrt[4]{48} = \sqrt[4]{16} \cdot \sqrt[4]{3} = 2\sqrt[4]{3}$

c) $\sqrt{20x^2 y^9} = 2\sqrt{5} \cdot x \cdot y^4 \sqrt{y}$

$= 2xy^4 \sqrt{5y}$

d) $\sqrt[3]{250c^{17}d^{12}} = 5\sqrt[3]{2} \cdot c^5 \sqrt[3]{c^2} \cdot d^4$

$= 5c^5 d^4 \sqrt[3]{2c^2}$

31) a) $\frac{5}{\sqrt{t}} = \frac{5}{\sqrt{t}} \cdot \frac{\sqrt{t}}{\sqrt{t}} = \frac{5\sqrt{t}}{t}$

b) $\frac{4}{\sqrt{6}-2} = \frac{4}{\sqrt{6}-2} \cdot \frac{\sqrt{6}+2}{\sqrt{6}+2}$

$= \frac{4\sqrt{6}+8}{6-4} = \frac{4(\sqrt{6}+2)}{2}$

$= 2(\sqrt{6}+2) = 2\sqrt{6}+4$

c) $\frac{n}{\sqrt[3]{4}} = \frac{n}{\sqrt[3]{2^2}} \cdot \frac{\sqrt[3]{2}}{\sqrt[3]{2}} = \frac{n\sqrt[3]{2}}{\sqrt[3]{2^3}} = \frac{n\sqrt[3]{2}}{2}$

33) $\sqrt{-16} = \sqrt{-1} \cdot \sqrt{16} = i \cdot 4 = 4i$

35) a) $9h^2 + 2h + 1 = 0$

$h = \frac{-2 \pm \sqrt{2^2 - 4(9)(1)}}{2(9)}$

$= \frac{-2 \pm \sqrt{4 - 36}}{18}$

$= \frac{-2 \pm \sqrt{-32}}{18}$

$= \frac{-2 \pm 4i\sqrt{2}}{18} = -\frac{1}{9} \pm \frac{2\sqrt{2}}{9}i$

$\left\{ -\frac{1}{9} - \frac{2\sqrt{2}}{9}i, \ -\frac{1}{9} + \frac{2\sqrt{2}}{9}i \right\}$

b) $(w+11)^2 + 4 = 0$

$(w+11)^2 = -4$

$w + 11 = \pm 2i$

$w = -11 \pm 2i$

$\{-11 - 2i, \ -11 + 2i\}$

c) $(b+4)^2 - (b+4) = 12$

$(b+4)^2 - (b+4) - 12 = 0$

Let $u = b+4;\ u - 4 = b$

$u^2 - u - 12 = 0$

$(u-4)(u+3) = 0$

$u - 4 = 0$ or $u + 3 = 0$

$u = 4$	$u = -3$
$b = u - 4$	$b = u - 4$
$= 4 - 4$	$= -3 - 4$
$= 0$	$= -7$

$\{-7, 0\}$

d) $k^4 + 15 = 8k^2$

$k^4 - 8k^2 + 15 = 0$

$(k^2 - 5)(k^2 - 3) = 0$

$k^2 - 5 = 0$ or $k^2 - 3 = 0$

$k^2 = 5$	$k^2 = 3$
$k = \pm\sqrt{5}$	$k = \pm\sqrt{3}$

$\{-\sqrt{5}, -\sqrt{3}, \sqrt{3}, \sqrt{5}\}$

37) a) $5x - 10 = 0$

$5x = 10$

$x = 2 \qquad (-\infty, 2) \cup (2, \infty)$

b) $2x + 3 \geq 0$

$2x \geq -3$

$x \geq -\dfrac{3}{2} \qquad \left[-\dfrac{3}{2}, \infty\right)$

39) a) (-3, 1)

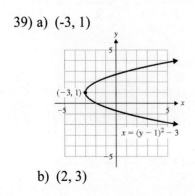

b) (2, 3)

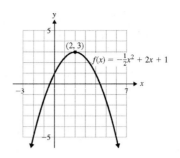

41) a) Let $\log_2 16 = x$

$2^x = 16$

$2^x = 2^4$

$x = 4$

b) Let $\log 100 = x$

$10^x = 100$

$10^x = 10^2$

$x = 2$

c) $\ln e = 1$

d) Let $\log_3 \sqrt{3} = x$

$3^x = \sqrt{3}$

$3^x = 3^{1/2}$

$x = \dfrac{1}{2}$

43) a) $5^{2y} = 125^{y+4}$

$5^{2y} = (5^3)^{y+4}$

$5^{2y} = 5^{3y+12}$

$2y = 3y + 12$

$y = -12 \qquad \{-12\}$

b) $\qquad 4^{x-3} = 3^{2x}$

$\ln 4^{x-3} = \ln 3^{2x}$

$(x-3)\ln 4 = 2x \ln 3$

$x \ln 4 - 3 \ln 4 = 2x \ln 3$

$x \ln 4 - 2x \ln 3 = 3 \ln 4$

$x(\ln 4 - 2 \ln 3) = 3 \ln 4$

$x = \dfrac{3 \ln 4}{\ln 4 - 2 \ln 3}$

$$\left\{\frac{3\ln 4}{\ln 4 - 2\ln 3}\right\};\{-5.13\}$$

c) $e^{-6t} = 8$

$$\ln e^{-6t} = \ln 8$$

$$-6t \ln e = \ln 8$$

$$-6t = \ln 8$$

$$t = -\frac{\ln 8}{6}$$

$$\left\{-\frac{\ln 8}{6}\right\};\{-0.35\}$$

45)

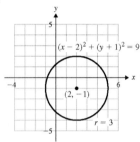

$(x-2)^2 + (y+1)^2 = 9$

$(2,-1)$

$r = 3$

47) a) $w^2 + 5w > -6$

$$w^2 + 5w + 6 > 0$$

$$(w+3)(w+2) > 0$$

$$(w+3)(w+2) = 0$$

$$w + 3 = 0 \quad \text{or} \quad w + 2 = 0$$

$$w = -3 \qquad w = -2$$

Interval A: $(-\infty, -3)$ Positive

Interval B: $(-3, -2)$ Negative

Interval C: $(-2, \infty)$ Positive

$$(-\infty, -3) \cup (-2, \infty)$$

b) $\dfrac{c+1}{c+7} \le 0$

Set the numerator and

denominator equal to zero

and solve for c.

$$c + 1 = 0 \qquad c + 7 = 0$$

$$c = -1 \qquad c = -7$$

Interval A: $(-\infty, -7)$ Positive

Interval B: $(-7, -1)$ Negative

Interval C: $(-1, \infty)$ Positive

$c \ne -7$ because it makes the

denominator equal to zero.

$$(-7, -1]$$

49) $S_6 = \dfrac{48\left(1 - \left(\dfrac{1}{2}\right)^6\right)}{1 - \dfrac{1}{2}}$

$$= \dfrac{48\left(1 - \dfrac{1}{64}\right)}{\dfrac{1}{2}}$$

$$= 96\left(\frac{63}{64}\right) = \frac{189}{2}$$

b) $S_6 = \dfrac{6}{2}\left(-10 + (-25)\right)$

$$= 3(-35) = -105$$

NOTES

NOTES

NOTES

NOTES

NOTES

NOTES

NOTES

NOTES

NOTES

NOTES

NOTES